Introduction to
Low Voltage Systems

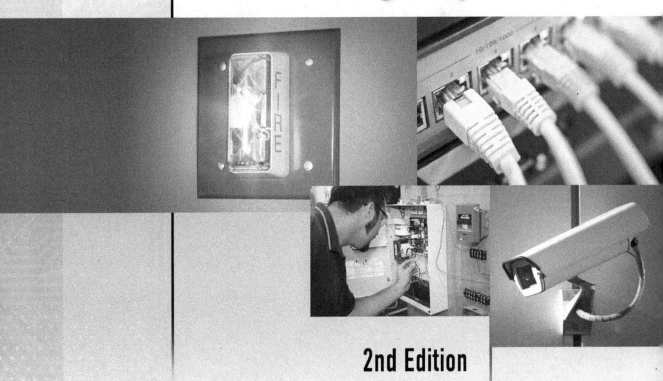

2nd Edition

Introduction to
Low Voltage Systems

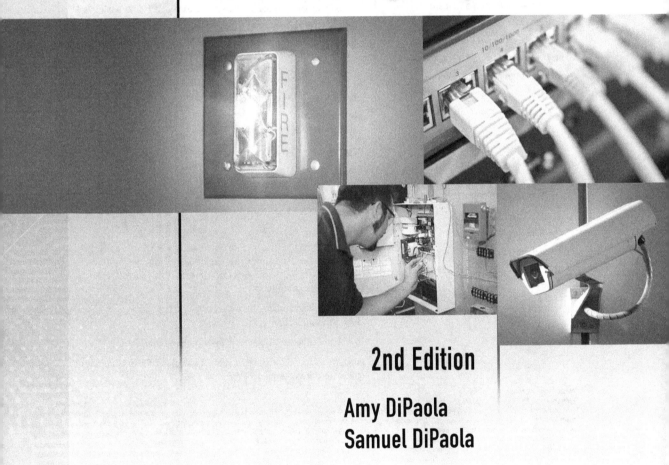

2nd Edition

Amy DiPaola
Samuel DiPaola

DELMAR
CENGAGE Learning·

Australia · Brazil · Japan · Korea · Mexico · Singapore · Spain · United Kingdom · United States

Introduction to Low Voltage Systems, Second Edition
Amy DiPaola, Samuel DiPaola

Vice President, Editorial: Dave Garza

Director of Learning Solutions: Sandy Clark

Acquisitions Editor: Stacy Masucci

Managing Editor: Larry Main

Senior Product Manager: John Fisher

Editorial Assistant: Andrea Timpano

Vice President, Marketing: Jennifer Baker

Marketing Director: Deborah Yarnell

Senior Marketing Manager: Erin Brennan

Marketing Coordinator: Jillian Borden

Senior Production Director: Wendy Troeger

Production Manager: Mark Bernard

Content Project Manager: David Barnes

Senior Art Director: David Arsenault

All cover images courtesy of iStockphoto.

Background Image: © CTR design LLC

Top left image: © Dean Turner

Center left image: © ryasick

Center right image: © Jakub Pavlinec

Bottom left image: © Lisa F. Young

Bottom right image: © michal kodym

For product information and technology assistance, contact us at
Cengage Learning Customer & Sales Support, 1-800-354-9706

For permission to use material from this text or product,
submit all requests online at **www.cengage.com/permissions**
Further permissions questions can be emailed to
permissionrequest@cengage.com

Library of Congress Control Number: 2011938604

Student Edition:

ISBN-13: 978-1-111-63953-2

ISBN-10: 1-111-63953-1

Delmar
Executive Woods
5 Maxwell Drive
Clifton Park, NY 12065
USA

Cengage Learning is a leading provider of customized learning solutions with office locations around the globe, including Singapore, the United Kingdom, Australia, Mexico, Brazil, and Japan. Locate your local office at **www.cengage.com/global**

Cengage Learning products are represented in Canada by Nelson Education, Ltd.

To learn more about Delmar, visit **www.cengage.com/delmar**

Purchase any of our products at your local bookstore or at our preferred online store **www.cengagebrain.com**

Printed in the United States of America
2 3 4 5 6   24 23 22 21 20

Contents

Chapter 1 The *National Electrical Code*

Sec 1.1 Purpose of the *National Electrical Code* 2
Sec 1.2 Important Definitions 4
Sec 1.3 Table of Contents and *National Electrical Code* Format 5
Sec 1.4 Important *National Electrical Code* Chapters and Articles for Power-Limited Systems 9
Sec 1.5 Standards and Standards Agencies 10
Sec 1.6 Listing and Labeling Laboratories 12

Chapter 2 Electrical Conductors and Cable

Sec 2.1 Cable Selection 16
Sec 2.2 Cable Construction and Insulation 18
Sec 2.3 Plenum versus Riser 24
Sec 2.4 Classified versus Listed Cables 28
Sec 2.5 Conductor Shielding 32
Sec 2.6 Electrical Properties of Cables 35
Sec 2.7 Types of Cable 47
Sec 2.8 Dissimilar Metals 57

Chapter 3 Grounding and Bonding

Sec 3.1 The Purpose of Grounding Electrical Systems 64
Sec 3.2 The Electrical Service Entrance 64
Sec 3.3 What Is Ground? 66
Sec 3.4 Ground Definitions and the *National Electrical Code* 68
Sec 3.5 The Grounding Electrode System 70
Sec 3.6 Ground Resistance and Supplemental Electrodes 71
Sec 3.7 Methods of Connecting EGCs, Grounding Electrode Conductors, and Bonding Jumpers 72
Sec 3.8 Lightning Protection Systems 75
Sec 3.9 Aluminum and Aluminum Grounding Conductors 75
Sec 3.10 Ground Faults 75
Sec 3.11 Bonding 77

Sec 3.12 Alternating Current Systems Less than 50 Volts (*National Electrical Code 250.20(A)*) *80*
Sec 3.13 Grounding Related to Other Articles of the *National Electrical Code* *81*

Chapter 4 Audio Physics

Sec 4.1 Decibels and Logarithms *90*
Sec 4.2 Sound Wave Physics *95*
Sec 4.3 Sound and Hearing *101*

Chapter 5 Audio Signal-Processing, Amplification, and Reproduction Equipment

Sec 5.1 *National Electrical Code* Requirements of Sound Systems *108*
Sec 5.2 Public Address Systems/Intercom *110*
Sec 5.3 Components and Electrical Properties of Sound Systems *111*
Sec 5.4 Speakers, Crossovers, and Their Electrical Properties *127*
Sec 5.5 Amplifiers, Signal Levels, and Their Electrical Properties *140*
Sec 5.6 Mixers, Preamplifiers, and Signal-Processing Equipment *150*
Sec 5.7 Additional *National Electrical Code* Requirements *156*

Chapter 6 Networking and Information Technology Equipment

Sec 6.1 Basic Networking, Architecture, and Topology *164*
Sec 6.2 Network Protocols and Their Functions *199*
Sec 6.3 Computer Network Addressing *206*
Sec 6.4 Media: Connections, Hardware, and Installation Techniques *208*
Sec 6.5 Other Concerns *213*
Sec 6.6 Low-Voltage Residential Network Applications *214*

Chapter 7 Power Supplies, Batteries, and Emergency Systems

Sec 7.1 Transformers *220*
Sec 7.2 Alternating Current Voltage Regulators *224*
Sec 7.3 Surge Protectors, Conditioners, and Filters *226*
Sec 7.4 Generators *230*
Sec 7.5 Static Uninterruptible Power Supply *233*
Sec 7.6 Direct Current Power Supplies *235*
Sec 7.7 Storage Batteries *243*
Sec 7.8 *National Electrical Code Article 700* *248*

Chapter 8 *Article 725* of the *National Electrical Code*; Classification of Circuits

Sec 8.1 Types of Electrical Circuits *256*
Sec 8.2 *Article 725* of the *National Electrical Code* *257*

Sec 8.3   Classification of Circuits and Class 1 *258*
Sec 8.4   Class 2 and 3 Circuits *National Electrical Code Article 725.121 261*
Sec 8.5   Power-Limited Tray Cable and Instrumentation Tray Cable *265*
Sec 8.6   Reclassification of Class 2 and 3 Circuits, Markings, and Separation
           Requirements *266*
Sec 8.7   Installation Requirements for Multiple Class 2 and 3 Circuits
           and Communications Circuits: *National Electrical Code
           Section 725.139 272*
Sec 8.8   Class 2 or Class 3 Circuit Conductors Extending beyond
           One Building *273*
Sec 8.9   Support of Conductors and Cables *273*
Sec 8.10   Calculate the Number and Size of Conductors in a Raceway *275*

Chapter 9   Fire Alarm Systems

Sec 9.1   The Intended Purpose of Fire Alarm Systems *280*
Sec 9.2   What Is a Fire Alarm System? *282*
Sec 9.3   Control Units and Alarm-Initiating Devices *291*
Sec 9.4   Notification Appliances *303*
Sec 9.5   Electrical Requirements of Power-Limited and Non–Power-Limited Fire
           Alarm Systems, *National Electrical Code Article 760 307*
Sec 9.6   Wiring of Fire Alarm Systems *315*
Sec 9.7   Troubleshooting *324*

Chapter 10   Fiber–Optic Cable and *National Electrical Code Article 770*

Sec 10.1   Introduction *328*
Sec 10.2   Basic Concepts of Light *328*
Sec 10.3   Optical Fiber Cable *329*
Sec 10.4   Applications *338*
Sec 10.5   Varieties of Fiber *339*
Sec 10.6   Types of Connectors *340*
Sec 10.7   Classifications of Fiber and *National Electrical Code Article 770 344*

Chapter 11   Telecommunications, and *National Electrical Code Article 800*

Sec 11.1   Telephone Basics *350*
Sec 11.2   Residential Cabling *355*
Sec 11.3   Commercial Systems *358*
Sec 11.4   The Private Branch Exchange (PBX) *361*
Sec 11.5   PBX Trunk Services *361*
Sec 11.6   Electronic Key Systems (EKS) *364*
Sec 11.7   Terminations and Color Codes *365*
Sec 11.8   *National Electrical Code Article 800* Communication Circuits *370*

Chapter 12 Security and Access-Control System Basics

Sec 12.1 Security Alarm Systems *378*
Sec 12.2 Wiring a Security Alarm System *385*
Sec 12.3 Access-Control Systems *389*
Sec 12.4 Wiring an Access-Control System *396*
Sec 12.5 Open Source Systems *400*
Sec 12.6 Electrical Code *402*

Chapter 13 Wireless Communications and *National Electrical Code Article 810*

Sec 13.1 A Brief History of Wireless Communications *406*
Sec 13.2 The Fundamentals of Wireless Communications *408*
Sec 13.3 Antennas *415*
Sec 13.4 Cellular Phone Communication *424*
Sec 13.5 Satellite Communications *432*
Sec 13.6 Wireless Computer Networks *444*
Sec 13.7 Cable Television System *452*
Sec 13.8 *National Electrical Code* Requirements for Radio and Television, *Article 810 454*

Chapter 14 Closed Circuit Television Camera Systems

Sec 14.1 Introduction *460*
Sec 14.2 The Purpose of Closed Circuit Television *460*
Sec 14.3 Closed Circuit Television Components *460*
Sec 14.4 Closed Circuit Television System Specifications *464*
Sec 14.5 Transmission Link *472*
Sec 14.6 Viewing Monitors and Video Formats *474*
Sec 14.7 Recording Devices *476*
Sec 14.8 *National Electrical Code Article 820* Requirements *480*

Glossary 485
Index 511

Preface

Introduction to Low Voltage Systems, 2E, is intended for beginners. As a prerequisite to low voltage systems, students should have already taken a basic course in ac/dc electricity. Many of the systems discussed throughout this book include concepts of series and parallel circuits, as well as a functional knowledge of capacitance, inductance, and transformers.

The primary focus of *Introduction to Low Voltage Systems,* 2E, is the functional basics of various systems and their connecting devices. *Introduction to Low Voltage Systems,* 2E, was not, however, written as an installation guide, nor should it be used as one. Instead, the intent of this book is to concentrate the basic theory, standards, and *Code* requirement of various systems into a single reference book for the beginning student. The types of systems in *Introduction to Low Voltage Systems,* 2E, include audio, video (closed circuit television [CCTV system]), security, telephone, fire alarm, computer networking, and wireless.

CODE REFERENCES

Code references listed throughout the book are taken from the 2011 edition of the *National Electrical Code* (NFPA-70) and the 2010 edition of the National Fire Alarm Code Handbook (NFPA-72).

TARGET AUDIENCE

This book is intended for use at the technical or community college level. It may also be used by electrical or electronics programs wishing to add a low-voltage component to their curriculum or by any low-voltage design programs already teaching low voltage systems.

FORMAT OF THE BOOK

All chapters begin with a list of objectives, followed by a chapter outline and a selection of key terms. Key terms are highlighted in bold throughout the chapter. The end of each chapter includes a selection of final questions to be gone over by students and the teacher after completion of assigned readings.

OUTLINE OF THE BOOK

Chapter 1 provides an introduction to the *National Electrical Code® (NEC)*. Students are introduced to the purpose of the *Code*, as well as the format and structure of the book.

Chapter 2 details metal cable construction, types of conductors, insulation requirements, and the electrical properties of cable. The electrical properties of cable include cable resistance, capacitance, characteristic impedance, frequency response, cross talk, signal attenuation, and shielding.

Chapter 3 provides an introduction to grounding and bonding and *Article 250* of the *NEC*. This chapter discusses the purpose and importance of ground, the grounding electrode system, and the differences between the grounded conductor, the equipment grounding conductor, and an ungrounded conductor. A description of a ground fault and the purpose of a ground-fault circuit interrupter are discussed and explained, as well as the purpose of bonding and the bonding conductor. Although it is true that building grounding systems are installed by electricians, knowledge of them is crucial to the safe installation of any system and must still be understood by all low-voltage electrical installers. The end of this chapter also discusses the grounding requirements of outdoor cables entering buildings, and the grounding of communications systems, as related to the 2011 *NEC Articles 800, 810,* and *820.*

Chapter 4 discusses the physics of audio, including sound and hearing, types of waveforms, frequency response, bandwidth, harmonics, octaves, decibels, logarithms, wave propagation, the inverse square law, and the definition and use of white noise and pink noise generators.

Chapter 5 discusses audio and the various components that make up an audio system. The chapter begins with the definition of permanently installed, portable, and temporary sound systems, as well as a discussion of public address systems and intercoms. The next part of the chapter discusses the components and electrical properties of sound systems, including microphones, microphone polar patterns, phantom power, impedance levels, balanced versus unbalanced signals, ground loops, wireless microphones, and connecting hardware. The second half of Chapter 5 discusses types of loudspeakers, filters, crossovers, speaker placement, echo and reverberation, amplifiers, mixers, preamplifiers, signal processing equipment, signal levels, impedance matching transformers, 25-/70-/100-V audio systems, isolated grounds, and the wiring requirements of audio systems as referenced by *NEC Article 640.*

Chapter 6 details the basics of computer networking. The chapter starts with a brief history of computer networking, leading into a discussion on the basic components that make up various types of computer networks. Topics include network architecture, topology, network access, methods of communication, and collision detection. From there, Chapter 6 discusses the seven layers of the OSI model, TCP/IP protocols and their functions, network addressing, network hardware (network interface cards [NIC], hubs, routers, switches, bridges), cabling and connecting hardware, and basic troubleshooting techniques.

Chapter 7 discusses the basics of power, including both ac and dc power supplies. The chapter starts with a discussion on isolation transformers and the need for ac voltage regulation, surge protection, power line conditioners, and filters. The second half of the chapter then moves to power generators (motor or engine driven), uninterruptible power supplies (UPS), rectification, dc power supplies, battery power sources, and *NEC 700, Part III,* which discusses the requirements of emergency power systems.

Chapter 8 takes an in-depth look at *NEC Article 725*, which details the classification of remote control and signaling circuits, including Class 1, 2, and 3 circuits. The chapter also discusses the comparison and use of power-limited tray cable as well as the use of instrumentation tray cable, as referenced in *NEC Article 727*. Additional topics include the installation requirements of Class 1, 2, and 3 circuits; installation options with communication circuits; support of conductors; and the number of conductors to be placed in a raceway.

Chapter 9 discusses the devices and components of a fire alarm system. The chapter starts with a description of a basic fire alarm system, the concept of supervisability, a detailed description of control units, alarm initiating devices (heat and smoke detectors), and alarm indicating devices (horns and strobes). The second half of the chapter covers the wiring requirements of fire alarm systems as discussed in *NEC Article 760*, which defines the specifications and wiring requirements of both non–power-limited fire alarm systems and power-limited fire alarm systems. The last section of the chapter details the specifics of fire alarm wiring, specifically the differences between Class A and B systems, as well as the various wiring styles related to initiation devices and notification appliances.

Chapter 10 details the theory and use of fiber-optic cable. The chapter starts with an introduction to fiber-optic theory and the basic concepts of light and light transfer. The discussion continues to fiber-optic cable construction, types of cables, single mode versus multimode, glass versus plastic, *NEC Article 770*, and types of fiber-optic connecting hardware splices, and fiber color codes.

Chapter 11 discusses the basics of telephone systems. The chapter starts with a brief history of the telegraph and the telephone in America and leads to a technical description of the local loop and the various components that make up a modern telephone system. Topics include the functional parts of the telephone, dual-tone multifrequency (DTMF), dialing ringers, ringer equivalence numbers (RENs), exchanges and area codes, cordless telephone systems, cabling, connecting hardware, *NEC Article 800* requirements, and a discussion of commercial systems, including the main distribution frame (MDF), the intermediate distribution frame (IDF), and the differences between a private branch exchange (PBX) and an electronic key system (EKS). Also included is a description and illustration of loop start versus ground start systems, as well as the terminations and color codes of 25-, 50-, 150-, and 300-pair telephone cables.

Chapter 12 details the basic components and wiring of security and access-control systems. The first half of the chapter deals with the theory of various security system sensors (passive and active infrared, ultrasonic, microwave/radar, motion detectors, acoustic and shock-type glassbreak sensors, photoelectric beams, and video motion detection), and the connection of devices to the control panel. The second half of the chapter discusses access-control systems, including card readers, the various types of card entry, request to exit (RTE), maglocks, electric strikes, reporting systems, keypads, and biometrics. Within each type of system, a detailed description of device wiring and the differences between open-loop and closed-loop connections are also discussed. The chapter ends with a discussion on integrated open source systems that operate over a computer network.

Chapter 13 discusses the fundamentals of wireless communication. Topics include the theory of basic wireless communication, carrier frequencies, modulation techniques, antennas, transmitters, receivers, and repeaters. The chapter comprehensively explores wireless with a discussion on cellular phone communication, satellite communication, wireless computer networks, and the requirements of *NEC Article 810* for radio and television communication.

Chapter 14 details the components that make up a closed circuit television (CCTV) system. The chapter begins with an introduction on the purpose of a CCTV system and leads into a discussion on the various types of CCTV cameras. Other topics include camera sensitivity, foot-candles versus lux, camera resolution, analog versus digital, lens formats, types of lenses, focal length, depth of field, field of view, *f*-stops, the shutter speed, transmission link, power-over-Ethernet (POE) standards, digital transmission, viewing monitors, video formats, recording systems, and the requirements of *NEC Article 820*.

SUPPLEMENTS

Lab Manual

A Lab Manual with at least one lab for each chapter is available. Objectives, required material, pre-questions and post-questions are provided. Some labs are analytical, and some require hands-on activity. ISBN: 9781111639549

Instructor Companion Site

An Instruction Companion Website containing supplementary material is available. This site contains Answers to Review Questions, testbanks, an image gallery of text figures, and chapter presentations done in PowerPoint. Contact Delmar Cengage Learning or your local sales representative to obtain an instructor account.

Accessing an Instructor Companion Website site from SSO Front Door

1. Go to: http://login.cengage.com and log in, using the Instructor e-mail address and password.
2. ENTER author, title, or ISBN in **the Add a title to your bookshelf** search box, and click on **Search**.
3. CLICK **Add to My Bookshelf** to add Instructor Resources.
4. At the Product page, CLICK on the **Instructor Companion site** link.

New Users

If you're new to Cengage.com and do not have a password, contact your sales representative.

ABOUT THE AUTHORS

Amy DiPaola is currently the owner and lead instructor of The Minnesota Electrical Training Center. She provides continuing education for the electrical and electronics industry, primarily for individuals needing to renew their state electrical licenses. Prior to opening her own business, Amy was an instructor for nearly 10 years with a technical college in Minneapolis, Minnesota. Amy gained her low-voltage experience as a project manager and design technician for the Federal Bureau of Investigation (FBI), assigned to the Intrusion Detection Program. Her work with the FBI included the building and fabrication of circuits, as well as the installation of security, access-control, and CCTV systems.

Samuel DiPaola is currently the training director for the Minnesota State-wide Limited Energy JATC, providing apprenticeship training for low-voltage electrical installers affiliated with the International Brotherhood of Electrical Workers (IBEW). Prior to becoming training director, Sam was the Director of Electronics at a technical college in Minneapolis, Minnesota. His industry experience includes seven years as a design technician for Emerson Process Management, previously known as Rosemount Engineering. His primary job responsibilities involved the building of automated test systems, as well as the development of test software used in the manufacturing of SMART temperature transmitters.

Prior to working in manufacturing, Sam gained his low-voltage experience with Muzac Inc., where he worked as an electronics technician of audio and video systems. His work also included the occasional design and installation of recording studios and theater systems.

Both Sam and Amy hold a valid Power-Limited License with the state of Minnesota and have been approved to teach low-voltage continuing education classes through the Electrical Licensing and Inspection unit of Minnesota.

ACKNOWLEDGMENTS

The authors would like to acknowledge the following people, who were instrumental in helping to guide this project to completion.

At Cengage Learning:

Dave Garza, thanks for your initial interest in this project.
John Fisher, thanks for all your work in helping us to stay on track, edit, and complete the final manuscript.
Stacy Masucci, thanks for being so supportive of this project.

Reviewers:

Charles W. Dale, Ed.D., Eastfield College, Mesquite, TX
Donald Farrell, Tidewater CC, Chesapeake, VA
Jane Moorhead, Mississippi State University, Starkville, MS
James E. Gordon, Clover Park Technical College
Bruce Rogol, College of Southern Nevada, Western Center, Las Vegas, NV
Steven Ray Vietor, Riverland Community College
Richard L. Schell, Luzerne Community College

Chapter 1

The *National Electrical Code*

Objectives

- Explain the purpose of the *National Electrical Code* (*NEC*).
- Describe and explain the administrative aspects and enforcement of the *NEC*.
- Recognize and explain important definitions, as covered by Article 100 of the *NEC*.
- Describe and explain the table of contents, structure, and layout of the *NEC*.
- Recognize which articles of the *NEC* specifically pertain to power-limited circuits.
- Recognize the use of standards and the various agencies established to maintain and enforce them.
- Recognize the use of listing and labeling and the agencies established to conduct product testing and safety evaluations.

Chapter Outline

Sec 1.1 Purpose of the *National Electrical Code*

Sec 1.2 Important Definitions

Sec 1.3 Table of Contents and *National Electrical Code* Format

Sec 1.4 Important *National Electrical Code* Chapters and Articles for Power-Limited Systems

Sec 1.5 Standards and Standards Agencies

Sec 1.6 Listing and Labeling Laboratories

Key Terms

annexes

articles

authority having jurisdiction (AHJ)

chapters

definitions

exceptions

index

informational notes

labeled

listed

listing and labeling laboratory

marginal notations

National Electrical Code (NEC)

National Fire Protection Agency (NFPA)

parts

qualified person

sections

standards agency

tables

units of measurement

1

SEC 1.1 PURPOSE OF THE *NATIONAL ELECTRICAL CODE**

In late August 1895, the first electrical generating station supplied electric power to Niagara Falls, New York, and its local industries. The station, which has long since been demolished, was the first large-scale multiphase power station to use commercially the unyielding pressure of Niagara Falls to provide alternating current systems rather than direct current systems. The revolution of high-voltage alternating current used for long-distance power transmission influenced the future of the electrical industry worldwide and set the pace for the growth of a billion-dollar industry.

Nevertheless, the newness of the electrical industry ultimately brought forth concerns of personal safety. Thomas Edison, a promoter of direct current power, was actively politicking the issue by publicly electrocuting elephants to prove the danger of alternating current. Although such activities proved to have great showmanship, they only managed to scare the uneducated masses. In truth, there was simply a general lack of qualified individuals at the time who really understood the dangers of electricity. Eventually, what finally drew the attention of a small group of prominent electrical industry members was a growing epidemic of fires. Ultimately, the epidemic was attributed to the expansive use and proliferation of electricity. To help combat the problem, the first committee meeting for the *National Electrical Code*® *(NEC*®*)* was held in 1896. Up to 1200 industry members met to set up guidelines for the electrical industry that would regulate all future installations of electrical apparatus. The committee's goal was to write *Code* that would potentially be adopted by electrical workers to ensure the safety of people and property.

The *NEC* is a safety manual. The first edition of the *NEC* was published in 1897, less than 20 years after Edison invented the electric lightbulb.

Today, the **National Fire Protection Agency (NFPA)** sponsors the *NEC*. In the early twentieth century, inventions and developments of the times were driving a new landscape of industry. By the late 1890s, North America was being flooded by immigrants from all over the globe. The plethora of working class individuals and the emergence of new technologies promoted an eruption of new industry and urban jobs. The challenge, however, was to find employees with the knowledge and expertise to ensure the clean and safe installation of apparatus and equipment. A need for clear standards that dictated safety began to emerge. To ensure the promotion of such standards and the safety of individuals, the NFPA has published the *NEC* since 1911. The publication is known as NFPA document 70 and is just one of the many documents published by the NFPA. In 1995, NFPA released the 1996 Edition that observed the 100th anniversary of the first *NEC* committee meeting.

In sponsoring the *NEC*, the NFPA revises the *NEC* every three years. This means that the NFPA solicits industry professionals interested in being part of *Code*-making panels to continually update and revise existing *Code*. The past panel consisted of more than 340 volunteer members who were broken down into

---

*National Electrical Code® and *NEC*® are registered trademarks of the National Fire Protection Association, Inc., Quincy, MA 02169.

20 *Code*-making panels and an 11-member correlating committee to oversee the developments and amendments to the *NEC*. The 2011 *NEC* is currently 870 pages long, the most detailed of any NFPA *Code* or standard.

National Electrical Code Article 90

Section 90.1 of the *NEC* defines the purpose of the *Code*. The purpose of the *Code* involves the following:

a. **Practical safeguarding:** the practical safeguarding of persons and property from hazards arising from the use of electricity
b. **Adequacy:** includes provisions that are considered necessary for safety, resulting in installations that are essentially free from hazard
c. **Intention:** the *Code* is not intended to be used as a design specification or as an instruction manual for untrained persons

Section 90.2 of the *NEC* details the scope of the *Code*.

Electrical systems typically include the installation of electric conductors, such as wires and cables, and electrical equipment, such as boxes, cabinets, enclosures, and raceways. The installation of conductors and equipment also include those connected to the electrical supply as provided by a main electrical utility company.

Additional electrical systems may include audio circuits (*NEC Article 640*), remote control and signaling circuits (*NEC Article 725*), fiber optic cables (*NEC Article 770*), and communication conductors. *Chapter 8* of the *NEC* details communications conductors, including telephone, alarm and security, antenna systems, television cables, and broadband networks.

The *NEC* also specifies specific installation structures such as buildings (public and private premises), mobile homes, recreational vehicles, and floating buildings, together with specific locations such as yards, lots, parking lots, carnivals, and industrial substations.

Fire investigations focusing on the electrical systems use NFPA document 70 as a guideline. Although the *NEC* has been adopted by all 50 states, many state, city, and local municipalities will overlay additional requirements to the *NEC*.

Code revisions used in one area may not be the same revision in other areas. For example, The Commonwealth of Virginia has just adopted the 2008 revision of the *Code*, whereas North Carolina has already adopted the 2011 edition of the *Code*.

As stated previously, NFPA document 70 is not intended as a design specification or as an instruction manual for untrained persons. However, this *Code* should be used as a guideline and minimum safety standard for construction plans and the installation of electrical apparatus and equipment. The difference between a safety standard and a design specification is that safety is concerned with saving lives and protecting property, whereas a design specification includes system performance. Although it may ensure safety, a safety standard alone does not necessarily provide the best system performance. A design specification, however, is not only safe but also ensures a system operates at peak levels of performance.

All electrical installations must be inspected. Inspection is performed according to the guidelines of the *NEC*; therefore, all electrical installations must be *Code* compliant or risk violations.

Special events, special occupancies, and special installations comprise *Chapter 5* of the *NEC*. Surprisingly, outdoor events such as concerts venues and sporting events venues are also covered here.

SEC 1.2   IMPORTANT DEFINITIONS

National Electrical Code Article 100 Definition; Approved, Authority Having Jurisdiction

The role of the **authority having jurisdiction (AHJ)** could not be more important. The AHJ's job is to enforce compliance with the *Code* on all installations. The AHJ is the electrical inspector.

National Electrical Code Article 100 Definition; Labeled

Equipment that is acceptable to the AHJ has to be proved able to withstand its environment and use by a listing agency, such as Underwriters Laboratories (UL). Equipment proved acceptable for use is then physically listed with the agency and **labeled** as an indication of its compliance.

National Electrical Code Article 100 Definition; Listed

Listed means that equipment, materials, or services have been tested as safe by an organization acceptable to the AHJ. An example of a listing laboratory is UL. Such an agency evaluates and tests a piece of equipment, guaranteeing that it meets required safety standards and that it is appropriate for public use. All equipment, materials, or services must be tested to ensure that they meet appropriate safety standards; they must also be found to be suitable for a specific purpose of use. After having passed a series of inspections and tests, the equipment, materials, or services are then placed on a list that is published by the listing agency, such as UL. The independent listing by such third-party inspectors verifies that the equipment, materials, or services are safe for public use.

National Electrical Code Article 100 Definition; Qualified Person

A **qualified person** must be familiar with the aspects of electricity, electrical construction and installation, and the operation of electrical equipment, but most importantly must also have had some form of electrical safety training.

Units of Measurement

National Electrical Code Article 90.9

When using the *NEC*, it is essential to understand how the *Code* uses **units of measurement**. References to measurements are always listed as International System (SI) of Units first, and the inch–pound conversion are listed in parentheses; for example, 2.5 m (8.2 ft). It is not always required to include the SI units on jobs where the inch–pound system dominates.

Understanding the *National Electrical Code*

Before maneuvering through and attempting to use the *NEC*, it is important to understand the style in which it is laid out and written. The *NEC* is made up of 11 components: chapters, articles, parts, sections, tables, exceptions, informational notes, definitions, marginal notations, annexes, and an index.

SEC 1.3   TABLE OF CONTENTS AND *NATIONAL ELECTRICAL CODE* FORMAT

Like any other non-fiction book used for educational purposes or for pleasure, the *NEC* has a table of contents. The table of contents is located in the beginning of the book. It is displayed in outline form using article numbers, titles, and roman numerals to break down each chapter. The table of contents is a good general reference guide to subject matter. In the 2011 edition of the *NEC*, the bottom of each page contains a number beginning with 70 (e.g., 70-23). The 70 indicates that this is NFPA document 70. Every page in the *NEC* begins with this page-numbering convention. The second number defines the specific page number of document 70. In this example, it would read "page 23 of NFPA document 70."

Chapters

Nine major categories in the *NEC* are known as **chapters**. Each chapter contains chapter-specific articles (Figure 1–1).

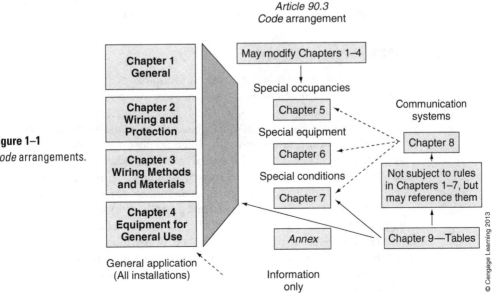

Figure 1–1
Code arrangements.

National Electrical Code Article 90.3

Chapters 1, 2, 3, and *4* of the *NEC* contain general applications that apply to all installations. *Chapters 5, 6,* and *7* apply to all special occupancies, conditions, and equipment that may modify various sections of *Chapters 1, 2, 3,* or *4. Chapter 8* is not subject to any rules in the previous chapters but may occasionally reference them, in which case they would apply. *Chapter 9* contains tables referenced in *Chapters 1* through *8.* The annex is not part of the *Code*; it is there for informational purposes only.

Articles

There are 147 articles contained in the 2011 edition of the *NEC*. **Articles** are individual subjects, each of which covers a specific subject matter. As you will notice as you look through the *NEC*, the last numbered article is *Article 840.* But there are not 840 articles contained within the *NEC*; as stated above, the *NEC* contains only 147 articles. As you page through the book, you will see that some article numbers have been skipped. This leaves room in the *NEC* for future expansion or for new technologies to be covered in later editions.

Article numbers directly relate to the chapter in which they are found. For example, *Chapter 1* contains *Articles 100* through *199, Chapter 2* contains *Articles 200* through *299,* and so on.

Parts

Many of the articles are broken into parts. **Parts** are used to subdivide articles into simpler topics of focus. Parts are displayed as roman numerals. Most articles contain three to four different parts; however, *Article 230* is a good example of an article subdivided many times. *Article 230* contains eight parts. When giving a *Code* reference, parts are not referenced as part of a correct answer.

Sections

Each part is broken down into sections. A **section** is assigned a corresponding article number and noted with a "dot" followed by the section number (e.g., *800.1*). This example indicates that we are referencing *Article 800, Section 1.* Sections increase chronologically, but like articles, some section numbers are skipped to allow for future expansion. When giving a *Code* reference, the section is the reference. *Code* sections may be broken down into subsections by letters in parentheses, such as *(A)*, then broken down further into numbers in parentheses, such as *(1)*, and subdivided even further into lowercase letters such as a, b, or c. When giving a *Code* reference you have to reference the subsection, just the section is not sufficient. Also, be sure when citing *Code* that you carefully identify the correct *Code* reference. Many times an article is broken down into several subsections and spans numerous pages; therefore, in finding the *Code* reference, it is easy to skip over the correct answer and reference a neighboring section.

Tables

Many *Code* rules are displayed within tables. **Tables** list rules in a logical and easy-to-read order. Many tables contain notes that simplify and more easily explain the rules listed within the table. Each table found within an article is article specific. When referencing a table within an article, the reference must include the word *Table* and the section number, for example, *Table 110.31*.

Exceptions

Exceptions are alternatives to specific sections of *Code*. There are many exceptions listed throughout the *NEC*. Often, the *NEC* states a rule to follow, and then clarifies that rule with an exception. In some cases, there may be a list of multiple exceptions, which is why it's important to read all of the exceptions first, to be sure of alternative options. Following are some examples of exceptions.

National Electrical Code Article 800.44

Section 800.44(A) explains that if we are running overhead communication wires that are originating from the outside and entering a building, then the wires have to:

1. Be located below any electric light or power conductors
2. Not be attached to the same cross-arm that carries the electric light and power conductors
3. Have sufficient climbing space to allow for servicing and other use
4. Be separated from the electric light and power conductors by 12 in. (300 mm) at any point in the span of the wires

Section 800.44(B) explains that overhead communication wires that originate from the outside and enter a building have to have a clearance from any part of a roof which they pass of not less than 8 ft (2.5 m). However, look directly below *800.44(B)* and you notice that there are three exceptions to this 8-foot rule.

Exception No. 1 states that over auxiliary buildings, such as garages, this rule does not apply. That means that over garages you do not have to provide an 8-foot clearance.

Exception No. 2 states that you can reduce the 8-foot rule to not less than 18 in. (450 mm) if you have not more than 4 ft (1.22 m) of communications wire passing over the roof, or if the communications wire is terminated at an approved raceway.

Exception No. 3 indicates that the 8-foot rule does not have to be followed if the roof you are running your communications wires above has a slope of not less than 4 in. (100 mm) in 12 in. (300 mm). In this case, a reduction from 8 to 3 ft (2.5 to 0.9 m) is allowed.

It is interesting to see that many times there are exceptions to rules. These exceptions allow for some other options in the installation. Notice also that the exceptions are in italic letters; it is easy to not see the italicized words and simply skip over them. Be sure to read all of the exceptions that apply to a specific section of *Code*; often, the exceptions are important and become fair game on most electrical examinations.

The verbiage used in *Code* articles is another important piece to understanding the *NEC*. There are two types of rules used in the *NEC*: mandatory and permissive.

A mandatory rule uses the words *shall* or *shall not*. The word *shall* means that if you are using the applicable code article, it is mandatory that you do it in a particular format. The words *shall not* mean that it is not allowed.

A permissive rule uses the phrase *shall be permitted*. This means that the *Code* article being used can be performed in a particular format, but it is not required. A permissive exception gives an option but is by no means mandatory.

Informational Notes

Informational Notes used to be called Fine Print Notes (FPNs). The 2011 edition of the *NEC* is now referring to them as Informational Notes. Informational Notes are included within the *NEC* for clarification purposes. As an example, take a look at *NEC Section 800.44*. At the end of the section is an Informational Note. Informational Notes are used to provide additional details to a section of *Code*, or to reference sources of other standards or codes which may apply.

Definitions

Like most textbooks you have used, the *NEC* contains a glossary of terms. These **definitions** are found in *Article 100*. All definitions found in *Article 100* apply to multiple *NEC* articles. It is often good practice to look for assistance in *Article 100* when reading a particular section of *Code* because *Code* definitions are included to clarify the meaning of unknown words and help reduce the possibility of reader confusion. The way *you* define a word may not be the same way the *Code* defines a word. Definitions found in *Article 100* should be referenced as *100* because section numbers are not assigned to that article. It is also important to note that *Article 100* does not contain all of the definitions found within the *NEC*. As a result, additional definitions may often be found printed within specific articles. In such cases, those definitions will be considered article specific. Article-specific definitions can usually be found in section 2 of such articles. As an example, *250.2* lists additional definitions for grounding and bonding not included in *Article 100*. Section 2 of many other articles lists article-specific definitions as well.

Highlighted Text or Marginal Notations

As previously mentioned, the *NEC* is revised every three years. When this is done, all changes to prior entries are indicated with highlighted text. When you page through the 2011 *NEC*, you will notice many lines of highlighted text. The highlights indicate that these lines have been changed from the previous 2008 edition. Completely new additions to the *NEC* are indicated by a marginal notation. A marginal notation is designated by the presence of a vertical line located on the left side of the text. This means that the entire article/section/subsection has been added, and is new to the *NEC* since the previous edition. A good example of this is *Article 840*, which is new to the 2011 *NEC*. Marginal notations are also used when the previous title of an article has changed; in such cases, it is at the left of the header title box. A revision may sometimes be as subtle as a single word change or as great as the addition of the entire section or article. A list of revisions is also published every *Code* change year and may be purchased individually from the NFPA.

Annexes

The **annexes** are located in the back of the book between *Chapter 9* and the *Index*. The annexes are a good reference source; however, they are not an official part of the *NEC*. The annexes are included for informational purposes only and are not accepted as a valid *Code* reference.

Index

The **index** is located in the back of the *NEC*. The index is like having your own little search engine. It is organized in alphabetical order with keywords found in the *NEC*. The index is the best reference for finding specific words and phrases in NFPA document 70. The index contains approximately 40 pages and should be used to locate specific *Code* rules.

SEC 1.4   IMPORTANT *NATIONAL ELECTRICAL CODE* CHAPTERS AND ARTICLES FOR POWER-LIMITED SYSTEMS

The following chapters and articles of the *NEC* are considered important for any power-limited technician to know about and understand. Not all of these articles are covered in Introduction to Low-Voltage Systems, but they are still worth noting.

- *Article 90 Introduction*
- *Article 100 Definitions, Part I*
- *Article 110 Requirements for Electrical Installations, Parts I and II*
- *Article 250 Grounding and Bonding*
- *Chapter 3 Wiring Methods and Materials*
- *Article 400 Flexible Cords and Cables*
- *Article 402 Fixture Wires*
- *Article 411 Lighting Systems Operating at 30 Volts or Less*
- *Article 500–510 Hazardous Locations*
- *Article 517 Healthcare Facilities, Part VI*

- *Article 520 Theaters, Audience Areas of Motion Picture and Television Studios, Performance Areas, and Similar Locations*
- *Article 530 Motion Picture and Television Studios and Similar Locations*
- *Article 540 Motion Picture Projection Rooms*
- *Article 640 Audio Signal Processing, Amplification and Reproduction Equipment*
- *Article 645 Information Technology Equipment*
- *Article 680 Swimming Pools, Fountains, and Similar Installations*
- *Article 650 Pipe Organ*
- *Article 700 Emergency Systems*
- *Article 701 Legally Required Standby Systems*
- *Article 725 Class 1, Class 2, and Class 3 Remote-Control, Signaling, and Power-Limited Circuits*
- *Article 760 Fire Alarm Systems*
- *Article 770 Optical Fiber Cables and Raceways*
- *Chapter 8 Communication Systems*
- *Chapter 9 Tables*

SEC 1.5   STANDARDS AND STANDARDS AGENCIES

The low-voltage industry is driven by codes and standards. A standard is a format that has been approved by a recognized **standards agency** and is established as an accepted practice by the industry. Standards exist for many fields in the realm of the low-voltage industry, most importantly programming languages, operating systems, data formats, communications protocols, electrical interfaces, wiring, and apparatus installation.

Standards are essential to system designers to ensure that various components from multiple manufacturers are compatible with each other and can be integrated with the system; otherwise, the system would not function. If standards did not exist, the installation of different components from different manufacturers would not be possible; such systems would then need to contain proprietary components from a single manufacturer, making the complex integration of various systems nearly impossible to achieve.

American National Standards Institute

American National Standards Institute (ANSI) was founded in 1918. It is a voluntary organization composed of more than 1000 members. Most members derive from the computer and telecommunications industries. ANSI creates technical standards for the electrical and computer industries.

Electronic Industries Alliance

Electronic Industries Alliance (EIA) is an association composed of trade representatives, mostly in the high-tech and communications industries, from the United States. EIA dates back to 1924, when it was originally known as the Radio Manufacturers Association. More recently, EIA has been solely responsible for the development of specific standards, such as the RS-232, RS-422, and RS-423 serial device connections, often used in computing for two-way communication between system devices.

Telecommunications Industry Association

Telecommunications Industry Association (TIA) is an organization made up of individuals and companies primarily representing the information technology and telecommunications industries. TIA was established in 1924 when a group of suppliers to the telephone industry organized an industry trade show. Ever since, TIA has been an industry leader through the development of trade shows, committee meetings, and standards writing. In 2000, the MultiMedia Telecommunications Association (MMTA) merged with TIA. As a result, TIA restructured to develop new departments focused on the emergence of international markets. TIA embodies the communications division of EIA.

International Organization for Standardization

An interesting fact about the name ISO (International Organization for Standardization) is that it is not an acronym; instead, the name derives from the Greek word *iso*, which means "equal." The ISO is an easily recognized, international organization made up of national standards organizations from several different countries. As an example, ANSI is a member of ISO, but it is the only member representing the United States. ISO has been instrumental in the development of computer standards and is most recognized for the research and standardization of the OSI model (Open Systems Interconnection), a standardized architecture for designing networks (see Chapter 6 in this book for more information).

Institute of Electrical and Electronics Engineers

The Institute of Electrical and Electronics Engineers (IEEE) was founded in 1884, with members directly out of engineering, scientific, and academic venues. The acronym for IEEE is pronounced *I-triple-E*. Most of the standards developed by IEEE are written for the computer and electronic industries specifically. Probably the most recognizable standard written by the IEEE is the IEEE 802 standard for local area networks. The 802 standard is widely followed and accepted in the field of computer networking and wireless communications.

SEC 1.6   LISTING AND LABELING LABORATORIES

A **listing and labeling laboratory** is an equipment, materials, and services testing agency that is capable of providing a wide range of testing and inspection services for industry. These tests should validate the functional use and safety of products and services before their being installed and used by the general public. Incidentally, any unauthorized modifications to a manufactured piece of equipment void out the original listing and labeling certification. Unless the manufacturer's instructions specifically tell you to make modifications, they are not allowed. If an inspector finds that an unauthorized modification has been made, the specific piece of equipment will be tagged and removed from public use. As an example, cutting off the third prong of an electrical plug will void out the original listing and labeling. In such a case, the equipment would need to be repaired prior to use or removed from service.

National Testing Association

National Testing Association (NTA) is a company composed of engineers, technicians, designers, and draftsmen all working together to assure the structural performance of building materials and components using industry-standard methods of evaluation. If no standard test method exists for a particular product or application, NTA develops a custom method to suit a company's needs.

Underwriters Laboratories

UL is an independent, nonprofit product safety testing and compliance organization. This agency has been testing products for safety since 1894, and each year more than 17 billion UL Marks are applied to products worldwide. UL is by far the industry leader for safety listing and labeling. Although UL specializes in the testing of electrical devices and programmable systems, they also provide quality-assurance processes for companies around the world.

Canadian Standards Association

The Canadian Standards Association (CSA) is a nonprofit, membership-based association serving business, industry, government, and consumers in Canada. CSA works in Canada and around the world to develop standards to enhance public safety and health.

CHAPTER 1 FINAL QUESTIONS

1. How many chapters are contained in the *NEC*?
   a. 8
   b. 9
   c. 147
   d. 840

2. The format of the *NEC* is made up of _____ parts?
   a. 6
   b. 7
   c. 9
   d. 11
   e. 13

3. How are changes to sections of *Code* from previous editions identified in the *NEC*?
   a. They are shown by highlighted text.
   b. They are listed in the annex.
   c. They are listed in the table of contents.
   d. They are shown by a marginal notation.

4. Who has the authority to enforce the *Code*?

5. Who sponsors the *NEC*?

6. Formal interpretations of the *NEC* are permitted to be made by _____.
   a. the authority having jurisdiction
   b. the local inspector
   c. the National Fire Protection Association
   d. Underwriters Laboratories

7. In the *NEC*, approved means _____.

8. What is the purpose of the *NEC*? Give a *Code* reference to support your answer.

9. What is the *NEC*'s primary system of units?
   a. English, inch-pound
   b. International System of Units (SI)

10. The *NEC* is revised and updated every _____.
    a. 2 years
    b. 3 years
    c. 5 years
    d. 10 years

11. What is the primary purpose of listing and labeling organizations?
    a. To validate the functional use and safety of products installed and used by the general public
    b. To provide product sales numbers
    c. To provide product instruction manuals
    d. To provide product safety training

12. The *NEC* should not be used as an electrical design manual. Why?

13. Communication Systems are covered in what chapter of the *NEC*?

14. Do the *Code* requirements for the communications chapter include the requirements of all the other chapters of the *NEC*? Explain.

15. The *NEC* is document number _____ of the National Fire Protection Association.

   a. 1
   b. 70
   c. 72
   d. 101

Chapter 2

Electrical Conductors and Cable

Objectives

- Identify different types of electrical conductors.
- Explain the physical properties of conductors.
- Recognize the various types of cable insulation and their applications.
- Categorize several classes of cable.
- Explain the electrical properties of a transmission line.
- Calculate attenuation and signal loss on a transmission line.
- Recognize the varieties, specifications, and uses of different types of cable.
- Describe the hazards of joining dissimilar metals.

Chapter Outline

Sec 2.1 Cable Selection

Sec 2.2 Cable Construction and Insulation

Sec 2.3 Plenum versus Riser

Sec 2.4 Classified versus Listed Cables

Sec 2.5 Conductor Shielding

Sec 2.6 Electrical Properties of Cables

Sec 2.7 Types of Cable

Sec 2.8 Dissimilar Metals

Key Terms

american wire gauge

attenuation

attenuation crosstalk (ACR)

bit rates

characteristic impedance

circular mil

coaxial

copper-clad

copperweld

crosstalk

delay skew

dissimilar metals

far-end crosstalk (FEXT)

multiconductor

near-end crosstalk (NEXT)

plating

plenum

power sum

riser

shielding

skin effect

standing waves

structural return loss

triaxial

twinaxial

twin-lead

twisted pair

velocity of propagation

wavelength

SEC 2.1   CABLE SELECTION

How do you know that you have selected the correct type of cable for a job? Because cable selection can cover a wide variety of topics ranging from the composition of the wire, type of insulation and jacketing, tensile strength of the metal, flexibility, temperature, impedance, capacitance, ampacity, frequency limit, size, number of conductors, shielding, attenuation, environmental effects, and the intended use of the product, just to name a few, the process can appear to be quite daunting. If you have ever spent time looking through a vendor's catalog, you have likely noticed that there are dozens of variations of cable, all having different part numbers and yet similar specifications. So how do you choose the right one? The answer is that it would be best to first view the individual details of the job and then ask what is needed, because choosing the correct cable has more to do with matching environmental considerations and job specifics to the intended use of the product than it does with simply searching for cable specifications.

All electrical circuits and products manufactured in the United States are required to be tested for safety and risk for fire by a variety of regulatory, standards, and listing agencies such as Underwriters Laboratory (UL) or the Canadian Standards Association (CSA). All tested products must first meet a variety of minimum required standards before they are given a classification and labeled accordingly, thus verifying them to be safe and reliable for public use.

The type of cabling needed for an installation must be determined by the rated classification, as labeled on a product, and the environmental specifics of the location. By first isolating the fine details of a job based on what is needed, and then matching cable specifications to a known criteria, the correct choice of cable becomes self-evident.

As an example, indoor cable, or cable intended for dry environments, would never be used in an outdoor setting because the exterior jacketing might not be resistant to moisture, heat, or radiation from the sun; therefore, over time it might breakdown or deteriorate, essentially destroying any intended fire-protective qualities originally engineered into the product by the manufacturer. In comparison, it would not be necessary to use a cable designed for exterior use in an indoor setting; although it is true that it would be safe and effective, the installation costs would be highly prohibitive and unnecessary (*NEC Article 310.10 (A–D)*). Cables are sufficiently engineered for specific purposes, and provided that they are installed and used correctly, as intended by the manufacturer, they are more than safe; any further derating or temperature correction over and above what is required by the *NEC* is not recommended.

National Electrical Code, Article 110.5– 110.11

Listing and labeling requirements, together with the classification of circuits and cables, as defined in *Articles 725, 760, 800,* and *820* of the *NEC*, are discussed in more detail later in this chapter, along with what defines the classifications and how they are applied.

Before choosing a specific type of cable for an installation, the job specifics must first be defined. The selection process for an appropriate cable is then based on individual size, shape, electrical specification, material makeup (e.g., copper vs. aluminum), and cost of the conductors. Examples of various cable types include solid versus stranded (Figure 2–1), single conductor versus multiconductor (Figure 2–2), multiconductor versus paired (Figure 2–3), shielded versus unshielded (Figure 2–4), copper versus aluminum (dissimilar metals), coaxial versus noncoaxial or twin-lead (Figure 2–5), and plenum versus nonplenum or riser. Let us first start with the basics of cable construction and then discuss specifications and types of classification later.

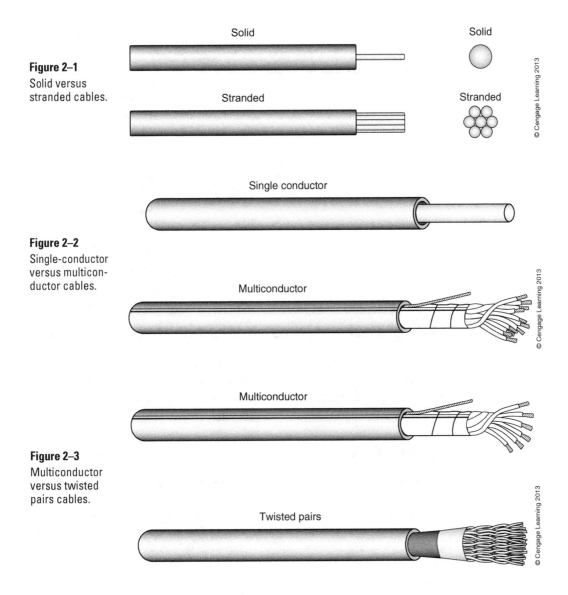

Figure 2–1
Solid versus stranded cables.

Figure 2–2
Single-conductor versus multiconductor cables.

Figure 2–3
Multiconductor versus twisted pairs cables.

Solid

Solid

Stranded

Stranded

© Cengage Learning 2013

Single conductor

Multiconductor

© Cengage Learning 2013

Multiconductor

Twisted pairs

© Cengage Learning 2013

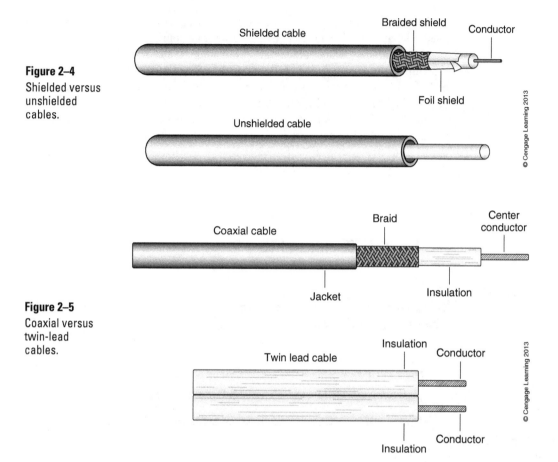

Figure 2–4
Shielded versus unshielded cables.

Shielded cable · Braided shield · Conductor · Foil shield

Unshielded cable

Figure 2–5
Coaxial versus twin-lead cables.

Coaxial cable · Braid · Center conductor · Jacket · Insulation

Twin lead cable · Insulation · Conductor · Insulation · Conductor

© Cengage Learning 2013

SEC 2.2   CABLE CONSTRUCTION AND INSULATION

Cable construction involves the material makeup of the conductors and the type of insulation and outer jacketing used to achieve a desired voltage, current, and temperature specification. Cable insulation also affects the level of capacitance per foot and the attenuation of high-frequency signals.

Types of Conductors

The conductor is defined as the metallic component of the cable or wire through which electrical current or electrical signals are transmitted. According to the *NEC*, there are three basic types of conductors available for use: copper, aluminum, and copper-clad aluminum. Other varieties include copper-weld, and copper-clad steel, which are not as widely used, but available.

Copper

Copper provides excellent conduction, being rated at 100% conductivity, and per size and foot-length it ranks highest in overall cost.

Copper-Weld

Copper-weld fuses a thin coating of hard-drawn, bare copper to a steel core. Although copper-weld is only 40% conductive at low frequencies, it has a significantly higher level of tensile strength, making it nearly impossible to stretch. Copper-weld is therefore an ideal wire for hanging between poles or supports where the added level of strength is required. Moreover, because of its high tensile strength, hard-drawn copper-weld is ideal for high-frequency applications, especially in the area of shortwave transmitting aerials and the construction of antenna systems. In such situations, the length of a single conductor is critical to the overall quality of a desired signal. Ideally, the length of an antenna element should not change once it has been tuned to a specific frequency. Having a high level of tensile strength makes it possible to produce a nearly perfect antenna system, one that is not prone to physical stress or change in characteristics over shifting temperatures.

Copper-Clad

Copper-clad is available in a variety of types. The most common example is copper-clad aluminum. However, copper-clad steel does exist.

Copper-clad aluminum fuses an outer layer of copper to a central core of aluminum. Although the conductor is nearly 90% aluminum, at high frequencies it performs as 100% copper, due to **skin effect**. Skin effect is discussed in the next section of this chapter. Copper-clad aluminum provides all the electrical benefits of copper at high frequencies but is one-third the weight and is less expensive. It is also sometimes installed as building wire, size 12 and larger.

Copper-clad steel fuses an outer layer of copper to a central core of steel. It is available at either 30% or 40% conductivity, depending on whether the temper is soft or drawn. The use of copper-clad steel is ideal where the electrical benefits of copper and the strength of steel are required. Grounding rods are commonly made out of copper-clad steel.

Aluminum

Aluminum, which is more commonly used by the electrical wire industry, provides only 61% the conductivity of copper at one-third the weight. At one point in the 1970s, aluminum wire was preferred because of its low cost, being nearly half as expensive as copper. However, currently, the costs of aluminum and copper have become comparable, and because of *NEC* regulations concerning dissimilar metals (the joining of unlike metals to each other or dissimilar splicing blocks and connectors), the joining of aluminum and copper becomes problematic, especially when considering future expansions or servicing. Dissimilar metals and the *NEC* regulations that apply are discussed in greater detail later in this chapter.

Conductor Coating

Once conductors have been manufactured, especially copper, they are usually coated with a thin layer of tin, silver, or nickel. The coating, referred to as **plating**, helps prevent a variety of insulations from attacking or adhering to the central conductor. There are three essential reasons why plating of conductors is necessary. First, plating helps to eliminate problems when soldering; second, it allows for a more simplified stripping and removal of insulation; and third, it helps to prevent deterioration of the central conductor at high temperatures.

Tin

Tin is the most common type of plating. It offers improved corrosion resistance and ease of solderability when compared with silver or nickel. Tinned copper is rated at temperatures up to 150°C (302°F).

Silver

Silver-plated conductors are used in high-temperature environments and for high-frequency applications. The high conductivity of silver helps to reduce high-frequency losses due to **skin effect**. At high frequency, current tends to travel along the edges of a conductor. Where this happens, the resistance of the conductor increases dramatically because the entire center of the conductor virtually disappears, leaving less surface area for current flow. Essentially, skin effect causes a round conductor to appear as a donut. Silver-plated copper is rated to temperatures up to 200°C (392°F).

Nickel

Nickel plating is used on conductors operating at temperatures higher than those for which silver is rated. Higher operating temperatures also cause copper to oxidize rapidly, effectively increasing the total resistance of the metal. Even without factoring in the losses caused by oxidation, the resistance of copper wire naturally increases with temperature. Nickel helps to reduce such effects. The one drawback to using nickel is that soldering is made more difficult. In such instances, nickel-plated wires can be joined more easily by braising rather than soldering. Wire braising uses a high temperature torch and is similar to welding. Nickel is rated to temperatures up to 260°C (500°F).

American Wire Gauge (*National Electrical Code Article 110.6, Conductor Sizes*)

In the United States, the system for specifying wire size is AWG, or **American Wire Gauge**. An increase of three gauge numbers doubles the area and weight of a conductor and simultaneously cuts the direct current (dc) resistance in half. In general, the larger the AWG size, the smaller the wire, except where size diameters

Figure 2–6

Conductor Sizes, *110.6.* Conductor sizes 18 AWG through 4/0 AWG are American Wire Gauge (AWG). Sizes 250 kcmil and larger are measured in circular mils (cmil), with k meaning 1000.

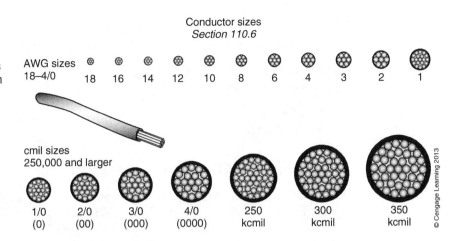

Conductor sizes
Section 110.6

AWG sizes 18–4/0

18   16   14   12   10   8   6   4   3   2   1

cmil sizes 250,000 and larger

1/0 (0)   2/0 (00)   3/0 (000)   4/0 (0000)   250 kcmil   300 kcmil   350 kcmil

© Cengage Learning 2013

exceed 1 AWG; then larger numbers equate to larger wire. Figure 2–6 illustrates the diameters of wire gauges as they range from 18 to 4/0 (0000). Larger than 4/0 the diameters are simply listed in cmil, or circular mil.

Wire gauges 12 through 4/0 are typically used by the medium- and high-voltage power industry for applications of power transmission from the power company to the individual consumer and as the internal electrical wiring for building structures. Sizes 18 to 12 AWG are used primarily for panel wiring and industrial control systems, whereas sizes 26 to 18 AWG are more commonly used within the communications, data, security, alarm, and low-voltage industries. Larger diameter wires can more safely handle greater levels of current flow and heat dissipation. The insulation material of a conductor must also be chosen accordingly, depending on voltage and temperature ratings needed, ampacity, and environmental conditions, to ensure operational safety and the prevention of possible fires.

Circular Mil

Circular mil (cmil) is a term used to define the cross-sectional area of a conductor. It is equal to area in a 1/1000-in. diameter circle. As cmil increases, size increases. Mcmil is an older term used to denote 1000 cmils; however, currently, kcmil is more commonly used as it fits more with the metric system of where k denotes kilo or 1000.

The cmil is the standard unit of measure for the cross-sectional area of a wire. A mil is a unit of measurement equal to 0.001 or 1/1000 of an inch. The diameter of a round conductor may be only a fraction of an inch; therefore, it is more convenient to express this fraction in mils, to avoid using decimals. For example:

$$0.025\text{-in. diameter} = 25 \text{ mil}$$

For a round conductor, the area in cmil is obtained by squaring the mil diameter; therefore, 25 mil squared equals 625 cmil. As stated earlier, conductors larger than 4/0 AWG are measured in direct cmil.

Example

What is the cmil area of a conductor that has a diameter of 0.25 in.?

Diameter in inches $\times$ 1000 = mil

0.25 $\times$ 1000 = 250 mil

mils squared = cmil

250^2 = 62,500 cmil

Stranded versus Solid Conductors

A conductor can be designed as a single solid wire or from a grouping of multiple strands. For example, a 22 AWG conductor can be made from a single 0.0253-in. diameter wire or from seven strands of 30 AWG (Figure 2–7).

Stranded wire usually is preferred over solid when wiring panels, controls, pumps, batteries, manufacturing test systems, or other components. Solid wire occasionally is used for long runs, most commonly by the telephone and data communications industry, but in most cases it does not work well. In many situations, making a good connection with solid wire, especially gauges 18 through 26, can be difficult, depending on the types of terminals being used. Often, the use of solid conductors runs the risk for breaking connections or internally breaking conductors if cables are stressed or overly flexed.

The telecommunications, data, voice, and video industries typically use solid twisted pair for all installations, but in these cases, they are connected to terminals that have been specifically designed for solid wire; stranded wire is not suitable for such installations.

Coaxial cable is unique in that it usually consists of a solid central conductor (stranded central conductors are also available) surrounded by a polyethylene insulator, with a concentric outer conductor, or shield, consisting of multiple braided strands; essentially, coaxial cables use a combination of solid and stranded wire. The specific uses of various types of cable and shielding are discussed in more detail later in this chapter.

Wiring Integrity

NEC, Article 110.7, Wiring Integrity, states that completed wiring installations shall be free from short circuits, ground faults, or any connection to ground other than required or permitted by the *Code.*

Figure 2–7

Comparison of solid versus stranded American Wire Gauge (AWG).

22 AWG
Solid

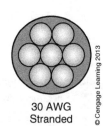
30 AWG
Stranded

© Cengage Learning 2013

National Electrical Code, Table 310.104 (A) Conductor Applications and Insulations

When discussing the construction of wire and cable, the type of insulation or outer jacketing may possibly be the most important topic. Remembering that the National Fire Protection Association (NFPA) is responsible for all entries to the *NEC*, and that safety and fire prevention ranks highest on the list of considerations when dealing with conductors of electrical current, the selection of a suitable insulation quickly becomes a primary concern.

The following factors govern the selection of a suitable insulation: stability; resistance to sunlight (ultraviolet); dielectric properties; electrical resistance; resistance to ionization, corrosion, ozone, high temperature, and moisture; mechanical strength; and flexibility. No single insulation is ideal for all of these factors. It is necessary, therefore, to select a cable that suitably matches the requirements of the particular job or installation.

Partial List of Various Types of Insulation and Their Properties

Ethylene-propylene-diene elastomer

Ethylene-propylene-diene elastomer (EPDM) is a high-temperature insulating rubber, ranging from –55°C to 150°C. It is flexible, abrasion resistant, and has good insulation resistance, together with moderate dielectric strength. In many situations, EPDM is replacing silicone rubber because it offers a higher cut-through resistance. EPDM is not resistant to oil, gasoline, kerosene, or most industrial solvents, and it should not be used in such environments.

Neoprene

Neoprene insulation offers a temperature range from –55°C to 90°C, depending on the type used. Neoprene is resistant to oil and sunlight and tends to be more stable in black, dark brown, and gray colors. Electrically, it does not insulate as well as other materials; as a result, it tends to be thicker to help compensate for the lower performance. Neoprene typically is used for separate lead wires and cable jacketing.

Rubber or Synthetic-Based Rubber

Rubber, natural or synthetic-based rubber compound, typically is used as insulation or jacketing. The majority of rubber compounds tend to have a temperature range from –55°C to 75°C, depending on their formulation.

Silicone

Silicone is a soft compound with low mechanical strength. Silicone can be used over temperatures ranging from –80°C to 200°C. It is also highly resistant to electricity, ozone, moisture, and radiation. Often, silicone is used in a gel form to plug up cable feeds between an exterior and interior run. The gel hardens in a matter of hours, providing good separation and isolation between dissimilar elements.

Polyethylene

Polyethylene provides good electrical insulation. It has a low dielectric value, which tends to remain quite constant over all frequencies. Polyethylene can be rated from stiff to hard, depending on its formulation, and it is lightweight, chemically inert, easy to strip, and highly resistant to moisture, gasoline, oils, and most solvents. Brown and black polyethylenes tend to provide greater resistance to weather. Polyethylene can be used in temperatures ranging from –60°C to 80°C.

Polypropylene

Polypropylene typically is harder than polyethylene, which makes it suitable for wall insulation; it also has a low dielectric constant. Depending on the formulation, polypropylene can have a temperature range from –40°C to 105°C; it is a durable and heat-resistive insulation.

Polyvinyl Chloride

Polyvinyl chloride (PVC) has thousands of formulation variations. It can be used, in most cases, over temperatures ranging from –40°C to 105°C. The various formulations differ greatly in pliability and electrical properties. However, they all tend to be highly resistive to moisture, sunlight, flames, oil, ozone, and chemical solvents. PVC is also best suited for frequencies in the audio range.

Teflon

Teflon has excellent electrical properties and is highly resistive to chemicals and high temperature. Teflon can usually be used in temperatures ranging from –70°C to 260°C. The downside of Teflon is that it should not be used in high-voltage situations or near nuclear radiation.

There are many more types of thermoplastic polymers and synthetic rubbers available that have a wide variety of specific uses and properties. *NEC Article 310.104, Table 310.104 (A–E)* also includes additional insulation types, together with acronym codes and suggested applications. Always be sure of the environment before choosing a specific type of cable. Resistance to oxidation, heat, cold, oil, weather, ozone, flame, radiation, water, acids, alkali, hydrocarbons, alcohol, and underground burial *(NEC Article 310.10 (F–G)* must be considered in all cases. Always read the fine print before choosing an insulation type.

SEC 2.3   PLENUM VERSUS RISER

Refer to *NEC, Article 100, Definition: Plenum* for more information.

Plenum

The space above a suspended ceiling where ductwork is run is known as the **plenum** (Figure 2–8). The plenum space has always been a convenient location

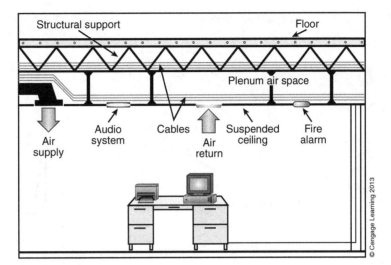

Figure 2–8
Plenum air space.

for a variety of wiring applications. However, in the event of fire, the use of these spaces may pose a serious health hazard. The main problem is that the building air exchange continues to run until the fire alarm panel triggers an alarm. Once the spread of fire reaches the plenum space, toxins may be released into the air from the melting insulation, and carried by smoke to the ventilation system. Few barriers exist to contain the smoke and flames, and within a matter of minutes, occupants can be overrun throughout all parts of the building.

Wiring installed within a plenum space requires plenum-rated cables. Plenum cables take a longer time to burn and are low smoke producing. The space below raised floors, where the under-floor area provides a pathway for ventilation, also requires the use of plenum cable. However, the underfloor pathway does not require plenum cable if it is an isolated chamber and not part of the main building air exchange.

Plenum cables are available as a low-cost alternative to using conduit in an environmental air space. Because the cost of rigid and semirigid conduit can be quite prohibitive in large commercial spaces, plenum cable provides a nice low-cost alternative for many low-voltage installations. In many instances, conduit is still preferable because of the physical protection against vandalism and abuse it provides. In locations where exposed wiring is not allowed, conduit is still a requirement for many applications.

Riser

A **riser** is an architectural term referring to the vertical column or shaft between floors in a building that is typically used for plumbing and air ducts. Figure 2–9 shows a riser backbone for a campus communication network. In the event of fire, both plenum and riser locations pose a potential health risk because smoke and poisonous gases are free to travel through unobstructed pathways around the building structure from floor to floor.

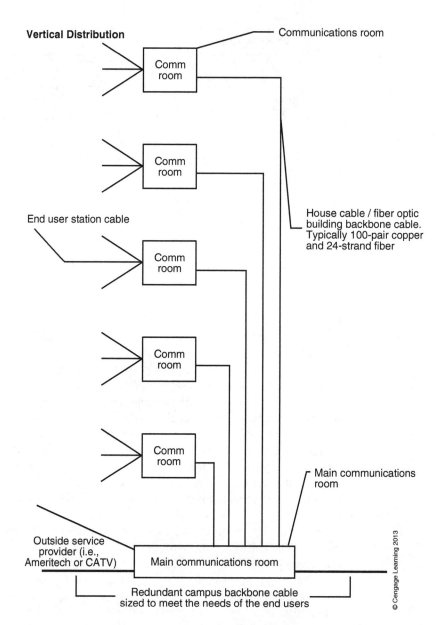

Figure 2–9
Vertical riser backbone. CATV, community antenna and television distribution.

The NFPA recognizes the potential risk for a hazard created by wire and cable in an environmental air space and has included regulations on the use of flammable insulations in all editions of the *NEC* since 1975. Today, the *NEC* states that all cabling not in conduit, installed in plenum spaces or risers, shall be listed as having adequate flame resistance and low smoke-producing characteristics. UL is one agency that tests for vertical flame, smoke, and emissions, certifying that manufacturer's cables meet the installation requirements of a plenum or riser space.

Abandoned Cable

Abandoned cable is classified as any Class 2, Class 3, power-limited tray cable (PLTC), power-limited fire alarm cable, optical fiber, audio, or communication cable not terminated at equipment and not identified for future use with a tag. All abandoned cable must be removed from a building space provided that it is accessible and not located within a concealed space. A concealed space is defined as a space closed off by the finish of a building. Obviously, the *NEC* does not require the removal of walls to facilitate the removal of all abandoned cable. The ruling applies only to exposed cables located within accessible parts of the structure.

Electrically, low-voltage wiring is not a major cause of fire, but in the event of a fire, the cables themselves act as a potential fuel source. Currently, a greater concern has to do with the large amounts of preexisting, nonplenum wire found in building structures, much of which has been disconnected electrically but has never been physically removed from the space. These abandoned cables pose a threat to life and safety if they ever combust. During a fire, such cables should be considered a potential toxic fuel source that not only help the fire to burn, but also spread dangerous fumes and gases through the environmental air spaces. As a result, the *NEC* now requires that abandoned cables be removed from a building or structure, provided that they are not terminated at both ends; tagged for future use; or concealed by the building structure, thereby making them inaccessible.

See definitions in *NEC*: *Article 725, Abandoned Class 2, Class 3, and PLTC Cable; Article 760, Abandoned Fire Alarm Cable; Article 770, Optical Fiber Cable and Raceways; Article 800, Abandoned Communications Cable; Article 820, Abandoned Coaxial Cable; Article 830, Abandoned Network-Powered Broadband; Article 640, Audio Signal Procession, Amplification, and Reproduction Equipment.*

Plenum versus Riser

Whereas plenum cables can be used for any application, riser cables can be used only in risers or below raised floors where the underfloor area is used for ventilation, as illustrated in *Table 725.154* of the *NEC*. Many companies, therefore, find it easier and more cost effective to use only plenum cables for all installations. By only purchasing plenum wire, a contractor often can offset the increased cost of plenum cable by buying in bulk, thereby receiving additional discounts from manufacturers.

The printing on the external jacket or internal marker tape identifies listed and classified plenum cable every 24 in. along the length of the cable (*NEC 310.120*). The jacketing for plenum cable typically is made from polyvinylidene diflouride, which in laboratory tests exhibits a low-flame spread and has low smoke-producing properties. Comparatively, riser cable usually is manufactured with a high-strength PVC outer jacket.

SEC 2.4   CLASSIFIED VERSUS LISTED CABLES

Refer to *NEC, Article 100, Definition: Listed*.

Wire and cables must be classified for a specific purpose of use, such as plenum, riser, commercial (general-purpose), or residential dwellings. Within each category, there are various types of cable ranging from Class 2 (CL2) or Class 3 (CL3); power-limited fire protective (FPL); non-power-limited fire protective (NPLF); communications (CM); multipurpose (MP); power-limited tray cables (PLTCs); and community antenna and television distribution (CATV).

To be classified, cables first must be tested through an approved agency such as UL or the CSA. All cables must pass a series of flame tests to qualify as suitable for a specific purpose of use. One such test is the UL910 test, often called the Steiner Tunnel Test, which evaluates flame spread and smoke emissions of cables and building materials as they are incinerated inside of a 25-foot horizontal chamber. The Steiner Tunnel was developed by UL engineer Albert J. Steiner nearly 20 years ago; it is currently located in Northbrook, Illinois. The listing of cables such as CMP (communications plenum) or MPP (multipurpose plenum) are then designated for such use only after the cable has exhibited the minimal amount of low-flame spread and smoke density, as specified by UL910 regulations. A printed label every 24 in. on the external jacketing or internal marker tape must then identify all listed and classified cables.

Another test, the vertical flame test, defined by the Multi-National Harmonized Communication Cable Standards CSA, C22.2, No. 0.3-M 1985, Test Methods for Electrical Wires and Cables, states that the damage by spread of fire for cables in cable trays is not to exceed a vertical flame or char of 1.5 m (4 ft 11 in).

Other similar vertical flame test examples include UL-1581 VW-1, FT4/IEEE 1202, IEEE 383, IEC323-3, and ICEA T-29-520. A comparison of each of these tests follows.

Vertical Flame Tests

UL-1581

UL-1581 test characteristics include 70,000 BTU/hour; burner position: horizontal, 3 in. (7.6 cm) from sample, 18 in. (45.7 cm) from tray base; tray dimension: 8 ft (2.4 m) in length, 12 in. (30.4 cm) in width, 3-in. (7.6 cm) side flanges; 20-min flame duration. Cable has failed test where the cable blistering has reached the top of the sample after cable has self-extinguished.

FT4/IEEE 1202

FT4/IEEE 1202 test characteristics include 70,000 BTU/hour; burner position: 20° up from horizontal, 2.95 in. (7.5 cm) from cable surface, 11.8 in. (30 cm) above floor; tray dimension: 9.84 ft (3 m) in length, 11.81 in. (30 cm) in width, 2.85-in. (7.2 cm) side flanges; 20-min flame duration. Cable has failed test where the cable char has exceeded a length of 4.94 ft (1.5 m).

IEEE 383

IEEE 383 test characteristics include 70,000 BTU/hour; burner position: horizontal, 3 in. (7.6 cm) from sample, 18 in. (45.7 cm) above tray bottom; tray dimension: 8 ft (2.4 m) in length, 12 in. (30.4 cm) in width, 3-in. (7.6 cm) side flanges; 20-min flame duration. Cable has failed test where the cable blistering has reached the top of the sample after cable has self-extinguished.

IEC 323-3

IEC 323-3 test characteristics include 70,000 BTU/hour; burner position: horizontal, 2.95 in. (7.5 cm) from cable surface, 23.6 in. (60 cm) above floor; tray dimension: 11.5 ft (3.5 m) in length, 19.7 in. (50 cm) in width, no flanges; 20-min flame duration. Cable has failed test where the cable charring has reached a height of 98.4 (250 cm) in. above bottom of the burner.

ICEA T-29-520

ICEA T-29-520 test characteristics include 210,000 BTU/hour; burner position: horizontal, 8.25 in. (21 cm) from cable surface, 12.25 in. (31 cm) above tray base; tray dimension: 8 ft (2.4 m) in length, 12 in. in width (30.4 cm), 3-in. side flanges (7.6 cm); 20-min flame duration. Cable has failed test where the cable blistering has reached the top of the sample after cable has self-extinguished.

Therefore, which type and classification of cable is needed for your installation? It depends on the National and Local Building and Fire Codes associated with the actual location of the installation site. For example, when it comes time to specify the flammability rating of a communications cable, markings such as CMP, CMR (communications riser), CMG (communications, general purpose), or CMX (communications, residential) are used to specify the many varieties. Cable classifications, together with flame test ratings, are printed along the outer jacketing for ease of identification. Listing and standards agency labels such as UL, CSA, Telecommunications Industry Association/Electronic Industries Alliance (TIA/EIA), or Institute of Electrical and Electronics Engineers (IEEE) is also included.

A typical jacketing label may appear as the following:

(UL) 24 AWG Type CMG 4PR FT4 ELT VERIFIED TIA/EIA 568-A CAT5
PATCH CABLE

The above example is a CAT5 network communications patch cable. ELT stands for electrical laboratory test. Let us now breakdown the cable classifications into more detail.

Types of Classified Cable

The following lists outline the various types of cables described within *Articles 725, 760, 800,* and *820.*

National Electrical Code Article 725

NEC Article 725				
	Plenum	**Riser**	**Commercial General Purpose**	**Residential**
Class 2	CL2P	CL2R	CL2	CL2X
Class 3	CL3P	CL3R	CL3	CL3X
Power-Limited Tray Cables	None	None	PLTC	None

© Cengage Learning 2013

Types CL2P and CL3P (Class 2 and 3 plenum) cables are listed as being suitable for use in ducts, plenums and environmental air spaces, as well as for having adequate fire-resistant and low smoke-producing characteristics.

CL2R and CL3R (Class 2 and 3 riser) cables are listed as being suitable for use in risers and shall also be listed as having adequate fire-resistant and low smoke-producing characteristics capable of preventing the carrying of fire from floor to floor.

CL2X and CL3X (Class 2 and 3 residential) cables are rated for use in dwellings or raceways and shall be permitted to have a total diameter of less than 0.25 in.

PLTC/CL3 can be rated for outdoor use, as well as being sunlight and moisture resistant. Cables installed in outdoor or indoor cable trays and raceways shall be listed as resistant to the spread of fire. Conductor sizes for PLTC shall range from 22 AWG through 12 AWG and must have an insulation rating suitable for up to 300 V. Conductor sizes for Class 3 cable shall not be smaller than 18 AWG. The insulation rating for Class 2 cable shall not be less than 150 volts, and for Class 3, not less than 300 volts.

National Electrical Code Article 760

NEC Article 760 covers the installation of wiring and equipment of fire alarm systems including all circuits controlled and powered by the fire alarm system. The following list defines the various types of cables and their uses covered under *Article 760*:

NEC Article 760				
	Plenum	**Riser**	**Commercial General Purpose**	**Residential**
Power-limited, fire-protective (FPL) signaling cable	FPLP	FPLR	FPL	None
Non-power-limited, fire-protective (NPLF) signaling cable	NPLFP	NPLFR	NPLF	None

© Cengage Learning 2013

Types FPL and NPLF cables are listed for general use only (not including ducts, plenums, or other space used for environmental air) and shall also be listed as being resistant to the spread of fire.

Types FPL plenum (FPLP) and NPLF plenum (NPLFP) cables are listed as being suitable for use in ducts and plenums and other space used for environmental air and shall also be listed as having adequate fire-resistant and low smoke-producing characteristics.

Types FPL riser (FPLR) and NPLF riser (NPLFR) cables are listed as being suitable for use in risers and shall also be listed as having adequate fire-resistant and low smoke-producing characteristics capable of preventing the carrying of fire from floor to floor.

Non-power-limited-fire alarm circuits shall be rated as Class 1 circuits. Power-limited fire alarm circuits shall be rated as Class 3 circuits.

National Electrical Code Article 800

NEC Article 800 covers telephone, telegraph (except radio), outside wiring for fire alarm and burglar alarm, and similar central station systems, and telephone systems not connected to a central station system but using similar types of equipment and methods of installation and maintenance. The difference between CM and MP cables is that MP consists of multiple conductors of various types in one cable, whereas a CM cable provides a single type of connection. The following list defines the various types of cables and their uses covered under *Article 800*:

NEC Article 800					
	Plenum	**Riser**	**Commercial, General Purpose**	**Under Carpet Wire and Cable**	**Residential**
Communications	CMP	CMR	CMG, CM	CMUC	CMX
Multipurpose	MPP	MPR	MPG, MP	None	None

© Cengage Learning 2013

Type CMP and MPP cables are listed as being suitable for use in ducts and plenums and other space used for environmental air and shall also be listed as having adequate fire-resistant and low smoke-producing characteristics.

CMR and MPR (multipurpose riser) cables are listed as being suitable for use in risers and shall also be listed as having adequate fire-resistant and low smoke-producing characteristics capable of preventing the carrying of fire from floor to floor.

Types CMG or CM and MPG or MP (general multipurpose) cables are listed for general use only (not including ducts, plenums, or other space used for environmental air) and shall also be listed as being resistant to the spread of fire.

Type CMUC (undercarpet communications cable) shall be listed as being suitable for undercarpet use and shall also be listed as being resistant to flame spread.

CMX cables shall be listed as suitable for use in dwellings or raceways and shall also be listed as being resistant to flame spread. CMX cables shall be permitted to have a total diameter of less than 0.25 in.

Communications cables shall have an insulation rating of not less than 300 volts.

National Electrical Code Article 820

NEC Article 820 covers CATV and radio distributions systems. The following list defines the various types of cables and their uses covered under *Article 820*:

NEC Article 820				
	Plenum	**Riser**	**Commercial General Purpose**	**Residential**
Community antenna television and radio	CATVP	CATVR	CATV	CATVX

© Cengage Learning 2013

Type CATVP (community antenna, television, and radio plenum) cables are listed as being suitable for use in ducts and plenums and other space used for environmental air and shall also be listed as having adequate fire-resistant and low smoke-producing characteristics.

Type CATVR (community antenna, television, and radio riser) cables are listed as being suitable for use in risers and shall also be listed as having adequate fire-resistant and low smoke-producing characteristics capable of preventing the carrying of fire from floor to floor.

Type CATV cables are listed for general use only (not including ducts, plenums, or other space used for environmental air) and shall also be listed as being resistant to the spread of fire.

Type CATVX (community antenna, television, and radio, residential) cables are rated for dwellings or raceways and shall also be listed as being resistant to flame spread. CATVX cables shall be permitted to have a total diameter of less than 0.375 in.

Hierarchy of Cable

NEC Figures 725.154 (G), 760.154(D), 800.154(b), and 820.154(b) display the flowcharts of cable substitution hierarchy. As stated earlier, plenum can be used in place of riser cable, riser cable can be used in place of commercial or general-purpose cable, and general-purpose cable can be used in place of residential. It should now be clear why contractors would rather choose to purchase plenum type cable exclusively and therefore not deal with the various substitution options at each level. By purchasing only types MPP, CMP, FPLP, CL3P, or CL2P cable, a contractor can not only accomplish any level of installation, but also can simplify ordering and reduce material costs in the process.

SEC 2.5   CONDUCTOR SHIELDING

Manufacturers offer a wide range of **shielding** for electrical cables. Shielding is designed to help eliminate or reduce the effects of interference and noise on technical circuits. (See *NEC Article 300.40, Insulation Shielding*.)

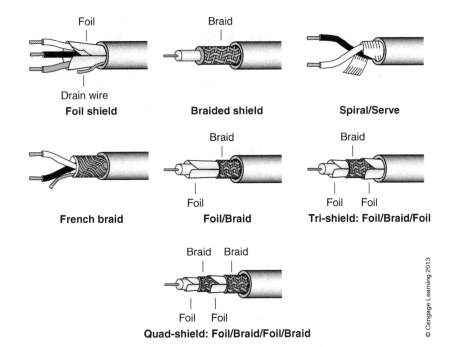

Figure 2–10
Various types of cable shield designs.

Shielding usually consists of aluminum foil, or it can be braided from individual strands of copper or aluminum. The shield can surround the insulation of a single conductor or a grouping of multiple conductors. The ideal braid can be made to cover from 60% to 100% of the insulated conductors, depending on the tightness of the weave. For added protection, a foil and braid combination is often used when designing most specialty-type cables. The shield then is typically covered over by an insulated outer jacket. The following sections describe typical shield designs (Figure 2–10).

Foil Shields

Foil shields consist of aluminum foil laminated to polyester or polypropylene film. The film backing provides additional mechanical strength to the aluminum, helping to prevent unintentional tearing or breaking of the foil as cables are flexed or stressed. A drain wire typically is included alongside the foil shield to help make ground terminations easier, thus providing a continuous discharge path for electrostatic energy. The downside of using foil shielding is that the dc resistance along the shield is greater than that of braided designs. Foil shields are more lightweight and flexible than braided shields, but they also have a shorter lifespan because of their low mechanical strength. Generally, foil shields are more effective on radio-frequency noise than braided shields because they provide 100% cable coverage.

Braid Shields

Braided shields consist of alternating weaves of copper (tinned or bare) or aluminum strands. One layer is woven in a clockwise direction, whereas the other is woven in the counterclockwise direction. Braids offer lower dc resistance than foil, and structurally provide greater strength to the shield and the cable. The cable also maintains good flexibility as a result of the braided design, and often lasts longer. The downside of a braided shield is that it can never be made to cover 100% of the cable; 60% to 95% coverage is typical. As a result, high-frequency noise signals eventually penetrate the weave. Braided shields are also bulkier, resulting in larger cable diameters, and in some cases, they are harder to terminate.

Spiral/Serve Shields

A spiral/serve shield consists of a wire wrapped in a spiral around an inner cable core. The spiral can provide up to 97% coverage, while still maintaining a high degree of flexibility and termination ease. The downside of spiral/serve shielding is that it can only be used in the range of audio frequencies, because the twisting creates a coiling effect that increases the total inductance of the cable.

French Braid Shields

A French braid shield is a double spiral that ties two spirals together through a single weave. The design provides improved flex life over a standard spiral and offers more flexibility than those of conventional braids. French braids also help to reduce microphonic and triboelectric noise.

Combination Shields

Foil Braid

Foil shielding is often used in combination with an outer braid. The combination offers greater protection from noise and interference, with increased mechanical strength as well.

Tri-Shield

Tri-shielding consists of a foil braid combined with an added layer of outer foil—foil, braid, foil. The second foil layer provides an additional barrier to unwanted noise signals and improves shield reliability.

Quad Shield

Quad shielding provides additional protection by adding a second layer of braid, essentially tri-shielding with an outer braid—foil, braid, foil, braid. (See *NEC Article 310.10 (E), Shielding.*)

The use of foil or braided shields helps to isolate conductors from stray electro-magnetic fields (EMFs) that can often be induced into a circuit, causing crosstalk between conductors, interference, and abnormal operation. Shielding also helps prevent electrostatic pickup of noise signals capacitively coupled from neighboring conductors through the insulation. Shields are usually connected to ground at the source side of the transmission line, providing a low-impedance pathway for induced and capacitively coupled noise currents. This helps to drain them these noise currents away from the circuit, thus the term *drain wire*. Low-voltage and solid-state control devices especially need to be shielded from EMFs, from sources such as radio waves, fluorescent lights, arcing of electric motors, generators, switches, and relays, because of their high degree of sensitivity.

Crosstalk

Crosstalk refers to the unwanted transfer of signals from one conductor to another through the process of electromagnetic induction or capacitive coupling. Neighboring conductors running through the same cable typically have this problem if they are not shielded from each other in some manner. Computer cables are available that provide separately shielded pairs within the same bundle, offering increased protection from crosstalk and stray EMF, if necessary.

An additional solution to inductively coupled noise would be to turn neighboring conductors 90° from each other. The 90° shift essentially polarizes the signals, and because inductively coupled noise can only be transferred by conductors having the same orientation, the chance of unwanted pickup is eliminated.

If it is not possible to turn neighboring conductors 90° from each other, then distance is required to help minimize the risk for unwanted interference. A separation distance of 2 to 3 ft is often required. The farther away interfering power or signaling cables are from each other, the better chance of containing and isolating them, helping to greatly reduce the pickup and transmission of unwanted noise throughout the system.

Shielding Low Frequency

Although aluminum shields work quite well to help eliminate high-frequency interference, they tend to have almost no effect on the low frequency, 60-Hz EMF, typically emanating from a basic power supply. Studies have found that aluminum and nonconducting plastics have virtually no effect on low-frequency radiation. The problem, however, usually can be solved by encasing the circuit in a steel enclosure or the connecting conductors in a galvanized rigid conduit or steel raceway. Steel enclosures are the best remedy for unwanted, low-frequency EMF. In situations where control circuits are near large power generators, or subjected to high levels of low-frequency radiation, steel or galvanized rigid conduit offers the best solution to the problem.

SEC 2.6   ELECTRICAL PROPERTIES OF CABLES

This section discusses the electrical properties of cables and wire, including wire resistance, capacitance, characteristic impedance, and frequency response.

Resistance of Wire

Resistance is defined as the opposition to current flow in a dc circuit. The properties that affect the resistance of a wire are type of conductor, length, and cmil.

Copper, because of its increased density, provides far more conductivity than that of an equally sized aluminum wire, based on temperature and material composition. As CM area increases (smaller AWG number), the resistance of the wire decreases; think of it as a larger pipeline that can provide more flow. Likewise, as cable length increases, resistance increases. The following formula is used to calculate the resistance (R) of a conductor in a dc circuit:

$$KL \div CM = R$$

where K represents ohms per mil-foot (mil-foot values are temperature dependent and different for all metals), L represents length of wire in feet, and CM represents the circular mil area of the conductor.

K values are material and temperature specific, representing the resistance of a conductor that equals 1 foot in length and 1 mil in thickness; remember that 1 mil represents an overall diameter of 1/1000 in. Different types of conductor, such as copper and aluminum, have specific ohms per mil-foot values that are not only based on the quality of the metal, but also on temperature. Increasing the temperature of a conductor usually increases the overall resistance. For this reason, resistance must always be calculated at specific temperatures of operation. *Chapter 9, Table 8* of the *NEC* lists wire gauges, showing dc resistance values for conductors of copper or aluminum at a temperature of 75°C; a formula is also included at the bottom of this table for calculating resistance at different conductor temperatures.

Voltage Loss at the Load

Voltage loss at the load can be calculated by Ohm's law, once the total resistance of the connecting cable is known.

$$I \times R = V$$

where I represents current flow.

Because cable resistance increases with length, the farther away the load is from the power supply, the greater the loss. Oftentimes a load may be hundreds or even thousands of feet from the main power supply, meaning that the actual conductive distance is double, because circuit current needs to travel out to the load and then back; therefore, a 500-foot run to the load turns out to be a 1000-foot conductive loop.

By substituting Ohm's law into the previous formula, the calculation can be rewritten as:

$$V_{line} = KLI \div CM$$

The voltage loss on any length of conductor can be found by simply knowing the total current flow in the loop. Because a voltage drop on the power lines represents a loss of power at the load, the expected load voltage can be found by subtracting the line loss from the source voltage.

$$V_{source} - V_{line} = V_{load}$$

A measured loss of voltage at the load also implies that the connecting wires are most likely consuming power. If the problem becomes too severe, the connecting wires can actually be dissipating a significant amount of heat, which is why cables need to be correctly sized for ampacity to help reduce the possible risk for fire.

The definition of *ampacity* is given in the *NEC* as *the current in amperes a conductor can carry continuously under the conditions of use without exceeding its temperature rating.*

The consumption of power on the connecting cables or transmission line results in heat dissipation and increased temperatures, caused by molecular friction, because circuit current flows through the calculated resistance of the conductors.

The formula Power $= I^2R$ shows how the total circuit current or conductor resistance can directly affect the power consumption and level of heat dissipation on the connecting wires. Cable ampacity must therefore be calculated using temperature ratings that are lower than the maximum temperature ratings of the connecting cable; otherwise, the potential risk for fire is increased.

When laying out a low-voltage system, such as a series of horns in an apartment building (Figure 2–11), it becomes necessary to calculate the maximum amount of load the system can handle based on the length of cable run from apartment to apartment, and the maximum possible current flow in the loop.

Knowing the number of annunciators that can be connected on a single line without having significant loss of end-line voltage is critical. The system power supply may be outputting 12 V at the front end of the circuit, but by the time the power reaches the last apartment, it may be less than 10 V. Why should we be concerned about a 2-V loss? Because the horn may require a minimum of 11 V to operate and 10 volts at the end of the line causes the last horn to malfunction, resulting in a potential safety hazard. The end-of-line resistance is also chosen at a specific value to achieve a specific level of current flow in the loop as a way of monitoring the integrity of the system for security purposes.

Figure 2–11

Voltage loss on long-distance runs.

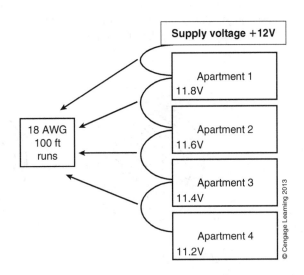

Skin Effect and Signal Attenuation

Although one would expect the value of cable resistance to remain stable in an alternating current (ac) circuit, the reality is just the opposite. Increasing signal frequency results in greater levels of ac resistance along the length of the cable. The phenomenon known as skin effect causes the conductive surface area of the cable to decrease as signal frequencies increase. At greater than 1 MHz, current flow tends to move along the edges of the conductor, effectively bypassing the center of the cable. Ultimately, the skin effect increases the total ac resistance or impedance of a cable by reducing the available conductive surface area as frequencies increase (Figure 2–12).

The effect is not very noticeable at low frequencies (dc to 100,000 Hz), but in the range of 1 MHz or greater, the problem can become quite severe, resulting in noticeable IR losses (voltage loss) and signal **attenuation** at the load. Attenuation is a measurement of signal loss, typically specified in decibels. As a result, specifications for communication cables usually list attenuation losses at different frequency levels, per 100 ft or 100 m of length.

As an example, a 1 MHz signal traveling down 1000 ft of solid conductor, 20 AWG, coaxial cable can lose up to 3 dB of signal voltage, or possibly more depending on the quality of the conductor and the material properties of the insulation or dielectric. Although 3 dB may not appear to be significant, in actuality, it represents a loss of more than 30% of the signal voltage and half the power being delivered to the load. Connecting any additional cable to the run would result in even more attenuation and possibly cause extreme degradation of signal at the end-line termination point. Chapter 4 discusses decibels and their calculations in more detail.

One possible solution for extreme attenuation of signal over the length of a cable would be to add a front-end amplifier to help boost the overall level and compensate for losses at the load. The amplifier needs to be attached at the front end where the signal is strongest, because attaching it too close to the load would only serve to amplify an already poor signal having a low signal-to-noise ratio. Although adding an extra amplifier increases the overall cost of the project, in some cases it may be the only solution to the problem.

Figure 2–12
Skin effect. (A) At low frequency, the entire cross-sectional area of the conductor is used. (B) At high frequency, the skin effect causes current to flow only along the outer edges of the conductor.

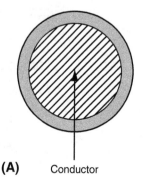

(A)   Conductor

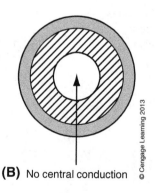

(B) No central conduction

© Cengage Learning 2013

Choosing the correct cable for the job usually depends on the length of run and the frequency of the transmitting signal, but there is also one more critical topic of discussion: the characteristic impedance of the cable. Not matching the load to the characteristic impedance of the cable results in even more losses and signal degradation. However, before jumping into a discussion of characteristic impedance, some explanation of what defines impedance and the components of a transmission line is necessary.

Impedance

Refer to *NEC Article 110.10, Circuit Impedance and Other Characteristics* for more information.

Impedance represents a combination of dc and ac resistance. Ac signals do not see the same level of resistance as those of dc. Compared with dc, ac currents react to the total level of inductance and capacitance on the line, which is why ac resistance is referred to as reactance. There are therefore two forms of frequency-dependent reactance: inductive reactance (X_L) and capacitive reactance (X_C) (Figure 2–13). As frequency increases, inductive reactance increases, whereas capacitive reactance decreases; the two are exactly opposite in nature and 180° out of phase with each other. If equal in magnitude, X_L and X_C completely cancel each other, leaving only dc resistance. The impedance, or Z, of the circuit, represents the vectored sum of the total inductive and capacitive reactance and dc resistance. Now let us apply these principles to a transmission line.

Components of a Transmission Line

Figure 2–14 represents the model of a short length of open-ended transmission line. Figure 2–14(A) represents a balanced line that has an equal level of impedance from either side of the line to ground, allowing for the cancellation of common mode signals that are traditionally the result of stray electromagnetic interference. A balanced line essentially guarantees that equal-level noise signals are induced on each side of the transmission line, because the overall proximity of the conductors to the electromagnetic interference is the same, and because the conductors are rated to identical specification and length. (Incidentally, the source of stray EMF may be from a nearby generator or simply the 60 Hz power radiating out of

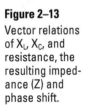

Figure 2–13

Vector relations of X_L, X_C, and resistance, the resulting impedance (Z) and phase shift.

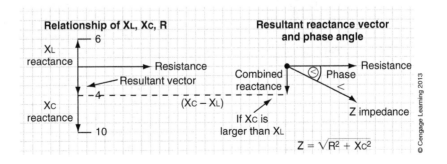

Relationship of XL, Xc, R

Resultant reactance vector and phase angle

$Z = \sqrt{R^2 + X_C^2}$

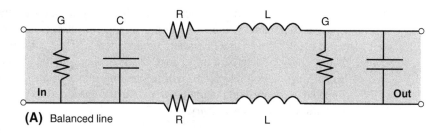

Figure 2–14
Model of a short
length of trans-
mission line.
(A) Balanced
line. (B) Unbal-
anced line.

(A) Balanced line

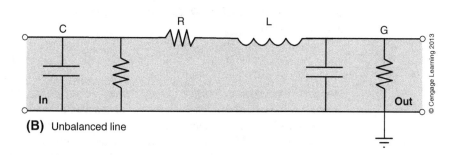

(B) Unbalanced line

the florescent lights.) Although both sides of the transmission line are 180°
out of phase with each other and balanced to ground, any signal common to
both conductors mathematically cancels out. Furthermore, any magnetic fields
radiating out from the conductors themselves, generated as a result of current
flowing down the line, also cancel for exactly the same reason, thus helping to
reduce possible line emissions that could otherwise interfere with neighboring
conductors.

Consider a 5 V signal traveling on a balanced line. One side of the line is
+2.5 V above ground, whereas the other side of the line is –2.5 V below ground,
thus measuring 5 V between. If 1 V of noise were induced equally into each
conductor, the new voltages would be +3.5 and –1.5 V. Notice that the difference between the lines is still the original 5 V. A balanced line remains immune
to induced noise because the level of signal between the two sides of the line
does not change.

Figure 2–14(B) represents an unbalanced line that is grounded on the lower
side, thus explaining why the resistance and inductance only appear on the top side
of the circuit. Unbalanced lines are unable to cancel out mutually induced noise
or electromagnetic interference because current flow on the ground side is flowing through zero ohms (the impedance of an ideal ground plane) and is unable to
develop the counteracting magnetic field needed to cancel out the radiated energy
around R and L (see Figure 2–14(B)) on the opposite side of the line.

For the unbalanced line, let us assume that the ungrounded side is sitting at
+5 V. Where a +1 V spike of EMF is induced into the line, the overall signal
level pushes up to +6 V. The ground side, however, remains at zero. Because the
ground side of the line is unable to shift by the needed 1 V, as was the case with
the balanced line, noise cancellation cannot occur. Therefore, the overall signal

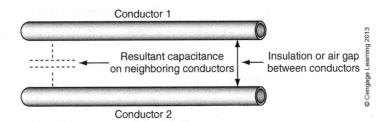

Figure 2–15
Conductor
capacitance.

Conductor 1

Resultant capacitance
on neighboring conductors

Insulation or air gap
between conductors

Conductor 2

level on the transmission line increases or decreases depending on the polarity of the induced noise. For this reason, unbalanced lines should use shielded or coaxial cable to help minimize the risk for electromagnetic interference.

Cable Capacitance

The capacitance on a transmission line, often called mutual capacitance, results from the electrostatic connection between two conductors and their separating insulation (Figure 2–15). The dielectric properties of the insulation, spacing, quality, and size of the conductors affects the level of capacitance per foot. Longer lengths of cable therefore exhibit greater levels of capacitance.

Bit Rates

For the data industry, high levels of capacitance on the transmission line may cause rounding of data edges and loss of signal definition. Where digital bits are not clearly square, they become unintelligible by the receiver, causing high bit error rates and loss of data (Figure 2–16). For this reason, data cable manufacturers often specify maximum transmission speeds as **bit rates**, or number of bits per second, over specific lengths of cable. Longer cable runs therefore have limited bandwidth and reduced bit rates attributed to them.

Cable Conductance

The "G" component (see Figure 2-14) represents the conductance of the dielectric or insulation. All insulations have different dielectric ratings that allow a certain amount

Figure 2–16
Data bits. (A)
Normal-shaped
data bits, no
distortion. (B)
Rounded data
bits. (C) Data
bits with over-
shoot.

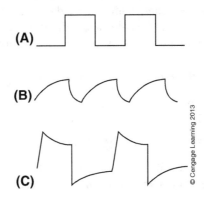

(A)

(B)

(C)

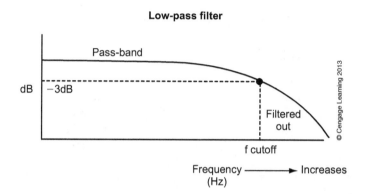

Figure 2–17
RC circuit,
low-pass filter
formed by cable
resistance and
cable capaci-
tance.

of conduction, small as it may be. There is no such thing as a perfect insulator, and any material eventually conducts if subjected to high enough levels of voltage.

Low-Frequency Transmission and Bandwidth

At low frequencies (dc to 50,000 Hz), levels of inductive reactance on transmission lines are insignificantly small compared with line resistance, and capacitive reactance is extremely high. Because L can be ignored, the cable now appears to be an RC low-pass filter (Figure 2–17), and characteristic impedance does not apply. For this reason, wire specifications for audio and communications cable list the amount of capacitance per foot or meter. A designer needs to know the value of capacitance on a specific length of cable to be able to calculate the low-frequency bandwidth. The bandwidth represents the acceptable range of transmittable frequencies over a specific length of cable. Transmitting frequencies above a certain range results in signal attenuation and loss of data.

Cutoff Frequency and Signal Attenuation

The maximum transmittable, sinusoidal frequency on a specific length of cable can be calculated by using the following formula:

$$\text{Maximum cutoff frequency} = 1/2\,\pi\,RC$$

where R is the total resistance of the cable plus the source resistance, and C represents the total capacitance of the cable. Simply multiply the length of cable in feet by the amount of capacitance per foot and you have the total capacitance of the transmission line. As cable length increases, so too does capacitance, and as circuit capacitance increases, the maximum transmittable frequency along the line decreases, attenuating and virtually eliminating all signal voltage above the cutoff frequency. The maximum cutoff frequency represents the upper cutoff of the low-pass filter, formed by connecting the source to the transmission cable. Any signals above the cutoff frequency are attenuated by more than 3 dB, which means that the end-line power being delivered to the load is less than half the input source power. Choosing a cable with low capacitance becomes critical where source resistance is high, transmission lines are long, and higher signal frequencies are desired.

High-Frequency Transmission

Once signal frequencies increase to greater than 100,000 Hz, inductive reactance on the transmission line increases significantly, whereas capacitive reactance decreases. As a result, cable resistance starts to become a more insignificant variable when compared with the higher levels of inductive and capacitive reactance. On short lengths of cable, resistance usually can be ignored. Longer runs may also have to consider losses due to skin effect, especially at very high frequencies, such as in satellite transmission, but for now the characteristic impedance of the transmission line is of greater concern.

Characteristic Impedance or Nominal Impedance

When cables are first powered up, the current does not flow instantaneously because of the distributed capacitance and inductance along the length of the transmission line. A surge of energy travels down the line charging the components. The **characteristic impedance** of the cable defines the cable as a charge resistance. Although the level of current is not zero, due to the charging capacitance along the line, and because it cannot be at maximum, due to the opposing inductance preventing any substantial flow, it is however, some finite measurable value. The characteristic impedance defines how much opposition the surge current faces as it moves down the transmission line. The transmission line can therefore be represented as an impedance value for all high-frequency transmissions, as shown in Figure 2–18.

The value of characteristic impedance for a line is determined by the geometry of the conductors, type of insulating material, and spacing. Figure 2–19 shows how the calculation differs when comparing a balanced line with that of a coaxial. Provided that the characteristic impedance of the line and connecting load are equal, the charge current and transmitted power dissipates 100% without reflecting back to the source; this is called matching the line to the load.

Matching the Line to the Load

High-frequency power that does not dissipate into the load reflects back to the source along the transmission line, just as waves of water on the ocean (see Figure 2–18(B)). The reflected waves of energy along the line are called **standing waves**; the data

(A) Characteristic impedance is equal to load.

Figure 2–18
(A) A matched transmission line resulting in 100% power transfer to the load and no standing wave. (B) An unmatched transmission line resulting in reflected power and a standing wave on the transmission line.

100%
power dissipates

50Ω ⟩ Ω internal

50Ω transmission line

50Ω
load

Signal
source

Power transfers to load
 with no reflection.
Transmission line appears transparent.
No standing wave.

(B) Characteristic impedance is not equal to load.

Load does
not receive
all of the
power.

50Ω ⟩ Ω internal

300Ω transmission line

Standing wave

50Ω
load

Incident power

Power reflects back to source due to
 mismatched line and load.
Standing wave appears on the
transmission line.

© Cengage Learning 2013

Examples of transmission lines

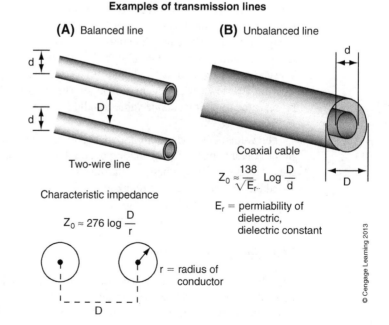

Figure 2–19
(A) Characteristic impedance of parallel conductors, balanced line. (B) Characteristic impedance of an unbalanced coaxial line.

(A) Balanced line

d

D

d

Two-wire line

Characteristic impedance

$$Z_0 \approx 276 \log \frac{D}{r}$$

r = radius of
conductor

D

(B) Unbalanced line

d

Coaxial cable

$$Z_0 \approx \frac{138}{\sqrt{E_r}} \, \mathrm{Log} \, \frac{D}{d}$$

D

E_r = permiability of
dielectric,
dielectric constant

© Cengage Learning 2013

communications industry refers to them as **structural return loss**. To prevent standing waves from occurring, the load must match the source impedance and the characteristic impedance of the line as closely as possible, thus guaranteeing 100% power transfer to the load without losses. In a purely resistive circuit, this is easy to achieve, but in most cases loads are inductive or capacitive, making the situation a bit more difficult to control.

In reality, it is impossible to match the line to the load perfectly because of the effects a changing signal frequency has on the load impedance. Often, wide ranges of complex frequencies are transmitted along the line, causing the load impedance to shift slightly in response. The shifting load impedance therefore results in standing waves reflecting on the line. A small amount of wave refection is tolerable, but if the load is severely mismatched from the characteristic impedance of the line, then large amounts of reflected energy or standing waves can result in the development of high-voltage nodes along the transmission line that can eventually break down and destroy the dielectric material of the cable. Additional symptoms of standing waves include greater levels of noise, increased heat dissipation, and possible signal ghosting.

Notably, the characteristic impedance of the line is not really a physical resistance in the circuit. Connecting a transmission line having 50 ohms (Ω) of characteristic impedance does not mean that the source sees an extra 50 Ω of resistance in the loop. Figure 2–18(A) shows that the source only sees the load impedance. The transmission line appears invisible provided the characteristic impedance of the connecting cable matches the load.

Velocity of Propagation

Velocity of propagation is a cable specification that refers to the charge delay of the transmitted signal as it moves down the cable to the load. The rate of transmission is compared with the speed of light, 300 million meters per second or 186,000 miles per second. Cable specifications usually specify the velocity of propagation as a percentage called the velocity factor. As an example, an 86% velocity of propagation means that the signal is traveling at 86% the speed of light, or 300 $\times$ $10^6 \times 0.86 = 258$ million meters per second. (Incidentally, the velocity factor is almost entirely dependent on the dielectric rating of the insulation material. Therefore, different cable velocity factors are usually the result of using different types of insulation.) By knowing the speed of transmission through the cable and the frequency of transmission, one could calculate the signal **wavelength**, which is necessary for calculating phase shift and delay. Where cable lengths are not exact multiples of a transmitting wavelength, phase shift and signal delay occur. The following formula is used to calculate wavelength:

$$\lambda = v_p/f$$

where λ is wavelength, v_p is velocity of propagation, and f is the frequency of the transmitting signal.

Taking the above propagation example and an 88 MHz transmission signal, the calculated wavelength is:

$$\lambda = 258 \times 10^6 \text{ m/sec}/88 \times 10^6 \text{ Hz} = 3.4 \text{ m per cycle}$$

What this means is that the signal frequency repeats on the transmission line every 3.4 m. Therefore, a 3 m cable causes the signal to reach the load at 317.64°, just short of a complete cycle, which may or may not be a problem, depending on the type of load.

$$3 \text{ m}/3.4 \text{ m} = x/360°$$

Solving for x gives 317.64°.

In some situations, such as mixing video signals, the length of cables is critical for preventing phase shifts, which causes a noticeable color imbalance in the image. Phase shift is usually a problem where multiple high-frequency signals are mixed together.

Low-frequency transmissions may exhibit similar problems, but only where the cables are long in length. As an example, let us calculate the wavelength of a 60 Hz signal on a cable with 86% velocity of propagation:

$$\lambda = 258 \times 10^6 \text{ m/sec}/60 \text{ Hz} = 4.3 \text{ million meters per cycle}$$

As shown, the transmission lines would need to be insanely long just to be able to see one complete wavelength of signal. A total of 4.3 million meters is equivalent to approximately 2672 miles. For this reason, transmission lines do not cause noticeable delays or phase shifts on low-frequency transmissions. Circuits can surely add delays or phase shifts to a signal, but 60 Hz transmission lines do not, unless of course, they are extended from New York to California, in which case the total line capacitance, resistance, and cutoff frequency have more effect on the end-line transmission, possibly showing up as severe signal attenuation and a loss of bandwidth.

Where the length of a transmission line is significantly small compared with the wavelength of the traveling signal, any noticeable delays or phase shifts are insignificant. The problem, however, becomes noticeable at high frequencies, when the wavelengths start decreasing to lengths more in comparison to the actual length of the transmission line. Figure 2–20 lists various frequencies together with wavelengths in meters.

As stated earlier, proper cable selection is about defining the job and then selecting a cable that matches specific criteria. What is important to understand is that circuit load impedances are always specified at specific operating frequencies: audio circuits are meant for signals in the 20 to 20,000 Hz range, video circuits up to 1 MHz,

Figure 2–20

Various frequencies and their associated wavelengths propagating at 300 million meters per second.

Frequency	Wavelength (meters)
1 kHz	300,000
10 kHz	30,000
100 kHz	3000
1 MHz	300
10 MHz	30
100 MHz	3
1 GHz	0.3
$f=c/\lambda$ $\lambda=c/f$	

© Cengage Learning 2013

radio and digital circuits greater than 1 MHz, and satellite systems greater than 1 GHz. Ac power cables are not concerned so much with frequencies greater than 1000 Hz and usually only specify the wire type, gauge, insulation, and maximum voltage rating in their specification data sheets. Communication cables are different from power cables in that their data sheets also specify the nominal impedance (another name for characteristic impedance); capacitance per foot; velocity of propagation; and the amount of signal attenuation (in decibels) over specific frequencies of operation, typically ranging from 1 MHz to 1 GHz, depending on the manufacturer's intended use of the product. It is important to understand all of these variables and concepts to choose the most suitable cable for the job.

SEC 2.7   TYPES OF CABLE

Single-Conductor Cables

Single-conductor cables comes in a variety of sizes (solid or stranded), voltage and temperature ratings, insulation types, and colors. Often called hook-up wire, these cables are most often used to interconnect circuits and control panels or computerized test grids, or both, and are typically found in an experimental or manufacturing environment.

Multiconductor Cables

Multiconductor cables (Figure 2–21(A)) are available in a variety of sizes (solid or stranded), voltage and temperature ratings, and insulation types; they also can be of the shielded or unshielded variety. The number of conductors in a cable can vary from 3 to 50, depending on the type of cable and application. Typical applications for multiconductor cables include communication, instrumentation, control, audio, and data transmission.

A specification example of a multipaired computer data cable is as follows:

Number of pairs	10
Wire AWG	24 (stranded 7 × 32)
Conductor type	Tinned copper, twisted pair
Maximum voltage and temperature	30 V, 80°C
Insulation	Polyethylene
Shielding	100% foil shield, 65% braided shield, 24 AWG stranded drain wire
Standard lengths	1000 ft or 304.8 m
Standard weight	90 lbs or 40.9 kg
Nominal dc Ω	78.7 Ω/km conductor, 7.9 Ω/km shield
Nominal outside diameter	0.427 in. or 10.85 mm
Nominal impedance	100 Ω
Nominal velocity of propagation	78%
Nominal capacitance (between conductors)	12.5 pF/ft or 41 pF/m
Nominal capacitance (between conductor and shield)	22 pF/ft or 72.2 pF/m

Twisted Pair Cable

Twisted pair cables (see Figure 2–21(B)) offer improved noise immunity over that of multiconductor cables. The twist in the cable helps to attenuate high-frequency noise transmission caused by the increased inductance. Because the conductors are grouped into pairs providing equal impedance to ground, they are often used by the telecommunications industry where balanced transmission lines are needed. The twist helps to provide common mode rejection and reduce noise and crosstalk between neighboring conductors. As a result, a twisted paired cable usually offers greater rates of data transmission or bit rates than multiconductor cable and has become the standard for communication cables in the industry. Twisted pair cables are available in a variety of sizes (solid or stranded), voltage and temperature ratings, and insulation types; and they can be of the shielded or unshielded variety.

Varieties of twisted paired cable include UTP (unshielded twisted pair; see Figure 2–21(C)), ScTP (screened twisted pair; see Figure 2–21(D)), and STP (shielded twisted pair; see Figure 2–21(E)).

Solid conductor UTP has become the standard for most data and telecommunication installations because of its relative low cost and ease of installation. Even though UTP is more subject to interference and offers a limited bandwidth, it still continues to be the cable of choice throughout the industry, especially where installers have to make upgrades to older installations and interface with existing wiring.

Figure 2–21

Types of conductors. (A) Multiconductor. (B) Twisted pairs. (C) UTP Category 5 (CAT5). (D) ScTP CAT5. (E) Shielded TP CAT5. (F) Coaxial. (G) Twinaxial. (H) Triaxial.

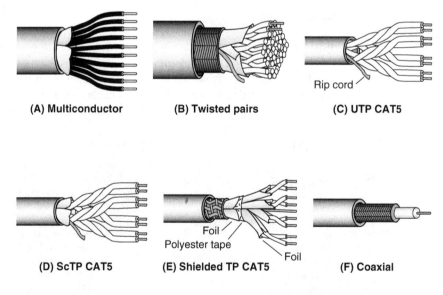

(A) Multiconductor

(B) Twisted pairs

(C) UTP CAT5

Rip cord

(D) ScTP CAT5

(E) Shielded TP CAT5

Foil
Polyester tape
Foil

(F) Coaxial

(G) Twinaxial

(H) Triaxial

© Cengage Learning 2013

The cost of upgrading an existing system to better cabling, which also requires different connectors and terminations if shielding is used, does not make sense and is rarely done. The characteristic impedance of UTP is typically in the range of 100 Ω.

ScTP provides a foil screening around the paired conductors to help reduce the risk for electromagnetic interference, but is unable to prevent interference caused by crosstalk.

STP, although it can provide the most protection against noise and EMF, is not widely used anymore because of its incompatibility with most of the new modular connectors and terminators currently used in the industry. Because the majority of telecommunication and data installations have historically used UTP, interfacing with STP becomes more challenging and difficult; it is often not used unless shielding is absolutely necessary. The characteristic impedance of most STP cables is typically in the range of 150 Ω.

Category 3, 4, 5, 5e, 6, and 7 Cable

Category 3

Category 3 cables (CAT3) are most commonly used for voice transmissions or older style, slow-speed networks having a maximum transmission speed of 16 Mbps. However, because of their limited bandwidth, they are not widely used by the computer industry anymore, except where maybe connecting computer modems to the main telephone line. Incidentally, CAT3 cable was not originally designed as telephone cable; it was meant for voice and data networks. Standard telephone cable consists of UTP, 22 AWG minimum. CAT3 cable has different specification requirements and twist rates than standard UTP, as defined by TIA/EIA-568, which lists the requirements and specifications for all category cables 1 through 6.

By stripping away the insulation from UTP, CAT3, and CAT5 cables, the differences can clearly be seen in how the pairs are bundled and twisted along the length of the cable (Figure 2–22). The various varieties of twisted pair cabling

Figure 2–22

Comparing the twist rates of unshielded twisted pair cable with Category 5 extended (CAT5e), CAT5, and CAT3 telephone wire.

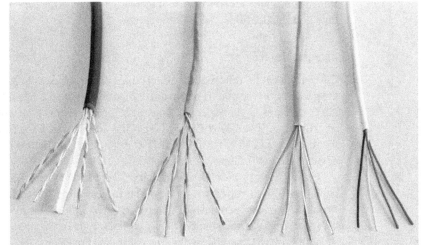

© Cengage Learning 2013

also require appropriate connecting hardware, because the style of the connector together with conductor size and pair spacing ultimately affects the characteristic impedance of the cable and quality of the connection.

Category 4

CAT4 was an interim design, operating at 328 ft (100 m) up to 20 MHz, and is no longer used, because the improved performance of CAT5 has served to dominate the industry.

Category 5

CAT5 cables are most commonly used by local area network (LAN) and Ethernet data networks. Composed of four pairs of 24 AWG, solid-copper UTP, they offer improved performance over traditional UTP cables because the rate of twist per pair differs, which helps to provide minimal crosstalk (referred to as **NEXT, or near-end crosstalk,** in the data industry), low interference from EMF, low attenuation, and increased bandwidth. To help ensure limited attenuation and high bandwidths, CAT5 cable runs are typically limited to a maximum of 328 ft (100 m) and 100 MHz.

Category 5e (CAT5e)

CAT5e, or Category 5 enhanced, was designed to be backward compatible to CAT5 for gigabit Ethernet connections operating at 100 m (328 ft) and 100 MHz, at maximum transmission speeds of up to 1000 Mbps (1 Gbps). Old style Ethernet communication used two pairs: one pair to transmit, and the other to receive. The gigabit Ethernet splits the signal into four parts and transmits it simultaneously down all four pairs. Each pair can transmit and receive duplex communication, such as on a telephone line where you and the caller are transmitting simultaneously on a solitary single-pair wire. The gigabit Ethernet, however, sends compressed data that use less bandwidth, and with four pairs transmitting and receiving simultaneously, it can offer much greater speeds of data communications.

Category 6 and 6a

Category 6 cable, referred to as CAT6, is a cable standard that was developed for the gigabit Ethernet and is backward compatible to the CAT5 and CAT5e standard. CAT6 features tighter specifications offering lower levels of crosstalk and noise transmission. The TIA/EIA 568C cabling standard for CAT6 provides performance of up to 250 MHz at 328 ft (100 m), and is suitable for networks operating at speeds above 1 Gbps.

  The construction of CAT6 contains four twisted pairs, 22–24 AWG copper conductors, with varied twist rates. The increased performance of CAT6 results from a better-insulated design. Compared to CAT5, CAT6 cable includes a central insulated spine in the shape of a cross. Each pair of the four pairs runs along a separate isolated chamber within the cable (Figure 2–23).

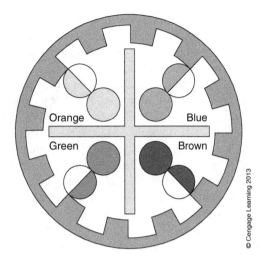

Figure 2-23
CAT6 cable
construction

Orange   Blue

Green   Brown

© Cengage Learning 2013

The added insulation and isolation of pairs helps to improve the overall cable performance. As a result, CAT6 cables are physically thicker and more expensive than CAT5 cables and take up more space in a raceway or cable tray.

Category 6a is referred to as Augmented CAT6. It is similar in design to standard CAT6 cable; however, it has tighter specifications, improved performance, and offers increased transmission frequencies, up to 500 MHz. From a distance point of view, CAT6 can operate up to 10 Gbps at 180 ft (55 m), whereas CAT6a can operate at that speed up to 328 ft (100 m).

Category 7 and 7a

Category 7 and Category 7a cable is a cabling standard for Ethernet that is backward compatible to CAT6 and CAT5. It has tighter specifications and offers improved performance over CAT6 cables. To achieve the tighter specifications, each of the four pairs, as well as the entire cable, are wrapped by a foil shield. As a result, CAT7 cables require special connectors that can accommodate the grounding of the internal shielding. Category 7 is designed to operate up to 600 MHz, on 10 Gbps Ethernets, at distances up to 328 ft (100 m).

Terminal Connectors

Terminal connectors for CAT5, CAT5e, and CAT6 cables use an 8-pin modular design referred to as an RJ-45 connector (RJ stands for "registered jack"). RJ-45 is similar to the RJ-11 and RJ-14 modules most commonly used by the telephone industry and on type CAT3 voice and data installations. RJ-11 and RJ-14 modules can typically be found on telephone equipment, modems, and telephone wall outlets. RJ-45 connectors can be found on most network interface cards (NICs) and computer network devices. Chapter 6 discusses the various color codes and pinout options for the RJ-45 connector, which is typically determined by the type of network installation and the speed of communication.

CAT7 terminations use hardware that can accommodate shielded cable, making them incompatible with most current systems, and except for the occasional new high-speed Internet services being developed, they are not widely used by many network designers or manufacturers.

It is important to use appropriate connecting hardware when installing twisted pair cable. As an example, CAT5 cable should never be used with CAT3 connectors; otherwise, signal performance is limited to CAT3 specifications. When installing RJ-45 connectors, it is also important to maintain the twist in all four pairs. Do not unwrap twisted pairs by more than half an inch. To do so would compromise the characteristic impedance of the junction and the quality of the data transmissions. For CAT6 cables, the specifications for installation are even tighter.

The difference in twist rates among the four pairs also causes the propagation delay of pairs to vary, because more twists of wire cause the total length of wire to increase along a single pair. As a result, the time it takes to transmit on one pair is different than the time it takes for the next. **Delay skew** represents the maximum difference in transmit times among all four pairs. The type of cable insulation directly affects the delay skew, because insulation dielectric directly affects cable capacitance and propagation velocity. The amount of delay skew for a specific cable must be checked to make sure the rates are fast enough between pairs; otherwise, communication glitches occur on some types of networks.

Ethernet standards contain several different physical layer specifications (see discussion in Chapter 6) that call for different types of cable, connecting hardware, and installation methods. The type of selected cable also determines the type of NIC to be used.

The majority of network designs use only two of the four pairs, one in each direction, allowing for full duplex or simultaneous bidirectional communications. High-speed Internet communications are now using all four pairs within the cable as a way to increase bandwidth and help reduce possible radiation emissions from the transmitting cable.

Specification Definitions for Communications Cable

The following sections provide definitions of the most commonly used communication and network cable specifications.

Near-End Crosstalk

NEXT results from high levels of transmission signal at each end of the line interfering with the reception of the intended signal. Consider also that the intended received signal has already been somewhat attenuated as a result of having to propagate down the length of the transmission line. The interference from near-end transmissions only serves to degrade the quality of reception further. High NEXT levels (in decibels) are desirable.

Far-End Crosstalk

FEXT, or far-end crosstalk, refers to signal distortion caused by the retransmission of a received signal at the far end of the line. The destination end of a transmission not only contains the intended signal but also the retransmitted receive signal from the other end. The mixing of transmitted and retransmitted data undoubtedly distorts the intended transmission, as well as generates unwanted errors.

Power Sum

Power sum (PSUM) is a test procedure where the same signal is applied simultaneously to three pairs of a cable, with the resultant measurement taken from the fourth pair. All pair combinations are tested and then averaged for a final power sum value, which is measured in decibels.

Attenuation Crosstalk Ratio

ACR, or attenuation crosstalk ratio, is a measurement similar to signal-to-noise ratio. High levels are therefore preferred. The perfect cable would have low attenuation and low crosstalk, thus allowing for maximum signal transmission down the line.

Delay Skew

Delay skew refers to timing differences among the four pairs, resulting from the varying twist rates of pairs and varying propagation delays. The timing of delay skew usually is measured in nanoseconds (a billionth of a second, or 10^{-9} second).

Structural Return Loss

Structural return loss is the digital equivalent to the standing wave ratio, typically caused by variations of characteristic impedance along the length of the cable. A mismatched system or transmission line also causes high structural return loss.

Specification Example of a CAT5e Cable

Number of Pairs	4
Wire AWG	24
Conductor type	Solid bare copper, twisted pairs, with rip cord*
Maximum voltage and temperature	30 Volt, 80°C
Insulation	Polyethylene
Shielding	None
Standard lengths	304.8 m or 1000 ft
Standard weight	10 kg or 22 lbs
Maximum dc ×	9 Ω/100 m
Nominal outside diameter	0.200 in. or 5.08 mm
Nominal impedance	100 Ω

(Continued)

Specification Example of a CAT5e Cable

(Continued)

Maximum capacitance (unbalanced)	66 pF/100 m			
Frequency, MHz	1	10	100	350
Maximum attenuation, dB/100m	2	6.4	21.7	44.3
Minimum PSUM NEXT, dB	65.3	50.3	35.3	27.2
Minimum PSUM ACR, dB/100m	63.3	43.9	17.1	NA
Minimum PSUM FEXT, dB/100m	60.8	40.8	20.8	9.9
Minimum structural return loss, dB	20	25	20.1	17

*The rip cord is a nylon fiber used to strip open the outer insulation.

© Cengage Learning 2013

Coaxial

There are two types of coaxial cable: flexible and rigid. The rigid varieties consist of a central wire mounted on isolative spacers inside of a concentric tubular conductor. Primarily used by broadcast and high-power transmitting stations, rigid coaxial cable is expensive because it must be filled with a moisture-resistant nitrogen, helium, or argon gas and then pressurized to between 3 and 35 psi before sealing. The gas provides isolation from exterior contaminates and insulation, which helps to equalize and stabilize the characteristic impedance of the cable along the length of the transmission line.

Flexible coaxial cables (see Figure 2–21(F)), most commonly used by the low-voltage communications, data, and video industries, consist of a central conductor typically surrounded by a polyethylene insulator, over which is placed a braided copper shield together with an outer jacket of insulation. Coaxial cables are available in a wide variety of material types, conductor sizes, and shielding configurations that are ideal for unbalanced systems. The concentric outer shield, combined with an internal layer of foil, provides maximum protection from possible radiation emissions and from interference caused by stray electromagnetic or electrostatic fields originating outside the cable.

The characteristic impedance of coaxial cable can vary between 50 and 150 Ω, depending on the type of insulation, size of the central conductor, and intended use of the product. Care must be taken not to bend, squeeze, or overly stress a coaxial cable, or damage and possible shifting of the characteristic impedance may result. Transmission cables tend to use low impedance lines, often in the range of 50 Ω, whereas video and data lines often use higher impedance levels, typically 75 to 100 Ω. Higher impedance levels help to minimize circuit currents, greatly reducing the possibility of unwanted emissions, interference, and crosstalk, while promoting wider bandwidths.

Different types of coaxial cable should not be intermixed within the same transmission line. As an example, types RG-6/U and RG62/U are not the same. First, RG-6/U is a 75 Ω cable, whereas RG62/U is 93 Ω. Other specification differences may include varying gauge sizes, capacitance per foot, and velocity of propagation. A broad range of insulation and shielding options are also available for each, depending on the type of circuit and its location; examples may include plenum or underground burial. For all of the reasons given, the mixing of dissimilar coaxial cable types within the same transmission line is not a good idea. In most cases, it can result in signal degradation and distortion at the load.

Applications for the use of coaxial cables include video broadcast, CATV, master antenna television (MATV), closed circuit television (CCTV), computer networking, and home entertainment systems. Where used for LANs, coaxial cable is run over relatively short distances (less than 10 miles), allowing for transmissions rates of 400 Mbps or more.

Transmission lines for coaxial cable can be of the baseband or broadband variety. Baseband, being least expensive, can only accommodate a single transmission signal along a given line. Broadband, as the name implies, provides a much broader range of transmission, using a technique known as multiplexing to increase the transmission capacity, often combining signals of voice, data, and video, at varying rates of speed, simultaneously along a single channel. Although broadband may sound like the better option, baseband is still widely used, not only because of cost but also because it is more secure from interference and therefore less complicated to set up and transmit over a network.

Specification Example of a Flexible Coaxial Video Cable, MATV Type RG-6/U

Wire AWG	21
Conductor type	Solid bare, copper-covered steel
Maximum voltage and temperature	30 V, 80°C
Insulation	Polyethylene
Shielding	97% copper braided shield
Standard lengths	1000 ft or 304.8 m
Standard weight	74 lbs or 33.6 kg
Nominal dc ×	105 Ω/km central conductor, 3.6 Ω/km shield
Nominal outside diameter	0.332 in. or 8.43 mm
Nominal impedance	75 Ω
Nominal velocity of propagation	66%

(Continued)

Specification Example of a Flexible Coaxial Video Cable, MATV Type RG-6/U

(Continued)

Nominal capacitance	20.5 pF/ft or 67.2 pF/m			
Frequency, MHz	1	10	100	1000
Nominal attenuation, dB/100m	1.3	2.6	8.9	32.1

© Cengage Learning 2013

Notice that the attenuation is worse at greater frequencies. This is because of the lower values of capacitive reactance (X_C) as frequencies increase, which, in turn, has the effect of decreasing the upper cutoff and narrowing the transmission bandwidth.

Twinaxial

Twinaxial cable (see Figure 2–21(G)) differs from coaxial in that it uses two central conductors, allowing for a low-noise, crosstalk-free, balanced transmission signal. Twinaxial is more expensive than coaxial cable and is most often used by the broadcast industry.

Triaxial

Triaxial cable (see Figure 2–21(H)) is coaxial cable with an added second layer of braided shielding. The second shield is separated electrically from the first by an added insulation barrier. Functionally, having two shielding layers allows for the broadcast of multiple signals over the same cable. Triaxial cables are most commonly used by the broadcast video industry; for example, to address the need to deliver dc power to the camera, intercom signals to the camera operator's headset, data feeds to the teleprompter, signal feeds to the monitor, and even automated robotic functions to the camera tripod and lens. To accommodate the multiple functions, signals are often multiplexed along the central conductor and isolated shields to increase the transmission capacity of the cable.

Twin-Lead Cables

Twin-lead cables (Figure 2–24) originally were designed for older style, balanced, television and radio antenna systems. In most cases, the characteristic impedance of twin-lead cables is 300 Ω, as determined by the conductor spacing and the gauge of the wire, 20 to 26 AWG being the most common. Notably, manufacturers of new style televisions, VCRs, and satellite receivers have stopped including 300 Ω inputs on many of their products, replacing it with the now standard 75 Ω coaxial "F" connector. Where it becomes necessary to connect a 300 Ω, twin-lead cable to a 75 Ω coaxial input, a matching transformer or balun is needed. Not using a matching device introduces an impedance mismatch into the circuit, causing signal loss and a degradation of quality.

Figure 2-24
A 300 Ω twin-lead cable.

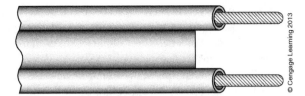

These days, with the proliferation of cable television and digital satellite, balanced 300 Ω transmission is beginning to disappear as more and more people are taking down their old-style television antennas, opting for the more preferable high-speed, low-noise digital systems. The future necessity for twin-lead cable may only survive in the realm of FM radio, HAM operators, and radio hobbyists.

Special Multipurpose Hybrid Cables

Special MP hybrid cables exist for the following applications: television cameras; CCTV; audio, communication, and instrumentation systems; and direct broadcast satellite (DBS). They offer a variety of intermixed cable types, such as shielded multiconductor, twisted pair, and coaxial, grouped and bundled inside of a single jacketing. As an example, television camera and CCTV cables exist that support a combinational variety of needs, including the transmission of system power, audio and video signals, and servomotor-operated, motorized lens control.

The grouping of various cable types into self-contained bundles is also common in the instrumentation and robotic control industry. Wire and cable manufacturers often are willing to work with customers to provide a variety of customized special orders provided they meet *NEC* requirements for safety and performance, and the regulations regarding the mixing of cable classifications. Specialized cables can often be quite expensive, so be ready to pay the extra manufacturing costs.

SEC 2.8   DISSIMILAR METALS

When connecting circuits, the mixing of **dissimilar metals** poses special problems. The use of solders, inhibitors, and chemical solvents must also be carefully considered to guarantee a high level of system performance.

National Electrical Code Article 110.14, Electrical Connections

Problems may arise when joining or splicing any two dissimilar metals, because the electrical properties of each can vary greatly because of the uniqueness of their atomic structures. Corrosion, being the primary concern, results where a connection is exposed to moisture or liquid, starting in motion a degenerative electrolytic process, referred to as galvanic action, which eventually precipitates into the electromechanical breakdown of the connection. Because most metals are not pure, they naturally begin to exhibit some form of corrosion where exposed to the environment, causing possible oxidation, pitting or crevicing, de-alloying, or hydrogen damage. The joining of dissimilar metals, however, accelerates the process dramatically.

The junction points of dissimilar metals also develop internal voltages and thermal currents caused by temperature gradients along the length of the conductors. Referred to as the Seebeck Effect, named after T. J. Seebeck, the scientist who discovered the phenomena in the 1820s, these thermal currents and voltages are often used in thermocouple sensors to measure temperature in heating, ventilation, and air conditioning (HVAC) and process control systems. Seebeck voltages, often ranging in millivolts, typically increase with temperature.

When comparing different types of metal, such as copper and aluminum, specifications regarding conductivity, density, and resistance differ greatly. Provided that temperatures remain equal along the lengths of conductors, there are not any noticeable adverse effects. However, the reality is that between any two points, temperature gradients do exist that result in the creation of thermal currents, unwanted junction voltages, and heat dissipation. The unwanted heat also results from friction as electrons flow from one type of metal to another, having different density and conductivity ratings.

Electrolytic corrosion, originally caused by moisture, further increases as temperature gradients along the surface of joined, dissimilar metals generate unwanted thermal currents and junction voltages. Over time, the degenerative process continues, causing the joint to pit, loosen, or possibly arc. In high current situations, heat buildup caused by increased resistance and the loosening and corroding of joints can often accelerate into a possible fire hazard.

The resistance of a loose, corroding connection can shift dramatically from an ideal of less than 1 Ω when tight, to greater than 10 or even thousands of ohms when loose. Now consider the added problem of repeated thermal expansion and contraction as power is routinely turned on and off from normal everyday use, resulting in the further loosening of pressure terminals and connectors. The cycle repeats endlessly until heat buildup and possible arcing of the corroding, loose conductors compromise the safety and integrity of the circuit, connector, lug, or splice. In addition, the increase of heat dissipation only serves to cause further oxidation of the surface area of the joined metals, causing even greater joint resistance and thus adding to the problem. As a result, the intermixing of dissimilar metals, such as copper and aluminum, or copper and copper-clad aluminum, is not allowed where splicing circuit conductors to each other or to connecting terminals that have not been identified for the intended use. Except for the case of actual thermocouple wire that has been specially designed and installed for the expressed purpose of measuring temperature for instrumentation control circuits, or in approved dissimilar metal applications such as the bimetallic components of a circuit breaker, the interconnection of dissimilar metals is not authorized, as stated in *Article 110.14* of the *NEC*.

Pressure terminals, splicing connectors, and soldering lugs must all be rated and identified as suitable for the material type of the conductor being used. Therefore, copper conductors must only be connected to copper-rated terminals, aluminum conductors must only be connected to aluminum-rated terminals, and copper-clad aluminum conductors must only be connected to copper-clad aluminum-rated terminals.

Solder, Fluxes, and Inhibitors

Solder, fluxes, and inhibitors shall be chosen as suitable for use, so as not to adversely affect the normal operation or safety ratings of conductors, insulation, or equipment. Many types of solder exist, in a range of mixtures of lead, tin, and silver, and their use depends on the type of conductors being joined. There are also two basic varieties of flux: rosin and no-clean.

National Electrical Code Article 110.11, Deteriorating Agents

Flux traditionally is made from pine tar and is used in soldering to promote the smooth flow and adherence of liquefied metal around the surface of conductors being joined. After soldering a joint with rosin core solder, the junction must be cleaned with a flux-removing compound; otherwise, over time, any remaining flux residue eats away at the joint, causing eventual corrosion and increased resistance. As stated earlier, a high-resistance connection can eventually heat up, loosen, oxidize, and become a potential fire hazard. In this specific case, however, the corroding joint is the culprit, because of chemical breakdown resulting from acidic residual flux, and not a loosening screw terminal or connector.

No-clean solder flux exists for situations where cleaning a solder joint would be inconvenient, especially where nearby insulations or sensitive materials may be adversely affected by the use of traditional flux cleaners and solvents. Because a variety of plastics and insulators often soften and deteriorate where they are made to come in contact with most industrial solvents, and also because many switch contacts and terminals are typically molded in plastic, the use of certain solvents and cleaners could adversely affect the normal use and safety ratings of these devices. Before using a specific cleaner or solvent, always be sure to read the warning label for a list of approved materials. To remedy the problem, the use of no-clean solder often is recommended, and in fact, the manufacturer does not typically authorize the cleaning of no-clean solder joints, because to do so would remove a residual protective coating deposited on the joint once soldering is complete. Always be sure to read the warning labels and be sure to know the consequences of intermixing certain materials and solvents.

Chemical inhibitors are also available and often are used when joining aluminum wires to terminal lugs. The inhibitors help to provide a moisture barrier, reduce oxidation, and promote conduction at the termination points. As a rule, always know the consequences of using chemical inhibitors, solvents, or cleaning agents, and read the warning labels carefully to ensure the proper use of products so they are not a detriment to conductors, insulating materials, and equipment.

CHAPTER 2 FINAL QUESTIONS

1. What type of conductor is rated for 100% conductivity?

    a. Aluminum
    b. Copper
    c. Copper-clad aluminum
    d. Copper-clad steel

2. What type of plating is most commonly used?

    a. Nickel
    b. Silver
    c. Tin

3. For conductors 1 AWG through 30 AWG, as the AWG increases, what happens to the physical size of the wire?

    a. It decreases.
    b. It increases.

4. Cross sectional area of a conductor is measured in _____.

5. When is the air space below a raised floor considered a plenum space?

    a. Always
    b. Never
    c. Only when the air flow is connected to the building air exchange
    d. Only when it is part of an Information Technology Center

6. A vertical shaft running from floor to floor in a high-rise apartment is an example of a _____.

    a. general-purpose space
    b. plenum space
    c. riser

7. What is the main risk of not removing abandoned cable in a building?

    a. It poses an electrical hazard to people in the building.
    b. It can cause a collapse of the suspended ceiling due to the added weight.
    c. During a fire, it will increase the total volume of smoke in the building.
    d. It can short-circuit other systems.

8. Type CMUC cable is suitable for what purpose?

9. The insulation ratings for Class 3, PLTC, FPL, and CM cables are rated for _____ volts.

    a. 120
    b. 150
    c. 300
    d. 600

10. What is the main purpose of conductor shielding?

    a. To physically protect the conductors from potential damage
    b. To increase the total capacitance of the cable for better conductance
    c. To lower the cable resistance and increase the signal output
    d. To help prevent electromagnetic interference and reduce noise

11. List three factors that affect the resistance of a conductor.

12. A copper wire at 20°C and 1624 cmil has a K value of 10.4 Ω per mil-foot. If the distance to the load is 600 ft (183 m), calculate the resistance of the loop.

13. Why does ac current not see the same level of resistance as dc?

14. As signal frequency increases, signal attenuation _____ .

15. As cable length decreases, signal attenuation _____ .

16. As cable length increases, cable capacitance _____ .

    a. decreases
    b. increases
    c. remains the same

17. As cable length increases, characteristic impedance _____ .

    a. decreases
    b. increases
    c. remains the same

18. As signal frequencies increase, cable resistance appears to _____ due to the skin effect.

    a. decrease
    b. increase
    c. remain the same

19. What causes a standing wave on a high-frequency transmission line?

20. What type of cable should be used to reduce noise on unbalanced lines?

    a. Coaxial
    b. ScTP
    c. UTP

21. How is a CAT6 cable physically different from a CAT5e cable?

22. What is the maximum transmission frequency at 100 m (328 ft) for CAT5e cable?

23. What is the maximum recommended transmission distance for Augmented CAT6 cable at 10 Gbps?

24. When is it appropriate to join copper and aluminum in a solder lug or terminal connection?

25. After soldering with rosin core solder, why is it important to clean the joint thoroughly?

Chapter 3

Grounding and Bonding

Objectives

- Describe the purpose of grounding and bonding electrical systems.
- Define important grounding and bonding terms and definitions.
- Define and describe the grounding electrode system.
- Describe the methods of connecting grounding conductors to grounding electrodes.
- Define a ground fault and explain the purpose of using a ground-fault circuit interrupter.
- Describe the requirements of grounding and bonding within various articles of the *NEC*.

Chapter Outline

Sec 3.1 The Purpose of Grounding Electrical Systems

Sec 3.2 The Electrical Service Entrance

Sec 3.3 What Is Ground?

Sec 3.4 Ground Definitions and the *National Electrical Code*

Sec 3.5 The Grounding Electrode System

Sec 3.6 Ground Resistance and Supplemental Electrodes

Sec 3.7 Methods of Connecting EGCs, Grounding Electrode Conductors, and Bonding Jumpers

Sec 3.8 Lightning Protection Systems

Sec 3.9 Aluminum and Aluminum Grounding Conductors

Sec 3.10 Ground Faults

Sec 3.11 Bonding

Sec 3.12 Alternating Current Systems Less Than 50 Volts (*National Electrical Code 250.20 (A)*)

Sec 3.13 Grounding Related to Other Articles of the *National Electrical Code*

Key Terms

bonded/bonding

bonding conductor

chassis ground

circuit common

earth ground

effective ground-fault current
path

equipment grounding
conductor (EGC)

ground

ground fault

ground-fault circuit
interrupter

grounded conductor

grounded/grounding

grounding electrode

grounding electrode
conductor

grounding electrode system

intersystem bonding
termination

solidly grounded

supplemental electrode

ungrounded conductor

SEC 3.1   THE PURPOSE OF GROUNDING ELECTRICAL SYSTEMS

The purpose of grounding electrical systems can be stated with one simple word—lightning. Because the earth's atmosphere regularly develops static electrical charges, which occasionally discharge in the form of lightning bolts, a protective, conductive pathway must be provided to help draw the energy away from buildings to safely dissipate it into the earth. Basic electrical theory states that electrical current always follows the path of least resistance. To protect a building or structure from lightning, a ground rod, or some other type of grounding electrode system, is added to provide the necessary means to safely dissipate static atmospheric charges into the earth. Not having a ground rod would mean that the lightning would be free to follow the next available path, which in most cases would be the electrical wiring connected to your house. A direct hit from a lightning bolt to an ungrounded electrical service could be catastrophic. If such an event were to occur, the conductive wiring would overheat, causing the surrounding walls and pathways within the structure to catch on fire. To reduce the risk, all electrical wiring entering a building or structure from the outdoors is required to be grounded.

According to the *NEC*, the term *ground* means "the earth," and all electrical systems that are **grounded** shall be connected to earth in a manner that limits the voltage imposed by lightning, line surges, or unintentional contact with higher-voltage lines, and stabilizes the voltage to earth during normal operation (*NEC Article 100* and *250.4(A)(1)*).

SEC 3.2   THE ELECTRICAL SERVICE ENTRANCE

The electrical service entrance is the point in the circuit where the electrical power lines are brought into the building or structure from outside by the utility. The voltage on the pole is considered high voltage, in the range of 7200 volts or more in some cases. In the United States, a transformer on the pole reduces the voltage

Figure 3–1
Service panelboard.

down to a single-phase, center-tapped (Neutral) 240 volts at the service entrance. The center tap (Neutral) is where the utility company's earth ground is connected. The service panelboard (Figure 3–1) is then divided into two phases, phase A on the left side and phase B on the right side, each measuring 120 volts to center or Neutral and ground. Incidentally, there are two types of panelboards, main lug and main circuit breaker.

A main lug panelboard connects phase A and phase B directly to the panelboard, without going through a set of main breakers. When installing a main lug panelboard, phase A and phase B typically connect through a main disconnect located outside the panelboard in a separate box, or on the outside wall of the building or structure, just prior to bringing the power lines into the building. Main lug panelboards also have output lugs located at the bottom of the cabinet to feed phase A and phase B to additional panelboards within the building.

A main circuit breaker panelboard differs in that it connects both phase A and phase B, as they enter the enclosure, to a main circuit breaker disconnect located on the inside, top portion of the panelboard. For most residential installations that don't require a main disconnect on the outside of the residence just prior to bringing power into the building, a main circuit breaker panelboard is typically used.

Within the panelboards, circuit breakers are then connected along the left- and right-hand sides, respectively, for phase A and phase B, for connection to each individual branch circuit running through the building or structure. The circuit breaker serves as the main disconnect for each branch circuit. The article of the *NEC* that covers the grounding and bonding of electrical systems is *Article 250*.

SEC 3.3   WHAT IS GROUND?

The *NEC* defines the term **ground** as "the earth." Therefore, any time the *NEC* makes a reference to ground, it means that the Neutral leg of the electrical supply is bonded to the earth (which happens only at the main service panelboard or a main disconnect switch) via a main bonding jumper and equipment grounding conductor to a grounding electrode as listed in *Article 250.52(A)*.

Aside from the term *ground*, or **earth ground**, there are two other terms related to ground that we must define: *Neutral*, or **circuit common**, and *chassis ground*. The schematic symbols for each type are illustrated in Figure 3–2.

Neutral, or Circuit Common

The circuit Neutral, also referred to as a circuit common, is a common point in an electrical circuit from which voltage measurements are taken. A voltage measurement is a 2-point differential measurement. When voltages are measured from circuit Neutral, or common, the black lead of the voltmeter is connected to the designated Neutral, or common, in the electrical schematic, and the red lead of the voltmeter is connected to a specific test point in the circuit. So, for instance, a voltage measurement taken at point A in a circuit would be notated as V_A. Because taking a single point voltage reading would be impossible, it is assumed in such a case that the black meter common would be connected to the Neural conductor or equipment grounding conductor (usually a bare conductor), and the red lead to point A. In some cases, the circuit ground may not be present at the test point; it all depends on the design of the circuit that is being examined. To be sure, you always have to check the schematic drawings to determine where voltage readings have been referenced. The designated circuit ground, or circuit common, is always indicated.

Chassis Ground

A **chassis ground** is indicated on a schematic where the electronic signal ground, or circuit common, of the equipment and the earth ground of the electrical system are attached to the outer metal enclosure of the equipment. The reason for a chassis ground is twofold. First, the chassis ground provides safety from ground faults and possible short circuits. And second, the metallic chassis provides a large conductive surface area for the electronic signal ground, or common, to help minimize ground resistance and reduce noise on the internal circuit. The third prong of the electrical outlet, referred to as the equipment grounding conductor, is always attached to the metallic chassis of any piece of equipment. All exposed metal is connected in a similar manner, to protect individuals from the accidental hazards of electrical shock.

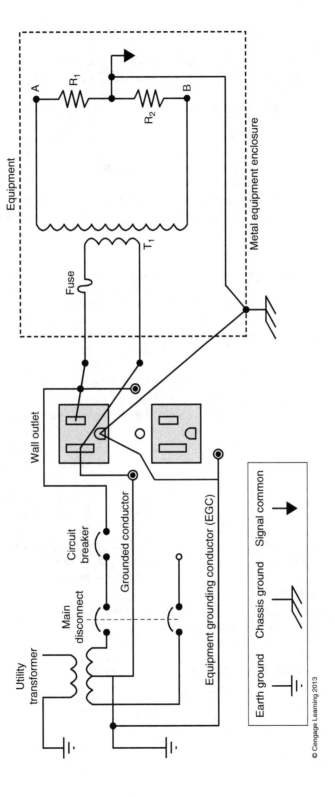

Figure 3-2
Ground symbols.

© Cengage Learning 2013

SEC 3.4 GROUND DEFINITIONS AND THE *NATIONAL ELECTRICAL CODE*

Grounded Conductor

The **grounded conductor**, Neutral, is an electrical system or circuit conductor that is intentionally grounded. In a single-phase, 120/240-volt system, the grounded conductor refers to the grounded center tap of the main supply transformer. Each branch circuit in a building is then made up of an ungrounded conductor (hot), coming from a designated circuit breaker, and the grounded conductor (Neutral), providing the transformer return. In the United States, the grounded conductor is white or gray.

Equipment Grounding Conductor

The **equipment grounding conductor (EGC)** provides the conductive path(s) to connect the normally non-current-carrying metallic parts of equipment together. In the United States, the EGCs are the bare copper or green wires connected to the third prong of the electrical outlet. The EGCs start at the main service panel, where the grounded conductors (white Neutral wires) and the grounding electrode conductor are bonded together. Once the EGCs leave the main service panel, they remain separate from and are never again interconnected with the grounded conductors. The main reason for the separation is that the grounded conductor is used to supply current to the load, whereas the EGCs are used to interconnect all the exposed metal, equipment racks, raceways, and equipment chassis in the building for protection against possible ground faults and short circuits. Accidentally interconnecting a white, Neutral conductor and a bare or green EGC with a branch circuit would cause the unwanted flow of current along all the exposed metal in the building, which could pose a potential shock hazard to anyone standing on the ground while simultaneously touching the outer metallic chassis of any piece of equipment. For this reason, the EGC and the grounded conductor shall only be bonded together at the main service panelboard.

EGCs for raceways and equipment shall be a minimum size based on *Table 250.122* of the *NEC*.

Figure 3–3 illustrates the differences between the grounded conductor and the equipment grounding conductor.

Grounding Electrode

The **grounding electrode** is a conducting object used to establish direct contact to the earth for the purpose of dissipating electrical surge currents and charges from lightning strikes. Examples include a metal rod, pipe, plate, ring, building steel, concrete-encased rebar, and a water pipe.

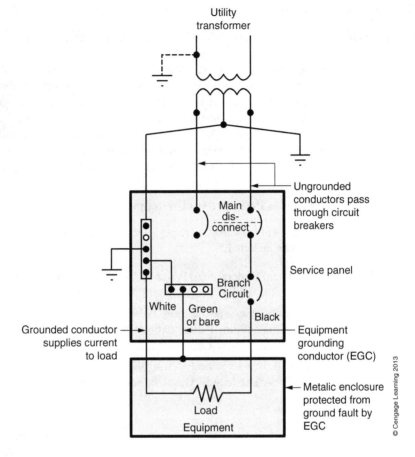

Figure 3–3
Grounded conductor vs. EGC.

Grounding Electrode Conductor

The **grounding electrode conductor** connects the system Neutral or grounded conductor (white or gray colored) and the equipment (exposed metal) to a grounding electrode or to a point on the grounding electrode system. The grounding electrode conductor provides the necessary connection to the grounding electrode.

Solidly Grounded

To be **solidly grounded** means that a connection to ground is made directly, without inserting an intermediary resistance or impedance into the connecting path.

Ungrounded Conductor

The **ungrounded conductor** is not connected to ground and is referred to as the hot power line, which provides current to the load. In the United States, the most common colors for ungrounded conductors are black and red.

SEC 3.5   THE GROUNDING ELECTRODE SYSTEM

The **grounding electrode system** is a system of interconnected electrodes at a building or structure that is used to provide a connection to the earth for the purposes of grounding an electrical system. Examples include a metal rod, pipe, plate, ring, building steel, concrete-encased electrode or rebar, and a metal, cold water pipe that extends at least 10 ft (3.04 m) beyond its entry point into the building. (Figure 3–4). When multiple electrodes are used at a single structure, they must be interconnected and bonded into what's referred to as the *grounding electrode system.*

Electrodes Permitted for Grounding

The following electrodes are permitted for grounding. The required specifications have been included.

1. Metal Underground Water Pipe—A metal underground water pipe in direct contact with the earth, 10 ft (3 m) or more, and electrically continuous.

2. Metal Frame of the Building or Structure—At least one structural member of the building that is in direct contact with the earth, 10 ft (3 m) or more, without concrete encasement.

3. Concrete-Encased Electrode—A concrete-encased electrode 20 ft (6 m) or more in length. The electrode shall consist of either bare steel or zinc

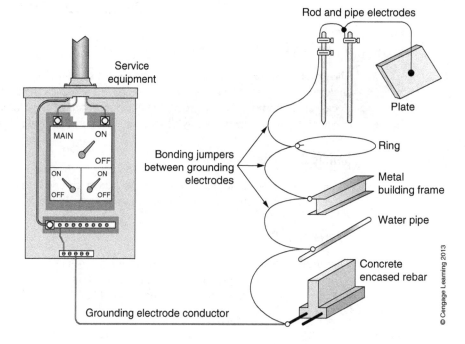

Figure 3–4
Grounding electrode system.

© Cengage Learning 2013

galvanized coated steel, or other electrically conductive coated steel rein-forcing bars or rods, with not less than ½ in. diameter.

4. Ground Ring—A ground ring that encircles a building or structure, not less than 20 ft (6 m) in length, consisting of bare copper, not smaller than 2 AWG. The ground ring shall be buried to a depth not less than 30 in. (762 mm).

5. Rod and Pipe Electrodes—A rod and pipe electrode not less than 8 ft (2.4 m) in length. A pipe electrode shall have a diameter not smaller than trade size ¾. A steel pipe shall have the outer surface galvanized or coated with some other type of suitable metallic coating for protection against corrosion. A rod-type electrode shall be stainless steel, copper, or zinc, and with at least a ⅝ in. diameter, unless listed. Rods shall be installed so that the entire 8 ft (2.4 m) in length is in contact with the soil. Where rock is encountered, the rod shall be driven at a 45-degree angle from vertical or shall be buried in a trench not less than 30 in. (762 mm) deep.

6. Plate Electrodes—A plate electrode shall expose not less than 2 ft² (0.185 m²) of surface area to exterior soil and be buried to a depth not less than 30 in. (762 mm). Bare or conductively coated iron or steel plates shall be at least ¼ in. (6.35 mm) thickness. Solid, uncoated, plate electrodes of nonferrous metal shall be at least 0.06 in. (1. 5 mm) thickness.

Grounding electrodes shall be installed as close as practicable to the service entrance. Rod, pipe, and plate electrodes shall be free from nonconductive coat-ings, such as paint or enamel, and if practicable, shall be embedded below the permanent moisture level, preferably in permanently damp soil.

Not Permitted as Grounding Electrodes

The following are not permitted to be used as grounding electrodes:

1. Metal underground gas pipes
2. Aluminum

SEC 3.6   GROUND RESISTANCE AND SUPPLEMENTAL ELECTRODES

The ground resistance of a single electrode shall have a resistance to earth of 25 Ω or less. Such a measurement cannot be taken with a simple ohmmeter. An ohm-meter only measures dc resistance, but when measuring earth ground resistance, measurements must be taken using a more complex, high-frequency ac meter to simulate the pulse currents of a lightning strike. To achieve such measurements, a special ground resistance meter, such as a Megger ground resistance tester, must be used. When such a meter is used, measurements are taken between multiple grounding rods staked at various distances from each other. The meter then sends out a series of high-frequency pulses, simulating lightning frequencies. The voltage drops between grounding stakes are then measured and plotted on a graph. The value of resistance is determined based on the shape of the curve.

Any single rod, pipe, or plate electrode not having a measured resistance to earth of 25 Ω or less shall be supplemented by the installation of an additional electrode. The **supplemental electrode** is used to help lower the total resistance into the earth by providing additional discharge paths for lightning. When multiple grounding electrodes are installed at a single structure, they shall be spaced not less than 6 ft from each other, and bonded together. Bonding conductors shall not be required to be larger than 6 AWG copper. The reason for the 6 ft (1.8 m) separation requirement is so that lightning does not potentially arc between the electrodes; it also prevents the radiation patterns into the earth from overlapping so that they can independently dissipate the incoming energy surge.

Metal Underground Water Pipes and Supplemental Electrodes

If used as a grounding electrode, a metal underground water pipe shall be supplemented by an additional electrode of a type previously discussed. In addition, attachment to the water pipe can only be made within the first 5 ft (1.5 m) of the water pipe's entrance to the building. Past the first 5 ft (1.5 m), the risks are too great that the water pipe may interconnect with plastic or some other type of nonconductive material, thereby removing connection to ground. Connection to the water pipe past the first 5 ft (1.5 m) is only allowed in industrial, commercial, or institutional structures or buildings that have qualified personnel supervising who can absolutely guarantee that the water pipe is electrically continuous to the outside. In addition, a 6 AWG bonding jumper shall be attached to the water pipe between the incoming and outgoing sides of the water meter, to prevent the accidental disconnection of ground in the event that the water meter is ever removed for servicing.

SEC 3.7   METHODS OF CONNECTING EGCs, GROUNDING ELECTRODE CONDUCTORS, AND BONDING JUMPERS

EGCs, grounding electrode conductors, and bonding jumpers have to be connected at equipment and to the grounding electrodes. A discussion of methods that can be used follows.

Methods of Connecting EGCs, Grounding Electrode Conductors, and Bonding Jumpers to Equipment

The following are permitted methods for attaching EGCs, grounding electrode conductors, and bonding jumpers to equipment (*NEC 250.8*):

1. *Listed pressure connectors* (Figure 3–5)
2. *Terminal Bars* (Figure 3–6)

Figure 3–5
Pressure
connector.

© Cengage Learning 2013

Figure 3–6
Terminal bars.

© Cengage Learning 2013

3. *Pressure connectors listed as grounding and bonding equipment*

4. *Exothermic welding processes (Figure 3–7)*

5. *Machine screw-type fasteners that engage not less than two threads or are secured with a nut*

6. *Thread-forming machine screws that engage not less than two threads in the enclosure*

7. *Connections that are part of a listed assembly*

8. *Other listed means*

Methods not permitted include fittings that depend on solder. Sheet-metal screws are also not allowed.

Figure 3–7
Exothermic weld.

© Cengage Learning 2013

Methods of Connecting Bonding Conductors and Grounding Electrode Conductors to the Grounding Electrode

Connection of **bonding conductors** or **grounding electrode conductors** to the grounding electrode shall be made by exothermic weld, listed lugs, listed pressure connections, listed clamps (Figure 3–8), or other listed means. Connections depending on solder shall not be used (*NEC 250.70*).

Clamps and Hardware—Listing and Labeling

All clamps and hardware used for grounding and bonding shall be listed and labeled. Ground clamps shall be listed for the material of the grounding conductor and the grounding electrode, and where buried, they shall be listed for direct burial or concrete encasement. Not more than one conductor shall be connected by a single clamp unless the clamp has been specifically listed for multiple connections. Hardware must always be installed according to the intended purpose of the product and always according to manufacturer's instructions (*NEC 250.70 and 110.3(B)*).

Figure 3–8
Listed clamps.

© Cengage Learning 2013

SEC 3.8   LIGHTNING PROTECTION SYSTEMS

Lightning protection systems are not to be used as a grounding means for the main electrical system within a building or for the grounding of other systems. All systems, devices, and cables are to maintain a minimum distance of 6 ft (1.8 m) from the lightning protection system. Bonding, however, is required between the lightning protection system rod, pipe, or plate and the main grounding electrode system of the building.

SEC 3.9   ALUMINUM AND ALUMINUM GROUNDING CONDUCTORS

Aluminum is not to be used as a grounding electrode. When placed in direct contact with the earth, aluminum corrodes and deteriorates within a short period of time, depending on the level of moisture and mineral content of the soil. Bare aluminum or copper-clad aluminum grounding electrode conductors are not to be installed where in direct contact with masonry or the earth or where subject to corrosive conditions. Where used outside, aluminum or copper-clad aluminum shall not be terminated within 18 in. (457 mm) of the earth.

SEC 3.10   GROUND FAULTS

What Is a Ground Fault?

A **ground fault** is an accidental or unintentional electrical connection between an ungrounded, hot conductor and the normally non-current-carrying conductors (EGC), metallic enclosures, metallic raceways, metallic equipment, or any other exposed metal within a building or structure. When a ground fault occurs, the exposed metal in the building becomes energized and serves as a potential shock hazard to anyone unlucky enough to make contact with it and the ground simultaneously. Before the days of 3-prong plugs, ground faults were a very real threat. Exposed metal would become energized, and the risk of electrocution was very high. Today, the risk has been minimized by interconnecting all the exposed metal to the EGCs, which provide the necessary pathway back to the power source in order to complete the circuit and short the main breaker of the branch circuit.

The grounding of electrical equipment is required by the *NEC* under *250.4(A)(2)*, which requires that the normally non-current-carrying conductive materials enclosing the electrical conductors or equipment, or forming part of such equipment, are connected together and to the electrical supply source in a manner that establishes an effective ground-fault current path. An effective ground-fault current path is essential to guarantee that the ground fault clears. Figure 3–9 illustrates a ground fault and the effective ground-fault current path back to the electrical supply source.

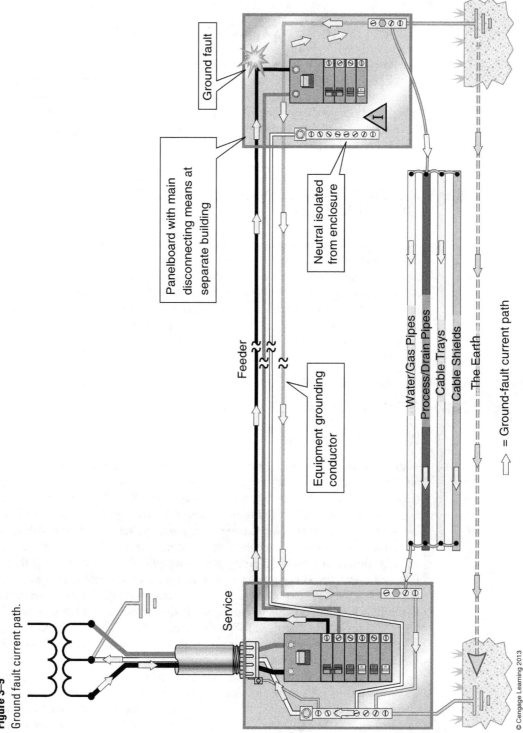

Figure 3–9
Ground fault current path.

Ground-Fault Circuit Interrupter

A **ground-fault circuit interrupter** provides a disconnect means at the power receptacle by monitoring the differential current traveling along the ungrounded and grounded power conductors. If the ground-fault circuit interrupter detects a difference of 5 mA between the hot and the Neutral power line, the circuit inside the ground-fault interrupter trips the branch circuit breaker, and power is disconnected. 5 mA represents the maximum harmless current intensity for the human body. Ideally, the hot and Neutral conductors should have the same level of current flow. A difference between them indicates that current is traveling through an alternate pathway, such as between the ungrounded conductor and the equipment enclosure, indicating that something is shorting.

SEC 3.11   BONDING

The Purpose of Bonding Electrical Systems

The main purpose of **bonding** electrical systems is to create a low-impedance pathway back to the service panel and the main power source to the building for the purpose of clearing a ground fault; this is commonly referred to as an effective ground-fault current path (*NEC 250.4(A)(3–5)*). Bonding between grounding electrodes is also required to neutralize any stray ground resistance in the earth, to prevent undesired voltage surges and spikes from occurring between multiple electrodes on the ground plane during a lightning strike. Figure 3–10 illustrates how bonding the electrical system completes the pathway back through the

Figure 3–10
System bonding.

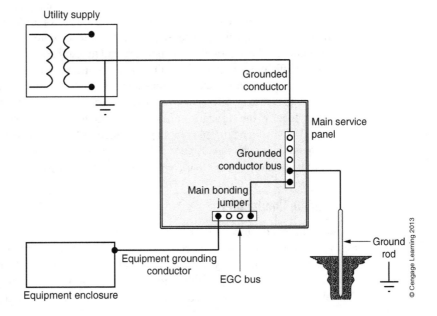

© Cengage Learning 2013

interconnection of the grounded conductor and the EGC by the main bonding jumper in the service panel.

As stated earlier, the equipment grounding conductor (EGC) is connected to all the exposed metal equipment racks and equipment enclosures within the building. Bonding ensures that a ground fault has an **effective ground-fault current path**, which is a low-impedance pathway back to the main service panel, through all the exposed metal in the building and the EGCs. The effective ground-fault current path guarantees that the circuit shorts from the point of the ground fault back to the electrical supply source. Provided that the low-impedance pathway back to the service panel is sufficiently low enough, the current surges above the 15- or 20-ampere rating of the circuit breaker to instantly open the branch circuit, thereby removing power to the ground fault.

Do Ground Rods Protect against Ground Faults?

The next question to ask is: Does a ground rod connected to the earth protect against ground faults? The answer is, not necessarily. The *NEC* states that *the earth shall not be considered an effective ground-fault current path* (*NEC 250.4(A)(5)*). The main reason is that the resistance to ground for any grounding electrode is required to be less than 25Ω. Although this is sufficient for protection against lightning, 25Ω may not be sufficiently low enough to guarantee a level of current high enough to trip a 15- or 20-ampere circuit breaker. A ground fault can only be cleared if the current can reach the break-over point of the circuit breaker. If the current flow in a ground fault never rises above the rating of the circuit breaker, power is never removed, and the current continues to flow unending. The danger here is that the exposed metal, where the ground fault is making contact, continues to remain energized, increasing the risk of a potential shock hazard and possible electrocution. In some cases, the resistance at the grounding electrode may be sufficiently low enough to short the circuit and trip the circuit breaker, but there are no guarantees, especially because ground resistance changes throughout the year, depending on the season and the moisture and mineral content of the soil. The only way to sufficiently protect a circuit from a ground fault is to have all the exposed metal bonded together and interconnected through the EGCs within the building in order to create an effective ground-fault current path back to the main source of power. Ideally, the resistance of the ground-fault current path should be $0 \ \Omega$, but there is always a negligible amount attributed to the wire resistance of the EGCs and bonding conductors. Figure 3–11 illustrates the problem and why the earth is not to be considered an effective ground-fault current path.

Bonding Conductors

Bonding conductors are used to interconnect multiple grounding electrodes, as well as for joining independent electrical or communications systems to the main grounding electrode system of the building. For most applications, a bonding conductor shall not be required to be larger than 6 AWG copper. TIA/EIA standard J-607-A gives the recommendations for the size of bonding conductors to

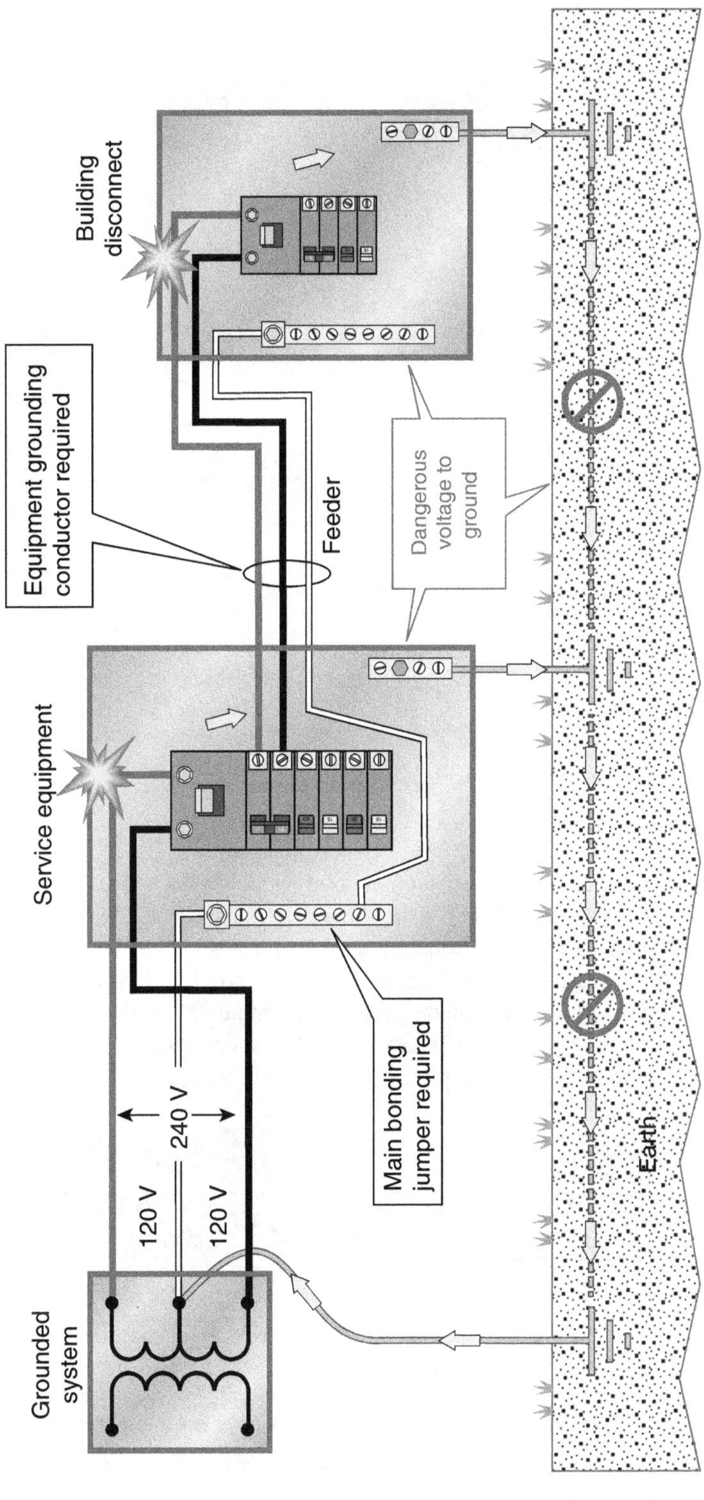

Figure 3–11

Earth is not an effective ground-fault current path.

© Cengage Learning 2013

Table 3–1

Sizing of the TBB

TBB length linear m (ft)	TBB Size (AWG)
less than 4 (13)	6
4–6 (14–20)	4
6–8 (21–26)	3
8–10 (27–33)	2
10–13 (34–41)	1
13–16 (42–52)	1/0
16–20 (53–66)	2/0
greater than 20 (66)	3/0

© Cengage Learning 2013

Figure 3–12
Methods of bonding service equipment.

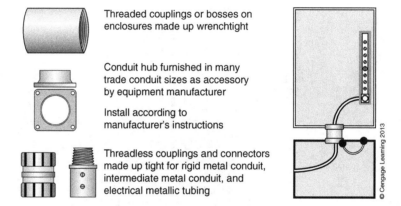

Threaded couplings or bosses on enclosures made up wrenchtight

Conduit hub furnished in many trade conduit sizes as accessory by equipment manufacturer

Install according to manufacturer's instructions

Threadless couplings and connectors made up tight for rigid metal conduit, intermediate metal conduit, and electrical metallic tubing

© Cengage Learning 2013

be used for a telecommunications bonding backbone (TBB) of a telecommunication system. Table 3–1 illustrates the recommended size based on length. In addition, methods of bonding service equipment shall be made through wrenchtight, threaded couplings, conduit hubs with bonding screws, threadless couplings made up tight for rigid metal conduit, intermediate metal conduit, and electrical metallic tubing (Figure 3–12). Good, solid connections are crucial to ensure a low-impedance return path to the electrical supply source.

SEC 3.12   ALTERNATING CURRENT SYSTEMS LESS THAN 50 VOLTS (*NATIONAL ELECTRICAL CODE 250.20(A)*)

Alternating current systems less than 50 volts shall be grounded under any of the following conditions:

1. When the main transformer supply exceeds 150 volts to ground
2. When supplied by transformers, if the transform supply system is ungrounded
3. When installed outside as overhead conductors

A very common circuit less than 50 volts ac is a standard, residential doorbell circuit. Based on these three conditions, it can be easily understood why a doorbell circuit is not required to be grounded. In the United States, a typical doorbell circuit is installed indoors, supplied by a 24-volt transformer, connected on the primary side by a grounded, 120-volt supply. The 24-volt secondary therefore does not require a connection to ground.

SEC 3.13 GROUNDING RELATED TO OTHER ARTICLES OF THE *NATIONAL ELECTRICAL CODE*

National Electrical Code Article 640 Audio Signal Processing, Amplification and Reproduction Equipment

When installing audio signal processing and amplification equipment, metal equipment racks and enclosures shall be grounded. Bonding shall not be required if the rack is connected to a technical power ground (*NEC 640.22*).

Wireways and auxiliary gutters shall be connected to equipment grounding conductors or to an equipment bonding jumper. Where the wireways or auxiliary gutters do not contain power supply conductors, the EGCs shall not be required to be larger than 14 AWG. Where wireways or auxiliary gutters do contain power supply conductors, the EGCs shall not be smaller than that specified in *Table 250.122* of the *NEC. (NEC 640.7)*.

National Electrical Code Article 800 Communication Circuits

Communication circuits entering a building from outside shall be connected through a listed primary protector. The primary protector shall be installed as close as practicable to the point of entrance (*NEC 800.90(B)*).

In addition, when entering a building or when terminating on the outside of a building, the metallic sheath members of communication cable shall be grounded as close as practicable to the point of entrance, or interrupted by an insulating joint or equivalent device (*NEC 800.93*).

NEC 800.100 states that bonding conductors or grounding electrode conductors shall be permitted to be insulated, covered, or bare, and copper or corrosion-resistant material, solid or stranded. The size of the bonding conductors or grounding electrode conductors shall not be smaller than 14 AWG and shall not be required to exceed 6 AWG. The current-carrying capacity shall not be less than the grounded metallic sheath members and protected conductors of the cable. For mobile homes, the grounding electrode conductor shall not be smaller than 12 AWG.

The length of the primary protector bonding conductor or grounding electrode conductor shall be as short as practicable, and in one- and two-family dwellings shall not exceed 20 ft (6 m). For mobile homes, the length shall not exceed 30 ft (9.1 m). Figure 3–13 shows a typical communication installation.

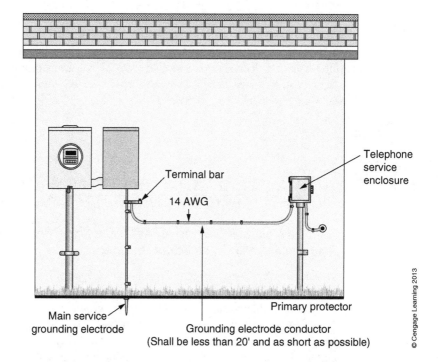

Figure 3–13
Grounded communication installation.

Telephone service enclosure

Terminal bar

14 AWG

Primary protector

Main service grounding electrode

Grounding electrode conductor
(Shall be less than 20' and as short as possible)

© Cengage Learning 2013

In one- and two- family dwellings, where it is not practicable to achieve an overall length of 20 ft (6 m), a separate communication ground rod, 5 ft (1.5 m) in length, and with ½ in.(12.7 mm) diameter, shall be driven into permanently damp earth. A separation of 6 ft (1.8 m) is required from any lightning conductors and from electrodes of other systems. Steam or hot water pipes or air terminal conductors shall not be used as electrodes for primary protectors or the grounded metallic members of cables. Figure 3–14 illustrates the requirements of the 20-foot (6 m) rule.

When separate electrodes are installed at a single building or structure, the communication electrode shall be bonded to the power grounding electrode. The bonding jumper shall not be smaller than 6 AWG copper. The interconnection of multiple electrodes at a single structure is required to help limit potential voltage differences between them and their associated wiring. Ground resistance between electrodes can develop high surge voltages as a result of stray ground currents from a lightning strike. Such a surge could potentially destroy all equipment connected to the communication electrode. The bonding jumper serves to reduce the potential voltage surge by providing a nearly 0 Ω parallel path, thus shorting out any resistance in the earth that may exist between electrodes.

Where exposed to potential physical damage, the bonding conductors and grounding electrode conductors shall be protected. Where installed in metal raceways, both ends of the raceway shall be bonded to the contained conductor or to the same terminal or electrode to which the bonding conductor or grounding

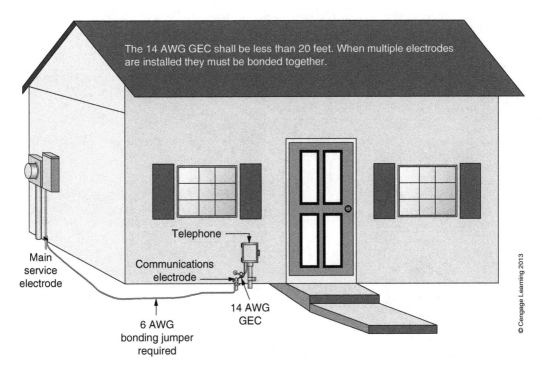

The 14 AWG GEC shall be less than 20 feet. When multiple electrodes are installed they must be bonded together.

Main service electrode

Telephone

Communications electrode

14 AWG GEC

6 AWG bonding jumper required

© Cengage Learning 2013

Figure 3–14
Exception to the 20 foot rule.

electrode conductor is connected. The idea is to make the raceway electrically continuous and interconnected with the bonding conductor or grounding electrode conductor from end to end.

Communication systems shall use the following as a grounding electrode (Figure 3–15):

1. An **intersystem bonding termination** as specified by *NEC 250.94*

2. The building or structure grounding electrode system as specified by *NEC 250.50*

3. The grounded metallic water pipe, within 5 ft (1.5 m) of the point of entrance to the building (*NEC 250.52*)

4. The power service accessible means, as covered in *NEC 250.94, Exception.* A 6 AWG copper conductor, with one end bonded to the grounded metallic raceway or equipment, with 6 in. (152 mm) or more made accessible to the outside wall

5. The nonflexible metallic power service raceway

6. The service equipment enclosure

7. The grounding electrode conductor or the grounding electrode conductor metal enclosure of the power service

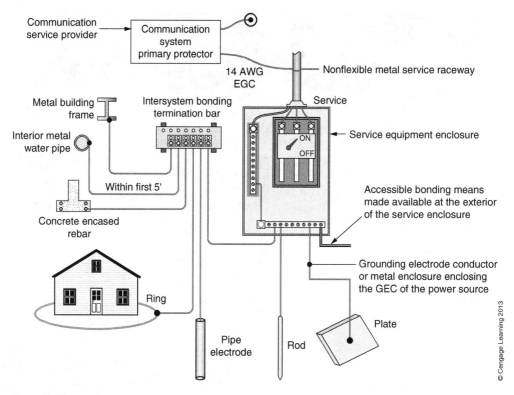

Figure 3–15
Communication system grounding means.

The bonding device shall not interfere with the opening of an equipment enclosure and shall be mounted on nonremovable parts. Bonding devices shall not be mounted on doors or covers, even if they are nonremovable.

Buildings or structures without intersystem bonding terminations or grounding means shall connect to a rod or pipe of 5 ft (1.5 m) in length and ½ in. (12.7 mm) diameter. The rod or pipe shall be driven into permanently damp soil. A separation of 6 ft (1.8 m) is required from any lightning conductors and from electrodes of other systems. Steam or hot water pipes or air terminal conductors shall not be used as electrodes for primary protectors or the grounded metallic members of cables.

Connection of the bonding conductor or grounding electrode conductor to the grounding electrode shall be made according to *250.70* of the *NEC*, as described earlier in Section 3.7 of this chapter. For indoor communication systems only, a listed sheet metal strap-type ground clamp, having a rigid metal base that seats on the electrode, shall be used. In addition, the material and dimensions of the strap shall be of a type that does not stretch during or after the installation (*NEC 250.70*).

National Electrical Code Article 820 Community Antenna Television and Radio Distribution (CATV)

Grounding Methods for *Article 820* of the *NEC* are nearly identical to those for *Article 800*, except for two major differences. Because CATV installations involve the use of coaxial cable, primary protectors are not required. Instead, *NEC 820.93* states that the outer conductive shield of the coaxial cable shall be grounded, and no other protective devices shall be required. Grounding of the outer conductive shield still requires it to be made as close as practicable to the point of entrance to the building.

The second difference involves the details of installing a separate CATV grounding electrode. Because the 20-foot (6 m) rule is still required in one- and two-family dwellings (30 ft [9 m] for mobile homes), *NEC 820.100(A)(4), Exception*, only references *250.52 (A)(5–7)*, which requires the installation of either an 8 ft (2.4 m) rod or pipe, or plate electrode. In addition, for buildings or structures without grounding means, although *800.100(B)(3)(2)* allowed for a 5 ft (1.5 m) electrode to be installed on communication systems, *820.100(B)(3)(2)* again only references *250.52(A)* for all possible electrode options when grounding a coaxial cable system. Essentially, an 8 ft (2.4 m) rod or pipe is required for grounding coaxial cable systems, and the 5 ft (1.5 m) electrode is only allowed on communications systems that do not use coaxial cables and are connected through listed primary protectors.

National Electrical Code Article 770 Optical Fiber Cables and Raceways

When installing optical fiber cables, *NEC Article 770* requires the grounding of all non-current-carrying metallic members of the optical fiber cables. *Article 770.93* states that either the metallic members of the cable have to be grounded as close as practicable to the point of entrance, or they must be interrupted by an insulating joint or equivalent device. Grounding methods for the metallic members of the optical fiber cable are then described in *770.100*, the requirements of which are identical to those in *800.100*, as previously discussed for communication systems. An additional option, *770.14*, states that the non-current-carrying conductive members of optical fiber cables shall be bonded to a grounded equipment rack or enclosure, as an alternative to the accessible grounding locations listed in *770.100(B)(2)*.

CHAPTER 3 FINAL QUESTIONS

1. A ground-fault current path is an electrically conductive path from the point of a line-to-case fault extending to the _____.
   a. ground
   b. earth
   c. electrical supply source
   d. equipment

2. Electrical systems that are grounded shall be connected to earth in a manner that _____.
   a. limits voltages due to lightning, line surges, or unintentional contact with higher voltage lines
   b. stabilizes the voltage-to-ground during normal operation
   c. facilitates overcurrent protection device operation in case of ground faults
   d. a and b

3. For grounded systems, non-current-carrying conductive materials shall be connected to earth so as to limit voltage-to-ground on these materials.
   a. True
   b. False

4. The earth can be considered an effective ground-fault current path.
   a. True
   b. False

5. Equipment grounding conductors and bonding conductors shall not be connected by _____.
   a. pressure connections
   b. solder
   c. lugs
   d. approved clamps

6. Sheet-metal screws can be used to connect equipment grounding conductors or bonding conductors to enclosures.
   a. True
   b. False

7. Ac circuits less than 50 volts shall be grounded if _____.
   a. supplied by transformers whose supply exceeds 150 volts
   b. supplied by transformers whose supply is ungrounded
   c. installed as overhead conductors outside of buildings
   d. All of the above

8. Interior metal water pipes located more than _____ from the point of entrance to the building cannot be used as part of the grounding electrode system.
   a. 2 ft (0.6 m)
   b. 4 ft (1.2 m)
   c. 5 ft (1.5 m)
   d. 10 ft (3 m)

9. Grounding electrodes that are driven rods require a minimum of _____ of contact with the soil.
   a. 6 ft (1.8 m)
   b. 8 ft (2.4 m)

   c. 10 ft (3 m)

   d. 12 ft (3.6 m)

10. Which type of grounding electrode requires a supplemental electrode and a bonding jumper?

   a. Building steel

   b. Metal underground water pipe

   c. Ring

   d. Rod

11. A bonding jumper for a supplemental grounding electrode is not required to be larger than _____ AWG copper wire.

   a. 4

   b. 6

   c. 8

   d. 10

12. The resistance to ground for a single grounding electrode shall be less than _____.

   a. 10 Ω

   b. 15 Ω

   c. 20 Ω

   d. 25 Ω

13. When multiple grounding electrodes are used, they shall be separated by not less than _____.

   a. 6 ft (1.8 m)

   b. 8 ft (2.4 m)

   c. 12 ft (3.6 m)

   d. 20 ft (6 m)

14. When multiple grounding electrodes are used, within a grounding system, they shall be bonded together.

   a. True

   b. False

15. A system or circuit conductor (one that carries current) that is intentionally grounded is referred to as _____.

16. What does it mean to be solidly grounded?

17. The following shall not be used as grounding electrodes:

   a. Aluminum

   b. Steel

   c. Metal underground gas pipes

   d. Both a and c

18. Which statement is true?

   a. Aluminum shall not be used as a grounding electrode conductor.

   b. Where outside, aluminum shall not be terminated within 18 in. of the earth.

    c. Bare aluminum or copper-clad aluminum grounding conductors shall not be used in direct contact with masonry or the earth where subject to corrosive conditions.

    d. Both b and c

19. An unintentional, electrically conducting connection between an ungrounded conductor of an electrical circuit and the normally non-current-carrying conductors, metallic enclosures, metallic raceways, metallic equipment, or earth is an example of _____.

20. All systems, devices, and cables are to maintain a minimum distance of _____ feet from the lightning protection system.

Chapter 4

Audio Physics

Objectives

- Describe the use of decibels and logarithms.
- Calculate the decibel as related to power and voltage.
- Define the threshold of hearing and the phon.
- Explain the inverse square law.
- Describe the mechanics of sound wave physics.
- Define frequency, dynamic range, bandwidth, octave, harmonics, phase, and coherence.

Chapter Outline

Sec 4.1 Decibels and Logarithms

Sec 4.2 Sound Wave Physics

Sec 4.3 Sound and Hearing

Key Terms

bandwidth	frequency response	pink noise
coherence	harmonics	radiation patterns
decibel	inverse square law	reflection
diffraction	octave	white noise
dynamic range	phase	
frequency	phon	

SEC 4.1   DECIBELS AND LOGARITHMS

Before we discuss audio physics, we first need to discuss briefly the use of decibels and logarithms.

Decibels

The **decibel (dB)** represents the common unit of measurement for sound. dB stands for 1/10 of a bel, on a logarithmic scale. Later in this chapter, we discuss how decibels can also be used to measure the magnitude of a changing circuit condition as related to amplification of power or signal attenuation. Let us first look at the decibel with respect to sound pressure and audio systems.

Sound level meters are calibrated in decibels and are used to measure the level of sound pressure or the intensity of sound within a given space. Zero decibel, however, does not represent the total absence of sound, but instead is equal to the threshold of hearing. The threshold of hearing refers to the lowest level of sound that can be detected by an average listener or by an individual with normal hearing ability. Mathematically, 0 dB is a zero reference level corresponding to a sound pressure level of 2×10^{-5} dynes/cm^2. The dyne is a unit of force in the CGS (centimeter–gram–second) system. Analogous units of force are the Newton (metric or International System of Units [SI] system) and pounds (lbs, English system).

In terms of intensity, this is equal to a power level of 1×10^{-12} W/m^2. The average jet airliner passing overhead is approximately equal to 130 dB, compared with a more quiet serene location that would measure closer to 35 dB. Locations appearing to be absent of sound typically measure 25 dB because of the simple movement of air molecules. Such levels of background noise in quiet rooms are considered ambient noise.

The following list provides typical examples of sound ranging from 0 to 195 dB and their associated power levels:

195 dB	Saturn rocket	20–40 million W
170 dB	Turbojet engine	100,000 W
160 dB	Jet engine, close up	10,000 W
150 dB	4-Propeller airliner	1000 W
140 dB	Threshold of pain	100 W
130 dB	75-Piece orchestra	10 W
	Jackhammer	
120 dB	Airport runway	1 W
	Piano	
110 dB	Power tools	0.1 W
100 dB	Subway	0.01 W
90 dB	Heavy traffic	0.001 W
80 dB	Factory	0.0001 W
70 dB	Busy street	0.00001 W
	Conversational voice	

(Continued)

60 dB	Average restaurant	0.000001 W
50 dB	Average office	0.0000001 W
40 dB	Low conversation	0.00000001 W
30 dB	Quiet office, whisper	0.000000001 W
20 dB	Quiet living room	0.0000000001 W
10 dB	Ambient noise	0.00000000001 W
0 dB	Threshold of hearing 200 $\times$ dynes/cm^2	0.000000000001 W

Notice that the threshold of pain is close to 140 dB; notably, however, permanent damage to the human ear could result if constant exposure to sound pressure levels greater than 100 dB are experienced.

Decibels and Power

The formula for calculating the decibel equivalent of a power measurement, whether acoustical (traveling through the air) or electrical (traveling through an electrical conductor or wire) is

$$dB = 10 \log(P/P_R)$$

where P represents the measured power, and P_R represents the reference power.

Without a known reference value, the calculation of a decibel cannot be achieved. In essence, a decibel measurement is nothing more than a logarithmic ratio, comparing the magnitude of some measured value with that of a known reference. Where calculating the decibel output of an electrical or electronic circuit, the power reference often is equal to 1 mW, or 0.001 W.

Decibels and Signal Attenuation or Signal Gain

In addition to power level calculations, the decibel can also represent the level of signal attenuation (signal loss) through a cable or a measured signal gain through an amplifier. Manufacturers commonly list the expected values of signal attenuation for a specific type of cable within their product data sheets. The example shown below represents the data for a typical RG-6U coaxial video cable. As stated below, attenuation values often are listed in decibels for every 100 m of cable.

RG-6U coaxial video cable attenuation specifications are as follows:

Signal frequency, MHz	1	10	100	1000
Nominal attenuation, dB/100 m	1.3	2.6	8.9	32.1

To use the table effectively, a technician must first know the frequency of the transmitting signal as well as the length of the actual cable run. The expected decibel values are then adjusted accordingly. As an example, for a 200 m length of cable, the listed losses in decibels must be doubled; likewise for a 50 m length of cable, the expected losses are cut in half.

Example

Using the cable data listed above, what is the expected signal attenuation for a 350 m length of cable at 10 MHz?

A total of 350 m of cable is 3.5 times longer than 100 m of cable. The expected losses are then

$$2.6 \text{ dB} \times 3.5 = 9.1 \text{ dB}$$

A loss of 9.1 dB can also be written as –9.1 dB.

A positive decibel number, therefore implies an amplification of signal or an overall gain.

As an example, if the specification data sheet for a signal amplifier lists a gain of +20 dB, then adding two amplifiers in series boosts the signal level by +40 dB. But what does +20 or +40 dB actually mean? And how can a loss or gain actually be measured and calculated? Doing so requires a decibel calculation based on relative voltage readings. A relative decibel calculation involves taking two signal level measurements, one before and the other after a changing condition.

Decibel Calculations Using Relative Voltage Measurements

Voltage measurements can be converted into decibels by using the following formula:

$$\text{dB} = 20 \log(V_2/V_1)$$

V_1 represents an initial voltage measurement across a specific resistance (before a changing condition), or it can represent a constant voltage reference. V_2 represents a voltage measurement from the same point in the circuit after a changing condition.

Notice that the only difference between the power decibel and the voltage decibel formula is the times 20 multiplier. Why 20? Because power is equal to

$$V^2 \div \text{Resistance,}$$

and because doubling the log is equivalent to a mathematical square, for example,

$$X^2 = 2 \log X,$$

any voltage ratio can be related to power simply by doubling the log.

Incidentally, the circuit resistance usually can be ignored provided that the resistance at the measurement point does not change, or that it is equal to the circuit resistance of the reference voltage. A typical example of how voltage may be used to calculate signal gain or attenuation is shown below.

If a measured signal voltage on a 600 Ω telephone line were to increase from 200 mV to 1 V, the decibel increase would be

$$20 \log 1 \text{ V}/200 \text{ mV} = +13.98 \text{ dB}$$

The affected increase would result in a gain of nearly 14 dB.

In this example, the 14 dB is a relative increase from 200 mV to 1 V. Relative measurements do not require a constant reference. For such a calculation, the reference is replaced with the initial voltage state before the measured change. Relative decibel calculations therefore require two measurements, not one, to calculate the magnitude of a change.

Logarithms

Logarithms typically are based on powers of 10; for example, the log of 100 is equal to 2. Why 2? Because when base 10 is raised to the power of 2, you end up with the number 100.

$$10^2 = 10 \times 10 = 100$$

Another example is that the log of 1000 = 3, because

$$10^3 = 10 \times 10 \times 10 = 1000$$

To calculate the log of any number between 100 and 1000, you need to use a log table or calculator. Logically, the answer is a decimal number between 2 and 3; for example, the log of 700 is 2.85, or

$$10^{2.85} = 700$$

Acoustical Power Measurements

If it is known that a jackhammer is producing a sound intensity of 5 W/m², then based on the previous formula, the decibel level is equal to

$$10 \log (5 \text{ W/m}^2 \div 1 \times 10^{-12}) = 127 \text{ dB}$$

The decibel notation system makes large intensities manageable, because 127 dB is 5,000,000,000,000 times larger than the threshold of hearing reference level of 1×10^{-12} W/m². Referring to the sound level as 127 dB makes it much easier to understand than it would be to say the sound is 5 trillion times larger than the threshold of hearing.

Let us now look at what happens when there are two jackhammers, each producing 5 W/m². If one jackhammer is equal to 127 dB, then two should be twice as loud, right? Wrong. The human ear responds to sound levels logarithmically, not linearly. The listener may notice an increase of sound level, but it is not twice as loud. Two jackhammers equal 10 W/m². The decibel equivalent is

$$10 \log (10 \text{ W/m}^2 \div 1 \times 10^{-12}) = 130 \text{ dB},$$

which is a total gain of only 3 dB.

What this means is that a doubling of power or sound intensity does not double the decibel output, but instead increases it only by 3 dB. Consequently, reducing the sound level to half the power reduces the output decibels by 3 dB. So where the jackhammer is muffled to a level of 2.5 W/m², then the output decreases to a level of 124 dB, because half the power only results in a loss of 3 dB.

How much of a change in decibels is noticeable by the human ear? The answer is that the average person does not notice a change in volume level of a complex waveform until it has increased or decreased by almost 7 dB. A 1 dB change of a pure single-tone frequency definitely can be detected, but when multiple frequencies are mixed together, such as in a complex musical arrangement, it becomes difficult for the human brain to detect and register all of the changing intensities. Because of this phenomenon, the human brain tends to average the perceived intensities of a complex waveform and mask the overall increase. What this means is that the average human ear would be hard-pressed to detect a 3 dB change or doubling of power. A highly trained ear may be able to detect a 5 dB change, but that is equivalent to just more than three times the power; this means that you would need three jackhammers, or possibly four, before the change in sound level starts to be noticeable. Because the human ear responds to sound logarithmically, not linearly, a doubling of power does not double the level of sound intensity; in fact, the average person hardly even notices the difference until the power levels get closer to four or five times larger. For this reason, large concert orchestras use 8 to 10 violinists to achieve the desired level of intensity, because one violinist would be lost in the background and two would hardly be noticed.

Inverse Square Law

Let us now look at what happens to the level of sound intensity when moving away from an object. Obviously, the closer you stand to a sound source, the louder it is; conversely, the farther you walk from the source, the lower or softer the intensity is. But how much softer is the sound? The **inverse square law** explains that all radiation patterns, whether they are light, sound, or radiofrequency signals, decrease by the square of the distance.

To apply this law, if you are standing 1 m (3.3 ft) from a source and measure 120 dB, and then you move to a distance of 2 m (6.6 ft), or double the distance, the sound pressure intensity is reduced by a factor of 4. Logarithmically, reducing the power by a factor of 4 is equivalent to a 6 dB reduction in volume intensity. Remember that losing half the power results in a 3 dB decrease, so then having only one-fourth the power means a loss of 6 dB. Think of one-fourth as being dividing by 2, twice.

The inverse square law states that the measured level of signal intensity decreases by 6 dB for every doubling of distance from the source. Likewise, at one-half the distance from the source, the measured level of signal intensity increases by 6 dB (Figure 4–1). The inverse square law will works for all types of waveforms radiating out into a three-dimensional space; examples include electromagnetic waves (radiofrequency), sound waves, and photons of light.

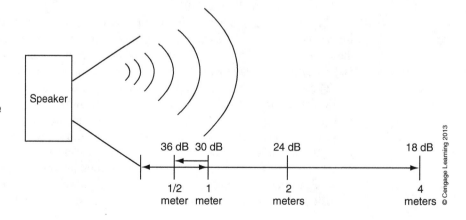

Figure 4–1
Inverse square law.

© Cengage Learning 2013

The following formula is used to calculate the expected loss or gain in signal level from a measured distance of d to the source. The formula also requires a comparative reference distance, which can be chosen from any arbitrary point. In most cases, however, a reference level is measured at a distance of 1 m from the source; all other signal levels can then be calculated and compared with this initial reading.

$$\text{dB loss or gain at distance } d = 20 \log (\text{Reference Distance}/d)$$

Example

Find the level of signal intensity 4.5 m from a signal source, when the reference intensity measures 25 dB/m² at 1 m.

Answer

$$\text{dB intensity at distance } d = 20 \log (1 \text{ m}/4.5 \text{ m})$$
$$\text{dB} = 20 \log 0.222$$
$$\text{dB} = -13.06 \text{ dB}$$

As a result, the level of sound intensity decreases by 13.06 dB. This means that the actual signal level, 4.5 m from the source, is

$$25 \text{ dB} - 13.06 \text{ dB} = 11.94 \text{ dB}$$

The answer to the problem is 11.94 dB.

SEC 4.2 SOUND WAVE PHYSICS

Sound Waves

Sound is produced by some form of mechanical vibration. Mechanical vibration causes the surrounding air molecules to vibrate back and forth, which in turn causes the air pressure to alternate between more dense and less dense. The compressions and rarefactions of air pressure radiate out continuously from an object in the form of waves, just as the waves of water form in the ocean. The peaks and valleys of a traveling sine wave graphically and mathematically illustrate the compressions and rarefactions of sound pressure. The waves that eventually reach the ear of a listener create the sensation known as sound (Figure 4–2).

Frequency

The **frequency** of sound is measured in Hertz (Hz). A quantity of Hertz represents the vibrational rate of a sound wave as it moves through a column of air. A 1000 Hz tone represents a sound wave that is vibrating back and forth 1000 times per second,

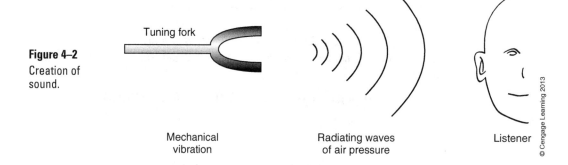

Figure 4–2
Creation of sound.

Mechanical vibration

Radiating waves of air pressure

Listener

or cycles per second, while radiating out from an object to the ear of a listener. Figure 4–3 illustrates one cycle of sound wave pressure.

The eardrum responds by vibrating at 1000 times per second in unison to the alternating air pressure. Musically, each note or tone of sound on a specific scale has an individual frequency pattern that not only makes it unique, but also distinguishable among all other tones vibrating simultaneously. Figure 4–4 illustrates the frequency ranges of various musical instruments.

Dynamic Range

Dynamic range represents the maximum span, or range, between the loudest and softest level of sound an object is capable of producing. An object can be a person, a musical instrument, or any mechanical device. Not all objects can produce the same levels or range of sounds. The human voice, for example, can often produce a broad dynamic range, as would be represented by comparing a whisper with a shout or a scream. Musical instruments, in contrast, all have specific dynamic ranges attributed to them, thus the occasional need for amplification. Think, for instance, of the comparison between a drum kit and an acoustic guitar.

Figure 4–3
Traveling sine wave, time, amplitude, and frequency (Hz).

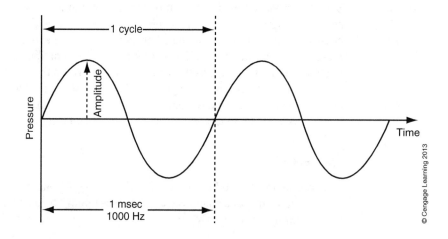

1 cycle

Amplitude

Pressure

Time

1 msec
1000 Hz

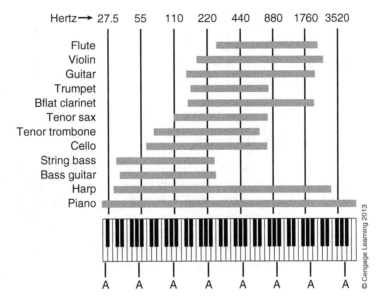

Figure 4–4
Various instruments and their associated frequency bandwidths.

Which one would you want your child playing with at 3:00 A.M.? Because of the wide dynamic range attributed to drums, they often are heard clearly over long distances, as compared with the strumming of an acoustic guitar, which, in most cases, would be lost in a sea of sound if not properly amplified.

Frequency Response

Frequency response represents the range of producible frequencies for an object, subject, instrument, or piece of audio amplification equipment. The frequency response of the human voice, for example, can vary greatly. A comparison of singers shows that a baritone cannot produce the same range of frequencies as a tenor because each is inherently limited by his or her natural ability; ultimately, they are unable to reach vocally beyond a specified range. From the hearing point of view, the frequency response of the human ear is somewhere between 20 and 20,000 Hz. Individuals who are hearing impaired often cannot hear the lowest and highest ranges. Most individuals have a difficult time hearing greater than 15,000 Hz and less than 100 Hz, although some more gifted individuals can actually perceive the full spectrum. Dogs and most other mammals can hear well into the range of 30,000 Hz, which explains why devices such as dog whistles appear to be silent to the human ear.

Bandwidth

Bandwidth represents the separation between the highest and lowest producible frequency within a system. For example, if a piece of audio equipment has a frequency response between 100 Hz and 1000 Hz, then the bandwidth is equal to 900 Hz: 1000 − 100 = 900.

Figure 4–5
Amplifier band-
width.

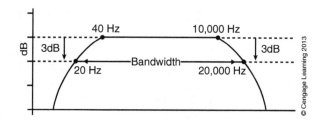

Amplifier Bandwidth

Amplifier bandwidth, or the bandwidth of any audio device, is represented by the separation between the highest and lowest producible frequencies, not having a signal loss of more than 3 dB. Outside of the bandwidth, voltages or signal levels decrease rapidly, making detection and reproducibility difficult to achieve. Earlier in this chapter, we discussed how a signal loss of 3 dB also represents the half power point, meaning that half the total power being delivered to the circuit or load has been lost. Figure 4–5 illustrates the typical frequency response of an audio amplifier.

Octave

An **octave** represents the doubling of a base frequency or half the value of a base frequency. Therefore, if 1000 Hz represents a reference base frequency, then 2000 Hz is the first upper octave, and 500 Hz is the first lower octave.

Multiple Octaves

Multiple octaves away from 1000 Hz can be calculated by multiplying or dividing the base frequency by 2^n, where n represents the level of octave to be calculated. Therefore, the second octave greater than 1000 Hz equals:

$$1000 \text{ Hz} \times 2^2$$
$$2^2 = 4$$
$$1000 \text{ Hz} \times 4 = 4000$$

Therefore, 4000 Hz is 2 octaves greater than 1000 Hz. Two octaves less than 1000 Hz is equal to

$$1000 \text{ Hz} \div 2^2$$
$$1000 \text{ Hz} \div 4 = 250 \text{ Hz}$$

The third, fourth, fifth, and so on octaves can be calculated by the same process, simply by changing the value of n in the exponent. Figure 4–6 illustrates a graph of octaves and harmonics.

The second octave also could be found by doubling the first upper octave.

$$2000 \text{ Hz} \times 2 = 4000 \text{ Hz}$$

The third octave can be found by doubling the second upper octave.

$$4000 \text{ Hz} \times 2 = 8000 \text{ Hz}$$

The formula can be repeated to calculate each subsequent octave. Either process works.

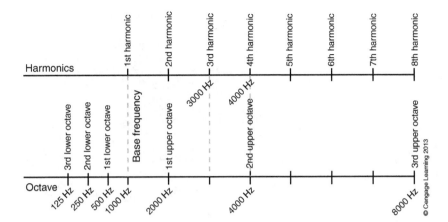

Figure 4–6
Octaves versus harmonics.

Harmonics

Harmonics represent direct multiples of a base frequency. If the base frequency is equal to 1000 Hz, then the second harmonic is 2000 Hz, the third harmonic is 3000 Hz, the fourth harmonic is 4000 Hz, and so on. The first harmonic is always equal to the base frequency.

The Human Voice

The energy response of the human voice is different between men and women. As shown in Figure 4–7, the average male voice tends to be heavier in low frequencies and more centered around 500 Hz. Notice, too, how 500 Hz also represents the greatest intensity of energy generated out of the male voice. In contrast, the average female voice pattern has little low-frequency content and also is centered up higher toward 1000 Hz.

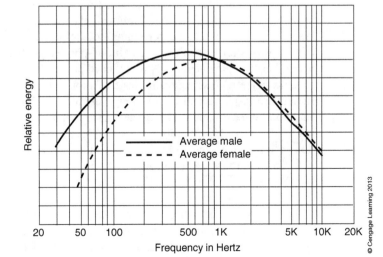

Figure 4–7
Average frequency response of male versus female voice.

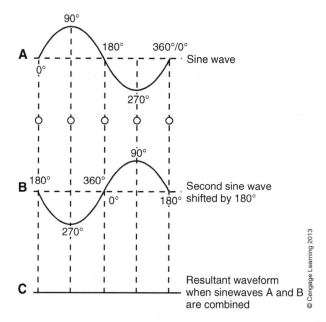

Figure 4–8

Phase cancel-
lation.

Phase and Coherence of Signal

Figure 4–8 represents two sine waves of identical frequency and amplitude. Sine wave A starts at 0° and completes one full cycle at 360°. Sine wave B starts at 180° and completes one full cycle after returning back to 180°. Comparing the starting points of each waveform shows that they are 180° out of **phase**.

As a result, phase cancellation results if both of these waveforms were to meet up with each other while traveling through the atmosphere, or inside of an electrical circuit, such as an amplifier or mixer. The two waves are considered to be out of phase or incoherent because waveform A reaches its maximum positive point (90°) at exactly the same time waveform B reaches its maximum negative point (270°). When combined, the two cancel out mathematically. Waveform C represents the zero output combinational phase cancellation after the two waves are mixed together. Out-of-phase signals are referred to as incoherent because they negatively affect each other when combined.

Coherence represents an in-phase relation between two signals. When two signals are in-phase and mixed, they positively combine to increase the total amplitude at the output, resulting in an increased signal level. Earlier in this chapter, we discussed how doubling the signal level, or signal voltage, results in four times the output power, which is equivalent to a 6-dB increase.

Figure 4–9 illustrates two complex audio waveforms. When comparing waveform A with waveform B, multiple points of coherence and incoherence can be seen as the two signals combine.

Coherence occurs when the two waveforms have the same instantaneous polarity, but at moments of polarity mismatch, the combination results in incoherence and cancellation of signal. Coherence or incoherence of signals occurs when sound waves traveling through space, a wire, or an amplifier are allowed to

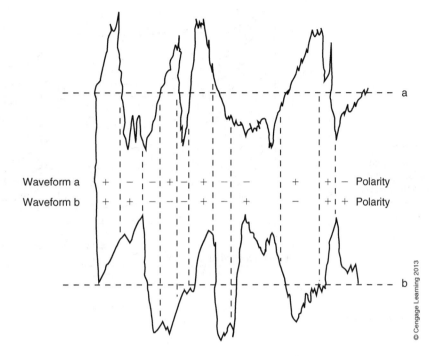

Figure 4–9
Coherence and incoherence of two complex waveforms.

Waveform a   +   −   −   +   −   +   −   −   +   +   −   Polarity

Waveform b   +   +   −   −   +   +   −   +   −   +   +   Polarity

a

b

© Cengage Learning 2013

combine and mix with each other. The effect can be problematic where sound can literally appear to die out in parts of a room, whereas at the same time increase a few feet away. Later in this chapter, and in Chapter 5, we discuss reflection, reverberation, and resonance to examine what can be done about this problem.

SEC 4.3   SOUND AND HEARING

As discussed, zero decibel has been defined as the threshold of hearing, which represents a sound pressure level of 10^{-12} W/m². The human ear does not hear all frequencies equally. For example, if a sound frequency of 3500 Hz is heard at a level of 0 dB, and the frequency then is reduced to 350 Hz (approximately 3 octaves and a major third below), the sound pressure level appears to decrease by nearly 17 dB, and the listener thinks you decreased the volume and changed the frequency. What this means is that for the average listener to be able to detect the 350 Hz tone with an equal intensity to the 3500 Hz tone, the signal would need to be amplified by an additional 17 dB.

Phon

A phon is equal to a decibel level of a 1000 Hz reference tone within a bandwidth of frequencies. As shown in Figure 4–10, the human ear's perception of loudness begins to flatten out as sound pressure levels increase. Notice that at a level of

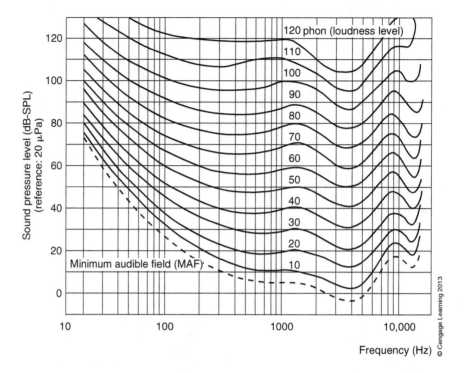

Figure 4–10

Robinson and Dadson's equal loudness contour of hearing graph.

© Cengage Learning 2013

80 phon (80 phon represents a level of 80 dB at 1000 Hz), 20 Hz must only be increased by 30 dB to be perceived as equal in level to a 1000 Hz tone. However, at 0 phon, a 20 Hz signal must be boosted by nearly 75 dB to have the same effect. Simply stated, as sound pressure levels increase across the entire bandwidth, the level of low and high frequencies do not need to be boosted as much, compared with when sound pressure levels are lower. As a result of this phenomenon, the loudness button on most audio amplifiers was designed to give a greater boost to lower frequency, and it should only be used at low volume or phon levels. However, when an amplifier is operating at much higher output levels, it is not necessary to boost the low frequencies as much, and the loudness button should not be used. As shown in Figure 4–10, greater phon levels do not need as much low-frequency boost for them to be perceived as equal in level to a 1000 Hz reference tone.

Understanding how the human ear perceives levels of sound pressure and frequency is crucial to balancing and recreating the original sound image. The main idea should be to perceive all frequencies within a certain bandwidth equally; otherwise, the overall sound image can appear to be more heavily weighted to one side of the spectrum or the other and not truly representing the original contours of sound.

Figure 4–10 represents the equal loudness contour of hearing graph that was originally developed by Fletcher and Munson, and later revised by Robinson and Dadson in the 1940s.

The graph represents the compiled data from an average sampling of human subjects and their measured perception of hearing. The sound pressure levels measured in decibels represent the minimum requirements needed for an individual to maintain a constant, perceived loudness level at all frequencies within the 20 to 20,000 Hz bandwidth. Let us start with the lower frequencies. By using 1000 Hz as the reference tone, it can be seen from the graph that 20 Hz occurs nearly 75 dB below on the zero **phon** level (dotted line in Figure 4–10). Zero phon represents a 1000 Hz reference level of 0 dB. For the observer to perceive the loudness or intensity of the 20 Hz signal as equal to that of the 1000 Hz reference, it must be amplified by almost 75 dB. Now look at 0 phon, 10,000 Hz. The problem still exists, but not as severely, because the level of sound pressure must now only be increased by 20 dB to maintain the same perceived level of intensity as the 1000 Hz reference. What this means is that the human ear has a much harder time perceiving low frequencies as compared with high frequencies, but it can hear quite clearly at midrange levels, somewhere between 1000 and 5000 Hz. In actuality, the human ear tends to be most sensitive to frequencies centered around 3500 Hz. Notice also that at each level of phon all the lowest points of the graph fall off at approximately 3500 Hz, indicating the frequency of maximum sensitivity to the human ear.

Bass and treble controls were developed on amplifiers to help alleviate the poor sensitivity of human hearing to low and high frequencies. Because perception levels can vary from person to person, allowances for personal adjustment had to be incorporated into the design of amplifiers. Remember that the Robinson and Dadson graph was designed originally around an average sampling of human subjects; therefore, allowances for people who do not compare with the measured average norms had to be taken into account.

White Noise Generators

White noise generators generate equal intensities of all frequencies in the range of 20 to 20,000 Hz. The effect is similar to the sound of a vacuum cleaner. Later in Chapter 5 of this book, we discuss how white noise generators are used to flatten the response of a room; they also are used as noise masking devices to help break the dead silence of a quiet room.

Pink Noise Generators

Pink noise generators are white noise generators with filters. Each octave band of frequencies, increasing from 20 Hz, is decreased by 3 dB. Therefore, 40 Hz is 3 dB less than 20 Hz, and 80 Hz is 6 dB less than 20 Hz. To understand this, you need to understand that a doubling of sound pressure intensities results in an overall power increase of 3 dB. For example, between 1000 and 2000 Hz, there are 1000 individual sound pressure intensities. Between 2000 and 4000 Hz, which represents the next octave, there are 2000 individual sound pressure intensities. As a result, every increase in octave results in a doubling of power, and the overall sound pressure energy increases by 6 dB. Because the range of intensities is wider between high-frequency octaves compared with low-frequency octaves, and because the human ear responds logarithmically to impulses, greater frequency octaves are perceived to be 3 dB louder. When listening to a white noise generator, the higher frequency octaves tend to overpower the lower frequency octaves, and the effect sounds like a high-frequency hiss, even though the actual intensity of energy is equal from one frequency to the next. Pink noise generators, however, sound more balanced and are filtered by 3 dB per octave to help level off the overpowering perception of high-frequency energy.

It may appear strange at first glance, but even though white noise generators are equal in energy per frequency, they actually sound 3 dB louder per octave. Pink noise generators, in contrast, are filtered and actually measure 3 dB less per octave, but they appear to sound equal or level from one octave to the next. Figure 4–11 compares the output response of the two generators over the 20 to 20,000 Hz bandwidth. To understand more, refer to Section 4.1.

Wavelength and Sound Traveling in Air

The wavelength (λ) of a sound wave can be defined as the velocity of sound in air, v (ft per sec), divided by the frequency in Hertz, f (cycles per sec).

$$v \div f = \lambda$$

The velocity of sound in air increases as temperature increases. At 70°F, the velocity of sound is approximately 1130 ft/sec.

Let us look at the wavelength at 100, 1000, and 10,000 Hz.

$$1130 \text{ ft/sec} \div 100 \text{ Hz} = 11.3 \text{ ft/cycle}$$
$$1130 \text{ ft/sec} \div 1000 \text{Hz} = 1.13 \text{ ft/cycle}$$
$$1130 \text{ ft/sec} \div 10{,}000 \text{Hz} = 0.113 \text{ ft/cycle, or } 1.36 \text{ in./cycle}$$

The length of a sound wave affects the **reflection** and **diffraction** of a signal as it moves past an object. Diffraction relates to the bending of a waveform around an object. When the wavelength of a signal is larger than the object in its path, the signal bends around the object and travels at an angle different from its original point of travel. Conversely, when the wavelength of a signal is smaller than an object in its path, the signal bounces off and reflects back at an angle equal to the angle of approach, similar to light bouncing off the surface of a mirror.

The effects of reflection and diffraction can cause distortions of the original sound image. As a result, the perception of signals heard from behind an obstacle appears to have a loss of high-frequency content and be heavier in low-frequency content.

Figure 4–11
(A) White noise.
(B) Pink noise.

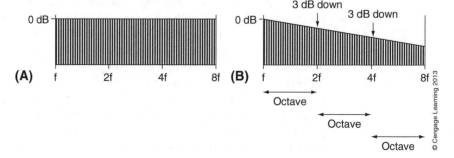

Because the higher frequencies bounce off and reflect away from their original point of travel, they never reach their final destination and appear to be nonexistent. Some of them may, however, reflect off the surrounding walls and eventually reach their intended destination; however, by that time, they are late and may be perceived as an echo or reverberation, and are greatly reduced in intensity. The size of the room and construction of the walls, ceiling, and floor materials greatly affect the type of reflections and reverberations returned or heard by the observer. This topic is explored in greater detail in Chapter 5 (see discussion of speaker placement).

Radiation Patterns of Different Frequencies

Wavelength, reflection, and diffraction cause the **radiation patterns** of differing frequencies to vary. Low frequencies tend to bend around most objects; as a result, they can easily cover and fill a three-dimensional space. It is nearly impossible to detect the point of origin for frequencies in the range of 20 to 80 Hz. Very high frequencies, in the range of 10,000 to 20,000 Hz, bounce around a lot and tend to travel in a more line-of-sight direction. High-frequency waveforms tend to be directional in nature, and for optimal detection, an observer should stand in the path of the traveling waveform.

CHAPTER 4 FINAL QUESTIONS

1. Calculate the decibel change when an output voltage increases from 2 to 6 V.
2. The input signal to a 1000-foot (304.8 m) cable run measures 25 V, whereas the load measures 17 V. Calculate the attenuation of the signal through the cable in decibels.
3. A total of 126 dB is equal to what level of power when referenced to 1 mW?
4. Define *dynamic range*. Give an example of an instrument with a wide dynamic range and one with a narrow dynamic range.
5. An amplifier has a frequency response from 400 Hz to 16 KHz. What is the bandwidth of the amplifier?
6. What is the frequency range of the human ear?
7. An amplifier has a base frequency of 500 Hz. Calculate the frequency of the first upper octave. Calculate the frequency of the fifth upper octave.
8. What is the sixth harmonic of 25 Hz?
9. What occurs when opposing signals mix through the same electrical connection, or meet within the same air space?
   a. The total signal power doubles.
   b. The signals add.
   c. The signals cancel and subtract from each other.
   d. It results in half the expected signal power due to cancelation.

10. When measuring electrical signal power, 0 dB represents
    _____.
    a. a total absence of power
    b. a gain or ratio of 1, where the output is equal to the input
    c. 0 volts
    d. 0 watts

11. Based on the perception of hearing and the phon, for the average listener, _____.
    a. treble frequencies need to be amplified more than bass frequencies at low listening levels
    b. base frequencies need to be amplified more than treble frequencies at low listening levels
    c. treble frequencies need to be amplified more than bass frequencies at high listening levels
    d. bass frequencies need to be amplified more than treble frequencies at high listening levels

12. The human ear is most sensitive at _____ Hz.
    a. 500
    b. 1000
    c. 3500
    d. 5000

13. How does a white noise generator differ from of a pink noise generator?

14. Calculate the wavelength of a 75 kHz sound wave traveling through a room temperature environment.

15. A 100 W amplifier is exchanged for a 600 W amplifier. The maximum output increases by _____ dB. Is the change noticeable?

16. Diffraction occurs when _____.
    a. the signal frequency is larger than an object in its path
    b. the signal frequency is smaller than an object in its path

17. Which statement comparing base frequencies to treble frequencies is true?
    a. Bass frequencies bend around objects and tend to fill a room, whereas treble frequencies are line of sight.
    b. Bass frequencies are line of sight, and treble frequencies tend to bend around objects and fill a room.

18. Give a real-world example of why you would use a white noise and a pink noise generator.

19. The reference intensity of a sound source measures 8 dB/m^2 at a distance of 1 m (3.2 ft). At a distance of 15 m (49.2 ft), the level decreases by _____ dB.

20. In the above question, the intensity of the sound source will then measure _____ dB at a distance of 15 m (49.2 ft) from the source.

Chapter 5

Audio Signal–Processing, Amplification, and Reproduction Equipment

Objectives

- Describe the mechanics and electrical properties of dynamic and condenser microphones.
- Explain microphone polar patterns and their uses.
- Describe the mechanics and electrical properties of loudspeakers.
- Explain crossovers and filters.
- Describe the electrical properties of mixers and amplifiers, their various signal levels, and types of available system connections.
- Describe the electrical properties of transformer-coupled audio systems, as well as their advantages and disadvantages.
- Describe and explain the use of various types of audio processing equipment.
- List the various specifics of audio as related to the *NEC*.

Chapter Outline

Sec 5.1 *National Electrical Code* Requirements of Sound Systems

Sec 5.2 Public Address Systems/Intercoms

Sec 5.3 Components and Electrical Properties of Sound Systems

Sec 5.4 Speakers, Crossovers, and Their Electrical Properties

Sec 5.5 Amplifiers, Signal Levels, and Their Electrical Properties

Sec 5.6 Mixers, Preamplifiers, and Signal-Processing Equipment

Sec 5.7 Additional *National Electrical Code* Requirements

Key Terms

active filter	cardioid	crossovers
audio mixer	coaxial speakers	damping
balanced	compliance	dynamic microphones
band-pass	condenser microphones	echo

equalizers	microphone paging	reverberation
field coil	midrange	shotgun
goosenecks	muting	signal level
graphic equalizer	Newtons	signal source
ground loop	notch filter	speaker efficiency
high-pass	omnidirectional	spectrum analyzer
horns	parametric equalizer	spider mount
integrated amplifier	phantom power	telephone paging
isolated ground	polar patterns	tip-ring-sleeve
lapel	power amplifier	transducers
lavaliere	preamplifier	tweeters
line pad	precedence	unbalanced
loudspeakers	proximity effect	unidirectional microphones
low-pass	radiators	woofers
matching transformer	RCA connectors	XLR

SEC 5.1   *NATIONAL ELECTRICAL CODE* REQUIREMENTS OF SOUND SYSTEMS

National Electrical Code Article 640—Audio Signal Processing, Amplification and Reproduction Equipment

Article 640 of the *NEC* covers permanently installed, distributed audio systems such as sound systems that would be found in restaurants, hotels, business offices, commercial and retail sales environments, churches, and schools. Also covered are portable sound systems; installation locations include residences, auditoriums, theaters, stadiums, movie theaters, and television studios. The article also covers temporary installation locations such as fairs, festivals, circuses, public events, and concerts.

National Electrical Code Article 640.4—Protection of Electrical Equipment

Like all electrical equipment, components of sound systems are required by *Code* to have sufficient hazard protection in the form of overcurrent/overvoltage devices. The use of such devices is meant to protect people using the equipment from electrical shocks resulting from the abnormal operation of sound system components. Overcurrent/overvoltage devices should also protect the facility in which the sound system is being installed from possible fires that could result in loss of life. In addition to electrical protection, equipment shall also be located or protected in such a manner that will guard against environmental exposure or physical damage.

National Electrical Code Article 640.5—Access to Electrical Equipment behind Panels Designed to Allow Access

When installing audio circuits, or any circuits for that matter, it is important that wires associated with the system be supported in such a manner so as not to block access to removable doors or panels. Wires and cables also should not be placed directly on top of suspended ceiling tiles. Suspended ceilings are listed and tested to support the weight of the tiles; they are not strong enough to support the weight of wires.

National Electrical Code Article 640.6—Mechanical Execution of Work

Pride should be taken when connecting the wires of audio components. All connections, regardless of whether they are terminated at equipment or tagged for future use, should be durably labeled so that servicing personnel are able to decipher the location of each wire and the component of the sound system which it serves (*NEC 640.6(D)*). Equipment, cables, and circuits shall be installed in a neat and professional manner to ensure that the wiring scheme of one system does not interfere with that of another. In particular, when the wires of one system are being serviced, upgraded, or removed, the wires of other systems must not be inadvertently damaged in the process. In general, the installation of any new wiring system should not be placed over or on any preexisting wiring; to do so would only cause clutter and confusion, and may introduce the possibility of short circuits to other systems during building renovations or system upgrades (*NEC 640.6(A)*).

Audio wires and cables shall be installed in a manner that does not cause damage to the conductors or insulation. Installers should choose an appropriate support means for all wiring and cables, one that does not adversely affect the system performance. It is never appropriate to attach wires or cables to the exterior of any conduit or raceway as a means of support. When attaching wires or cables to support hardware, care must be taken so that they are not strapped too tightly to constrict the interior of the cable or stapled in such a manner that could cause penetration of the central conductors; wires also should not be hung from structural components in such a manner that could cause sag or interfere with other systems, system wires, or prevent normal building use. Normal building use may include the opening and closing of a door or window (*NEC 640.6(B)*).

With regard to suspended ceilings, wires and cables should not be attached to the suspension wires of the suspended ceiling. It is not acceptable to use the suspended ceiling as a means of support. Instead, audio wires should be installed on separately approved mounting hardware or inside of isolated conduits or raceways. In addition, the accessible portion of abandoned cable shall be removed unless previously tagged for future use.

SEC 5.2   PUBLIC ADDRESS SYSTEMS/INTERCOMS

Telephone Paging

Telephone paging is defined as an audio signal sent from the main phone system of a building structure to an audio system equipment rack for distribution through the network of building speakers.

Microphone Paging

Microphone paging results from connecting a microphone directly to an audio system equipment rack for distribution through the network of building speakers. The microphones can be permanently installed or be portable. A portable system has microphone connection jacks mounted on the walls in various locations around the building for use as needed. The user connects the microphone through premade microphone cables of various lengths, as needed.

Muting

Muting controls allow for selected inputs of amplifiers or mixing consoles to be shut down during a microphone or telephone page. The intent is to interrupt or reduce the intensity of a music-programming source during an announcement or page.

Precedence

Precedence is defined as a signal that has priority over all other signals. An example of a signal with precedence would be a fire alarm signal or emergency paging signal; in such circumstances, a fire alarm signal or emergency paging signal would have precedence over all other input signals to a mixer or amplifier. All signals not having precedence, such as music programming, are muted or shut down during an emergency broadcast.

Bells, Tones, and Alarms

Many industrial or professional audio systems provide the option of supplying bells, tones, or alarm signals to audio speaker networks. Bells or tones often are used as a prepaging signal to draw attention to an upcoming broadcast. In such cases, a bell or tone sounds 1 or 2 seconds ahead of an important page or announcement to draw the attention of the listener. School systems often use timed bells that are programmed throughout the day to announce the start, end, and changeover of classes or periods.

SEC 5.3   COMPONENTS AND ELECTRICAL PROPERTIES OF SOUND SYSTEMS

Microphones

Microphones come in two basic varieties: dynamic or condenser. Available options may also include wireless. In this section, we discuss the theory of operation, frequency response, and microphone polar patterns. Understanding such topics is critical to choosing the correct type of microphone for a specific job.

Transducers

A **transducer** is a device that converts one form of energy into a secondary form. An example would be the mechanical energy of one system being converted into the electrical energy of another. A microphone is an example of an audio transducer. It converts the mechanical sound pressure of the atmosphere into an electrical signal. A loud speaker does just the opposite; it converts the electrical signals of an amplifier into mechanical sound pressure. Both microphones and **loudspeakers** are examples of audio transducers. Figure 5–1 illustrates a transducer.

Dynamic Microphones

A **dynamic microphone** operates under the principal of induction, as sound pressure variations cause a coil of wire to move within a magnetic field. When a coil of wire moves through a magnetic field, a voltage is induced and current flows. The level of current changes in proportion to the rate of motion. As a coil moves faster through a magnetic field, the induced voltage and current flow increases. The strength of the induced signal is also related to the number of turns of wire in the coil and to the strength of the magnetic field. Induction also occurs if the magnetic field moves or varies, and the coil remains stationary. Either way, whether the coil is moving through the magnetic field or whether the magnet is moving across the coil, induction and current flow occurs. Motion is required; without motion there can be no current.

Figure 5–2 illustrates the effect of induced current flow through the coil of a dynamic microphone. The coil, referred to as the voice coil, connects to the microphone diaphragm. A magnetic field surrounds the coil. As an individual speaks, sound pressure vibrations within the atmosphere cause the diaphragm to move.

Figure 5–1
Model of a transducer.

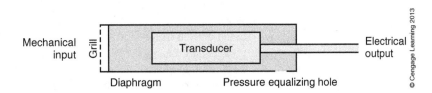

Mechanical input — Grill — Transducer — Electrical output

Diaphragm   Pressure equalizing hole

© Cengage Learning 2013

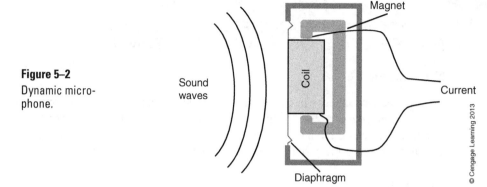

© Cengage Learning 2013

Figure 5–2
Dynamic microphone.

The coil then is forced to move in unison within the magnetic field, thereby inducing an electric current, one that represents an exact electrical copy of the mechanical vibration.

When an individual stops speaking into the diaphragm, the coil stops moving and the current stops flowing. No current flow is equal to an absence of sound. Remember, however, there is always some minimum level of ambient sound because of the simple movement of air molecules against the microphone diaphragm. Even the smallest, most subtle vibration of ambient sound is detected and amplified by the device. Some microphones are more sensitive than others, and the level of output will depend greatly on the type of diaphragm, size of the coil, and strength of the magnetic field. Specifications of different styles and types of dynamic microphones must be compared to know which works best for the intended use and the required bandwidth.

Condenser Microphones

A **condenser microphone** is made from a capacitor. A capacitor is a device capable of storing electrical charge in an electrostatic field between two metal plates. An insulator separates the metal plates so that the electrostatic field can exist. In early times, capacitors were referred to as condensers, hence the name condenser microphone. The diaphragm of a condenser microphone attaches to one of the plates of the capacitor. As sound pressure from the atmosphere pushes on the diaphragm, the spacing between the metal plates changes. The overall capacitance increases as the plates move closer together and decreases as the plates move farther apart. As capacitance changes, so too does the quantity of electrostatic charge stored inside the device.

A condenser microphone requires the use of an external power source to operate. Figure 5–3 illustrates the internal design of a condenser microphone. The external source supplies a direct current (dc) bias voltage to the plates of the capacitor. The bias voltage sets up an initial charge on the device. Atmospheric sound pressure variations on the diaphragm then alter the level of stored charge as the spacing of the plates varies. The variations of stored charge represent a direct analog to the changing sound pressure. In most cases, a battery inside the microphone supplies

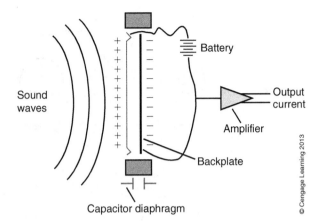

Figure 5–3
Condenser
microphone.

the bias voltage. An alternate supply method would be to use phantom power, as described in the next section.

A condenser microphone also requires the use of a preamplifier circuit, typically located inside of the microphone housing. The amplifier circuit is needed to detect the subtle changes in electrostatic charge as sound pressure variations alter the capacitance of the device. A condenser microphone has greater sensitivity compared with a dynamic microphone because of the built-in preamplification circuit. The preamplifier circuit also receives its power from the internal power supply or from the phantom power.

Phantom Power

Phantom power is an external power source for condenser microphones, typically supplied by the mixing console through the connecting wires of the microphone. The phantom power acts as a replacement for the internal batteries of the microphone. Most manufacturers use a level of 48 V dc. A button at each individual channel of the mix console is available for turning the phantom power on or off, as desired. The microphone never loses power because the electrical wall outlet powers the mix console. Batteries going dead halfway through a performance or presentation is no longer an issue when using phantom power.

Microphone Sensitivity and Polar Patterns

Microphone **polar patterns** graphically illustrate the sensitivity of a microphone. A polar response is measured by rotating the microphone through a 360° arc, while keeping the sound source constant, both in level and direction. Figure 5–4 illustrates the various types of microphone patterns that are available. The most common varieties are omnidirectional, bidirectional, unidirectional or cardioid, super cardioid, hypercardioid, and shotgun.

Omnidirectional Microphone. The **omnidirectional** microphone is sensitive to sounds in almost a perfect 360° pattern around the diaphragm (Figure 5–5). Sounds originating from virtually any direction can be detected from the

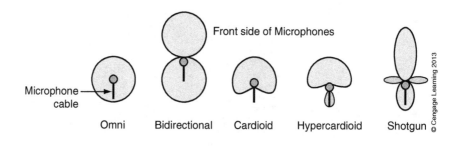

Figure 5–4
Microphone
polar patterns.

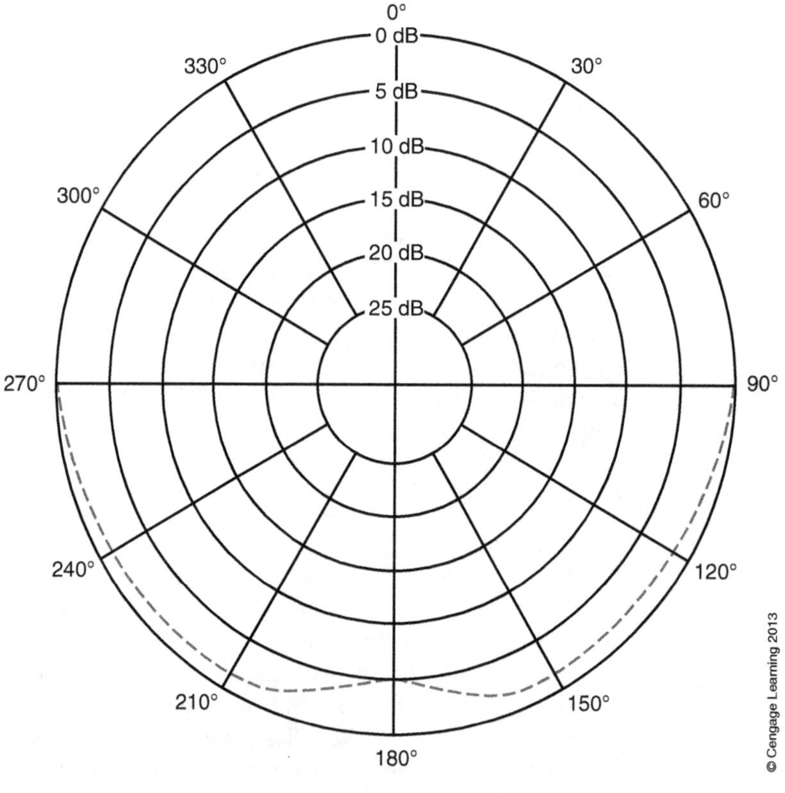

Front side of Microphones

Microphone cable

Omni　　Bidirectional　　Cardioid　　Hypercardioid　　Shotgun

© Cengage Learning 2013

Figure 5–5
Omnidirectional
polar pattern.

0°
0 dB
330°　　　　　　　　30°
5 dB
10 dB
300°　　　　　　　　　60°
15 dB
20 dB
25 dB
270°　　　　　　　　　90°

240°　　　　　　　　　120°

210°　　　　　　　　　150°
180°

© Cengage Learning 2013

front, sides, and rear. In addition, sound ports usually are designed into the diaphragm cover to allow signals entering from the rear of the device to reach the diaphragm.

Bidirectional. The bidirectional microphone pattern (Figure 5–6) is sensitive only to sounds originating from the front and rear of the diaphragm. The microphone does not detect off-axis signals entering from 90° or 270°.

Unidirectional or Cardioid. The **unidirectional microphone**, often called the **cardioid** microphone (Figure 5–7), detects sounds well in the forward direction but has a greatly reduced sensitivity toward the rear of the diaphragm.

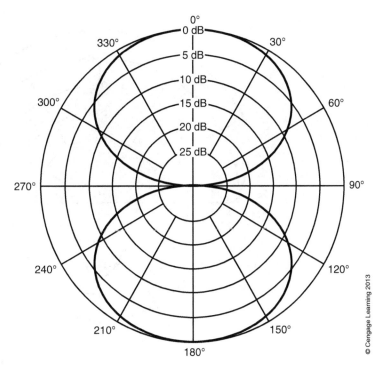

Figure 5–6
Bidirectional polar pattern.

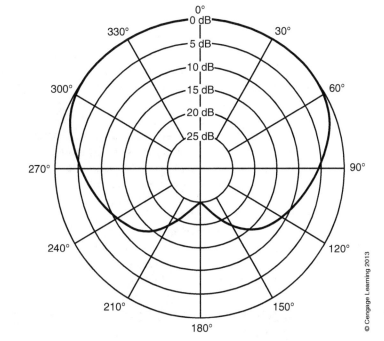

Figure 5–7
Cardioid polar pattern.

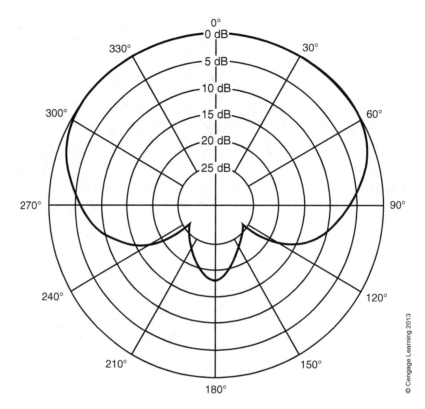

Figure 5–8
Supercardioid polar pattern.

© Cengage Learning 2013

Signals approaching from the rear, at 180°, are reduced by as much as 25 dB. Notice also that the sensitivity of the 90° and 270° axis is reduced by nearly 6 dB, and falls off more quickly as the angle approaches 180°. It is important to understand that the microphone is not completely deaf to off-axis signals. However, they are greatly reduced in decibel level, as indicated by the polar pattern.

Supercardioid and Hypercardioid. The supercardioid (Figure 5–8) and hypercardioid (Figure 5–9) patterns are far more selective toward the front axis of the diaphragm, and tend to more greatly attenuate signals that enter from the back of the pattern, somewhere between 90° and 270°. Compare the shape of the cardioid with the supercardioid as they are superimposed over each other on the same polar graph (Figure 5–10). The reduced sensitivity along the rear and sides of the diaphragm can be seen more easily. Supercardioids, hyper-cardioids, and **shotgun** patterns typically are used in situations where extreme selectivity of signal along the front axis of the diaphragm is required. Noises occurring at the rear of the diaphragm or off-axis are reduced significantly and essentially not detected. The shotgun pattern (Figure 5–11) is the most selective microphone design because it barely allows a 60° window of detection along the front of the diaphragm.

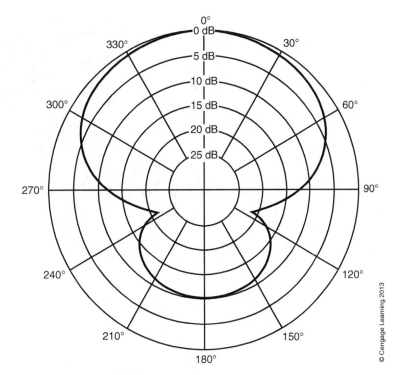

Figure 5–9
Hypercardioid
polar pattern.

© Cengage Learning 2013

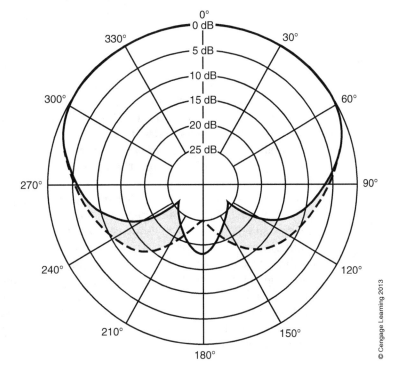

Figure 5–10
Comparison of
cardioid and
supercardioid.

© Cengage Learning 2013

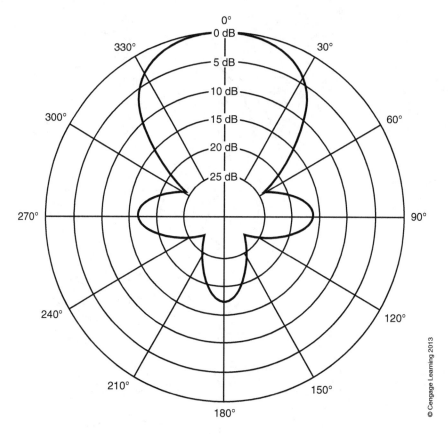

Figure 5–11
Shotgun polar pattern.

© Cengage Learning 2013

Choosing a microphone sensitivity pattern depends greatly on the environment and the quality of ambient noise surrounding the desired signal. In loud environments, it is more useful to use hypercardioid microphones because they greatly reduce the level of unwanted, off-axis sounds and help to minimize interference from other sources. However, if you are interested in detecting the overall sound image of an environment, such as the busy sounds of an outdoor mall, in a 360° pattern, then an omnidirectional microphone makes more sense.

Also available are more complex microphone designs that use dual diaphragms. A dual-diaphragm microphone often uses one diaphragm for low frequencies and the other for high frequencies, while also combining the effects of differing polar patterns. As described in Chapter 4, low-frequency signals traveling through three-dimensional space tend to be omnidirectional, whereas those of high frequency are straighter, line-of-sight, and directional. A designer often can achieve various degrees of frequency response across the entire spectrum simply by deciding on the correct polar pattern for each diaphragm. The overall sound image then represents the combination of the two diaphragms as they mix electronically.

Proximity Effect

Many cardioid microphones exhibit a phenomenon known as the **proximity effect**. The proximity effect causes an increase in bass response as a microphone moves closer to a sound source. Although an increase in bass may be beneficial in some cases, it may not always be desired; this is why some cardioid microphones have switchable high-pass filters built into them to help counteract the effect.

Microphone Impedance

Low-impedance microphones operate from **balanced** circuits, meaning that they are ungrounded and measure an equal level of impedance from either side of the line to ground. The balanced connection is desirable when long lengths of cable are required between the microphone and the mixing console or amplifier, to help reduce the amount of noise and interference from stray electromagnetic fields or electrostatic interference. Typical low-impedance microphones measure between 150 and 600 Ω. Impedance is defined as resistance to ac. Do not expect to measure 600 Ω when you connect a dc ohmmeter to the voice coil of a dynamic microphone or speaker. Dc resistance and ac impedance are similar to each other, but they do not measure the same. Also, if you attempt to connect a dc ohmmeter across the plates of a condenser microphone, you measure an open, because capacitors do not pass dc current. Ac impedance is a more complex measurement, and a higher level of electronics is necessary to perform the process. For this reason, it is more advisable to obtain the impedance level for the microphone from the specification data sheet listed inside the manufacturer's instruction manual.

High-impedance microphones operate from grounded circuits. They are unbalanced and typically measure between 10,000 and 35,000 Ω. High-impedance microphones are not inferior to low-impedance microphones; rather they are just limited to shorter cable lengths (typically <8 ft [2.4 m]) to help improve the frequency response of the device and also to keep noise levels to a minimum. If cable lengths were to increase significantly (greater than 8 ft [2.4 m]), the microphone then would exhibit a poorer frequency response because of the high impedance of the device combined with the increased capacitance of the connecting cable. Figure 5–12 illustrates why this is necessary. As cable length increases, so too does the total capacitance of the wire connected between the microphone and the input of the amplifier. The impedance of the microphone combined with the capacitance of the cable forms a low-pass filter. The increased capacitance of long cable runs lowers the passband of the low-pass filter, greatly attenuating the high-frequency content of the signal. The overall sound image then appears to be heavy in bass and dull or muffled in quality when compared with the original source. For this reason, the length of cables should be kept to an absolute minimum when using high-impedance microphones. High-impedance microphones are also more susceptible to noise pickup because they are not operating from a balanced line as do the low-impedance variety.

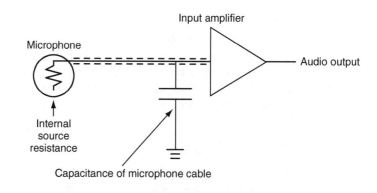

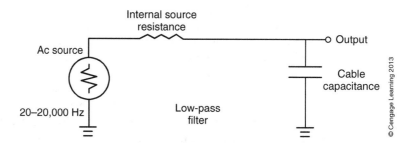

Figure 5–12
The internal source resistance of a microphone and the capacitance of the microphone cable form a low-pass filter network at the input of the amplifier.

Balanced versus Unbalanced

A balanced line is one that uses two circuit conductors, operating from an ungrounded or floating source, having equal impedance from either side to ground (Figure 5–13(A), (C)). The balanced line allows for the cancellation of identical but opposite radiating magnetic fields, thereby helping to minimize crosstalk and interference to neighboring conductors. The surrounding shield of the balanced line also helps to guard against any unwanted stray electromagnetic radiation or electrostatic currents that may try to impose themselves on the circuit. In most cases, the shield connects to one end of the circuit, typically at the source, so as not to introduce any ground loops into the system. An unbalanced line differs in that it uses a single central conductor and a surrounding grounded shield. In this case, the grounded shield acts as a circuit conductor and must be connected at both ends to complete the circuit (see Figure 5–13(B), (D)).

The benefits of a balanced line are shown in Figure 5–13(C). The two circuit conductors are transformer coupled, floating, and surrounded by an outer, grounded shield (see dotted line in Figure 5–13(C). The shield helps to ground out any stray electromagnetic or electrostatic fields that would otherwise impose unwanted noise on the circuit conductors. The circuit currents, flowing in opposite direction through two central conductors, have a balanced level of impedance to ground. As a result, the radiating magnetic fields from both conductors oppose each other equally and cancel. Shielded, balanced lines are ideal for low-level signals, such as microphones, which could be easily overwhelmed and corrupted by unwanted noise.

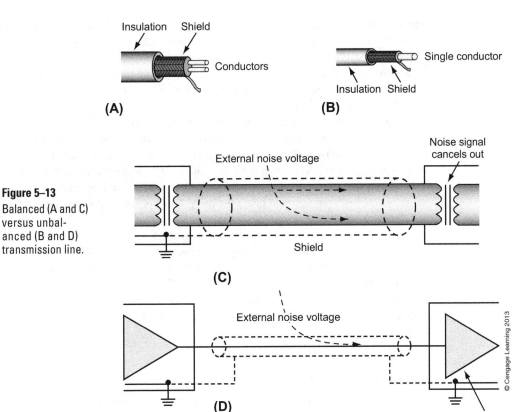

Figure 5–13
Balanced (A and C) versus unbalanced (B and D) transmission line.

In comparison, the **unbalanced** line uses a single-ended input from a grounded source to a grounded load. Because one side of the circuit is grounded and unable to balance off the main conductor, a magnetic field is radiating out from the main line, thus acting as a potential source of interference to other conductors. A shielded or coaxial cable should be used in such instances to help reduce the effects of unwanted radiation and to prevent the pickup of stray electromagnetic field and electrostatic interference (see Figure 5–13(D)).

Balanced lines are also superior because they can connect directly to a differential amplifier, thereby preventing the amplification of common-mode noise signals (see Chapter 7). In addition, an input transformer allows for impedance matching between an unequal source and load. When a mismatch of impedance occurs between a source and load, maximum power transfer does not occur, thus causing a loss of power as the signal moves through the circuit.

Ground Loops

Ground loops result from unwanted ground currents flowing between two low-impedance ground points, typically a result of 60 Hz power being referenced to the earth at multiple locations. The presence of unwanted currents along the ground plane can develop stray ground voltage and unwanted noise along the connecting ground resistance. Once present, the ground loop can also act as an inductive pickup for any other stray magnetic or electromagnetic fields in the environment. When present, a ground loop generates noise spikes, voltage fluctuations, and a 60 Hz hum within a circuit.

Ideally, a ground reference is supposed to measure zero Ω and be infinitely large to help dissipate stray voltages, but the reality is that there can be as much as 25 Ω of resistance between two points on a common ground plane, thereby causing unwanted loss of voltage and interference with normal circuit operations. To help minimize ground loops, systems should strive to use single-point grounding because a single point of reference on a ground plane typically does not build up enough noticeable resistance to develop or cause stray voltages.

Cable shields occasionally can generate ground loops in circuits if they are terminated at both ends, thus joining two ground points separated by a measurably high level of impedance. A remedy to the problem would be to disconnect the shield at the far end, thus breaking the loop. By maintaining a ground connection only at the source, the shield is still able to perform its useful function without simultaneously providing alternate current paths for unwanted stray currents within the environment.

Wireless Microphones

Wireless microphones operate on a radio frequency between the amplifier and the microphone. They transmit the signal over a select range of frequencies somewhere between 518 and 550, 630 and 662, 740 and 772, 790 and 822, and 838 and 879 MHz. Some designs have the ability to auto-select between available channels to help cut down on interference from other transmitters that may already be on that frequency.

The benefit of using a wireless microphone is that it allows the user more freedom of movement, without having to be permanently tied to a cable. The downside of wireless microphones is that they can accidentally pick up signals from other transmitters, such as other wireless devices operating on the same frequency. Some interference sources may include citizens band (CB) radios, taxicabs, and wireless headphones. Wireless devices also have a limited range; that is, the speaker cannot walk too far away from the base station without risking the degradation of signal quality. The problem is similar to tuning in a distant radio station. In such cases, the signal is weak and noisy.

The reason for the term *wireless* is that the transmitter of a wireless microphone is battery operated and encapsulated inside the housing of the microphone. A wall outlet powers the receiver or base station, which connects by cable to the amplifier. The radio frequency signal between the transmitter and receiver is the wireless connection. The communication signal is strong and clear, provided that it is not picking up interference from other sources operating on the same frequency. The solution to interference is to find a different channel, one that has no other signals operating within range of your location. For this reason, base stations and microphones have channel selection switches; the user must decide which to use for a specific location.

Lavaliere or Lapel Microphones

The **lavaliere** or **lapel** microphone is a type of wireless condenser microphone. As referenced by the name, it is often clipped to the lapel of a shirt or worn somewhere close to an individual's chest (Figure 5–14). The placement of the device demands a unique frequency response, one that has been optimized by the manufacturer to be flat across a wide range of low and high frequencies, more truly representing the actual voice of the speaker. If a lavaliere microphone is held too close to the mouth and spoken into directly like a traditional microphone, it produces a distorted frequency response, not having the correct mixture of low- and high-frequency content. For this reason, it must always be worn close to the user's chest to generate optimum results.

Lavaliere microphones are small and portable. As shown in Figure 5–14, the microphone wire travels down from the diaphragm to the wireless control unit, which typically is worn on the user's hip or slipped into a pocket.

Microphone Frequency Response

The frequency response can very greatly from one microphone to the next. Some have bandwidth ranges of 20 to 20,000 Hz, whereas others may only be useful between 200 and 10,000 Hz. Deciding which microphone is more suitable for the job depends more on the actual frequency of the sound source. You first need to know the quality of the source before deciding on what type of a microphone will do the best job. In some cases, an inexpensive microphone may be all that is required; therefore, spending a lot of extra money for more bandwidth may not be necessary or worth the investment.

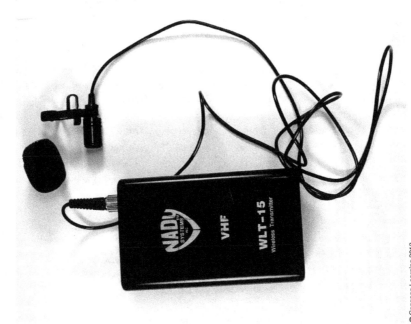

Figure 5–14
Lavaliere microphone.

© Cengage Learning 2013

Figure 5–15
Balanced XLR connector.

© Cengage Learning 2013

Microphone Hardware

Connectors

There are three basic varieties of connectors for microphones: XLR, 1/4-in. tip-ring-sleeve, and 1/8-in. miniconnectors. Balanced, low-impedance microphones most likely use an **XLR** connector (Figure 5–15) or a 1/4-in. **tip-ring-sleeve** connector (Figure 5–16). Notice that these connectors contain three poles: two represent the opposite poles of a balanced ac circuit, and the third represents the surrounding shield. When dealing with the two poles of a balanced circuit, it is important not to accidentally mix up the high side with the low. To do so may cause microphone phase reversal, especially when multiple microphones are recording the same event. When phase reversal occurs, two microphones picking up the same event electronically cancel at the mixer or preamplifier. Such a cancellation results in a shallow or deficient sound quality compared with the original source. In most cases, a unique color code identifies the high and low side of a balanced circuit. Typically, a white wire represents the high conductor, and a black wire represents the low conductor. Where connecting XLR or tip-ring-sleeve connectors among a variety of devices, it is important first to check the installation or instruction manuals to verify the pin out of the high and low conductors. Not all manufacturers use the same format. When building microphone cables, it is important to use the same pin out on each end of the cable. Doing so guarantees that phase reversal does not occur.

One-quarter–inch tip-ring-sleeve connectors (often referred to as stereo or phone plugs) also are used to connect stereo headphones or to interconnect stereo signals

Sleeve   Ring   Tip

Figure 5–16
A 1/4-in. tip-ring-sleeve connector.

© Cengage Learning 2013

between multiple devices of an audio system. When used in a stereo configuration, the tip and ring connect to the high side of the left and right channel conductors, whereas the sleeve connects to the circuit common. In this type of configuration, two unbalanced circuits share the same connector.

High-impedance microphones often use a two pole, 1/4-in. monoplug, or a 1/8-in. miniplug. The 1/8-in. miniplug looks similar to the 1/4-in. plug but is smaller. There are also metric miniplugs, often sold in a 3.5 mm format, together with a variety of other connectors known as microplugs. The metric variety looks similar to the 1/8-in. plug but is a slightly different size. Miniplugs and microplugs also are available in a three-pole, stereo format, offering the tip-ring-sleeve configuration for connection to stereo headphones or balanced circuits. Most consumer electronic devices now use the miniplugs and microplugs as a standard to connect headphones and a variety of peripheral devices. The smaller size is more practical on pocket-sized electronic equipment.

Two-pole connectors often are referred to as monoplugs, because they can only be used on a single unbalanced line (Figure 5–17). Mono, two-pole plugs have a tip-sleeve configuration. In the majority of cases, the sleeve connects to the grounded shield, and the tip connects to the main circuit conductor.

Windscreens

Windscreens are often made of foam, but they can also be made from cloth, fur, or any type of material that can be used as a mechanical filter. The idea is to help shield the microphone diaphragm from sudden gusts of wind or atmospheric pressure. Because microphones are highly sensitive to pressure, the slightest breeze or gust of wind can have the same effect as someone tapping on the diaphragm with a finger, causing intense levels of interference to the actual sound image. It is always good practice to use a windscreen when recording in an outdoor setting.

Shock mounts

Shock mounts act as shock absorbers to help reduce the effect of jarring and bumping as a microphone is moved from place to place during a recording session. The slightest amount of pressure to the exterior housing of a microphone often can generate unwanted noise and interference to the device. The use of a shock mount helps to relieve such stresses. There are many designs of shock mounts typically intended for specific styles of microphones. Always check with the manufacturer for a complete listing of accessories and microphone options.

Figure 5–17
A 1/4-in.
monoplug.

© Cengage Learning 2013

Goosenecks

Goosenecks are flexible metal conduits, available in various lengths for mounting and connecting a microphone to a podium or countertop (Figure 5–18). The flexibility of the device allows the user to position the microphone at multiple angles to achieve optimum performance. A secondary benefit to having the neck made of flexible metal conduit is that it provides an added level of shielding to the internal wires, as well as a physical barrier to help protect the device from possible damage.

Muting Switches

Muting switches on microphones, when pressed, connect ground to a separate muting control wire inside the microphone cable. The control line terminates back at the microphone mixer or **preamplifier**. Audio mixers or preamplifiers capable of muting contain special muting control inputs, which, when grounded, shut down or mute a specific input channel on the device. Muting controls are most commonly used to temporarily reduce or shut off a music-programming signal within a desired zone or location of a building during a period of microphone or telephone page.

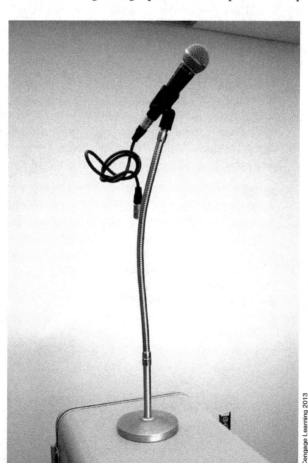

Figure 5–18
Gooseneck.

© Cengage Learning 2013

SEC 5.4   SPEAKERS, CROSSOVERS, AND THEIR ELECTRICAL PROPERTIES

National Electrical Code Definition 640.2—Loudspeaker

A loudspeaker converts an electrical current back into a mechanical vibration to reproduce sound. The term *direct **radiator*** is also used to describe a loudspeaker because it can directly affect the movement of air molecules, thereby generating sound waves and moving wave fronts within the atmosphere.

A basic speaker consists of a voice coil, a magnetic field, and a mechanical cone or diaphragm (Figure 5-19). The voice coil is mounted inside of a stationary magnetic field, typically generated by the presence of a permanent magnet. Before the 1950s, long-lasting permanent magnets were not readily available. A dc current was made to pass through a secondary outer coil, referred to as a **field coil**, to generate the stationary magnetic field. Today, however, permanent magnets are more widely available and easier to manufacture, which is why they have essentially replaced the older-style field coil designs.

To generate an audible sound, signal currents from the amplifier flow through the voice coil. In most cases, the voice coil has 4 to 8 Ω impedance. The reactive qualities of the coil attempt to resist the current flow by generating a counteracting magnetic field. The stationary magnetic field of the permanent magnet and the shifting magnetic field of the voice coil either attract or oppose each other; their interaction depends on the polarity of the signal current through the coil. The physical displacement of the voice coil generates a vibrational pattern within the speaker diaphragm, one that is an exact physical replica of the electrical waveform passing through the amplifier.

The moving voice coil is mounted on a free-floating membrane called a **spider mount**. The spider mount acts as a shock absorber to help dampen the coil movement

Figure 5–19
Cross section of a speaker assembly.

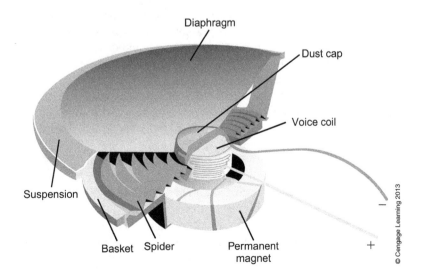

during the absence of current flow. Obviously, if the current stops flowing, then the coil should not be moving. The spider mount helps to prevent unwanted bounce during the absence of signal. The level of **damping** is important; it affects the transient response of the diaphragm and the residual bounce on the cone. The damping level also affects the resonant frequency of the speaker.

During the manufacturing process, one end of the voice coil is glued to a conical-shaped diaphragm. The diaphragm moves in unison with the voice coil as signal currents flow. The vibrational pattern of the diaphragm subsequently forces air molecules to move within the surrounding atmosphere. An observer detects the movement of air molecules as variations of pressure on the eardrum, and therefore senses the presence of sound.

Speaker Resonance and Impedance

Speakers come in various sizes, and because voice coils are reactive in nature, their impedance values must be optimized for specific resonant frequencies. Because low frequencies are associated with longer wavelengths, they have to be applied to larger diameter speaker diaphragms for optimum performance and reproducibility. Likewise, high-frequency signals are associated with shorter wavelengths that must be applied to smaller diameter diaphragms. The diameter of the speaker cone is directly related to the optimum frequency that can resonate on the diaphragm. The idea is similar to tuning a pipe organ. The length of a specific pipe has to be cut to an exact resonant length to recreate the desired musical note or tone. Speaker diaphragms are slightly different in that they can reproduce a broader range of frequencies, above and below their resonant value. The result of reproducing a broad range of frequencies across the speaker diaphragm also has the effect of shifting the inductive reactance of the voice coil. Because frequency and inductive reactance are directly proportional, an 8 Ω voice coil only measures 8 Ω at a specific resonant frequency. If frequency values decrease, then so too does the reactance and impedance of the coil. If frequency values increase, then impedance likewise follows. An exact 8 Ω level is only maintained at the specific resonant frequency. A slight shift above or below does not matter because the overall average of a complex audio waveform maintains the desired 8 Ω over a specific period of time. Excessive shifts, however, can become problematic and eventually load down or damage the amplifier or speaker diaphragm if they persist for too long.

Woofers are large-diameter cones, typically larger than 8 in. They are used to recreate low-frequency bass signals. **Midrange** cones are somewhat smaller, typically between 2 and 8 in. (50.8 and 203.2 mm), and are used to recreate a midrange of frequencies, between 500 and 8000 Hz. **Tweeters** are smallest in diameter and more often resemble a flat diaphragm rather than a cone; typically, they are less than 2 in. (50.8 mm) in diameter. Tweeters are used to reproduce high-frequency content, often referred to as *treble*.

A typical speaker enclosure includes two or three sizes of diaphragms. Some enclosures contain a woofer and a tweeter, whereas others contain all three: woofer, midrange, and tweeter. Having a three-diaphragm speaker system more accurately recreates the full range of sounds than does a two-diaphragm system.

From the impedance point of view, an 8 Ω woofer only measures 8 Ω at a specific resonant bass frequency, whereas an 8 Ω tweeter only measures 8 Ω at specific resonant treble frequency. So, how are frequencies separated electronically to ensure that they reach the optimum-sized diaphragm within the speaker enclosure? The answer is crossovers and filters.

Crossovers

Crossovers are filter networks that are used to separate and direct a complex mixture of audio frequencies to their required size speaker diaphragm. For example, a **low-pass** filter is used to pass low-frequency bass signals to a woofer, a **band-pass** filter is used to pass midrange frequencies to a midrange diaphragm, and a **high-pass** filter is used to pass high frequencies to a tweeter. Figure 5–20(A–C) graphically illustrates the frequency response of low-, high-, and band-pass filters.

The next question to ask is: What would happen if midrange to high-range frequencies were accidentally sent to the woofer? Some type of sound surely would be generated, right? The answer is yes, but not as efficiently, because the large-sized diameter woofer cannot vibrate as effectively as the smaller-sized midrange

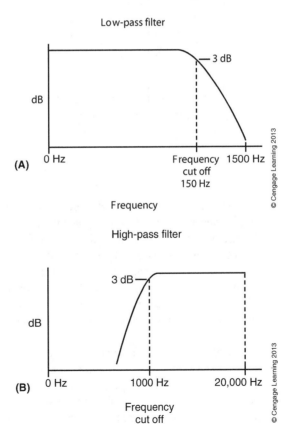

Figure 5–20
(A) Low-pass filter. (B) High-pass filter.

(Continued)

Band-pass filter

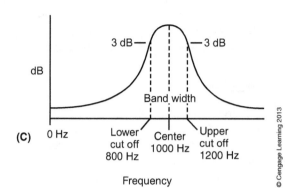

Figure 5–20 Cont'd
(C) Band-pass filter.
(D) Notch filter.

Notch filter

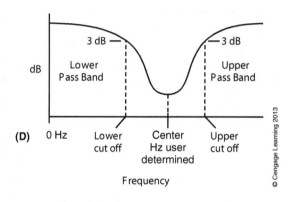

or tweeter. Higher frequency rates are usually well outside of a woofer's resonant range. A secondary problem would be that the value of voice coil impedance would increase dramatically from its optimum 8 Ω, thereby decreasing the amount of total current flow and power transfer to the load.

It would be more dangerous, however, to send low-frequency bass frequencies to a midrange or tweeter coil. Such a situation would cause the reactance of the voice coil to plummet, virtually short-circuiting the amplifier output. As a result, current values would increase dramatically, possibly burning up the output of the amplifier and tweeter.

Having a crossover ensures two important outcomes. First, the woofer, midrange, and tweeter better maintain the optimum average impedance of 8 Ω required by the amplifier; properly matching the amplifier to the load ultimately prevents the accidental overload and possible destruction of the device. Second, the crossover helps to recreate a cleaner quality of sound because the speaker diaphragms only receive frequencies within their required resonant range, ones that can be reproduced more reliably and efficiently.

Active versus Passive

Two types of crossovers are available: active and passive. Active crossovers are amplified filter networks that are adjustable and often contain some form of internal electronic feedback to help stabilize the signal. The user can adjust the gain, or level of output, and the cutoff frequency of each filter range. This makes them versatile, allowing the user to fine-tune the network, ensuring that the correct signals are getting to the right speaker diaphragm and with the right intensity of **signal level**. Passive crossovers are not adjustable and typically are designed from a combination of inductors and capacitors that are built into a network of low-, band-, and high-pass filters. The user cannot change the parameters of a passive crossover network without literally unsoldering a component and exchanging it for a different value. In most cases, the design of filter networks can be quite complex. Individuals not having the required knowledge or tools available to them should not attempt to perform such modifications; otherwise, damage to devices and circuits may occur. As a result, a passive crossover network is fixed, nonvariable, and usually mounted inside of a speaker enclosure where it is not tampered with by curious individuals (Figure 5–21).

Active crossovers are external systems, separate from the speaker, and usually are mounted next to the mixer or amplifier, inside the sound equipment rack. The output for each section of crossover must connect to an individual amplifier; a crossover containing low-, mid- and high-frequency separation, therefore, needs three separate amplifiers, each driving an individual speaker of proper size and power rating.

Speaker Compliance

Speaker **compliance** indicates the ease with which a speaker diaphragm moves and is measured in **Newtons** per meter (N/m) . It can be measured by dividing the

Figure 5–21
Passive cross-over mounted inside of a speaker enclosure.

© Cengage Learning 2013

diaphragm displacement by the applied magnetic force. A highly damped speaker has little compliance because the diaphragm movement per unit of applied force is small. If the speaker diaphragm is not damped, then the level of compliance reaches a maximum at the resonant frequency of the diaphragm. If damping level is increased, compliance is reduced; as a result, the resonant frequency of the speaker diaphragm also increases.

Speaker Efficiency

Speaker efficiency is measured as the ratio of output power to input power. Input power is electrical, whereas output power is radiating, acoustical sound pressure. A speaker's efficiency usually is measured (in decibels) at a distance of 1 m (3.2 ft), and it is derived from a level of electrical input power of 1 W; therefore, if 1 W input power generates 90 dB acoustical sound pressure, at a distance of 1 m (3.2 ft), then the efficiency is 90 dB. Using the inverse square law, you can now calculate the amount of acoustical power as the distance is doubled or quadrupled from the speaker.

Coaxial Speakers

Coaxial speakers are special designs that have two or three diaphragms mounted inside one speaker housing, driven by one voice coil (Figure 5–22). The different sized diameter cones are mounted on top of each other, terminating back to one central point, where they are mechanically connected to the voice coil. One voice coil can now be used to move three separate diaphragms simultaneously, thereby

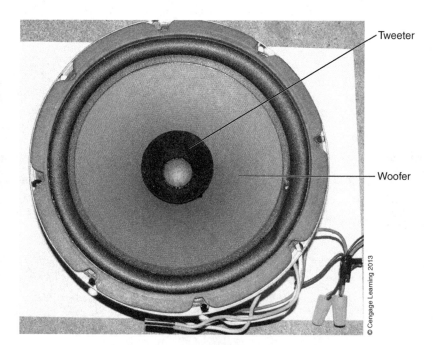

Figure 5–22
Coaxial speaker, combination midrange and tweeter.

Tweeter

Woofer

© Cengage Learning 2013

generating a full range of sound frequencies within one space, all radiating out from a central point.

Coaxial speakers are ideal for the installation of ceiling speaker arrays. They provide all the necessary midrange for paging and also have the ability to reproduce the high- and low-frequency content of the wider band music programming.

Horns

Horns are examples of indirect radiators. The speaker diaphragm, known as a compression driver, is mounted on an acoustic transformer, mechanically shaped as a horn or megaphone, flaring out in diameter as the length moves away from the center radiator. The throat of the acoustic transformer is considerably smaller in diameter than the speaker diaphragm (Figure 5–23). This creates a value of high acoustic impedance that more closely matches the mechanical impedance of the speaker, allowing for maximum power transfer.

The sound pressure disperses over a progressively larger cross-sectional area as it moves down the length of the horn, eventually reaching open air. Because the cross-sectional area increases gradually, the sound wave transfers more efficiently from the original high-pressure/high-impedance output to a low-pressure/low-impedance wave front. As a result, the sound pressure levels emitted by horns are much greater than those of standard direct radiators.

Although horns are more efficient than direct radiators, they tend to have narrower dispersion patterns, especially at high frequencies; and because high-frequency content is more line of sight, it tends to get lost as it radiates out into the atmosphere. Unless you are standing directly in front of the radiating pattern, the signal may not be detected. For this reason, multicellular horns have been developed at various dispersion angles which can simultaneously radiate high-frequency content across a wider range, thereby providing greater coverage, which helps to eliminate spatial signal losses. The advantage of multicellular horns is that they allow the user to easily reconfigure the dispersion angle by plugging up individual cells or ports as needed to better fit room patterns. Notably, horns having a single-output cell or fixed radiator dispersion angle output greater levels of high-frequency content along the central axis, while falling off slightly toward the edges. Therefore, multiple, single-cell, horns may be needed to complete the coverage area, depending on the shape of the room and output pattern desired.

Figure 5–23
Horn.

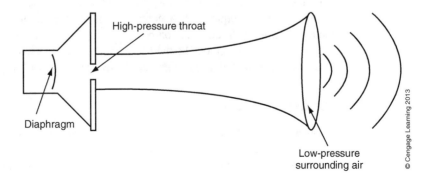

© Cengage Learning 2013

Filters

Filters are designed as line-level signal-processing devices, inserted between the input of the mixer and input of the **power amplifier**, to either pass a specific range of frequencies along to the next stage or to block or prevent a specific range of frequencies from getting to the next stage. There are four basic varieties of filters: low-pass, high-pass, band-pass, and notch filters.

Low-Pass Filter

The low-pass filter passes a band of frequencies on the low end of the spectrum until the −3 dB cutoff is reached (see Figure 5–20A). Any frequency past the −3 dB cutoff point is effectively filtered out from the signal. In the standard single-pole filter design, represented by a single resistor in series with a capacitive load, the frequencies drop off at a rate of −20 dB per decade past the cutoff frequency. A decade represents 10 times the cutoff frequency. Remember also that the −3 dB point represents the half-power point; at −3dB, half the signal power has been eliminated.

High-Pass Filter

The high-pass filter passes a band of frequencies on the upper end of the spectrum (see Figure 5–20B). Compared with the low-pass filter, the high-pass filter has the −3 dB cutoff point on the low side of the passband. Any frequencies below the cutoff roll off at a rate of −20 dB per decade; a single-pole filter design consists of a capacitor in series with a resistive load.

Band-Pass Filter

The band-pass filter can be thought of as a high-pass filter combined with a low-pass filter (see Figure 5–20C). The low-pass cutoff is higher up from the high-pass cutoff. The range of frequencies within the passband, between the upper and lower cutoff frequency, represents the bandwidth of the filter. All frequencies inside the bandwidth are passed on to the next stage of the circuit, thus the name, band-pass.

Notch Filter

The **notch filter** is the opposite of the band-pass filter, and it removes any frequency not within the passband (Figure 5–20D). Therefore, the only frequencies that make it through the notch filter are those outside of the bandwidth, above and below the cutoff frequencies. A notch filter is used in graphic equalizers to help remove unwanted frequencies or noise within the signal. If an unwanted noise signal is appearing at 5000 Hz, then a notch filter can be used to eliminate just that frequency, thereby helping to clean up the overall quality of the intended signal.

Speaker Placement

Speaker placement often can be a complex issue. When deciding on the best location for a speaker, the conversation often turns to issues of echo, reverberation, the quality of the room and furnishings, the size of the space, the type of programming, and the optimum listening level for the environment. For this reason, each room design is unique. Always start with the size of the room, then decide what the average listening level needs to be, based on the type of programming. Room furnishings may also come into play, depending on the quality of sound you are attempting to achieve.

Echo and **reverberation** are a result of reflection. A desired sound wave may take a direct path to the ear of the observer, but it may also bounce around the room, off the walls, floor, ceiling, and furnishings, before eventually arriving at the final destination. An echo is an example of an early reflection, arriving just after the direct sound wave. Reverberation is considered a later reflection because it arrives some time after the echo, due to the more indirect pathway it takes from the source to the ear of the observer (Figure 5–24).

A key to reducing the amount of echo and reverberation in a space is to reduce the number of reflective surfaces. Wood or tile floors, glass windows, or any smooth, hard surfaces only help to contribute to the number of reflections. To remove excessive reflections within a space, you must soften room surfaces; to do so may involve the installation of acoustic baffles. Acoustic baffles are designed to absorb sound, thereby reducing the number of possible reflections. Baffles are often installed on walls and the ceiling or in the corner spaces of a room. When

Figure 5–24
Echo versus reverberation.

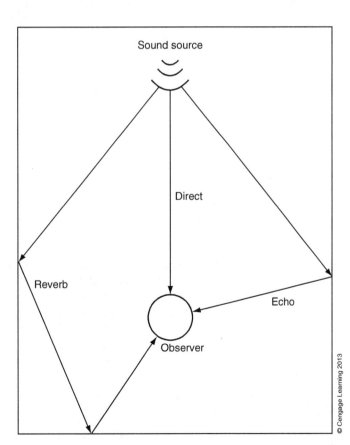

© Cengage Learning 2013

purchasing acoustic baffles, be sure that they are made of a safe, fireproof material, one that does not give off toxic fumes or vapors when burned. Fire inspectors insist on it.

When deciding on the placement of speakers, a designer must know the speaker efficiency rating as well as the dimensions of the room, the height of the ceilings, and the desired listening level to be achieved in decibels. Armed with such information, a designer then can make a reasonable assessment of the situation by using the inverse square law, as described in Chapter 4. The chart shown below is based on a speaker with an efficiency rating of 90 dB at 1 W/m; this means that if you are supplying a 1 W signal to a speaker and you are standing 1 m (3.28 ft) from the device, the sound pressure level measures 90 dB. The inverse square law then can be used to calculate the expected decibel level at various distances from the speaker. Notice that doubling the distance from the speaker reduces the signal level by 6 dB. Doubling the output power, however, increases the signal level by 3 dB. To help simplify the layout of a design, you can recalculate the chart below for any speaker efficiency. A designer then only needs to plot out the placement of speakers within a given space, based on the height of the ceilings, dimensions of the room, and desired listening level; this is assuming that all the major reflective surfaces within the space have already been dealt with, and acoustic baffles have been installed. The size of the power amplifier is determined based on the total number of speakers and the sum of their individual power dissipations. The required size of a power amplifier is discussed in more detail later in this chapter.

Output Power	1 m (3.28 ft)	2 m (6.56 ft)	4 m (13.12 ft)	8 m (26.24 ft)
0.25 W	84 dB	78 dB	72 dB	66 dB
0.5 W	87 dB	81 dB	75 dB	69 dB
1 W	90 dB*	84 dB	78 dB	72 dB
2 W	93 dB	87 dB	81 dB	75 dB
4 W	96 dB	90 dB	84 dB	78 dB
8 W	99 dB	93 dB	87 dB	81 dB
16 W	102 dB	96 dB	90 dB	84 dB

*Speaker efficiency rating.

© Cengage Learning 2013

When using multiple speakers for coverage, the only other factor to consider is the dispersion angle of the speaker diaphragm. By knowing the dispersion angle of the diaphragm, a designer can decide how close to place speakers within a given space to effectively cover the room with minimal overlap, hotspots, or dead zones. Hotspots imply that portions of the room are too loud, whereas dead zones refer to a lack of signal or coverage. In most cases, speaker placement should be planned out on graph paper, which allows the designer to see possible trouble spots ahead of time. In some instances, a variety of speakers and speaker enclosures may be needed, based on the size and shape of the room and the desired listening level.

Subwoofers

Subwoofers are large-diameter bass speakers. For most spaces, a single subwoofer is required, and the placement of the device is not critical because bass frequency tends to fill a space quite effectively in an omnidirectional manner. In some cases, sub-woofers are built into the pedestal of a table, placed behind furniture, or mounted up and out of the way, often out of view. Because low-frequency bass does not give the impression of directionality, it simply does not matter where the subwoofer is placed. As long as the output port of the subwoofer is not being blocked or obstructed, the device can be placed just about anywhere within a space. Subwoofers do, however, require large power amplifiers to operate because of their low efficiency. It is not unheard of to operate a subwoofer from a 1000 W amplifier. This may sound like a lot, but remember that a doubling of power only increases the output level by 3 dB; and 10 times the power only produces a gain of 10 dB, because it is a logarithmic calculation (see Chapter 4). In addition, a crossover is needed to operate a sub-woofer. The cutoff frequency of the low-pass filter typically is at 100 Hz. This means that all frequencies below 100 Hz pass along to the subwoofer.

Speaker Enclosures

Another factor to consider is the type of speaker enclosure, whether it is a wall mount, ceiling mount, or corner mount, and how this may affect the overall signal level within the space. The ideal sound source is a point source that radiates out into three-dimensional space in all directions. If the sound source were then moved against a wall and the sphere of radiation were bisected, forcing the total sound energy to radiate out into half-space, then the measured sound pressure level would double, because of the added concentration, and be 6 dB greater. Sound pressure is analogous to voltage, not power; in such cases, the decibel level is 6 dB instead of 3 dB. A doubling of voltage increases power by 6 dB, whereas a doubling of power is equal to only a 3 dB increase (see Chapter 4).

If the sound source were then moved into the corner, the intersection of the two walls once again cuts the source in half, thereby increasing the sound intensity another 6 dB. The last case would be to move the source toward the floor or ceiling, to intersect three surfaces; once again, the source would be cut in half, this time by the floor or ceiling, and the signal level would increase by another 6 dB.

The previous theoretical description of a sound source works only provided the source is an ideal single-point source; that is, one that can radiate out equally in all directions. But a loudspeaker is not an ideal sound source. The speaker enclosure and the speaker housing often prevent or block the radiation pattern from moving toward the rear of the device. By drawing the speaker diaphragm schematically, the radiation pattern can be shown to be at a finite angle, rather than radiating out into full space (Figure 5–25).

In addition, the radiation pattern of a typical speaker can be shown to be based on frequency. Low frequencies ultimately wrap around and become omnidirec-tional, appearing behind the speaker. Midrange frequencies travel off more to

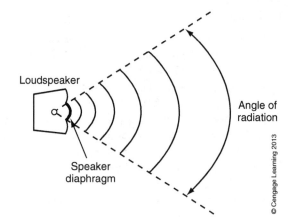

Figure 5–25
A loudspeaker
angle of radiation.

the sides, whereas the high-frequency content only appears in a narrow angle
of radiation toward the front of the diaphragm, moving in a direct line of sight
(Figure 5–26).

So then, what really happens when a speaker is placed against a wall? Do all
the frequencies experience a 6 dB increase? The answer is no. The increase is
only noticed in the range of low frequencies because they are the ones reaching
the rear of the device and which are able to reflect off the surface of the wall. The
presence of the wall barely affects midrange frequencies, and does not affect high
frequencies at all. The next step is to move the speaker into the corner. Moving
the speaker into the corner produces a similar effect, but in this case the angle of
the corner walls, now at 45°, helps to boost the midrange response more than the

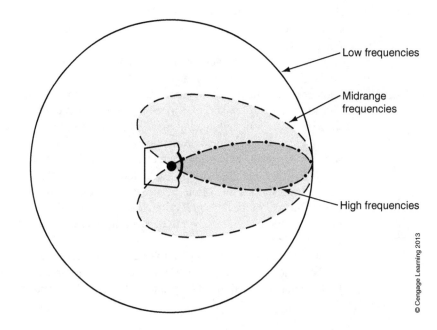

Figure 5–26
Radiation pat-
tern on a typical
speaker.

previous example. In the corner position, the bass sees an additional 6 dB gain, whereas those frequencies in the midrange only now start being affected. Once again, however, high-frequency content does not see any noticeable change in level caused by the positioning of the loudspeaker. The last example places the speaker in the corner, near the floor or ceiling. As in the previous examples, the new position has the effect of increasing the bass frequencies by another 6 dB. But now there is a more significant increase in the level of midrange frequencies because the reflective area of the added floor or ceiling surface cuts the dispersion angle and helps to concentrate more low and midrange frequency content toward the front of the loudspeaker. As stated before, speaker placement does not affect the level of high-frequency content, because the radiation angle of high frequency is narrow to begin with and travels in a more direct line of sight out from the diaphragm (Figure 5–27).

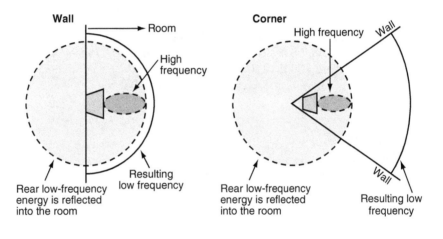

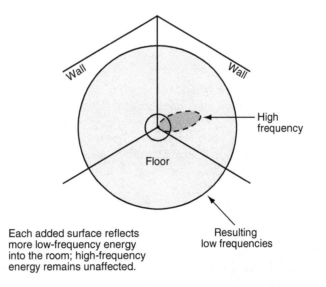

Figure 5–27
Speaker placement and the resulting radiation patterns: wall mount, corner mount, and floor or ceiling mount.

SEC 5.5 AMPLIFIERS, SIGNAL LEVELS, AND THEIR ELECTRICAL PROPERTIES

The next section discusses the various signal levels and electrical properties of amplifiers. Knowledge of signal levels is important to the interconnection of system devices so as not to cause a potential mismatch of impedance or damage to the system.

Signal Levels

There are four ranges of signal level in audio equipment: microphone level, line level, power amplifier input level, and power amplifier output level.

Microphone Level

Microphone levels typically are between 300 μV and 2 mV, or about –60 to –40 dBm. (The dBm is referenced to 1 mW as 775 mV is dissipated across a 600 Ω load.) So if a 150 Ω microphone outputs a signal strength of –60 dBm, how much voltage is measured? Transpose the following formula and solve for V:

$$10 \log (V^2/R)/1 \text{ mW} = \text{dBm}$$
$$10 \log (V^2/150 \text{ Ω})/1 \text{ mW} = -60 \text{ dBm}$$
$$\log (V^2/150 \text{ Ω})/1 \text{ mW} = -6 \text{ dBm}$$
$$(V^2/150 \text{ Ω})/1 \text{ mW} = 1 \times 10^{-6}$$
$$V^2/150 \text{ Ω} = 1 \times 10^{-9}$$
$$V^2 = 150 \times 10^{-9}$$
$$V = 387.3 \text{ μV}$$

Compared with dynamic microphones, condenser microphones tend to have greater output levels because they have built-in preamplifiers.

Line Level

Line-level signals are meant for any intermediate signal processing or wave shaping before final amplification. All signal processing among mixers, preamplifiers, or any other signal-processing equipment, such as crossovers, equalizers, or sound-effect processing equipment input and output signals at line level. Typical line levels for home recording equipment or musical instruments measure between 100 and 300 mV, or about –10 dBm. Professional recording studios and broadcast studios use higher levels, typically between 1.2 and 2 V. These levels translate into +4 or +8 dBm, respectively.

Volume indicator meters (VU meters) typically are calibrated so that 0 VU corresponds to +4 dBm for professional audio equipment. Telephone and broadcast equipment calibrates 0 VU at +8 dBm, and consumer audio equipment calibrates 0 VU at –10 dBm. It is important to know whether audio equipment is rated as consumer, professional, or telephone broadcast where attempting to mix or inter-connect signals from various sources, because their calibrated reference levels vary. As a result, broadcast audio equipment outputs louder or higher signal levels compared with consumer home audio equipment, which needs more amplification.

Line Pads

Line pads are available to help reduce signal levels, if necessary, to avoid overload-ing the channel inputs of mixers or amplifiers. Distortion occurs if signal levels are excessive. Most mixers have pad switches that can reduce a signal level −20 or −40 dB, as needed, to maintain impedance levels and line balance. The use of such switches is dependent on the types of source or signals. Because microphone outputs can vary by as much as −20 dBm, and line-level signals can usually fall any-where between −10 and +8 dBm, adjustments occasionally may have to be made.

In-line pads are also available that can connect as passive resistor networks to microphone or line-level cables to achieve a desired signal attenuation. Such networks are specifically designed to maintain an impedance match between two points in the system without disrupting a balanced or unbalanced signal. Most in-line pads are designed around required parameters and often must be made to order, depending on the amount of attenuation desired, impedance values at the input and output of the network, and type of line grounding. Just as with filter net-works, a more detailed level of electronics is necessary for proper design.

Headphones and Headphone Amplifiers

Power amplifier output levels are much greater than line levels. As a result, the 100 V, 70 V, 25 V, 16 Ω, 8 Ω, and 4 Ω outputs of a power amplifier must never be directly connected to a pair of headphones. The excessive amounts of output voltage and wattage being delivered by a power amplifier could cause perma-nent damage to eardrums and possible loss of hearing to the person wearing the headphones. To achieve optimum signal level for headphones, it is usually recom-mended that line-level signals be routed through separate headphone amplifiers, each of which provides multiple outputs for auxiliary connections. Be sure to check with the manufacturer for a list of configuration options and the number of desired output ports when deciding on a suitable headphone amplifier. The outputs and inputs of multiple amplifiers usually can be daisy-chained together to increase total output capacity. Thus, two 4-channel headphone amplifiers can be connected together to provide seven available outputs.

Specifications on headphones do vary. In most cases, headphone impedances range from 30 to 75 Ω, depending on the manufacturer, and typically have a sen-sitivity rating of 90 dB/mW. Headphones do not need a lot of signal level; 1 mW, which corresponds to about 274 mV across 75 Ω, is generally more than sufficient.

Line Input Connectors

Unbalanced line inputs either use a standard **RCA connector** (Figure 5–28), 1/4-in. monoplugs, or 1/8-in. miniconnectors (Figure 5–29). Balanced line inputs are connected through XLR or 1/4-in. tip-ring-sleeve connectors. Connections

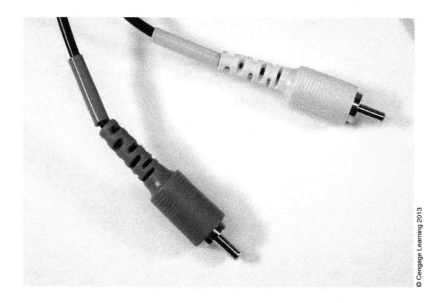

Figure 5–28
Unbalanced
RCA connector.

may also use screw terminals and spade connectors. Be sure to check with the manufacturer of the mixer or amplifier to be sure of the connection style and available options. Often, they can be ordered as desired.

Power Amplifier Input Levels

Power amplifier input levels typically are required to be in the range of 1 to 2 V, or +4 to +8 dBm. The manufacturer's specifications usually specify optimum

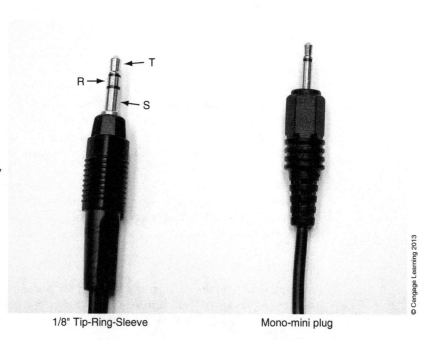

Figure 5–29
A 1/8-in. tip-ring-sleeve,
and a mono-
miniconnector.

1/8" Tip-Ring-Sleeve   Mono-mini plug

input levels needed for proper operation. The inputs can either be balanced or unbalanced, depending on the type of power amplifier, and use either XLR, 1/4-in. phono plugs, or spade connectors and screw terminals for making input connections.

Power Amplifier Output Levels

(Refer to *NEC Article 640.9(C), Wiring Methods* for exact specifications.)

Power amplifier output levels vary, depending on the position of the volume control. For direct 4, 8, or 16 Ω outputs, the maximum output voltage driving the speaker is dependent on the maximum output power rating of the amplifier. In such cases, the formula

$$\text{Wattage} = \text{Voltage}^2 \div \text{Resistance}$$

can be used to calculate maximum output voltages.

An alternative to using the direct 4, 8, or 16 Ω output of an amplifier would be to use the Class 2 or Class 3 output. The Class 2 or Class 3 output of an audio amplifier is provided by an internal transformer. The primary of the internal transformer is usually connected to the direct, 8 Ω output. The secondary then provides a Class 2 or Class 3 input signal to the primary of the speaker transformer, installed some distance away.

The Class 2 or 3 transformer inside the amplifier is inherently limited, meaning that it it is designed to operate up to a determined limit, set by the manufacturer. The internal limiting guarantees that the system will be protected against a possible short circuit or from the hazards of a possible transformer overload. According to *NEC Chapter 9, Table 11(A)*, an inherently limited, Class 3 power source cannot exceed 100 VA and 100 V. (Noninherently limited Class 3 sources, which require external overcurrent protection, are limited to 150 volts). A Class 2 source cannot exceed 100 VA and 30 V. Any source supplying a voltage greater than 150 V, and more than 100 W, must by default be considered a Class 1 circuit and wired accordingly. The classification of circuits is explained in more detail in Chapter 8.

The standard voltage output for Class 2 audio is 25 V, and for Class 3 is 70 or 100 V. Transformer outputs allow for variable impedance loads; in addition, they help to reduce the line loss and signal attenuation over long cable runs. The listed transformer output voltages represent maximum levels during full volume or full rated power. For example, the output of a 25 V transformer never develops more than 25 V across the load during full power. The output voltage can always be less, depending on the position of the volume control, but never more; the same holds true for the 70 and 100 V audio outputs.

To comply with the *NEC*, a 25 V output must be connected through Class 2 wiring, and a 70 or 100 V output must be connected through Class 3 wiring. As stated earlier, the maximum power for such outputs cannot exceed 100 VA. This means that a single Class 2 or 3 cable run must not be connected to more than 100 W of speakers. Any additional speakers, over and above the 100 W limit, must be connected through a separate Class 3 cable run. By default, any amplifier outputs exceeding 150 V or 100 W, including directly connected 4, 8, or 16 Ω outputs, must be wired and connected as Class 1 circuits. See Chapter 8 for a description of Class 1, 2, and 3 wiring.

Why are 25 V, 70 V, or 100 V **matching transformers** needed on industrial or professional audio equipment? The answer is given in the next section. (Refer to *NEC Article 640.9 (D), Use of Audio Transformers and Autotransformers* for more information.)

Maximum Power Transfer and Speaker Systems

Most consumer audio equipment is designed around an 8 Ω load, which is why most speakers operate from 8 Ω of impedance. Maximum power transfer occurs when the impedance of the load matches the impedance of the source. Let us see why.

Example 1. A source voltage of 10 V, with an internal impedance of 8 Ω, is driving a 1000 Ω load. How much power is delivered to the load?

The total current flow can be calculated by

$$V_{Supply} \div R_{Total}$$

$$10 \text{ V} \div 1008 \text{ Ω} = 9.92 \text{ mA}$$

And the total load power can be calculated by

$$I_{Total}^2 \times R_{Load}$$

$$(9.92 \text{ mA})^2 \times 1000 \text{ Ω} = 98.4 \text{ mW}$$

Now, let us change the load to 8 Ω and recalculate the load power. The total current flow is:

$$10 \text{ V} \div 16 \text{ Ω} = 625 \text{ mA}$$

And the total load power is:

$$(625 \text{ mA})^2 \times 8 \text{ Ω} = 3.125 \text{ W}$$

For the last calculation, change the load value to 1 Ω. The total current flow is:

$$10 \text{ V} \div 9 \text{ Ω} = 1.11 \text{ A}$$

And the total load power is:

$$(1.11 \text{ A})^2 \times 1 \text{ Ω} = 1.235 \text{ W}$$

Notice that the load power reaches a maximum level only when the load impedance matches the source impedance. The 1 and 1000 Ω load both have the effect of reducing the total amount of output power reaching the load. Maximum power transfer only occurs when the source and load impedance are matched. Such power losses also have the effect of decreasing the overall efficiency of the system.

Although the impedance of the load is important for transfer of power, a more significant problem may be the resistance of the connecting wires over a long-distance connection. To illustrate the point, let us connect an 8 Ω speaker to an amplifier through 500 ft (152.4 m) of two-conductor cable. To start with, the actual circuit length doubles because the current must not only travel out to the load but also back to the amplifier; this means that from the circuit point of view, a 500 ft (152.4 m) run actually connects through 1000 ft (304.8 m) of wire. A 1000 ft (304.8 m) length of 18 AWG copper wire typically has a total resistance of 6.39 Ω, based on the type of wire, gauge, temperature, and length of the conductors. By now you should have noticed a problem. If an 8 Ω source connects to an 8 Ω speaker through 6.39 Ω of wire, then where is the majority of power being delivered? The answer is that the amplifier and the connecting wires are dissipating more power than the load. In this example, the source impedance, connecting wire, and speaker form a series voltage divider. The first rule of any series voltage divider is that the largest voltage level or power dissipation in the loop is measured across the largest resistance. In this case, the source impedance and the connecting wires make up the majority of the circuit impedance, not the load.

Let us now look at how much power actually is being delivered to the speaker from the source.

Assume that the source measures 10 V, through 8 Ω of internal resistance, connecting to an 8 Ω load through 6.39 Ω of wire. The total current flow in the loop is:

$$10 \text{ V} \div 22.39 \ \Omega = 446.63 \text{ mA}$$

Total power being delivered to the speaker is:

$$(446.63 \text{ mA})^2 \times 8 \ \Omega = 1.6 \text{ W}$$

Total power dissipation on the internal source impedance and speaker wires is:

$$(446.63 \text{ mA})^2 \times 14.39 \ \Omega = 2.87 \text{ W}$$

From the above example, it can be seen that the amplifier and the external wires are receiving more than 64% of the total power, and the speakers only 36%; this is highly inefficient because the connecting wires and the internal resistance of the amplifier are receiving the majority of the power.

Audio Transformer Outputs

To solve the previous problem and achieve maximum power transfer, you need to drive the load through a set of 25, 70, or 100 V matching transformers. A 25, 70, or 100 V system has an audio transformer (refer to *NEC Definitions 640.2, Audio Transformer* for an exact definition) mounted inside of the amplifier, connecting between the direct 8 Ω output and the speaker. The manufacturer sizes the transformer accordingly to accommodate maximum output power and desired voltages: 25, 70, or 100 V. Speaker wires are connected to appropriately labeled output terminals, depending on the type of matching system; in some cases, all three levels are available through a multitapped output transformer.

Figure 5–30
Speaker mounted matching transformer.

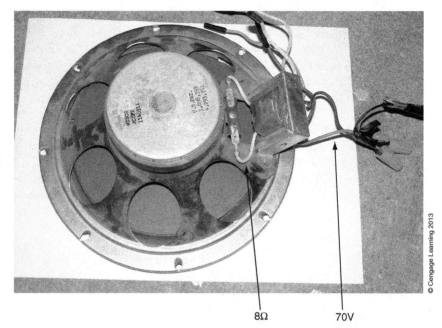

8Ω   70V

Speaker Transformers

At the load end of the circuit, the connecting wires terminate to the primary side of a matching transformer, mounted inside the enclosure of the 8 Ω speaker (Figure 5–30). The matching transformer is necessary to achieve maximum power transfer between the 25, 70, or 100 V amplifier outputs and the 8 Ω speaker.

The input to the speaker transformer is labeled as 25, 70, or 100 V, depending on the type of system being used. The type of system is critical for two reasons: first, the interconnection of Class 1, 2, or 3 systems is not allowed (see discussion in Chapter 8); and second, the interconnection of differing systems may be destructive to speakers, amplifiers, or the connecting wires.

The speaker transformer (Figure 5–31) has multiple input taps to choose from, each representing a different output rating. Some common transformer options include: 25 V Class 2 transformer, having 10, 5, 2, 1, 1/2, 1/4, or 1/8 W taps; and 70 V Class 3 transformer, having 10, 5, 2, 1, 1/2, or 1/4 W taps.

As one might expect, the 10 W tap provides more power to the speaker than the lower level taps and therefore sounds louder. The secondary winding of the transformer, labeled 8 Ω, connects directly to the speaker.

The downside of transformer-coupled audio is that inductive reactance can shift dramatically over the 20 to 20,000 Hz bandwidth. As a result, low-frequency bass signals can cause excessive output currents to flow, possibly overloading and damaging the output of the amplifier during long signal durations. For this reason, a transformer-coupled amplifier usually provides internal current limiting together with fast-acting, short-circuit protection to help prevent overheating and ensure the safety of the system.

Figure 5-31
Multitapped
70 V speaker
transformer.

The bandwidth of most transformer outputs may also be limited and unable to recreate the entire 20 to 20,000 Hz bandwidth of the amplifier. In such cases, when it is more desirable to achieve maximum bandwidth, it would be better to avoid using matching transformers, and instead connect the output directly to the 4, 8, or 16 Ω output.

Transformer outputs are also not efficient. A typical 70 V audio transformer tapped at 10 W may draw as much as 14 W of power from the amplifier, depending on the manufacturer and quantity of losses inside the core. For this reason, the connection of ten 10 W transformers to an amplifier output may draw as much as 140 W, not 100. Forgetting to factor transformer inefficiencies and losses into the total power draw of the amplifier may result in possible overheating and destruction of the device. For this reason, it is necessary to de-rate power amplifiers to 70% their maximum output power to ensure that losses and transformer inefficiencies are accounted for. In the above example, a 140 W or greater amplifier would be required to handle the 100 W transformer load. Having the speaker transformers rated for wattage rather than impedance makes it easier for the installer to simply count the number of speakers, and multiply by the rating to obtain the total wattage of the load.

Example

Calculate the minimum requirements of a power amplifier needed to safely drive a load consisting of ten 70 V transformers, all tapped at 2 W.

Answer

The amplifier must be able to deliver 20 W of power to the 10 speakers, and be able to account for possible transformer losses, if any. Because manufacturer's ratings on transformers may or may not include losses, it would be safer to assume that they do not. Therefore, to ensure the useful life of the amplifier, the output power should never be pushed past 70% of the maximum rating; that is, if the amplifier is rated for 100 W, the load should never attempt to draw more than 70 W.

In this example, the load requires 20 W. Divide 20 by 70%.

$$20 \div 0.7 = 28.57 \text{ W}$$

Based on the above calculation, a 30 W or greater amplifier would be needed. De-rating the amplifier not only helps protect it from periods of high current draw often caused by low-frequency content shifts, but also guarantees that there is enough available power to cover any circuit inefficiencies. Continuously pushing an amplifier to its maximum rating is not recommended, as to do so may ultimately bring about the untimely destruction of the device.

Transformer Volume Control

External, wall-mount volume controls are made from 25, 70, or 100 V adjustable autotransformers (Figure 5–32). The volume is adjusted by moving the selector switch to a different position on the autotransformer coil. A single volume control has three connections: input, output, and common. The common is shared between the input and the output. Stereo controls are also available, but mono remains the most common for industrial applications.

A single volume control can adjust the volume for a single speaker or an entire zone of speakers. The maximum power level a control can handle is listed in the manufacturer's specification data sheet for the product. In simple terms, if a volume control is rated for 20 W, then there must not be more than 20 W of load connected to the output. It is not necessary to de-rate the volume control. A 20 W control can safely drive a 20 W load without risk of damage to the device.

Figure 5–32
Transformer volume control.

© Cengage Learning 2013

Integrated Amplifier

An **integrated amplifier** is a combination of a preamplifier and a power amplifier. The integrated amplifier is a self-contained audio-processing device that provides multichannel input selection, tone controls, linkage to the power amplifier, and output to the speakers. (Refer to *NEC Definitions 640.2, Mixer-Amplifier* for an exact definition.)

Power Amplifier

The power amplifier is the last stage of amplification between the signal input and the output speakers. Power amplifiers can be multichannel or monaural, and they must be rated to deliver all the necessary power required by the load. As stated earlier, because of low-frequency draw, transformer losses, and inefficiencies, power amplifiers should be de-rated to 70% their maximum output power to help reduce the risk for overloading and device malfunction. For example, if a network of speakers requires 70 W of power, then the amplifier should be rated for a maximum output power of 100 W, to help ensure the long-term reliability of the device.

Power amplifiers provide wide-band amplification across the entire audio spectrum, from 20 to 20,000 Hz. They do not offer any tone control or signal-processing functionality other than adjustable volume control for the main signal outputs. Power amplifiers are also designed for connection to 4, 8, or 16 Ω loads, or for a maximum output of 25, 70, or 100 V, through a properly rated matching transformer. The output connections on the back of the amplifier indicate the type of connections available for use. Where making output connections, be sure to check the instruction manual provided by the manufacturer. Jumper wires are sometimes necessary depending on the output configuration and type of grounding scheme.

Telephone Paging Signals

(Refer also to *NEC Article 640.3, Locations and Other Articles.*)

Telephone paging signals are sent out from the central phone system of a building to the various inputs of mixers or preamplifiers driving the audio network. The telephone system uses a balanced 600 Ω line input level; that is, proper connection requires that you either need to have a balanced 600 Ω low-impedance input on the amplifier, or you must first connect the signal through a line-matching transformer, one that converts between a balanced, low-impedance line and an unbalanced, high-impedance line. Typical high-impedance line inputs range between 10,000 and 35,000 Ω. Maximum signal level and power transfer between the phone line and the amplifier system is not achieved unless the matching transformer is used.

In addition, directly connecting a balanced phone line to an unbalanced line input also creates a grounding problem between the two systems. Direct connection causes the grounded side of the unbalanced input to short out half of the balanced telephone line. The matching transformer provides the necessary ground isolation between the two sides, as well as the impedance match between systems.

SEC 5.6 MIXERS, PREAMPLIFIERS, AND SIGNAL-PROCESSING EQUIPMENT

Mixers

Audio mixers (Figure 5–33) combine line-level input signals from multiple sources, either to a single output channel or to multiple output channels, as selected by the user. The functionality of a mixer is to also adjust tone controls of individual signals that can be routed through selectable processing equipment or to audio effects generators as desired by the audio engineer. The output of a mixer typically is fed to the line input of a preamplifier or power amplifier for final amplification.

Mixers also provide separation between individual audio signals so that they do not interfere with or load each other when combining. Essentially, there are two reasons why you would never want to connect separate signals in parallel across the same amplifier input without a mixer.

First, parallel connections provide no separation between individual branches; that is, if one of the inputs were to ever short out, then all of the inputs would be disabled simultaneously.

Figure 5–33
Audio mixer.
(A) Top view.

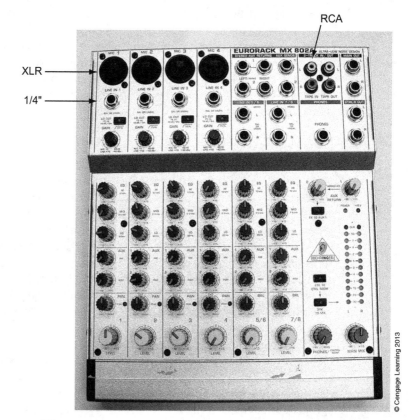

RCA

XLR

1/4"

© Cengage Learning 2013

(A)

(Continued)

Figure 5–33 Cont'd
Audio mixer. (B) Rear view, including phantom power switch.

AC POWER IN

USE THE SUPPLIED POWER SUPPLY ONLY !

ON
POWER

ON
PHANTOM

(B)

Rear view

© Cengage Learning 2013

Second, the input impedance of the amplifier would decrease dramatically because of the parallel combination of **signal sources** with the input channel of the amplifier. For example, if the internal resistance of all five signal sources each measures 10,000 Ω, then connecting them in parallel with the amplifier input would reduce the total input impedance by a factor of 5; five, because an individual signal source sees the other four sources as parallel resistances to the 10,000 Ω input impedance of the amplifier. The total input impedance of the amplifier then would be 10,000 ÷ 5, or 2000 Ω. Because a source of 10,000 Ω is expecting to see a load of 10,000 Ω, the resulting 2000 Ω would severely attenuate all five signal sources and maximum power transfer would not be achieved.

Audio mixers often are chosen based on the number of input channels needed for the job. The total number of individual signal sources to be amplified determines the number of channels. Channel inputs can be of the high- or low-impedance variety, and the engineer typically can adjust individual inputs between microphone or line level through a series of selector switches on the control panel. Be sure to check with the manufacturer on all the available options because not all models have the same features.

Preamplifier

Preamplifiers are line-level signal amplifiers that provide tone control and level adjustment for input signals before final-stage amplification. Output connections can be either 150 or 600 Ω balanced, or 10,000 to 35,000 Ω unbalanced. Preamplifiers are not designed to drive 8 Ω loudspeakers. The maximum output power for most preamplifiers is well below 10 W, which is adequate for line-level signal processing but not nearly enough for the high current demands of most 8- or 12-in. loudspeakers. (Refer to *NEC Definitions 640.2, Audio Amplifier or Pre-Amplifier* for specific definitions.)

Equalizers

Equalizers are used to help filter out or increase specific frequencies within the 20 to 20,000 Hz audio spectrum. They process line-level signal inputs and usually are inserted into the system just before final power amplification. There are two basic types of equalizers used in audio signal processing: graphic and parametric.

Graphic Equalizers

Graphic equalizers use a combination of **active filters** separated by equal bandwidth, sometimes as narrow as 1/6 of an octave along the entire audio spectrum, to help adjust and change the frequency response of a line-input audio signal. The graphic equalizer gets its name from the front panel display that graphically illustrates the adjusted frequency response of the output signal (Figure 5–34). Slider switches are placed left to right on the display panel, each representing a specific frequency band along the audio spectrum; they can be moved up or down as needed to either increase or decrease a specific band by as much as 6 to 20 dB, depending on the specifications of the equalizer. When complete, the placement of slider switches along the front panel displays the shape of the adjusted curve, illustrating the new frequency response of the audio signal, across the 20 to 20,000 Hz bandwidth. Graphic equalizers are used together with white or pink noise generators to adjust the frequency response of a room.

Parametric Equalizers

Parametric equalizers are most often used by studio engineers. They offer continuous variability and adjustment to frequencies and bandwidths, thus allowing the user to fine-tune the signal precisely to any desired shape or curve. Multiple filter types, whether they are band-pass or notch, can be selected together with the central frequencies, output levels, and bandwidths. The user can then independently adjust each parameter to obtain any desired signal response or outcome. The parametric equalizer is not as user friendly as the graphic equalizer, and a more detailed knowledge of signals and filtering is needed to operate the device.

Amplifier with graphic EQ

Figure 5–34

Multi-input amplifier with graphic equalizer.

Graphic equalizer

Signal-Processing Equipment

Signal-processing equipment provides special-effects processing for line-level audio signals before final amplification. Signal-processing devices usually are placed between the preamplifier and the power amplifier. Most integrated amplifiers, consisting of a preamplifier/power amplifier combination, provide a linkage between the preamplifier and the power amplifier for the expressed purpose of inserting a signal-processing device or equalizer into the signal loop. If the linkage is removed, then the electrical connection between the preamplifier and power amplifier is broken, causing a total loss of signal at the output.

When designing sound images and room acoustics, the following processing devices typically are used: echo, delay, and reverberation. Echo and delay processors add user-defined echoes or timing delays to audio signals. Reverberation is used to create the impression of space within a small environment. Small rooms can be made to sound larger, depending on the amount of reverberation and signal processing. Surround sound, concert hall effects, and amphitheaters usually can be simulated with a variety of echo, delay, and reverberation processing. Programmable processors allow the user to choose from a variety of standard effects, all of which can be adjusted further by the user to achieve a desired outcome.

Digital-Processing Equipment

Digital processing provides a nice low-cost alternative to analog. Instead of purchasing individual components, all of which have to be physically mounted, wired, and grounded to an equipment rack (refer to *NEC Definitions 640.2, Equipment Rack* for an exact definition) or cabinet, it is far more efficient and easier to send line-level audio signals to a central computer. The central computer can literally provide all the benefits of a professional sound studio in one simple, neat, compact, portable package. Through a variety of software, a virtual studio can be assembled quickly. In most cases, a computer terminal can now replace a variety of analog devices such as compressors, limiters, graphic equalizers, filter networks, echo, reverberation, delay, harmonizers, effect processors, crossovers, sound meters, and analyzers. Virtual mixers and preamplifiers also can be obtained through a variety of musical instrument digital interface (MIDI) and virtual studio packages, now available for either PCs or Macintosh-based computers. Additional hardware is needed for the interconnection of audio inputs and outputs to the computer motherboard; but once connected, all the benefits of the virtual sound studio are at the fingertips of the operator, through a simple mouse click or keyboard command.

Sound Level Analyzers/Spectrum Analyzers

A sound-level **spectrum analyzer** graphically illustrates the full range of signal levels along the entire audio spectrum, from 20 to 20,000 Hz. The analyzer is designed to measure the frequency response of a room. An omnidirectional microphone is placed on the top of the analyzer to detect the sound pressure levels of the room. A series of band-pass filters inside the analyzer separate and drive the graphical display. A bar graph on the display of the analyzer illustrates the level of detected frequencies picked up by the microphone. In most cases, a pink or white noise source is used to level the room. The frequency response of the sound level analyzer is also designed around the frequency response of the human ear. The graphic display illustrates the level of sound as it would be heard by an individual. In addition, weighted filter networks A, B, or C are selectable on the device, corresponding to the 40, 70, and 100 Phon listening levels, as shown in the Robinson and Dadson contour of a hearing graph (see Chapter 4).

Signal Sources

Signal sources provide the input signals to audio systems. They can range from telephone paging, microphone, music, video programming, computers, security alarms, fire alarms, or a series of scheduled bells or tones.

Music or video sources can come from a satellite, radio tuner, compact disc player, digital video disc player, audio or video cassette players, or a computer.

Security or fire alarm control panels often control audible alarm signals. Notably, however, combinational systems involving the interconnection of fire alarm systems and music or paging systems, typically are not allowed unless otherwise permitted by the authorities having jurisdiction (see *NEC 640.3(J)* for more information). In cases where the use of combinational systems is allowed, devices typically are rated for 2 hours of fire resistance, as referenced in *NEC 760. NEC 640.25* also states that loudspeakers and enclosures installed in fire-rated walls or ceilings must be appropriately rated for the purpose of installation, to maintain the fire resistance rating of the building.

Bells and tones are generated from signal sources that are usually programmable. They often provide a variety of signal tones and event sequences throughout the day, as desired by the user.

Before connecting any input device to a system, electrical specifications must be checked and verified. First, determine whether the source is providing a balanced or an unbalanced signal output. In addition, how many channels are needed? Are they stereo or monaural? Then verify the input impedance, signal levels, and types of connectors needed, while also taking into account the type of mixer or amplifier to be used. As stated earlier, signal sources must be correctly matched and terminated to sound systems to achieve optimum signal reproduction.

Completed System Block Diagram

Figures 5–35 through 5–37 illustrate the flow chart of a completed audio system, including all elements and signal levels discussed throughout this chapter.

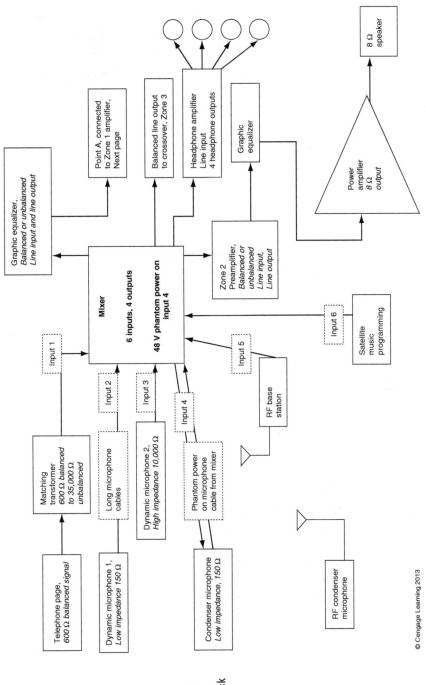

Figure 5-35
Completed
system block
diagram.

© Cengage Learning 2013

Figure 5–36

A 70 V speaker connection block diagram.

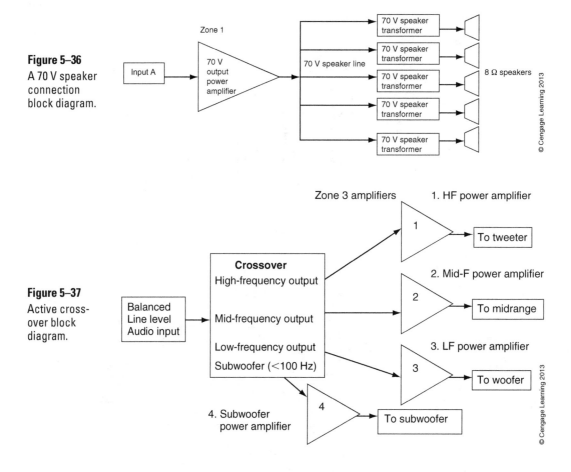

Figure 5–37

Active crossover block diagram.

SEC 5.7 ADDITIONAL *NATIONAL ELECTRICAL CODE* REQUIREMENTS

The *NEC* further covers audio systems with several requirements for installations.

Systems near Water

When installing an audio system near a body of water, either natural or artificial, the following restrictions apply (see *NEC 640.10 (A) and (B)*):

Audio equipment supplied by a branch circuit cannot be installed within 5 ft (1.52 m) of the inside wall of the body of water and must be protected by a ground-fault circuit interrupter.

If an audio system is powered by a Class 2 supply, the placement of wires must meet the requirements and specifications of the manufacturer.

Number of Conductors

When installing audio systems, the number of conductors allowed in a single raceway must comply with *NEC Chapter 9, Table 1* (*NEC 640.23*). (*NEC Annex C* is also a good reference for determining the number of conductors in a raceway.) An example of conduit fill using *NEC Chapter 9, Table 1* is included at the end of Chapter 8.

Bonding

When wiring metal equipment racks, the rack itself and equipment must be bonded. Equipment racks typically are bonded in a star pattern configuration, from a bonded connection at the main service panel for the building. The purpose of bonding is to prevent electric shock. All metal boxes, raceways, cabinets, and mounting racks are required to be bonded together to prevent people and technicians from the hazards of electric shock, the source of which could be unwanted current flow on exposed metal.

In addition to bonding, the metallic chassis of mounted equipment is also internally connected to the EGC through the third prong of the power cord to prevent accidental ground faults. Nonmetallic equipment enclosures will be powered through a two-prong power connection, and therefore will not require connection to the third prong. Figure 5–38 illustrates a typical grounding scheme for equipment racks.

Isolated Ground

Another type of ground connection involves the technical circuit. In many instances, the grounding of a technical circuit is separated from a chassis or equipment ground as a way to isolate the circuit from possible noise on the existing ground plane. In cases where the technical circuit ground is separated from the existing chassis or equipment ground, a suitable ground replacement must be made. In such cases, an **isolated ground** must be established.

An isolated ground is a single-point grounding scheme set up by a system installer. In cases where an isolated ground is used, ground straps are removed or lifted from technical circuits, isolating them from the existing equipment ground or chassis ground; this is often the case for telecommunications circuits, networking, and audio/video circuits. The required ground connection to the technical circuit is then replaced by connecting an insulated bonding conductor between the technical circuit and a grounding bus bar. The grounding bus bar finally terminates to an earth ground through a single-point connection. The benefit to using an isolated ground is to help isolate sensitive equipment from possible building noise on the existing equipment ground.

In cases where the technical circuit is grounded through an isolated ground connection, the chassis of equipment and equipment racks are still bonded though the main building's bonding jumper. Figure 5–39 illustrates the wiring of an isolated circuit ground.

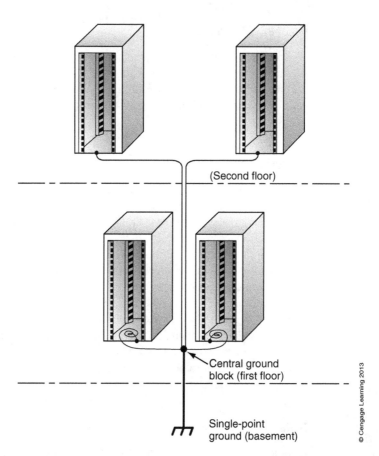

Figure 5–38
Star-grounded typical equipment racks.

(Second floor)

Central ground block (first floor)

Single-point ground (basement)

© Cengage Learning 2013

The bus bar for an isolated ground typically is made from a 1/4-in. thick, 2-in. wide (6.35 mm thick, 50.8 mm wide) copper or aluminum plate. Keep in mind that if aluminum is used, you will need to follow the rules for connecting dissimilar metals, *NEC 110.14*. The length of the grounding bus often varies depending on the number of equipment grounding conductors attached to the bar. Each bonding connector connects to a different equipment rack or piece of equipment within the system through a star configuration. A single-point ground then is achieved by bonding from the grounding bus plate all the way back to the main service panel or intervening panelboards with an insulated bonding conductor not smaller than 6 AWG copper. To ensure that the technical circuit is isolated, the bus bar also is mounted on an insulator. In addition, bonding conductors connecting between technical circuits and the bus bar should be equal in length and of insulated, 14 AWG copper. Having equal length conductors helps to ensure that a zero-point ground connection is made to the bus bar, thus forcing the elimination of unequal ground resistance between components. Having unequal levels of ground resistance between various pieces of equipment and the bus bar often adds to the level of system ground noise. Using insulated bonding conductors ensures that the bonding conductors maintain their isolation from the existing chassis or equipment grounds, so as to not corrupt the integrity of the isolated ground connections.

Figure 5–40 illustrates an isolated ground connection.

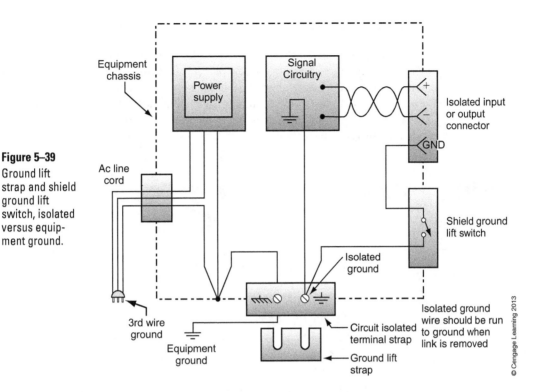

Figure 5–39
Ground lift strap and shield ground lift switch, isolated versus equipment ground.

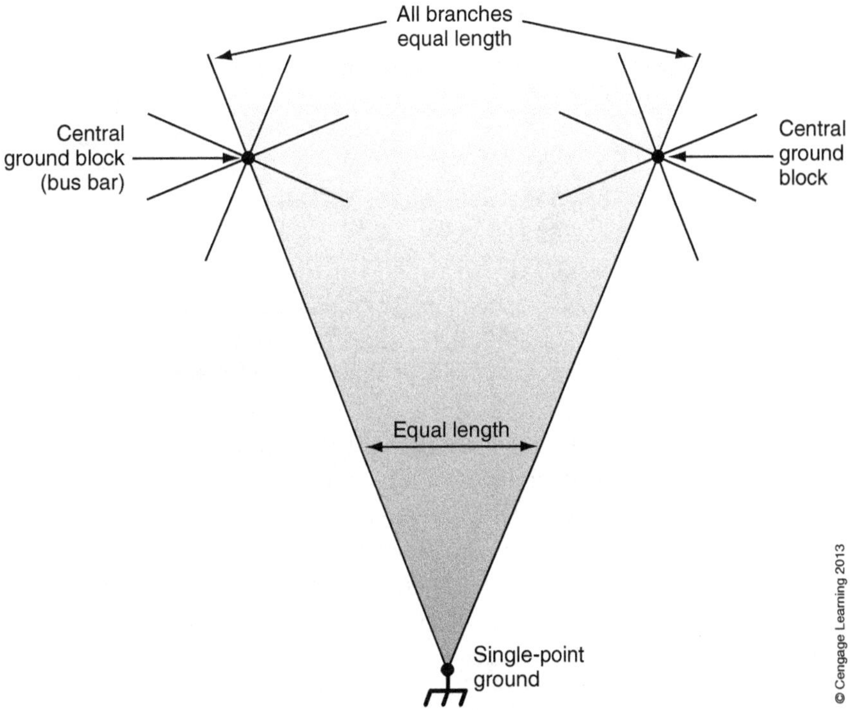

Figure 5–40
Technical star ground.

Additional Concerns

Nonmetal racks must not allow access to Class 1 or 3 circuits or primary circuit power without the removal of a safety device (see *NEC 640.43*). Equipment racks must be wired in a neat and professional manner. The wiring of equipment must not deny access to any of the devices housed within the rack itself. Any wiring that leaves the confines of the rack must be secured by approved hardware, to avoid risk for damage to the cable.

CHAPTER 5 FINAL QUESTIONS

1. What article in the *NEC* covers audio signal processing, amplification, and reproduction equipment?
2. A dynamic microphone requires power.
   a. True
   b. False
3. A condenser microphone is _____.
   a. electromagnetic in nature and does not require power
   b. electromagnetic in nature and requires an external power source
   c. electrostatic in nature and requires an external power source
   d. electrostatic in nature and does not require power
4. What is the typical impedance of a low-impedance microphone?
   a. 75–150 Ω
   b. 150–600 Ω
   c. 600–1200 Ω
5. What is the impedance of a balanced telephone paging system?
   a. 75 Ω
   b. 150 Ω
   c. 300 Ω
   d. 600 Ω
6. An unbalanced, high-impedance microphone cable should not be longer than _____ ft.
7. The polar pattern of a microphone determines the _____.
   a. frequency response of the device
   b. impedance of the device at various distances
   c. sensitivity pattern of the device within a 360° rotation
   d. maximum power limits to prevent damage to the device
8. How can you determine that there is a ground loop on an audio circuit?
   a. The circuit has tripped a ground-fault interrupter.
   b. The speaker fuse has blown.

   c. There is a presence of noise and hum in the output.

   d. The output power is significantly low.

9. Which type of audio connector is balanced?

   a. XLR

   b. RCA

10. How does the physical size of a woofer, midrange, and tweeter relate to the resonant frequency of the device?

11. If a speaker has an efficiency rating of 70 dB/m, using the inverse square law, calculate the dB/m at a distance of 8 m (26.2 ft).

12. What type of device is used to direct specific frequencies to a speaker driver?

   a. A balun

   b. A crossover

   c. A line pad

   d. A transformer

13. What type of filter will effectively filter out all frequencies above the −3 dB cutoff point?

14. Explain the difference between a band-pass and a notch filter.

15. What type of device is used to reduce signal voltages on an audio pre-amplifier input?

   a. A balun

   b. A crossover

   c. A line pad

   d. A transformer

16. Which statement is not true about a 70 V audio transformer?

   a. A 70 V audio transformer provides an impedance match between the amplifier output and the speaker.

   b. The output of a 70 V audio transformer is calibrated in wattage.

   c. The output of a 70 V audio transformer will reach 70 volts when the amplifier volume control is at maximum.

17. Calculate the minimum wattage for a power amplifier if the connecting load requires six 10 W, 70 V transformers; four 2 W, 70 V transformers; and fifteen 1 W, 70 V transformers.

18. What type of device is used to connect a balanced 600 Ω source to an unbalanced 10,000-Ω input?

19. The output of an audio preamplifier provides what level of signal voltage to the amplifier?

20. Define *reverberation*. How does it compare with an echo?

21. What piece of test equipment is designed to measure the frequency response of a room?

22. Explain the benefits of an isolated ground for most types of technical circuits?

23. The equipment grounding conductor connected between equipment and the grounding bus bar shall not be smaller than _____ AWG.

24. The bonding conductor connected between the isolated grounding bus bar and the main service panel, should not be smaller than _____ AWG.

25. Audio systems supplied by branch circuits are not to be installed within _____ feet from the inside wall of a body of water and must be protected by ground-fault circuit interrupters.

Chapter 6

Networking and Information Technology Equipment

Objectives

- Compare a wide area network and a local area network and explain their differences.
- Describe some popular network architectures.
- Describe the purpose of the Open Systems Interconnection model and list the various levels.
- List various Transport Control Protocols and Internet Protocols and their functions.
- Explain the process of addressing computers on a network.
- Categorize the cabling used to send data between multiple points of a network.
- List and describe the various hardware components used in building a network.
- Explain a variety of installation techniques that ensure a clean install.
- Describe the purpose and use of various operating systems and device drivers.
- List the various methods and type of connections needed to access the Internet.

Chapter Outline

Sec 6.1 Basic Networking, Architecture, and Topology

Sec 6.2 Network Protocols and Their Functions

Sec 6.3 Computer Network Addressing

Sec 6.4 Media: Connections, Hardware, and Installation Techniques

Sec 6.5 Other Concerns

Sec 6.6 Low-Voltage Residential Network Applications

Key Terms

bridge	lan extender	shielded twisted pair (STP)
bus topology	local area network (LAN)	simple mail transport protocol (SMTP)
client/server	modem	
digital subscriber line (DSL)	network	simple network transfer protocol (SNTP)
domain name server (DNS)	open systems interconnection (OSI)	star topology
ethernet		
fiber distributed data interface (FDDI)	operating system (OS)	token ring network (TRN)
	parallel communication	topology
file transfer protocol (FTP)	physical addressing	transport control protocol (TCP)
firewall	ping	tree topology
firewire®	post office protocol 3 (POP3)	uniform resource locator (URL)
gateway	protocol	universal serial bus (USB)
hub	repeater	unshielded twisted pair (UTP)
integrated services digital network (ISDN)	router	user datagram protocol (UDP)
	serial communication	wide area network (WAN)
internet protocol (IP)	server	

SEC 6.1   BASIC NETWORKING, ARCHITECTURE, AND TOPOLOGY

Networking involves the intercommunication between computers. Network **topology** involves the manner in which multiple computers are physically connected. Network media refers to the method of transporting communication signals. Three common methods include copper wire, optical fiber, and wireless. The desired speed of communication is determined based on which is used. This chapter discusses the basics of computer networking and how systems are interconnected locally and over long distances.

A Brief History of the Network

A **network** is made up of a collection of individual computers, computer-processing equipment, printers, or communication devices, which are linked together by interconnected routes or pathways, for common use by all participants. The first networks, known as **Ethernets**, were developed in the 1970s by Xerox, and ran at maximum speeds of 2.94 Mb/ps. Xerox soon joined with Intel and Digital Equipment Corporation to develop a standard for a 10 Mb/ps Ethernet, known as DIX (Digital, Intel, Xerox). During the same period, the Institute of Electrical and Electronic Engineers (IEEE) was developing the 802 task force that was working to establish technical standards for **local area networks (LANs)** and metropolitan area networks (MANs). Figure 6–1 illustrates a LAN.

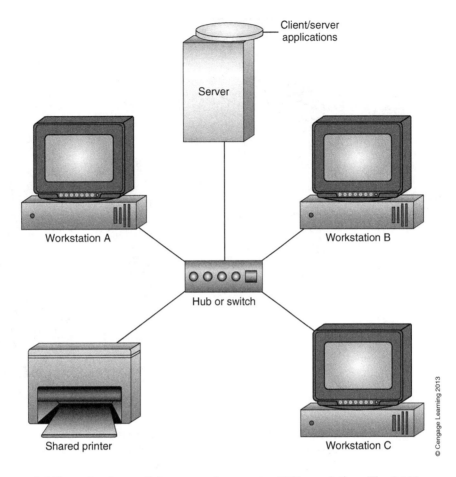

Figure 6–1
Local area
network.

LANs evolved out of the personal computer (PC) revolution. The LAN succeeded in linking the communications of multiple users for file exchanges, exchange messaging (e-mail), and the sharing of resources such as printers, information databases, and file **servers**. A LAN refers to the linking of users in a small geographical area, such as an office complex, school, or college campus.

MANs, now more commonly known as **wide area networks (WANs)**, interconnect multiple LANs over a wide geographical area, such as a city or state. WANs achieve a communication link over a long distance through a network line brought in by a service provider. The service provider can provide access to WANs and the Internet by a variety of methods. Examples include a dial-up telephone **modem**, **integrated services digital network (ISDN)**, high-speed digital coaxial cable, **digital subscriber line (DSL)**, satellite, optical fiber, or by wireless. Figure 6–2 illustrates a WAN.

By 1985, IBM had developed the IBM **token ring network (TRN)**, which linked computers through a dedicated system of data relay. The ring functioned by passing a bit stream, referred to as the token, from station to station. Every station in the loop was required to pass on the circulating information until the token returned back to the source, thereby ensuring that all participants in the network

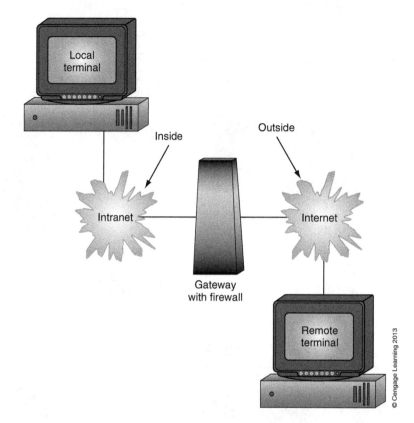

Figure 6–2
Wide area
network.

had equal access to the circulating data. Figure 6–3 illustrates a TRN. Addressable data stream packets were then developed as a security measure to ensure that individual stations around the ring only had access to data packets addressed to them; the token, containing the remainder of unopened data packets, then was passed on to the next designated station in the loop.

The token ring process would be similar to a mail carrier delivering the neighborhood mail to only one house on the block; each house then would only be responsible for opening their individual mail, as the bag moved from house to house. Eventually, all the letters would make it around the block to the required destinations. The bag then would be emptied of any outgoing letters and refilled with new incoming letters as soon as it returned to the originating mail stop.

The IBM TRN was so successful that it was quickly adopted by the IEEE 802 task force and standardized. It remained a popular network design through the mid-1990s.

As the physical topologies of network connections were improved and adapted to permit greater speeds of communication, ease of connectivity and a more reliable method of cabling were needed. Customers also insisted on a more telephone-based cabling structure, rather than the bulky and more expensive coaxial cable, which typically was being used on early Ethernet and TRNs. Ultimately, **unshielded twisted pair (UTP)** was adopted as an industry standard because it was less expensive and easier to install.

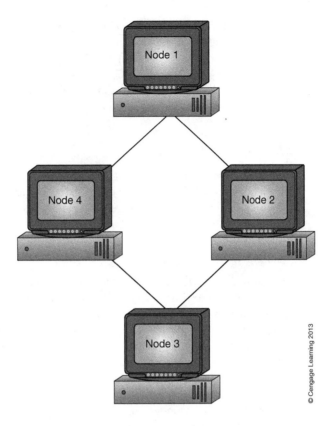

Figure 6–3
Token ring
network.

© Cengage Learning 2013

In 1987, **fiber distributed data interface (FDDI)** networks were developed that used optical fiber to transport token ring data at speeds of up to 100 Mbps. Nevertheless, within a few short years of its release, and because of the high cost of optical fiber, shielded and unshielded copper twisted pair became the desired industry cable of choice. As a result, the copper distributed data interface (CDDI) was organized to develop standards and specifications for the more commonly used UTP and **shielded twisted pair (STP)** cables.

FDDI did not, however, disappear altogether. It still exists as the main backbone for many installations requiring long-distance runs; examples include universities and large corporate complexes. FDDI is also commonly used to interconnect two or more LANs, because optical fiber inherently transmits a higher quality signal over greater distance than does copper. We look more closely at optical fiber in Chapter 10.

The expansion of TRNs ultimately was slowed by the development of the 10Base-T Ethernet in 1987. 10Base-T refers to a communication speed of 10 Mbps on a twisted pair line. Such networks can reliably transmit over a 100 m run. The downside of TRNs was their inability to provide quality control as network traffic increased, because the addition of more stations on the ring would ultimately slow down the entire network. A proprietary system of permissions to specific files also was not in the nature of TRNs; data were simply passed on from one computer to the next, with no concern as to whom the recipient might be or whether the recipient had an actual right to the data. Ethernet solved the problem by transmitting out only to desired stations within a network, via a **hub**, digital switch, or router. As a result, data moved faster and in a more direct manner without having to visit each individual station along the way. Although 4, 16, and even 100 Mbps were possible through token ring designs, the eventual 10Base-T and later 100Base-T and 1000 Base-T Ethernets proved to dominate the industry. This dominance was cost-related in many respects, because token ring was a proprietary system and expensive to implement.

Access to the Internet

As stated earlier, an individual needing to gain access to a WAN or to the Internet can do so through a variety of means: dial-up modem, DSL, ISDN, cable modem, satellite, optical fiber, or by wireless. In most cases, the choice is determined by price, speed, and availability of service.

Dial-up Modem

A dial-up modem is a device that allows a computer to communicate over an analog telephone line. The word *modem* is a conjunction of *mod*ulation and *dem*odulation. Modulation places an intelligent signal onto a communication carrier for transmission. An intelligent signal may be voice, data, or video. The carrier then carries the signal through a designated channel to the recipient of the communication. In a wired system, the channel is the connecting wire; in a wireless system, the channel is the air. Demodulation is the opposite of modulation; it removes the intelligent signal from the carrier so that the recipient can interpret the message. A dial-up modem also provides a digital-to-analog conversion so that the analog phone line can use and pass the signal.

The maximum speed of most dial-up modems is 56,000 bps. Limitations on connection speed are also introduced into the system by the telephone line. Often, Internet connection speed over a telephone line is limited because of low bandwidth and traffic on the line. The type of information being transmitted also affects the rate of data flow, because signals heavy in video and graphics ultimately take up more space and slow down the system. In most cases, 56,000 bps is not even possible, depending on how far the connection is from the telephone company switch gear. The closer an individual is to the telephone company's switch, the better. Longer transmission distances add unwanted capacitance, resistance, and inductive loading to the telephone line that inhibit the passage of high-speed communications. Figure 6–4 illustrates a dial-up modem network connection.

Modems also are limited by the type of data they can transmit. Some have the ability to handle voice and fax, as well as data. Another option may include data compression to help optimize and increase data transfer rates. When purchasing a modem, be sure to read all the specifications carefully, as well as the available options; otherwise, issues of hardware incompatibility may arise.

Digital Subscriber Line

DSL offers the next alternative to a standard dial-up modem because it allows high-speed computer data to share the telephone line with the analog voice

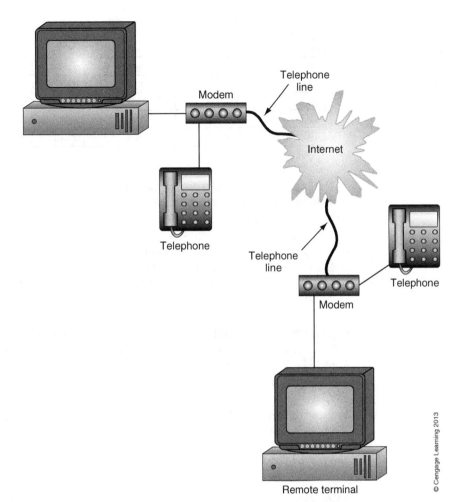

Figure 6–4
Modem
connection to
the Internet.

Telephone
line

Modem

Telephone

Internet

Telephone

Telephone
line

Modem

Telephone

Remote terminal

© Cengage Learning 2013

communication. The service provider of a DSL system installs a special DSL modem that filters and splits the analog voice signal from the high-speed digital data. A passive low-pass filter, referred to as a plain old telephone system (POTS) splitter, is used to separate the voice signal from the digital data at frequencies just less than 20 kHz, which represents the upper threshold of human hearing. The DSL uses all frequencies above 25 kHz for computer communication and connection to the Internet. Splitting the signals also ensures that if the digital section of the DSL modem ever fails, the voice section continues to operate.

DSL can operate up to 50 times faster than dial-up modems, at nearly 1.5 Mbps. However, just as with the dial-up modem, maximum speeds are unlikely to be realized if the connection is located far from the telephone company, or telco. Line loading and line-conditioning equipment installed by the telephone company to optimize the voice signal often inhibit high-frequency transmission at great distances, not to mention the added capacitance and resistance of the cable.

© Cengage Learning 2013

Figure 6–5
Digital
subscriber
line (DSL)
specifications.

Type of DSL	Upstream rate	Downstream rate	Distance
ADSL	640 kbps	1.5 Mbps	18,000 ft (5486.4 m)
SDSL	2.3 Mbps	2.3 Mbps	22,000 ft (6705.6 m)
RADSL	1.0 Mbps	7.0 Mbps	18,000 ft (5486.4 m)
HDSL	1.5 Mbps	1.5 Mbps	12,000 ft (3657.6 m)
VHDSL	16 Mbps	52 Mbps	4,000 ft (1219.2 m)
ISDL	144 kbps	144 kbps	35,000 ft (10,668.0 m)

DSL may also introduce noise or hum to the voice section of the telephone line. To help prevent such interference, low-pass filters are added to phone jacks that do not use the DSL connection. Although the telephone must still dial out to make a voice call, the DSL is always connected and requires no dial-up to access the service provider. The DSL connection is on 24 hours a day, 7 days a week.

DSL comes in six varieties, all varying in speed, upstream and downstream symmetry, and maximum distance. The chart in Figure 6–5 details the six varieties of DSL.

Asymmetrical Digital Subscriber Line. ADSL, or asymmetrical DSL, provides an upstream rate of 640 kbps and a downstream rate of 1.5 Mbps. As a result, uploads from an individual's computer are slower than downloads. ADSL also is limited to a distance of 18,000 ft (5486.4 m) from the telephone company switch box.

Symmetric Digital Subscriber Line. SDSL, or symmetric DSL, allows upstream and downstream rates to be equal, to a maximum of 2.3 Mbps. SDSL does not, however, allow simultaneous access to the telephone line. Maximum distance of SDSL is limited to 22,000 ft (6705.6 m) from the telephone company switch box.

Rate-Adaptive Digital Subscriber Line. RADSL, or rate-adaptive DSL, is a variation of ADSL that allows the modem to vary the speed of transmission based on the length and the quality of the line. RADSL is asymmetric, allowing maximum downloads rates of 7 Mbps and maximum upload rates of 1 Mbps. The maximum distance from the phone company switch box is limited to 18,000 ft (5486 m).

High-Bit-Rate Digital Subscriber Line. HDSL, or high-bit-rate DSL, is symmetric, offering a maximum data rate of 1.5 Mbps in each direction. The downside is that it requires three telephone lines: two for the DSL connection, one being the uplink and the other the downlink; and the third line for the regular telephone. HDSL has a maximum transmission distance of 12,000 ft (3657.6 m) from the telephone company switch box.

Very-High-Bit-Rate Digital Subscriber Line. VHDSL, or very-high-bit-rate DSL, provides the fastest data rates, 52 Mbps for download and 16 Mpbs for upload. But to achieve such speeds, the maximum distance from the phone company switch box cannot be greater than 4000 ft (1219.2 m).

Integrated Services Digital Network Digital Subscriber Line. ISDL, or ISDN DSL, is for existing users of ISDN who choose to use their existing equipment. Although the rates of ISDL are the slowest of all DSL options, coming in at 144 kbps each direction, it does offer the greatest transmission distance, 35,000 ft (10,668.0 m) from the phone company.

Termination of a DSL to a computer is to either a **universal serial bus (USB)** port or a registered jack (RJ)-45 port, depending on the type of DSL transceiver being used. The service provider often refers to the transceiver as an ADSL Termination Unit-Remote (ATU-R).

Integrated Services Digital Network

ISDN is a dial-up service to the Internet. It does not offer a dedicated or always-on connection to a network like DSL. In some respects, ISDN is similar to a standard dial-up modem connection, except that the ISDN has the ability to share the line simultaneously with the telephone, fax, and computer. A standard dial-up modem connection can only provide access to one user at a time. The service provider of ISDN provides a network terminal adapter, which is nothing more than a special type of modem that allows the integration of voice, fax, and computer on a single telephone line. ISDN allows all devices to share the phone line simultaneously.

To accomplish a shared line, the ISDN terminal adapter provides the use of a bearer channel (B channel) that automatically scales back the bandwidth usage of the line to allow the fax or telephone access to 64 kbps on incoming calls. In this way, the computer does not actually disconnect, but instead slows down to achieve the integration of services on the line.

Two levels of service exist through ISDN:

Basic rate interface (BRI): consists of two 64 kbps B channels and one 16 kbps D channel. The BRI can provide up to 128 kbps when other devices are not using the B channels.

Primary rate interface (PRI): consists of 23 B channels and one 64 kbps D channel in the United States, or 30 B channels and 1 D channel in Europe.

The B channels carry voice, data, and special services, whereas the D channel, or Delta channel, carries control and signaling information.

The local telephone company in most urban areas can provide ISDN service. The cost, however, is quite high compared with that of DSL and is based on usage. ISDN is simply not a competitive option to DSL, because of the lower transmission rates and greater costs, and also because it cannot provide an always-on, dedicated line to the user.

Cable Modem

The cable television (CATV) service provider in most regional areas provides cable modem Internet service. The modem signals travel on a two-way hybrid fiber or coaxial cable, often at speeds ranging from 320 kbps to 10 Mbps. Subscribers, however, typically achieve maximum download speeds of only 1.5 Mbps,

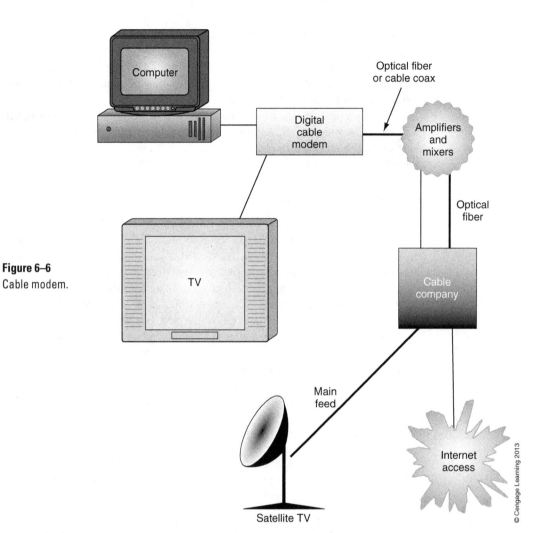

Figure 6–6
Cable modem.

as a result of having to share bandwidth with other subscribers on the system. Figure 6–6 illustrates a cable modem network connection.

The cable modem provides two outputs, one for the television and one for the PC. The modem itself is capable of receiving and processing information at rates of up to 30 Mbps, thousands of times faster than a standard dial-up telephone modem. The obvious advantages to using a cable modem service are higher speeds and greater bandwidth. Cable service is also often more readily available than DSL. DSL must often be brought into a location at additional charge, whereas cable television service already has an existing infrastructure in most locations.

Satellite

The theory of satellite communication is explained in Chapter 13. From a computer networking point of view, a direct broadcast satellite (DBS) connection is capable of downloading Internet service at speeds of up to 45 Mbps. An uplink to a DBS satellite is not possible, though, because the system is capable of providing only a one-way signal from the satellite in the sky to a ground base station. To achieve a network uplink, the system still needs a secondary telephone connection, which, as described earlier, is considerably slower. Figure 6–7 illustrates a satellite network connection with a secondary uplink connection. Satellite uplinks can only be made by the service provider. The data follow a point-to-multipoint path; that is, multiple recipients have the ability to download signals originating from a single ground station.

Satellite systems do exist, however, that offer the ability to uplink and downlink data, to and from an Internet Service Provider. In most cases, such systems are quite expensive, though, often running as high as $150.00 per month.

The main advantage to using satellite is that it provides a wireless high-speed connection from just about anywhere in the world. All that is needed is a dish, a

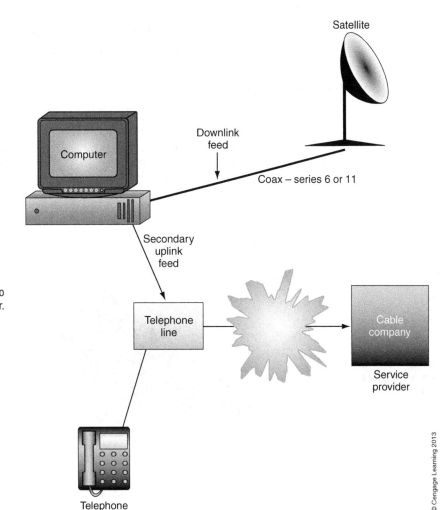

Figure 6–7

Satellite downlink with secondary uplink via the telephone line to service provider.

© Cengage Learning 2013

receiver, and an unobstructed, line-of-sight view to the orbiting transceiver in the sky. For rural areas, it is often the only available choice if high-speed network connections are required.

Optical Fiber

Optical fiber remains the best choice when considering a hard-wired connection to the Internet. Fiber can transmit data at rates in the gigabits and over long distances without having to worry about electromagnetic interference (EMI) or signal attenuation. The theory and specifics of optical fiber are covered in Chapter 10. The cost of optical fiber can be somewhat prohibitive, which is why most people opt for DSL or cable modems. Many new construction projects are now bringing optical fiber into the home. Although business and industry have been using optical fiber for years, it only recently has become more affordable for residential use.

Wireless Networks

Wireless networks can operate on cellular, Bluetooth, infrared, or satellite. The details and specifics of wireless networks are discussed in Chapter 13.

Network Interface Card

The network interface card (NIC), as shown in Figure 6–8, is the link between the computer and the network. The card slides into the expansion slot of the computer. The main function of the card is to interpret data passing between the microprocessor and the network. The card also manages the flow of data to ensure that bidirectional communication is logically taking place on both sides of the link.

Figure 6–8
Network interface card.

© Cengage Learning 2013

The NIC also provides a unique identifier number known as the media access control (MAC) address. The number is built into the electronics of the card. A secondary address, known as the **Internet Protocol (IP)** address (see later), combines with the MAC address to form a logical address, enabling the computer to communicate to any device or computer on the network.

A Byte

A byte refers to a digital word, consisting of an 8-bit binary count, ranging anywhere from 0000 0000 to 1111 1111. Computers "talk" to each other through binary switching. A 0 refers to a logic low, off, or no, whereas a 1 refers to a logic high, on, or yes. The entire word of 8 binary bits can also represent the decimal equivalent of a base-10 number. Because computers can only deal with binary math, any numbers must be ultimately reconverted back to base 10 decimal for the operator. As an example, 0010 1001 represents a decimal equivalent of 41. Bytes are read from right to left. The farthest bit to the right, bit 0, is the least significant digit (LSD), whereas the farthest bit to the left, bit 7, represents the most significant digit (MSD). The base 10 decimal equivalent of each bit in a binary word would look like this: 128(bit 7), 64(bit 6), 32(bit 5), 16(bit 4) 8(bit 3), 4(bit 2), 2(bit 1), 1(bit 0). To obtain the decimal total of a binary word, simply add the decimal equivalent of each bit indicating an ON position, or a value of 1. In our previous example, 0010 1001, add bits 0, 3, and 5 to obtain a decimal equivalent of 41; $1 + 8 + 32 = 41$.

Computer programmers use digital, binary logic to write programs and make computers work. How fast a binary word can clock or switch its way through the computer or the network determines how fast the computer can complete a desired task or even send and receive information on the Internet.

Networking Challenges

The ultimate goal of any network is to have sufficient bandwidth, allowing for fast communication, high performance, and reliability. The original Ethernets of 10 Mbps have mostly been upgraded and improved to the more common levels of 100 Mbps, and now even tens of Gbps, for very-high-speed digital communication, translating to well over 10 billion bps.

Today, the challenge of most networks has more to do with the management, support, and security of many disparate technologies. Different types of equipment, operating at varying speeds, through different styles of communication, across a variety of media types, makes for a continuously changing and demanding marketplace. Networks, therefore, must be reliable, relatively problem free, and yet remain flexible enough to adapt to the continual changes and challenges of the industry. Security cannot be overstated. As more and more networks are interconnected, the protection and security of information systems from outside attack and hacking becomes a central concern to the overall connectivity and reliability of the system. Internal attacks must also not be overlooked; because most security breaches result from insiders who have access to security passes and codes, these breaches can effectively and efficiently undermine the normal everyday operations of any networked system which naively has been assumed to be secure.

Methods of Communication

Multiple possibilities exist for sending and receiving data bits or binary words between any two points of a communication network. Possible communication methods may include a **serial communication** link, a USB, a high-speed **FireWire**®, or a parallel connection. The chosen method typically is based on the size and width of the data highway, and the style and type of connecting hardware available to devices. The data highway refers to the number of physical connections between two points of the communication link, together with the total distance and maximum travel time. Possible communication methods are explained and compared below.

Serial Communication

Serial communication sends data bits one bit at a time between two points in a computer network or system. The process is slow, because binary words cannot be decoded or processed by the system until the entire word has arrived at the opposite end. The process is similar to catching balls out of an automatic pitching machine. Now imagine that each ball has an individual letter written on it; then you are told to catch the next eight balls before you can read the intended message. This is how serial communication works. A serial port and serial cable are shown in Figure 6–9.

At the receive end of the system, data bits are stored into memory locations called registers and reassembled to their maximum readable size. When the entire binary word has been received, it then is sent to another registry location for processing. The previous register then is cleared and made ready for the next available bit stream to be received and assembled. The process ultimately continues

Figure 6–9
Serial connector,
RS-232.

© Cengage Learning 2013

until all arriving data bits have been reassembled into complete binary words and processed for meaning.

Universal Serial Bus

USB is a high-bandwidth data communication standard that can support multiple devices within a system. Figure 6–10 shows a USB port and cable. The maximum number of devices is limited to 127. There are currently three USB standards available for data transmission: USB 1.1, USB 2, and USB 3. The USB 1.1 standard, released in 1998, offered two transmission speeds, low and high. Low speed supports data rates up to 1.5 Mbps and high speed, up to 12 Mbps. The popularity of USB quickly replaced the serial data port as the preferred method of connection to devices. As demand for more transmission speed increased, the USB 2 standard, released in April 2000, pushed the maximum data rate to 480 Mbps. The third and most recent USB standard, USB 3, was released in November 2008. Also known as SuperSpeed, USB 3 was designed for a maximum data rate of 5 Gbps.

Maximum cable lengths for USB 1.1 and USB 2 should not exceed 16.4 ft (5 m). For USB 3, it is recommended that cable lengths not exceed 9.8 ft (3 m).

Power capacity for USB is rated in unit loads. For USB 1.1 and USB 2, 100 mA is defined as a unit load. For USB 3, 150 mA is defined as a unit load. Devices are divided into two categories, low power and high power. Low-power devices operate from one unit load. High-power devices can operate up to a maximum of 5 unit loads for USB 2 and 6 unit loads for USB 3. The maximum current draw

Figure 6–10
Universal serial bus connector.

© Cengage Learning 2013

for USB 2 is 500 mA per port, and for USB 3, 600 mA per port. Devices operating above such levels can draw more power through the use of a dual USB cable that divides the required load between two ports.

FireWire

FireWire, also known as i.Link, is a high-bandwidth, serial communication standard developed by IEEE (IEEE 1394). A FireWire cable is shown in Figure 6–11. The connection allows a maximum of 63 devices, with a data transfer rate of 400 Mbps. Although FireWire is ideal for the transfer of video and audio information, it is also commonly used to connect hard drives and computer storage devices because of its high bandwidth and speed. Maximum cable lengths for FireWire are limited to 14 ft (4.3 m). Any attempt at increased lengths significantly decreases the overall transfer speeds and increases the number of data errors on the bus.

Parallel Communication

Parallel communication sends 8 bits, simultaneously, between two points in a communication system. The total transfer time is therefore eight times faster than serial communication because an entire binary word can be sent all at one time.

The connecting cable for parallel communication is physically larger than a serial cable because a separate pathway or wire is required between the send and receive points of the system for each binary bit. Without the presence of a separate pathway for each bit, multibit transmission would not be possible. A parallel port and cable are shown in Figure 6–12.

Figure 6–11
FireWire cable.

© Cengage Learning 2013

Figure 6–12
Parallel port
connection.

Seven Layers of the Open Systems Interconnection

For a network to be functional, information from one software package on one computer must move through a physical medium to communicate with a software package on a secondary computer. As an example, if a Microsoft Word® document exists on a file server, and a remote computer connected to the network would like to open the existing document, the two computers ultimately have to communicate with each other to transfer the data. Early on, when the computer industry was attempting to iron out details as to how the interconnection and communication would or should take place, many computer manufacturers were attempting to come up with their own proprietary systems, ultimately raising issues of incompatibility. It therefore became obvious early on that whatever level of communication was adopted, there had to be some form of standardization to work from as a way to make the physical interconnectivity of different devices plausible. As a result, a conceptual model, or framework, was developed by the International Organization for Standardization (ISO) in 1984. Subsequently, the **Open Systems Interconnection (OSI)** reference model was created, which is now considered the primary architectural model for intercomputer communications. The OSI reference model divides the movement of communications between computers into seven tasks or layers. A group of tasks then is assigned to each layer, which makes them independently self-contained operations of the overall communication system. Conceptually dividing the model into seven self-contained tasks helps to reduce the complexity of the network. Smaller tasks ultimately are easier to understand and manage. The seven layers are listed as follows:

Layer 7: Application
Layer 6: Presentation
Layer 5: Session

Layer 4: Transport

Layer 3: Network

Layer 2: Data Link

Layer 1: Physical

Keep in mind that the OSI model is only a framework of how communication should take place, and it is not a working system in and of itself; ultimately, manufacturers do not need to conform to such standards. But having the OSI model makes the conceptualization of the network simpler to understand. Engineers can work more efficiently on modular functions, and provided their layer effectively communicates with adjacent layers, the overall interoperability of the system will function reliably, even between inherently dissimilar systems.

Layers 5, 6, and 7 are called the upper layers of the OSI and deal with application- and software-related communication issues. Application layer 7, the highest layer, is closest to the user end of the system. The computer operator interacts directly with layer 7.

Layers 1 through 4 are considered the lower layers of the OSI and deal with the data transport, the physical network, cabling, and hardware. The actual physical connection to the network medium is closest to layer 1, the physical layer. The implementation of communication is accomplished through a combination of software and hardware as the communication works its way through the individual layers. Figure 6–13 illustrates the layers of the OSI model.

Communication between computers is made possible through a set of **protocols**, which are a formal set of rules governing the transfer and exchange of data through the network medium and individual layers of the OSI. A wide variety of protocols exist, such as LAN or WAN protocols, network protocols, or routing protocols. As an example, **router** protocols exist in the network, layer 3 of the OSI. Their purpose is to provide an organized exchange between routers to select the best available pathway for network traffic. Because the actual network appears more as a spider web than a road, there are always multiple pathways that can accomplish the same destination outcome. The router protocols attempt to simplify the data traffic as it moves through the network, helping to prevent crashes or redundancy of information.

Figure 6–13

Open Systems
Interconnection
(OSI) model.

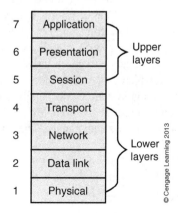

Communication and the Open Systems Interconnection Model

Communication starts at layer 7, the application layer, as the user interacts with a software application, such as e-mail, that has to communicate with other systems resource files, printers, or network resources, or has to simply share resources between network locations. The application layer provides direct interaction of the software package necessary to implement network communications. Its functions typically include the identification of communication partners, determination of available resources, and the synchronization of data flowing back and forth on the network.

For any type of communication to take place, the identity of partners must be clearly known, resources must be made available, and the logical synchronization of data must occur in an organized manner for it to be clearly and intelligibly transferred between desired locations. Let us use a telephone conversation as an example. To make a call, you have to know who you are calling—identification of partners. To be able to make the call, the phone must not already be in use by someone else—available resources. And finally, a clear protocol of speaking and passing information back and forth across the phone line must be implemented so as not to cause confusion (I talk, you listen; you talk, I listen)—synchronization of data. Protocols are implemented at each level of the OSI to help solve these three crucial steps in the communication process.

The exchange of information along the OSI is implemented from one layer to the next, in a top-down, bottom-up method, as information winds its way between communicating parties. If system A wishes to communicate with system B, data transfer from the application layer 7 to presentation layer 6, of system A.

The Presentation Layer

The presentation layer converts the data from a user format to a computer or Internet format. Common data formats are the American Standard Code for Information Interchange (ASCII) and the Extended Binary Coded Decimal Information Code (EBCDIC). As an example, in the ASCII format each key symbol on a standard keyboard has an associated binary computer code for communicating with the computer. From layer 6, the data are transferred to layer 5, the session layer. This layer also is important for encryption, because both systems need to be on the same page when encoding and decoding the data. Back to the phone example: Are we speaking English, German, Spanish, or French?

The Session Layer

The session layer establishes, manages, and terminates the communication session between prospective parties. It can be thought of as the local operator of the telephone system. The session layer is the last level of the upper OSI. From there, the data are transferred to the transport layer, which represents the lower level of the OSI and the physical connection of the computer to the network.

The Transport Layer

The transport layer accepts data from a session and transports it across the network. The transport layer is responsible for dividing the data stream into smaller, addressable chunks, ensuring the data are delivered error free. The transport layer also manages flow control so that multiple parties are not communicating simultaneously across the network. **Transport control protocols (TCPs)** are discussed in more detail in the next section. From the transport layer, data are then sent to the network layer 3. This is where the operator actually sets up and tears down the actual connection. Think of it as the old-style plug-and-pull operator switch stations at the telephone company.

The Network Layer

Network layer 3 of the OSI controls the address, destination, and routing of messages between nodes or locations on the network. Layer 3 also determines the way that the data are sent to a recipient device, such as over an Ethernet or phone line, or through a token ring or FDDI, and provides the IP addresses necessary to achieve such transmissions.

Internet Protocol

IP is a routing protocol that contains the addresses of the sender and the intended network destination. IP addresses can be permanently or temporarily assigned, depending on how the computer is connected to the network. For most dial-up connections through telephone modems, a new IP address is assigned to the user by the Internet Service Provider (ISP) at the start of each session. Corporate connections, however, assign each individual computer on the network a permanent address. The network administrator managing the local network is responsible for setting up the IP addresses for each individual user, or station. To illustrate this point, think about an envelope when you write a letter. The envelope has written on it the name and address of the recipient, as well as the name and address of the sender. IP uses this type of addressing style on every packet sent from a device. The source and destination addressing use either the IP address or network address. The next stage of communication along the OSI is layer 2, data link.

The Data Link Layer

The data link layer provides the reliable transfer of data across the physical network. Depending on the data link layer, the protocols can include physical addressing, network topology, error notification, sequencing and framing of data, and flow control. **Physical addressing**, which is different from network addressing, defines how devices are addressed at the data link layer; it is similar to a zip code as opposed to an actual house address. Network topology refers to the type of physical connection between computers, such as a bus, ring, or star pattern. Topology is discussed further in the next section of this chapter. Error notification alerts the upper layers of the ISO that transmission errors have occurred. Flow control manages the flow of data on the network so as not to overwhelm the network with more data than it can handle. The data link layer communicates next with the physical layer, which represents the last stage between the computer and the network.

The Physical Layer

The physical layer defines the electrical and mechanical functions and protocol specifications for activating, maintaining, and deactivating the physical link between computers. The physical layer is responsible for the following characteristics: voltage levels and the timing of transitions, data rates, and distance of transmission lines and their physical connection. Each bit (0 or 1) is represented by a specific signal value.

Encapsulation

From the physical layer, the information from system A is transferred across the network to system B. Once it arrives at system B, the information works its way up the OSI model, starting at physical layer 1 and progressing through successive layers until reaching application layer 7 and the user of system B. The user of system B can now respond, causing the process to repeat in reverse order, from system B, application layer 7, back down to physical layer 1, across the network to system A, in to physical layer 1 and up the OSI chain to application layer 7, where it originated. This entire process is called *encapsulation*. Figure 6–14 illustrates the

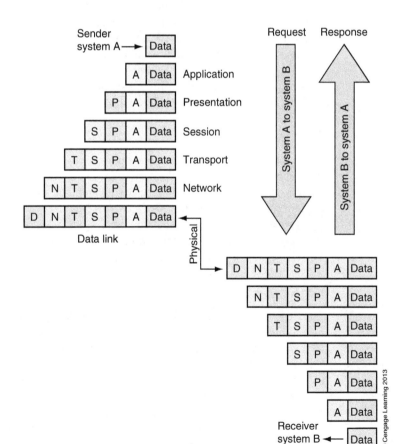

Figure 6–14

Encapsulation of the Open Systems Interconnection (OSI).

encapsulation process. Each layer of the OSI model corresponds to its counterpart on the receiving device. Therefore, the physical layer talks to the physical layer, the data link layer talks to the data link layer, and so on up the line. Information that is encapsulated at layer 7 is not truly exposed or open until it gets up to the receiver's layer 7. A toy many toddlers play with can illustrate the process. Remember from childhood the assortment of five or six different-colored cups that fit inside of each other? Starting with the big cup, you have to place the cups inside of each other; you put one inside the big cup, then another inside of that one, and another inside of that one, until you are out of cups. Simply put, this is what encapsulation does. The littlest cup would then represent the data sent by the application.

Network Topology

LANs are designed to connect a group of computers within a building or group of buildings. Individual workstations usually are connected through a central computer system called a network operating system (NOS), which allows multiple users to access the same software, printer, or even document. Network administrators manage the individual user's rights, allowing access to certain documents by some individuals, while preventing similar access by others. The rights to change or alter a document also must be administered, thus ensuring the security and safety of the system and stored data. As an example, an instructor has rights to post and change student grades on an Internet Web site, but the students only have read-access rights, allowing them to see the grades but not have the permissions necessary to alter any of the posted documentation. The physical geometry of the network structure can take on many forms.

Star Network

A star network connects individual workstations or computers through a central device that acts as a hub to the surrounding satellites. Each of the network connections is considered a node to the central hub. The **star topology** works well for organizations needing a centralized system where multiple workstations must have access to a common communication line. The downside of the star topology is that if the central device were to fail, the entire network would be disabled, preventing the free access and exchange of any data on the system.

Ring Network

The ring network connects each node of the network in series with the next through a connecting cable. Communication is achieved by passing a short message in the form of binary data packets, called a *token*. The token passes continually around the ring, from node to node, in the form of a message relay. When a station wishes to send data, it changes the token to busy and alters the data packets before passing on the message. Each station receiving the token checks to see whether the station is the intended destination; if so, the message is opened and processed. The processed message, now altered, continues along the path, passing from node to node until eventually reaching the original destination. If the intended destination of the message is for a node other than the receiving node, the token then is simply passed on to the next station until eventually making its way back to the originator. The originator, once receiving the reply, releases the token from busy to free, and then passes it on.

Token passing ensures that only one message is on the network at a time. It is an effective way of managing network traffic, helping to prevent congestion or multiple messages in the loop. The downside of TRNs is that the failure of any node on the network crashes the entire system, preventing the return of tokens. Also, every additional node added to the network has the effect of slowing down the entire network. Because each node must examine the token before passing it on, the time it takes for the token to make a complete loop around the ring increases. The originator of the busy token likely sees reply delays as more and more participants are added to the loop.

Bus Network

The bus network connects each node of the network to a common line. It is a broadcast environment where all devices on the bus hear all transmissions simultaneously. There is no central device such as in the star design. A token can still be passed on a bus, but instead of in a ring, the token would be passed up and down the line from computer to computer. Special timing and addressing provisions must also be put into effect so as to prevent multiple messages from being sent along the bus simultaneously. At any time, there should be only one message appearing on the bus. Figure 6–15 illustrates the star, ring, and bus configuration of network topology.

Combinational Topology

Combinational network topologies exist such as the star-wired ring, star-wired bus, or tree. The star-wired ring connects all nodes of the ring through a central hub or multistation access unit (MSAU), which ensures the continual use of the network by all participants in the event of failure by a specific node. In the traditional ring design, a single failure can take down the entire network. With the star-wired ring design, all failures are bypassed by the central hub or MSAU so as to continue the passing of the tokens on to the next destination. Figure 6–16 illustrates the star-wired ring topology configuration.

Another type of combinational topology is the star-wired bus. The star-wired bus does not use a host computer as the master central connection; instead, all nodes are terminated to a central hub or multiport repeater. The distance to or from any node then is considered only one hop away through the central repeater. **Tree topology** is nearly identical to **bus topology**, except multiple branches from nodal points are possible.

Collision Detection

On bus-type networks, the method for preventing multiple transmissions is a process called carrier sense multiple access with collision detection (CSMA/CD). By using CSMA/CD, bus networks can monitor the data bus for data. If the bus is not

Topology

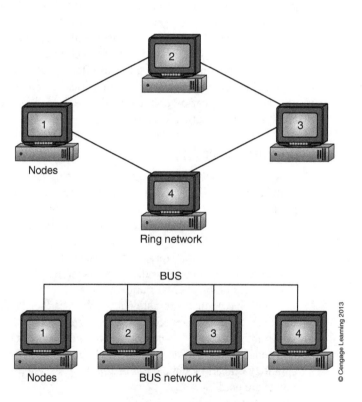

Figure 6–15
Star, ring, and
bus, network
topology.

© Cengage Learning 2013

busy, a data packet is sent out on to the bus having a specified address or destination, and only a node having the correct address can access the data. Problems can occur, however, if one node begins a transmission after another node has already started a transmission. Because it takes a specific amount of time for data to transfer down the line, there is always the possibility for one computer to think there

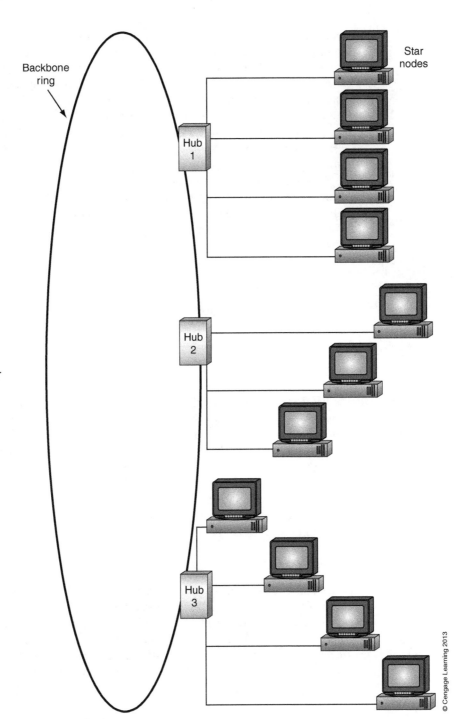

Figure 6–16
Ring-wired star
network.

Backbone
ring

Hub
1

Hub
2

Hub
3

Star
nodes

© Cengage Learning 2013

are no data on the line, when in fact there are, but they have not yet reached the listener. Two or more nodes transmitting simultaneously results in what is termed a *data collision.* A collision immediately corrupts the data on the line and causes errors. At this point, all participants in the collision must stop and wait a predetermined variable amount of time before retransmitting their data; this way, they do not retransmit at the same time. CSMA/CD sets up the required protocols for managing collisions and the transmission and retransmission of data on the bus.

Bus designs using CSMA/CD tend to move data faster because there is no token to pass and participants can simply transmit at will, provided the line is not busy or already transmitting. Collision rates will increase as more nodes are added and bus traffic is heavy, which ultimately can slow down the network as the number of retransmissions caused by errors begins to increase.

Token ring systems inherently avoid collisions; however, they can take longer because each participant must wait for the token to access the network.

Transmission Methods

Communication along a LAN or network can take place in one of two common ways: baseband or broadband. (See *NEC Article 830, Network Powered Broadband Communication Systems,* and *Article 830.2, Definition, Network Interface Unit.*)

Baseband

The baseband method is used by most LANs and Ethernet systems. Transmissions usually consist of high-speed serial packets of digital data along one shared channel, meaning that the combinational transmitted or received signals from the transceiver share the same cable connection. Protocols must be in place to ensure that the transmitter and receiver, residing on the NIC of the computer, are not operating simultaneously; otherwise, data errors occur. Baseband is cheap, easy to connect, and commonly uses either coaxial or twisted pair cables. Typical uses for baseband would be digital voice or data transfers from one computer to the next.

Broadband

Broadband differs from baseband in that it uses multiple channels to communicate. The signals are analog in nature; they transmit and receive on separate radio frequencies, and they must also pass through a specially designed modem to properly negotiate the protocols for the sending and receiving of data packets. A cable modem is a perfect example of broadband modem. Inside a cable modem, frequencies are allocated to different data streams. One frequency is for the digital video, whereas another is for Internet traffic.

National Electrical Code Article 830. Broadband communications are considered asymmetric because they transmit and receive along separate channels at different frequency rates; baseband is considered symmetric because the communication remains at a constant rate for both the transmitter and receiver along the same pair of wires or length of coaxial cable. Typical uses for broadband are data, voice, video, or closed circuit television. Cable television systems inside of hotels are examples of broadband networks, which not only supply customers with in-room movies but also a variety of services and games.

***National Electrical Code Article 830.179* as Related to Broadband Equipment and Cables.** Cable to be used for broadband application must be listed as being suitable for the purpose in which it is being used by a listing laboratory such at Underwriters Laboratories. Cable being used for medium-power network broadband applications must be rated for a minimum of 300 V and should be marked as type BMU, BMR, or BM, with type BMU being suitable for outdoor underground use. Cable being used for low-power network broadband applications must be rated for a minimum of 300 V and should be marked as type BLU, BLX, or BLP, with type BLU being suitable for outdoor underground use.

***National Electrical Code Article 830.90* as Related to Protection.** Primary electrical protection is to be provided on all network broadband communication conductors that are not grounded. Primary protectors must be located within the city block where the service is being provided. Fuseless protectors may be used where the currents on the protected cable are safely limited to its current-carrying capacity. Fused protectors can be used where requirements cannot be met by the means explained earlier. Fused-type protectors must provide a fuse in series with each conductor to be protected. The location should be an approved method in *NEC Article 830.90(B)* and should be within a practical distance from the grounding point.

Unicast, Broadcast, and Multicast

Transmissions along an Ethernet can be of the unicast, broadcast, or multicast variety. Unicast transmissions involve the transmission of data from one individual to another. The connections involve a single transmitting station and a single receiving station. Broadcast transmissions are sent out by a single station to all participants on the network. Broadcasts do not discriminate as to destination or recipient. Multicast transmissions involve a list of select recipients within the broadcast domain. Although there may be a variety of multicast address lists to choose from, multicast transmissions do not have to be limited to a single multicast address. Compared with unicast, broadcast and multicast provide a more efficient means of data transmission across the network.

Velocity of Propagation as Related to Collision and Carrier Sense Multiple Access with Collision Detection

The classical Ethernet originally used a thick, 50 Ω, coaxial cable for connection to the bus. One hundred nodes typically were allowed over a 500 m (1640.4 ft) cable length. The data transferred along the bus at a rate of 10 Mbps. The system was called 10Base-5, which stands for 10 Mbps, baseband, along a 500 m of cable. Thick Ethernet cable is now typically reserved for the backbone of the bus, referring to the main cable run where nodes connect. Smaller networks and clusters of nodes in larger networks usually were connected through thin Ethernet cable, consisting of RG-58/U coaxial cable (RG stands for radio guide and U means universal). Although the thinner coaxial cable was considerably cheaper for the shorter node connections, it was more limiting because the cable runs had to be reduced to lengths of not much more than 200 m (656.2 ft). Such connections were referred to as 10Base-2, which meant 10 Mbps, baseband, along a 200 m (656.2 ft) length. Ethernets later adapted to twisted pair lines, similar to what the telephone industry was using, now known as CAT5 cable, for the connection of 10Base-T systems, which stands for 10 Mbps, baseband, on twisted pair cable. Lengths of 10Base-T connections were restricted to 100 m (328.0 ft) between switches, hubs, or routers. Figure 6–17 shows the comparison between the thick and thin Ethernets.

The reason for cable run limitations on Ethernets or LANs can be found by calculating the velocity of propagation along the cable length, and then relating it back to the CSMA/CD network protocol.

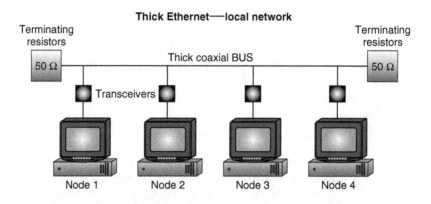

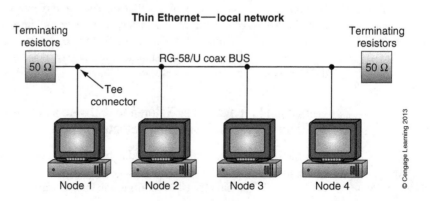

Figure 6–17
Thick Ethernet versus thin Ethernet.

© Cengage Learning 2013

As discussed in Chapter 2, assuming the velocity factor for a typical coaxial cable is 66%, the travel distance over time can then be calculated for any signal traveling along the cable. For example:

$$d = vt$$
$$= v_f ct$$
$$= 0.66 \times 300 \times 10^6 \text{ m/sec} \times 9.6 \text{ } \mu\text{sec}$$
$$= 1900 \text{ m (6234 ft)}$$

where d represents distance traveled, v represents velocity of propagation along the cable (see Chapter 2), t represents propagation time, v_f represents velocity factor, and c represents speed of light in meters per second.

From the above example, imagine that station A has already started transmitting along the network, whereas station B is listening. The signal has not yet reached station B because the distance between stations may be more than 1900 m (6233.6 ft). Station B assumes that no one is transmitting and erroneously begins a second transmission. In the event of such an occurrence, a collision occurs, thereby forcing all transmitting participants to stop what they are doing, discard the now erroneous data on the network, and wait. How long must they wait? The worst-case scenario exists for when nodes A and B are connected on opposite ends of the network. The wait time must therefore be double the normal propagation time because node A or B would not only need to wait out the normal propagation length of the network, but also wait for any error detection to be returned and recognized. Any software error detection would therefore have to consider the round trip for transmitting nodes to recognize that a problem may exist.

If the length of cable is known, then the maximum wait time can be calculated by rearranging the formula. For the next example, let us use an Ethernet bus with a length of 800 m (2624.7 ft), on a coaxial cable with a velocity factor of 66%. The propagation delay is

$$v = v_f c$$

With a given distance, the time can be calculated as

$$t = d/v$$

For an 800 m (2625 ft) cable, the propagation delay is

$$t = 800 \text{ m}/(0.66 \times 300 \times 10^6 \text{ m/sec}) = 4.04 \text{ } \mu\text{sec}$$

The maximum time to detect an error would then be double or 8.08 μsec. With a transmission speed of 10Mbps, this would translate into 80.8 data bits (8.08 μsec × 10 Mbps = 80.8 bits), which is quite fast when you consider the amount of total data being transmitted at any one time; data typically are transferred in the range of 12,000 total bits.

Remember too that it requires 8 bits to make a byte; this means that roughly 10 transmission bytes are involved in the collision and therefore corrupted by the error. In the event of an error, transmitting nodes must stop and wait out the required round trip propagation time. After the wait time, they then again have to listen to the line for any further traffic before attempting the retransmission of

previous data. If an excessive number of errors are detected, communication protocols typically add on additional wait times to specific nodes to attempt to alleviate
the occurrence of continued data crashes.

Network Devices

Networks can be connected through a variety of devices on the physical layer as
a way to either interconnect multiple networks, which would otherwise not communicate, or to extend cable distances beyond maximum limitations. Following is
a list of commonly used devices and their definitions.

Repeater

A **repeater** is a network device that amplifies and retimes incoming data signals
before sending them out to their final destination. As cable lengths increase, signal
quality can become degraded, causing voltages to attenuate or fall off, whereas
data bits may also lose shape, become rounded, or even overly skewed, often
taking on a more triangular shape than the ideal square. Excessive skewing causes
data bits to become blurred, making it more and more difficult for the microprocessor to distinguish between a 0 pulse and a 1. Electrical or atmospheric conditions such as lightning and thunderstorms may also influence the quality of the
signal. Multiple repeaters may be used at various points on long cable runs, as
necessary, to help reshape, reamplify, and retime signals along the network to
ensure error-free transmissions. System timing requirements dictate how many
repeaters may be used on a specific segment of the network.

Hub

A hub connects multiple user stations through multiple ports. The hub does not
make any decision or determination on the data and instead acts as a multiport
repeater. Because the hub functions as a central relay station having a low level of
intelligence, collisions result as more and more ports are connected. This device
fits into layer 1 of the OSI model.

Bridge

A **bridge** is a network device that is used to connect and disconnect computer
clusters operating in two separate domains. For example, consider a situation
where there are 15 computers operating in the accounting department of a college
or university, and another 25 computers operating in the chemistry department
across the street. The basic problem is that the accounting staff does not really
need to know what the chemistry department is up to; likewise, the chemistry
staff does not need to know about the daily operations of the accounting department. For all practical purposes they can probably exist as separate networks
because they rarely need to communicate with each other. The question then
remains: Why should 40 computers exist as a single network, when essentially
they would be much better off operating separately? By operating as smaller,
separate networks, they produce lower levels of network congestion, have fewer

collisions and errors, and ultimately are faster and more efficient. But what happens when accounting has to consult with the chemistry department about an overdue charge for 500 test tubes? At some point, they must communicate with each other, and a network bridge can solve the problem. Figure 6–18 illustrates the concept of a network bridge.

A bridge has the ability to separate computer domains from each other and reconnect as necessary, based on communication requests from either side. For example, if a computer in domain A would like to communicate with any other computer in domain A, the bridge remains off. As a result, domain B is not aware of the additional network traffic flowing in domain A. However, when a computer in domain A wishes to communicate with a computer in domain B, the bridge responds by turning the necessary connection on, thus allowing the transfer of data to flow between the two sides. By having the ability to switch the interconnection of domains on or off, a bridge can directly control and manage the amount of network traffic, and ultimately prevent excessive collisions and possible network errors. For such switching to take place, however, the bridge must be intelligent enough to recognize addresses. Addresses typically are either programmed into a bridge manually by a network administrator or designed to be updated and stored into the internal memory automatically. The type of bridge being used determines the setup

Figure 6–18
A network
bridge.

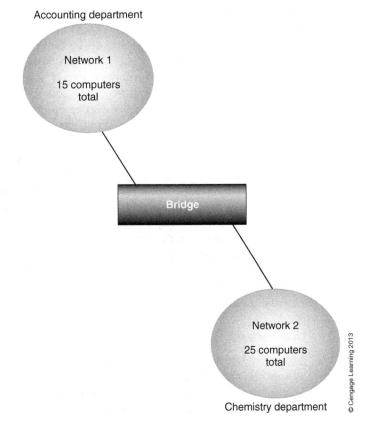

Accounting department

Network 1

15 computers
total

Bridge

Network 2

25 computers
total

Chemistry department

© Cengage Learning 2013

process. As a result, bridges are considered smart devices compared with repeaters or hubs, which lack the overall intelligence to make any kind of network decision as to where data should flow and to whom. These devices are defined at the data link layer of the OSI model. The actual connection switching occurs at this level.

Router

A router is essentially a network bridge that is used to link entirely separate networks. A bridge connects clusters of computers within the same network, whereas a router is used to connect isolated networks. As an example, a router would be used to link separate entities, such as the University of California to the University of Minnesota. The router still operates from a list of programmed addresses; however, the address is targeting a location inside the network layer of the OSI by using IP addressing instead of using the data link layer, where CSMA/DC manages network traffic.

Switches

A switch is an Ethernet or network device that can send incoming data packets to one of several address destinations. A switch can be thought of as a multiport bridge. Some switches have higher level functions that allow a network administrator to turn ports on and off. Switches can also separate LANs so that they act like they are on their own switch, creating a virtual local area network (VLAN). A VLAN has the ability to separate computers into separate networks as a way to impose security restrictions on specific terminals or to help reduce contention between nodes attempting to transmit simultaneously.

There are three varieties of switches available: fragment-free, cut-through, and store-and-forward. The fragment-free style uses the calculation of propagation over the Ethernet to determine which section of a transmitted data frame may be corrupted. This works out to be the first 64 bytes of any Ethernet frame. The cut-through switches operate faster because they forward on data to their destination as soon as the destination address has been determined. The store-and-forward variety stores the entire frame of data in internal memory before sending it on to the required destination. This allows the frame to be checked for errors using a mathematical process. At the end of each frame is a frame check sequence (FCS) that is an answer to the same mathematical process performed when the frame was sent. By comparing the two, the system determines whether the frame has been corrupted during transmission. As a side note, this process is done by all devices within an Ethernet network when a frame enters layer 2 of the OSI model. A switch can operate from data link layer 2. Because a bridge operates on layer 2, essentially a layer 2 switch is a multiport bridge.

LAN Extenders

LAN extenders are remote-access multilayer switches that are used to filter out and forward network traffic coming from a host router. Although they can be used to help filter out unwanted traffic by address, they are unable to segment or create security **firewalls**. Firewalls require passwords and user IDs to gain access to the network.

Gateways

Gateways are used to interconnect incompatible computer systems, e-mail systems, or any network devices that would otherwise be unable to communicate directly with each other. Examples would include Macintosh to IBM PC, or even different varieties of mainframe computer. Another example would be when different LANs have different protocols; for example, a TCP/IP network talking to an IPX/SPX network.

Network Software Management

Once the network has been designed and the topology constructed, the management of network software becomes the primary issue, together with the availability and user rights of shared resources on the network. There are two basic designs for managing network software: peer-to-peer and **client/server**.

Peer-to-Peer Networks

Peer-to-peer networks are meant for smaller systems where each node of the system has equal rights and access to any shared software, resource, or device on the network. Individual nodes are able to control their network availability with respect to what resources may be shared or used by other nodes. The construction of peer-to-peer networks is relatively simple and inexpensive to create. A typical setup includes less than 10 computers.

Client/Server

A client/server network designates one node of the system as the main file server and system resource for all other connecting nodes of the network. The main server typically is dedicated as a network resource, and normally would not be used as a separate individual workstation; this is the only downside of the client/server model. There are, however, many benefits in using client/server networks; they are inherently provide a more orderly and organized model for managing system resources and shared files.

A client/server network also has the ability to monitor system usage and software licensure. Software packages typically are sold based on quantity of maximum users. For example, if specific software requires a maximum of 25 users at any one time, the centralized control of the main server can more easily keep track of and manage client requests, ensuring the proper enforcement of licensure agreements. As a result, users occasionally may find themselves locked out of certain software resources on the network until they are released for use and made available by the main server. Connecting nodes or workstations of the client/server model may also have the ability to share personal resources with each other, from workstation to workstation if necessary.

Multiple software packages residing on the main server must also be able to run simultaneously to various nodes in the network, all based on client usage and demand. The main server must therefore have multitasking abilities to accommodate the various requests. As a result, **operating systems (OSs)** such as Unix or

Microsoft Windows NT® and XP® were designed specifically for the management and daily operations of the client/server network, thus allowing multiple programs to run simultaneously on various nodes without the risk for interference. These OSs are known as network operating systems (NOSs).

The client/server network also has the ability to manage and keep track of alterations to files, thus helping to reduce the possibility of error and confusion. As an example, if a specific file is being updated or changed by a client while other network participants are requesting use of the document, the requesting parties are instead given read-only rights by the server and told that the document is currently locked and unavailable for alteration. Because complete user rights to a document can only be given to one network participant at a time, any requesting participants must ultimately wait for the document to be released back to the main server by the current user before any additional changes can be made. The main server also can permanently lock out rights to certain clients such as students requesting class grades. In such circumstances, the main server allows students to read only current grade data, without giving them the ability to make changes or alterations. Students could, however, have access to personal files where they would have rights to update user information such as address, phone number, and so forth. In all such cases, network participants would have to provide a log-in ID and password to the main server to identify and verify user rights and privileges to available documents and software packages residing in the system.

As stated earlier, client/server networks are inherently more organized and able to handle and manage system resources more effectively than peer-to-peer networks because of the centralized control of the main server.

Point-to-Point Links

Point-to-point links consist of leased lines, often from the telephone company, which are used by network carriers or service providers to connect customers across a WAN to remote networks. Leased lines are quite expensive and more often priced based on transmission length and bandwidth requirements. The benefit of having a point-to-point network link is that the line is not shared. The customer is paying for a dedicated line, which is made available by the network provider 24 hours a day for the continuous transfer of data, as needed.

Circuit Switching

The alternative to expensive, dedicated, leased lines is to connect to network circuitry only as needed and then disconnect when not in use. The concept of circuit switching is similar to making a standard telephone call, but instead using a digital line called an ISDN. A system router makes the connections available, as required by the customer, and then disconnects, or terminates, the connection once the data transfer has been completed. A customer pays less because the network connection is never tied up for any long period. The line is therefore made available to many customers throughout the day, provided that they do not intend to monopolize large amounts of time transmitting or receiving data. The downside of circuit switching is that the connection must be redone every time the address of the destination changes. Another alternative to dedicated leased lines or circuit switching would be to use packet switching.

Packet Switching

Packet switching involves the transfer of datagrams across the network by best available means; that is, the network itself determines the delivery pathway without directly connecting or involving the original sender. Back in the mid-1970s, packet-switching protocols were being developed by an organization known as the Defense Advanced Research Project Agency (DARPA). The main goal of the organization was to develop a system where incompatible computer systems would not only be able to send and receive data, but also would be able to make the transfer regardless of downed lines or intermittent connection across the network or WAN. The U.S. Department of Defense realized early on that they needed to develop a highly redundant and secure data delivery model that would be able to negotiate and quickly sidestep any sporadic outages of power or telephone service, possibly the result of a terrorist or nuclear attack. Such an event of unthinkable magnitude ultimately could have had a crippling effect on the nationwide communication infrastructure, which at the time was largely dependent on the circuit-switching protocols of the analog telephone system. Defense planners therefore envisioned their ideal delivery model as one that would be able to use multiple pathways and connections, at will, through the linking of intelligent hardware, ensuring the safe and guaranteed transfer of information, especially in times of greatest risk or need. As a result, a connectionless data transfer system was soon established.

Connectionless implies that the original sender of a data packet does not ever actually connect to its destination node address. Instead, the originator simply relays a data packet to a network router. The network router then is responsible for negotiating the transfer of data by first storing the data packet, which includes the destination address, within its internal memory, and then sending it on only after having chosen the best available route to the final destination. As a result, packet-switching networks often are called store-and-forward networks. Once the network router has taken control of the data packet, the original sender is released from the connection and remains uninvolved with the rest of the transmission.

Packet switching ultimately was designed as a way to route and reroute datagrams across the network, by best available means, to guarantee their reliable and safe delivery. The odds are remote that a terrorist group could ever effectively disable all the interconnected and cross-connected pathways of a WAN. By using intelligent routing hardware and packet switching, a datagram is virtually guaranteed to reach its final destination, unless the primary connection to the access server or router is severed, which would effectively isolate data packets and permanently prevent them from gaining access to the Internet or WAN. In most cases, the primary access to the Internet is available through either a dial-up telephone connection or a secondary cable provider. So long as the primary Internet connection remains available, any data placed on the network appears to have an almost limitless variety of pathways to choose from while negotiating its route between nodes A and B; this is similar to finding a route from Chicago to California through the network of national roads and highways. Imagine a car intelligent enough to decide on its own how to get from point A to B; the *how* does not matter just as long as the machine is successful. Packet switching accomplishes such an outcome. So what are datagrams?

Datagrams

Datagrams are data packets that have been organized into frames representing compartmentalized sections of binary code, each designed to carry specific information about the transmission, together with the actual data. Each compartmentalized section of a frame is referred to as a field. Figure 6–19 illustrates the concept of a datagram. The example shown is an Ethernet frame. Depending on the type of network protocol being used, the individual fields within the frame may represent all or some of the following transmission details: preamble, used to synchronize the system; start and stop delimiters, indicating when to begin and end a transmission; destination address; source address; data type and length; control information; error check; and the actual data. Each field is limited to a specific size based on its coded purpose. In most cases, the data field occupies the largest part of the packet.

Types of Data Errors

The occurrence of data errors on a digital network may result from one or all of the following: latency, jitter, bandwidth, or lost packets.

Latency

Latency involves the time delays between a transmitting node and a receiving node; that is, the time it takes for data to be processed and transferred across the network.

Jitter

Jitter involves time shifts between data packets, which ultimately can create a shaky or jerky quality, especially in video streams where the necessity for smooth, high-speed data flow is required to recreate the desired image.

Figure 6–19
Datagram.

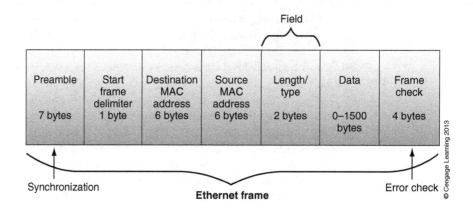

Low Bandwidth

Bandwidth represents the size of the pipeline. Low bandwidth may cause data to be delayed, increase network traffic, and ultimately produce greater rates of collision and error; this is similar to what would happen if a three-lane highway were reduced to a single lane.

Lost Packets

Lost packets occur when a device such as a hub, router, or switch is unable to keep up with the incoming network traffic. Losses occur when data arrives too fast or too often, causing the network device to drop or throw away data packets, especially as internal memory becomes overloaded and unable to store or buffer the incoming stream. The conceptualization of lost packets can be compared with overfilling a glass of milk; eventually, the vessel becomes full, and any further attempts to force more into the container results in overflow and spillage. As network memory becomes overloaded or full, data spillage occurs and packets are lost.

The increasing need for most organizations to achieve greater network bandwidths, higher transmission speeds, and lower error rates is primarily because of current high demands for streaming video and real-time video conferencing, which has now become a daily routine and a necessary business tool for many networked corporations around the world. The convergence of voice, data, and video signals across a single network connection has significantly increased traffic density for all participants, making traffic management and network performance a primary concern of most network providers and analysts.

SEC 6.2 NETWORK PROTOCOLS AND THEIR FUNCTIONS

Network protocols provide three basic functions: They determine the size of the data packet, organize the frame layers, and determine the requirements for data control, helping to provide error-free transmissions. Examples of IPs for data packets include X.25 Packet-Network Protocol, frame relay, asynchronous transfer mode (ATM), and TCP/IP.

X.25

The X.25 Packet-Network Protocol dates back to the 1970s, when the need to develop wide area connectivity across public networks became necessary for packet-switched networks. X.25 was established as a WAN protocol, and structured and modeled around the first three layers of the OSI. The first layer determines the link to the OSI physical layer; the second layer, called the frame layer, links to the data link layer of the OSI; and the third layer, referred to as the packet layer, involves the network layer of the OSI.

There are three general categories of X.25 network devices: data terminal equipment (DTE), data circuit-terminating equipment (DCE), and packet-switching exchange (PSE). DTE corresponds to devices that communicate across the X.25 network as end-system devices. Examples include terminals, network hosts, and personal computers. DCE provides the communication link through devices such as modems and packet switches as a way to interconnect the end-line hardware and the data flow. PSEs provide the switching mechanisms between devices to transfer data across the network.

Frame Relay

Frame relay was developed as a WAN protocol to operate only in the first two layers of the OSI, the physical layer and the data link layer. Originally developed to operate across the ISDN, frame relay often is described as a streamlined version of X.25. Both protocols are similar in that they can each send and receive variable-length data packets and also accommodate statistical multiplexing techniques to help improve access to the network and overall use of bandwidth. Frame relay, however, does not provide a rigorous error check field. For this reason, frame relay should only be used on low-error or low-collision networks. Because frame relay only operates on the first two levels of the OSI, it tends to offer greater efficiency and less overhead than X.25, which requires extra packet fields to accommodate and manage network layer 3.

The actual data fields of the X.25 or frame relay protocols can vary anywhere between 128 and 4096 bytes. Longer data packets are preferred because the ratio of data bytes to control bytes is greater, making the transmission more efficient. Longer data packets are at a slight disadvantage, however, because they require larger sections of internal memory and take more time to transmit across the network from router to router. As a result, transmission delays may be noticeable or even significant as packet sizes increase. Also, notably, data packets do not necessarily arrive at their final destination in their original order. Depending on the limitations of network memory, traffic levels, number of collisions, type of network gateway being used, or even communication between incompatible protocols, such as X.25 and ATM, packets may become scrambled or broken apart, or both, eventually needing to be sorted and reassembled at the final destination. The system therefore requires a greater degree of network intelligence to process and decode the flow of data. For communication to occur, data packets ultimately must arrive at their final destination in the correct order and in the correct format.

Frame relay falls into two general categories: DTE and DCE. But unlike X.25, frame relay does not use PSEs because it is unable to accommodate network layer 3 of the OSI.

Asynchronous Transfer Mode

ATM is used for high-speed data transmissions of voice data and video, well into the gigabits per second, typically over long-distance fiber-optic lines. ATM is well suited for the delivery of broadband over the ISDN. The data are sent out in small 53-byte packets, having a speed advantage over X.25 and frame relay because the smaller, shorter bursts of data packets use up less memory and travel faster across the network. The ATM architecture only corresponds to the physical layer of the OSI and part of the data link layer, which is why it is able to provide greater network speeds and efficiencies over X.25 and frame relay. Transmitting ATM through fiber-optic cable also is a benefit because fiber provides superior quality over long-distance runs, with a minimal amount of loss or error. The greater transmission speeds of ATM allow for real-time transmissions of live digital video and audio streams. The advantage of not having to wait as long for data to be received, sorted, and reassembled significantly increases the real-time quality of the broadcast signal, allowing for greater levels of detail and clarity in the final product.

Transfer Control Protocol/Internet Protocol and Related Protocols

Two of the most commonly used protocols for accessing the Internet and World Wide Web (WWW) are TCP and IP. TCP/IP works as a suite of protocols providing connectivity among communicating systems across the Internet. The protocols originally were developed by the U.S. military in 1969 for use on a four-node network referred to as ARPANET (Advanced Research Projects Agency Network). Since then, they have grown significantly beyond their initial conception into a much broader range of protocols that currently have the ability and flexibility to handle a wide variety of applications. Ultimately, these protocols provide the communication link among any similar or dissimilar mediums and data structures, which makes TCP/IP extremely reliable. Ultimately, these protocols provide the communication link among any similar or dissimilar mediums and data structures, which makes TCP/IP extremely reliable. To further expand on this topic, a description of the structural basics of TCP/IP is first required, and then we can review a variety of other protocols which may be required to operate within the TCP/IP framework.

Transfer Control Protocol/Internet Protocol

TCP/IP forms a hierarchy of protocols; it is a protocol suite, similar to the OSI, but centered on a four-layer model instead of seven layers. The four layers of TCP/IP are as follows:

Layer 4: Application/process

Layer 3: Transport

Layer 2: Internet

Layer 1: Network access interface

Note that the top layer (application/process) has combined the session, presentation, and application layers into a single entity entitled Application and Process. The transport layer and Internet layer of the TCP/IP suite map directly to the transport layer and network layer of the OSI model, respectively. The last layer of the TCIP/IP suite maps to the bottom two layers of the OSI model (data link and physical).

The functionality of TCP/IP is split between layers 3 and 4 of the OSI. IP starts working on network layer 3, whereas TCP resides in transport layer 4. Because TCP/IP operates above the physical and data link layer, any compatibility issues between dissimilar mediums or data systems become a nonissue. As a result, TCP/IP has the ability to provide linkage to any communication system, as long as the physical and data link layers are properly designed and working appropriately. Therefore, the type of system or data becomes irrelevant because TCP/IP does not concern itself with layers 1 or 2; ultimately, it can provide linkage to the Internet through a variety of network connections and over any style of LAN or WAN.

Internet Protocol

IP provides two main functions to the Internet: It sends out continuous streams of datagrams to the required network nodes, and it also manages the removal of lost or recirculation packets in the system. IP is considered connectionless because it does not concern itself with the results of a transmission, nor does it care about the order or error checking of data packets. Datagrams are simply moved through the system without keeping track of their final outcome as they propagate from node to node; any error checking, sorting, or data rearranging eventually is taken care of in the transport layer by TCP.

The management of lost packets becomes a secondary function of IP. Lost packets, if not properly managed, potentially could migrate around the network forever, without ever arriving at their final destination. Remember that data packets do not necessarily take a direct route across the network, nor are they required to take the same route each time. For this reason, there can be a great number of packets, broken apart and out of order, continuing to circulate the network endlessly. Lost packets, if not properly controlled, ultimately can increase the level of network traffic, resulting in additional congestion, loss of bandwidth, greater rates of collision, and error. To solve this problem, IP has been given the ability to control a data field inside the packet frame referred to as the time-to-live count. As a packet attempts to "hop" from one node to the next, IP reduces this count by one. Once the time-to-live count has reached zero, the system automatically discards and terminates the packet.

Transfer Control Protocol

TCP operates from the transport layer at both ends of the communication link. Its capabilities include data stream transfer management, flow control, full-duplex operation, and communication multiplexing. TCP is mainly responsible for the sorting and reassembling of data packets; error correction; and the detecting, requesting, and resending of lost packets. Full-duplex operation means that TCP can send and receive packets simultaneously, whereas multiplexing refers to the ability to communicate multiple conversations simultaneously over a single connection. Because TCP is not operating at the network level, it does not concern itself with the destination routes of the various packets; instead, it simply operates from the end points of the communication link, spending the majority of its time sorting and reorganizing packets as they arrive. TCP ensures that all packets have been received and are in order before handing them up to the next layer. If data errors or lost packets are detected, then the retransmission of specific bytes is requested through means of acknowledgments. The sending computer waits for confirmation of receipt; if the confirmation does not arrive, then the transmission is sent again. This process does not happen forever; there is a time associated with retransmission called a "time out" that tells the sender when to stop trying. If a sending computer does not receive a confirmation known as an acknowledgment and the time-out value has expired, an error code is reported to the upper layers of the sending computer. Typically, the error is related to "connection lost." TCP is the connection-oriented protocol in the TCP/IP suite; that is, it is responsible for ensuring a connection is established between hosts. The connection is maintained by the acknowledgments during a transmission. When paired with the ability of IP, the communication can traverse many different networks to a destination.

User Datagram Protocol

User Datagram Protocol (UDP) is a connectionless protocol that does not send back any type of acknowledgment to the originating transmitter regarding the status of data packets. A UDP transmission therefore must rely on upper-layer protocols for reliability because data checks are not built into the communication.

Port numbers are assigned to TCP and UDP segments to distinguish the presence of multiple applications running from a single device. The port number contained within the TCP or UDP data segment is used to identify the application to which the data packets belong. Standardized port numbers have been assigned to applications to keep the level of confusion to a minimum so that differing and multiple implementations of the TCP/IP suite can interact and operate together.

Application and Process Layer

The application and process layer provides all the necessary protocols needed for a Web browser, such as Internet Explorer, to communicate with a Web server. The following items are examples of application and process layer protocols: Hypertext Transport Protocol (HTTP), **File Transfer Protocol (FTP)**, Telnet, **Simple Mail Transport Protocol (SMTP)**, and **Post Office Protocol 3 (POP3)**.

Hypertext Transport Protocol. Network Web browsers, such as Internet Explorer, Safari, and Mozilla Firefox, are all examples of application programs that use HTTP to communicate across the Internet. HTTP communicates through a programming language known as HTML, or hypertext markup language. HTML has all the functionality of a word processor and also allows links to other Web pages residing on the same or even different servers. HTML essentially provides an Internet Web-browsing program with the ability to read and display text files and graphics.

Although HTML may be suitable for reading or displaying Internet text or graphics on a Web browser, there are often times when a Web user needs to interact with a specific site rather than just be a passive observer. Two additional programming languages exist, Java and Flash, each of which offers its own level of specialization and functionality to Web browsing through HTTP.

Java was developed as an interactive Web-browsing programming language. Because HTML can only display the contents of a Web page, Java allows the user the ability to fully interact with or even make changes to the contents of the site.

A common example would be an online shopping page where buyers would be required to enter in their name, address, and payment option to purchase a desired item. Without the ability to interact with the Web site, online ordering would not be possible.

Flash takes Java to another level by providing the functionality of interactive Web browsing and also allowing for the design and presentation of multimedia-type programming across the Internet. Flash has the ability to produce high-quality graphics and animations, including audio and video, which then can be displayed and played by visitors to a Web site. Although Java has the ability to create some graphics and simple animations, the overall process would be quite daunting and cumbersome to programmers if they attempted to design a Web page as fully inter-active as one created with Flash. Flash, being more highly adaptable and functional than Java, is simply a better programming tool, especially when a multimedia experience on the Internet is desired.

File Transfer Protocol. FTP is often used to upload or download files to or from network servers. FTP is built into most Unix- and Windows-based OSs and Web browsers. When files need to be transferred between servers or network nodes, FTP is engaged to perform the desired task and often is transparent to the user. FTP is how files move around the network or Internet.

Telnet. Telnet was designed as a communication link between a server or main-frame computer and a remote terminal, primarily to perform administrative tasks such as turning switches on or off, configuring routers and hubs, or setting user permissions, just to name a few. By having the ability to log in and alter network parameters from a remote location, the network administrator has the freedom to control the system from anywhere, through the use of a single keyboard connection and monitor. The process, however, is not user friendly and demands a high level of command line knowledge and insight into computer code to be able to commu-nicate effectively with the main server. Telnet commands look much like the old disk operating system (DOS) or present-day Unix/Linux commands. Each com-mand must be typed in separately, and the user must then wait for a response from the main server or device before attempting to transmit any additional commands.

Although Telnet is still available, allowing individuals the ability to access, change, and move data from a remote location, in recent years it has been losing popularity due to the more graphically based and secure protocols on the Internet. Keep in mind that Telnet was originally developed back when network connections were more text-based, connection-oriented systems. The use of flashy graphic interfaces and connectionless packet switching was somewhat nonexistent because the main priority of the day was simple two-way communication and information transfer, modeled after the public telephone system. The term *Telnet* is actually a shortened form of *tel*ephone *net*work. It also would have been considered a great luxury of the time to use higher level graphic displays because they would have required far more memory to operate, gobbling up precious network bandwidth in the process, and ultimately bogging down the entire system. As computer speeds increased substantially from the low megahertz of the past to current levels of 1 GHz or more, the availability of reliable, more affordable memory chips became a reality, allowing the higher level graphic interfaces of HTTP to flourish. The landscape of computer networking was permanently changed once users were able to choose between the high learning curve and burdensome code aspects of DOS and UNIX, and the simplicity of point-and-click, graphical user interface-based operating systems and Internet Web browsers.

Simple Mail Transport Protocol. SMTP allows for the sending of e-mail and messages from a user station to an e-mail server. The process occurs in one direction. As a result, SMTP cannot request mail, only send it. To retrieve mail, the user must use POP3.

Post Office Protocol 3. POP3 defines the method and protocol for retrieving e-mail from an e-mail server; both protocols, POP3 and SMTP, work together to send and receive electronic messages across the Internet. Figure 6–20 illustrates the process.

There are also a variety of other protocols, which are explained briefly below.

Simple Network Transfer Protocol. Simple Network Transfer Protocol (SNTP) exists as a means of monitoring computers, routers, switches, and network usage.

Figure 6–20
E-mail protocols.

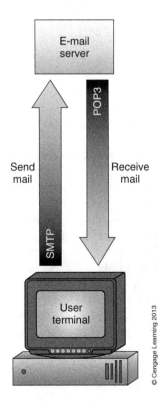

Internet Control Message Protocol. Internet Control Message Protocol (ICMP) is used for IP errors and control messages.

Address Resolution Protocol. Address Resolution Protocol (ARP) is used to map IP addresses to MAC addresses in the physical layer (layer 1 of the OSI).

Internet Mail Access Protocol. An alternative to POP3 is Internet Mail Access Protocol (IMAP). IMAP adds features and abilities which allow a user to manipulate and search messages that are still on the server, rather than simply retrieve them.

Applications using TCP/IP are constantly being developed as the growth and emergence of the Internet continues. Teleconferencing, voiceover IP, streaming audio, and video have all become commonplace in the network environment, and new applications are appearing daily to take the place of the older, outdated modes of communication. The list of protocols shown here for TCP/IP simply represents the current, most commonly used variety. Remember that the strength of the Internet resides in its versatility and the power of those connectionless data packets, which ultimately can adapt to any change as the growth of Internet usage moves forward into uncharted territories.

SEC 6.3   COMPUTER NETWORK ADDRESSING

Internet Addressing

All available resources or files on the Internet must have a unique address known as a **uniform resource locator (URL)**. The URL provides the necessary protocols for finding and accessing Web sites or information files residing on the Internet. URLs are registered through network providers to ensure that they are unique and available for use, because two individual Web sites are not allowed to have the same URL.

An Internet computer address using Internet protocol version 4 (IPv4) consists of a 32-bit binary number, broken into 4 groups of 8 bits called octets, specifying the network it belongs to and its host address on that network. The 4 groups of 8 bits are displayed in base-10 format and expressed as dotted decimal. The following number represents an example of a possible Internet address: 145.22.78.82. Mathematically, the maximum number of possible combinations for a 32-bit address is 4,294,967,296, which is equal to 2^{32}, 32 binary characters, 2 states each. This may seem like a lot, but as the size of the Internet continues to grow at a rapid pace, available address space eventually will start to fall short of demand. To help alleviate this problem, Internet protocol version 6 (IPv6) has increased the allocated size of the address field to 128 bits, thus ensuring the continued growth of the Internet for many years to come.

To make Internet addressing easier to remember, URLs are written in word format rather than using a seemingly meaningless series of numbers. For example, let us say the above Internet address, 145.22.78.82, represents the location on the

Web for a local pet shop. Instead of having to remember the four segments of numbers, the URL can be coded as http://www.petshop.com/newpets/blacklabs. html. For this process to work, however, there has to be a conversion between the words and the numbers. The translation is accomplished by a network computer called a **Domain Name Server (DNS)**. The DNS is updated regularly with a current list of computer addresses presently connected to its network. Because a domain server cannot possibly have the address list for every computer connected to the Internet, there must then be a process for finding or searching out unknown computers. This process is defined by the Domain Name System; both the Domain Name Server and Domain Name System are referred to as DNS. They really are the same, except they have different names. The overall system is defined to convert or resolve friendly computer names to IP addresses. Each DNS server has the ability to communicate with other DNS servers. All organizational LANs contain a DNS server for outside access to the local names. A request for resolving the name of an IP address from an outside computer is answered by the DNS server. If the name-to-IP mapping is unknown, then a process of recursion can be used where the DNS server asks other DNS servers whether they have the mapping; all this is done on behalf of the computer that is asking. Most organizations point their DNS servers to the Internet's root DNS servers, which can resolve names for the entire Internet space.

The fictitious address for the pet store, http://www.petshop.com/newpets/ blacklabs.html, breaks down in the following manner: "http://" (or sometimes possibly ftp://) represents the protocol resource to be used when communicating; "www" represents the pointer to a computer or a resource on the World Wide Web; "petshop.com" represents the domain name of the desired resource; and the remainder of the address, "/newpets/blacklabs.html," represents the hierarchical name or pointer to a specific file (in this case, blacklabs.html) located on the pet store server. The URL also could point to CGI programs, Java or Flash programs, graphic files, streaming media, or any other available Internet resources.

Firewalls

Often, private organizations must be connected to the Internet but they do not want to have to worry about outside networks or users accessing confidential files or restricted parts of their internal network, or intranet. The intranet represents an internal network of a private company or organization that can only be accessed by individuals internal to the organization or by those having recognizable user names, log-in IDs, and passwords. This usually is accomplished through a special type of software referred to as a firewall. A firewall provides an individual access to a private intranet only if the desired authentification codes are provided. Firewalls also are necessary to help prevent computer hackers from gaining access to and control of a network or a computer for malicious or destructive purposes, such as the theft of valuable information or planting of viruses. The firewall creates a two-way controlled environment where individuals can communicate safely, inside and outside of the organization, while also preventing access to unauthorized individuals.

SEC 6.4   MEDIA: CONNECTIONS, HARDWARE, AND INSTALLATION TECHNIQUES

Media Types and Evolution of the Ethernet

Media type categorizes the cabling used to send data between two points in a network. As the Ethernet evolved from its original 1 Mbps to the current levels, well into the gigabits, the type of cabling used to connect either the main network backbone or a hub to individual work stations also needed to evolve. Increased data speeds and the physical number of data packets being transmitted over the network, and later the Internet, meant that cabling had to improve together with network topology, OSs, and communication protocols to accommodate greater frequencies. The table listed in Figure 6–21 illustrates the

Figure 6–21

Common network cable specifications. MMF, multimode fiber; SMF, single-mode fiber.

Flavor	Cable type	Segment length* (meters)	Year of IEEE ratification	Speed
1BASE-5	UTP	250 (820.2 ft)	1987	1 Mbps
10BASE-5	Thick coax	500 (1640.4 ft)	1983	10 Mbps
10BASE-2	Thin coax	185 (606.9 ft)	1985	(Ethernet)
10BROAD-36	Coax (broadband)	3600 (11,811.0 ft)	1985	
10BASE-T	Cat 3 UTP	100 (328.0 ft)	1990	
10BASE-FP	Fiber	1000 (3280.8 ft)	1993	
10BASE-FB	Fiber	2000 (6561.7 ft)	1993	
10BASE-FL	Fiber	2000 (6561.7 ft)	1993	
100BASE-TX	Cat 5 UTP	100 (328 ft)	1995	100 Mbps
100BASE-T4	Cat 3 UTP (4 pairs)	100 (328 ft)	1995	(Fast Ethernet)
100BASE-T2	Cat 3 UTP (2 pairs)	100 (328 ft)	1997	
100BASE-FX	Fiber	2000 (6561.7 ft)	1998	
1000BASE-T	Cat 5 UTP	100 (328 ft)	1999	1000 Mbps
1000BASE-CX	Coax	25 (82.0 ft)	1998	(Gigabit Ethernet)
1000BASE-SX	62.5/125 Fiber	260 (8530.0 ft)	1998	
(850 nm)	50/125 Fiber	525 (1722.4 m)	1998	
1000BASE-LX	62.5/125 Fiber	550 (1804.4 m)	1998	
(1300 nm)	50/125 Fiber	550 (1804.4 m)	1998	
	SMF	3000 (9843 ft)	1998	
10GBASE-SR	MMF (850 nm serial)	65 (213.2 ft)	2002	10 Gbps LAN
10GBASE-LX4	MMF (1300 nm WDM)	300 (984.2 m)	2002	(10-Gigabit Ethernet)
10GBASE-LR	SMF (1310 nm serial)	10,000 (32,808.3)	2002	
10GBASE-ER	SMF (1550 nm serial)	40,000 (131,233.6)	2002	
10GBASE-SW	MMF (850 nm serial)	65 (213.2 ft)	2002	10 Gbps WAN
10GBASE-LW	SMF (1310 nm serial)	10,000 (32,808.3 ft)	2002	(10-Gigabit Ethernet)
10GBASE-EW	SMF (1550 nm serial)	40,000 (131,233.6 ft)	2002	

Single mode optical fiber or multi-mode fiber

*Segment length is the backbone cable length for 10BASE-5 and -2. For others, it is the distance from hub to station (or other attached device). Gigabit ethernet distances over fiber depend on the type of fiber and transmitter used. Specifications for 10-gigabit Ethernet are preliminary and not yet approved by standard-setting groups.

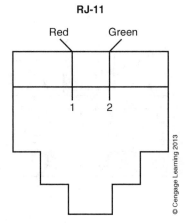

Figure 6–22
RJ-11 connector.

evolution of the Ethernet through the years. It concerns the type of cabling, type of network, transmission speed, and cable length. Fiber-optic cable is discussed in more detail in Chapter 10.

Modular Wire-Mapping Options

The categories of twisted pair cable and their specifications are discussed in Chapter 2. At this point, we need to discuss the modular wire-mapping options available for RJ-11, RJ-14, RJ-45, and RJ-61X connectors.

RJ-11 is commonly used as a single-channel, modular, telephone connection. The red and green wires, corresponding to line one, are connected as shown in Figure 6–22.

The RJ-14 connector, shown in Figure 6-23, was developed as a two-line modular telephone connection, using all four wires on the telephone line: red, green, black, and yellow. Notice that the red and green wires are occupying the same center location of the connector as in the RJ-11, whereas the black and yellow wires are placed on either side. This allows for compatibility in the event an

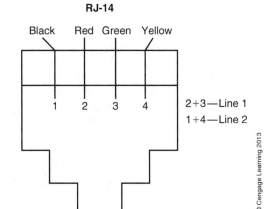

Figure 6–23
RJ-14 connector.

Figure 6–24

Electronic Industries Alliance/ Telecommunications Industry Association (EIA/ TIA) 568A RJ-45 connector configuration.

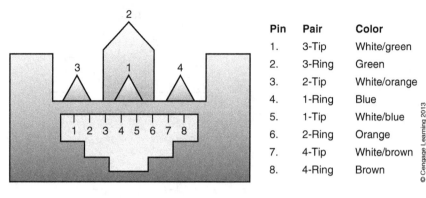

Pin	Pair	Color
1.	3-Tip	White/green
2.	3-Ring	Green
3.	2-Tip	White/orange
4.	1-Ring	Blue
5.	1-Tip	White/blue
6.	2-Ring	Orange
7.	4-Tip	White/brown
8.	4-Ring	Brown

© Cengage Learning 2013

RJ-14 connector is ever connected to a single-line telephone; the red and green wires always make the proper connection, thus always guaranteeing the use of line one.

The RJ-45 module was developed as an eight-pin connector, to be used with four-pair CAT3, CAT5, or CAT5e cable. There are two wiring options for the RJ-45 connector: Electronic Industries Alliance/Telecommunications Industry Association (EIA/TIA) 568A or 568B. Figure 6–24 illustrates the 568A configuration; Figure 6–25 illustrates the 568B configuration.

The only difference between the two configurations is the positioning of pairs 2 and 3, which are swapped. Installations requiring AT&T specifications commonly specify the 568B wire-mapping method, whereas most other organizations will use the 568A configuration. It is important to know which connection is being used to properly interconnect equipment. Because equipment on one side of the cable may specify 568A, and equipment on the other end may specify 568B, problems may result if the installer is not observant or aware of the difference; such a mistake would result in confusion and miscommunication on the network.

RJ-61X is another eight-pin, four-pair design, referred to as USOC (Universal Service Order Code), which was set up by the Federal Communications Commission

Figure 6–25

Electronic Industries Alliance/Telecommunications Industry Association (EIA/TIA) 568B RJ-45 connector configuration.

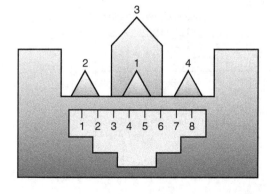

Pin	Pair	Color
1.	2-Tip	White/orange
2.	2-Ring	Orange
3.	3-Tip	White/green
4.	1-Ring	Blue
5.	1-Tip	White/blue
6.	3-Ring	Green
7.	4-Tip	White/brown
8.	4-Ring	Brown

© Cengage Learning 2013

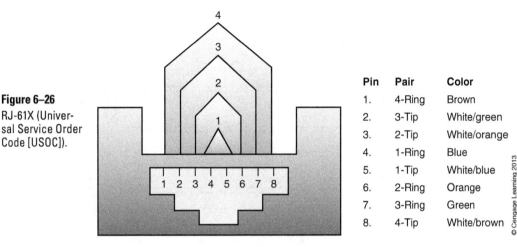

Figure 6–26

RJ-61X (Universal Service Order Code [USOC]).

Pin	Pair	Color
1.	4-Ring	Brown
2.	3-Tip	White/green
3.	2-Tip	White/orange
4.	1-Ring	Blue
5.	1-Tip	White/blue
6.	2-Ring	Orange
7.	3-Ring	Green
8.	4-Tip	White/brown

© Cengage Learning 2013

for voice and low-speed data transmission. Figure 6–26 illustrates the wire mapping of the connector. Notice how the pairs stack over each other. As a result, such a wiring configuration results in far more crosstalk and interference if used on a high-speed network, which is why USOC has been relegated to low-frequency transmissions and slow-speed data applications.

Connector Troubleshooting

The troubleshooting of connectors usually concerns the following: shorted or open lines, reversed pairs (red being where green should be, and green being where red should be), transposed or crossed pairs (pair 1 occupying the position of pair 2, and vise versa), or split pairs (a wire from pair 1 is swapped with a wire from pair 2). Any of these misconnections results in a faulty transmission.

The testing of terminated cables, once connectors have been assembled, usually can be accomplished through the use of line beepers to verify pair mapping from end to end, or by connecting automated loss testers, which not only check for proper conductor mapping, but also record cable length, signal attenuation, near-end crosstalk (NEXT), and far-end crosstalk (FEXT). Figure 6–27 illustrates the concept of NEXT and FEXT. Cable ties and clamps that are too tight may also cause signal degradation by changing the characteristic impedance of the cable, increasing levels of crosstalk, or if severe, actually causing internal shorts or opens along the length of the cable.

The following issues should be considered when performing any type of installation:

1. Always use more cable than needed. Leave plenty of slack, and be sure to include a service loop to be able to connect to the line at a future date, if necessary.

2. Test every part of a network as you install it. Even if it is brand new, it may have problems that will be difficult to isolate later.

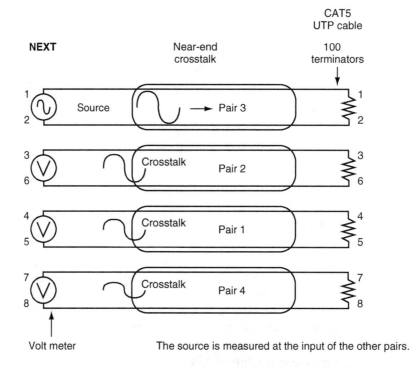

Figure 6–27
Near-end cross-talk (NEXT) versus far-end crosstalk (FEXT).

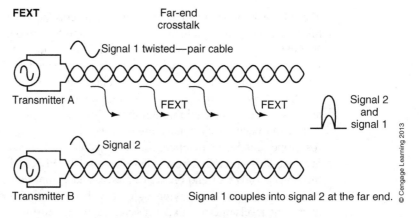

3. Stay at least 3 ft (0.9 m) away from fluorescent light boxes and other sources of electrical interference.

4. If the cable must run across the floor, be sure to cover it with an approved, durable cable protector; also be sure that the cable is properly rated for the specific location of installation.

5. Label both ends of each cable. Any unused cables must be tagged for future use, or any accessible portions of the cable must be removed from the building.

6. Use cable ties (not tape), wiring harnesses, or any style of listed wiring supports, trays, or raceways to help better organize cables appropriately in a desired location.

7. Do not put any undo stress or strain on the conductors or cables and be sure they are clear of any doors or moving devices.

8. Be neat and organized.

SEC 6.5 OTHER CONCERNS

Device Drivers

Almost all new hardware requires you to install a device driver. A device driver is an interactive program that tells the OS how to control, or "drive," the device or hardware. For example, if you physically install a new CD-ROM drive in your computer, chances are it will not work once you boot up the computer. This is because you have not installed the information to tell the computer to look for this particular device. The information that is needed should be packaged together with the CD-ROM drive and needs to be installed once the drive is physically in place. The driver software tells the computer to look for the CD-ROM drive and identify it for use. Currently, most OSs are what we call "plug and play"; that is, you can pretty much plug in any device and the computer recognizes it. The latest version of Windows comes with thousands of device drivers built into the OS software; thus hardware installation has become much quicker and easier.

Ping

Ping is a standard troubleshooting utility available on most NOSs. Besides determining whether a remote computer is "alive," ping also indicates something about the speed and reliability of a network connection. Ping is like the game of Ping-Pong (table tennis); you hit the ball to the other side and it should be hit back. When you use ping, you are actually sending a request directly to another computer simply to respond.

Operating Systems

The OS is the focus of a computing experience. It is the first software you see when you boot up the computer, the software that guides you through your applications and sees that the computer is safely shut down when you are ready to end your session. Most contemporary OSs use graphical user interfaces (GUIs), meaning they use pictures and icons to interact with the user. The OS is the software that allows programs to be launched and used together while organizing and controlling the hardware.

Of course, not all computers have OSs. The computers that control operations in your car do not need a user interface such as an OS. The most common of all OSs is the Windows family of interfaces.

OSs can be grouped generally into four different groups based on the types of computers they are working for and the types of software application they support.

The first group is the real-time operating system (RTOS). These operating systems are used specifically in industrial settings to run controls or for instrumentation purposes. This operating system has little interaction with the user because it is running a specific task with a specific outcome in mind. An RTOS executes an operation in a given amount of time over and over again, making it almost application-like in itself.

The second group is the single-user, single-task OS. This OS allows only one user to do one thing at a time. A good example of this OS is usually found on a handheld computer.

The third group is the single-user, multitasking OS. This OS allows one user to do many different tasks virtually at the same time. This OS typically is found on a desktop or laptop computer, where more multiple processes run simultaneously. For example, an OS that allows the user to print a word-processing document while surfing the Web would qualify as a multitasking OS.

The fourth group is the multiuser, multitasking OS. It allows multiple users to access a computer's resources at the same time. To ensure proper operation, this OS comes with a series of checks and balances that automatically "balance" out the system so that a problem with one user does not affect the entire network of users. Examples of this style of OS are Unix, Windows XP®, Windows Vista, Windows 7, and Windows server 2008®.

Most OSs running network applications today require multitasking and networking capabilities.

SEC 6.6   LOW-VOLTAGE RESIDENTIAL NETWORK APPLICATIONS

Applications for networked low-voltage systems in a residential location include PCs, stereo gear, telephones, automated lighting, intercoms, surveillance cameras, and security and alarm systems. This is obviously not a complete list but illustrates many of the most commonly applied applications.

The term *structured wiring* now exists, which refers to the integration and interconnection of various low-voltage, networked devices, all of which may ultimately communicate through a centrally located household computer. By providing special (multimedia and control) outlets and terminations that consolidate telephone jacks and data jacks together with video, audio, and control cables, into a single, unified housing, the future management, maintenance, troubleshooting, and upgrade of such complex systems can be more easily accomplished. The inherent logic of terminating all convergent systems to a centrally located junction box should be obvious.

CHAPTER 6 FINAL QUESTIONS

1. What is the difference between a LAN and a WAN?
2. What is the purpose of the network interface card (NIC)?
3. How does a dial-up modem compare to DSL?
   a. DSL cannot access the Internet if the telephone line is being used for voice transmission.
   b. DSL can operate up to 50 times faster than a dial-up modem and share the telephone line.
   c. Dial-up modems provide a filter to separate out the voice signal from the digital data.
   d. Dial-up modems provide wider bandwidths than DSL.
4. Compare a dial-up modem to ISDN. Which has the ability to share the line simultaneously with a telephone, fax, and computer?
5. Why may it not be possible to achieve maximum speeds when connected to a dial-up modem or DSL?
6. What is the difference between symmetrical and asymmetrical DSL?
   a. Asymmetrical offers the same upstream and downstream speeds.
   b. Symmetrical offers the same upstream and downstream speeds.
   c. Symmetrical is faster than asymmetrical.
   d. Asymmetrical is faster than symmetrical.
7. A token ring network topology _____.
   a. only sends specific data to network nodes based on the address of the workstation
   b. operates from a dedicated system of data relay, from one network node to the next
   c. is the most secure style of network topology
   d. broadcasts the same data to everyone on the network simultaneously
8. Ultimately, the speed, performance, and reliability of a network depends on the _____.
   a. number of bits per second being transmitted
   b. operating system of the workstation
   c. amount of bandwidth
   d. size of the data file
9. How many bits are there in a byte?
   a. 4
   b. 8
   c. 16
   d. 32

10. In the binary number 1101 0110, which number is the LSD? Which number is the MSD? What is the decimal equivalent?

11. Which layers of the OSI deal with the data transport, physical network cabling, and hardware?

    a. 1–4

    b. 5–7

12. The layer of the OSI that is closest to the user end of the system is

    _____.

13. What is the purpose of a network protocol?

    a. To define the size of the data packet, organize the data layers, and provide data control

    b. To determine the physical connection to the Internet and the number of network nodes

    c. To determine the level of network security between network nodes

    d. To determine the type of transmission on the network, such as uni-cast, broadcast, or multicast

14. A successful communication link requires the identification of commu-nication partners, a determination of available network resources, and

    _____.

    a. high network bandwidth

    b. a point-to-point link

    c. the synchronization of data over the network

    d. a client/server network

15. Which layer of the OSI converts the data from a user format to a computer or Internet format?

    a. Data Link

    b. Network

    c. Application

    d. Presentation

16. Which layer of the OSI manages and terminates communication between parties?

    a. Data Link

    b. Session

    c. Presentation

    d. Transport

17. Which layer of the OSI assigns IP addresses?

    a. Data Link

    b. Network

    c. Transport

    d. Session

18. Which network topology causes the entire network to fail if the central computer goes down?

19. Which transmission method uses multiple channels to communicate?
    a. Baseband
    b. Broadband

20. Which is faster: serial or parallel communication? Why?

21. A USB 2 port defines a unit load as _____, and can operate to a maximum of _____ unit loads.
    a. 100 mA, 5
    b. 100 mA, 6
    c. 150 mA, 5
    d. 150 mA, 6

22. What does 10Base-5 mean?
    a. 10 Mbps, baseband, along 500 m of cable
    b. 10 Gbps, baseband, along 500 m (1640.4 ft) of cable
    c. 10 Mbps, baseband, along 500 ft (152.4 m) of cable
    d. 10 Gbps, baseband, along 500 ft (152.4 m) of cable

23. How does propagation delay affect Internet speed and collision detection?

24. Which type of network device helps prevent voltage attenuation across a long cable run?
    a. Bridge
    b. Repeater
    c. Router
    d. Switch

25. Which type of device has a low level of intelligence and acts as a multiport repeater?
    a. Bridge
    b. Hub
    c. Router
    d. Switch

26. What type of device allows a large network to divide into a collection of smaller, independent networks?
    a. Bridge
    b. Hub
    c. Router
    d. Switch

27. Which type of device links separate isolated computer networks?
    a. Bridge
    b. Hub
    c. Router
    d. Switch

28. Which is faster, a cut-through-type network switch or a store-and-forward? Why?

29. What type of network allows every node to have equal rights and access to any shared software, resource, or devices?
    a. Peer-to-peer
    b. Client/server

30. What type of network is inherently more orderly and organized, and allows multiple programs to run simultaneously on various nodes without risk of interference?
    a. Peer-to-peer
    b. Client/server

31. Data packets over an IP network will arrive at their intended destination in exact order. True or false?

32. Why is Internet Protocol (IP) considered connectionless?

33. Explain full-duplex operation as it pertains to TCP.

34. Which protocol is used to retrieve e-mail?

35. What type of protocol allows the viewing of files attached to e-mail?

36. How are binary addresses converted into alphanumeric Web addresses?

37. What is the difference between an RJ-14 and an RJ-45 connector?

38. What is the main difference between EIA/TIA 568A and 568B?

39. What is the purpose of a device driver?

40. What is a network ping, and how is it used?

Chapter 7

Power Supplies, Batteries, and Emergency Systems

Objectives

- Explain transformer operation.
- Describe and explain the uses of various types of transformers.
- Describe the methods of alternating current voltage regulation.
- Identify the uses of surge protectors, power conditioners, harmonic filters, and noise filters.
- Describe and explain the differences between motor- and engine-driven generators.
- Describe the purpose of an uninterruptible power supply (UPS).
- Describe and explain the differences between a double- and a single-conversion UPS.
- Describe and explain the purpose of a rectifier.
- Describe and explain the differences between linear and nonlinear power supplies.
- List the determining factors for choosing a power supply.
- List the basic elements of a battery and the functional differences between primary and secondary cells.
- Describe and explain the differences between lead-acid and nickel-cadmium storage batteries.
- Describe and explain the particulars of *National Electrical Code Article 700, Part III,* as related to emergency system power sources.

Chapter Outline

Sec 7.1 Transformers

Sec 7.2 Alternating Current Voltage Regulators

Sec 7.3 Surge Protectors, Conditioners, and Filters

Sec 7.4 Generators

Sec 7.5 Static Uninterruptible Power Supply

Sec 7.6 Direct Current Power Supplies

Sec 7.7 Storage Batteries

Sec 7.8 *National Electrical Code Article 700*

Key Terms

autotransformer	generator	regulator
battery	harmonic filters	secondary cell
buck-boost transformer	isolation transformer	surge protector
common-mode signal	lightning arrestor	uninterrupted power
constant-voltage transformer	power line conditioner	supply (UPS)
regulator	primary cell	varistor

SEC 7.1   TRANSFORMERS

Isolation Transformer

An **isolation transformer** is used to isolate a piece of electrical or electronic equipment from the main power line. A 1:1 relationship exists between the primary and the secondary coils of the isolation transformer. Although the output has the same voltage and current flow as the input, it is electrically isolated, having only a magnetic coupling between the two sides of the transformer. As a result, the secondary voltage is floating, and no longer grounded, as compared with the Neutral line on the primary side. By lifting the ground from the secondary, the hot line also disappears, making either side of the secondary output safe to touch with one hand. A shock hazard exists only on the secondary if a complete circuit is made by simultaneously touching both sides of the coil.

Notice also that because the secondary is now floating as an isolated loop circuit (Figure 7–1), it is impossible to draw current to the hot side of the primary because no physical connection exists between the secondary and the primary

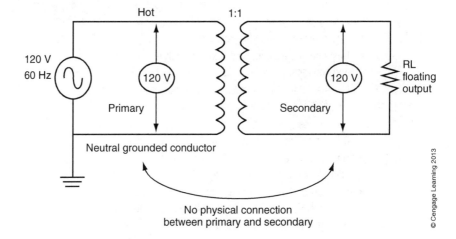

Figure 7–1
Isolation transformer.

© Cengage Learning 2013

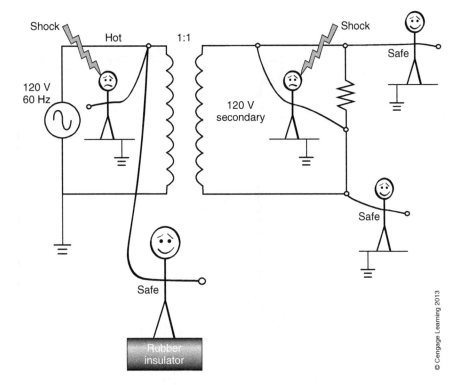

Figure 7–2
Shock hazard.

© Cengage Learning 2013

for current to flow. Essentially, it looks like an open circuit. Figure 7–2 illustrates the concept of shock hazard. In all cases where the individual is getting shocked, there is a complete loop connection between each side of the power line. All of the nonshocked individuals are floating.

Incidentally, the hot power line is "hot" with reference to ground; because the Neutral line is staked at a permanent ground reference by the power company, the hot side is always hot, representing a possible shock hazard to anyone standing on the ground. This is why birds can sit on the hot side of the power line without being electrocuted: They are not touching ground and hot at the same time. Figure 7–3 illustrates the concept. However, people walking on the ground who mistakenly touch the hot side of the power line are shocked unless they first isolate or float themselves by placing an insulator between their feet and the ground.

A 2-in.-thick (50.8 mm thick) piece of rubber is a good example of how technicians could float themselves and safely remove the risk for shock from the hot line, as long as they do not accidentally step off the rubber or touch anything grounded while working on a high-voltage piece of equipment. A better idea would be to power the equipment through an isolation transformer and discard the rubber insulator altogether.

A secondary benefit to using an isolation transformer is that common-mode noise signals present on input do not transfer to the output. A **common-mode signal** is characterized as a signal present on both sides of the power line, measuring equal in level to ground. Such signals, however, have no differential

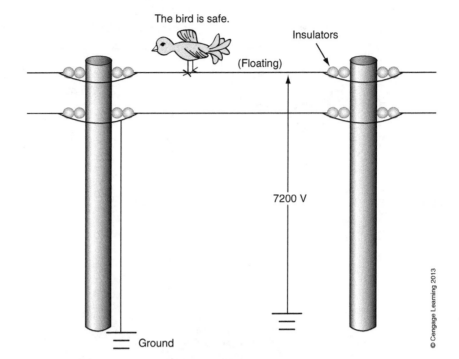

Figure 7–3
Floating circuit,
bird on a wire.

quantity when measuring directly between the power line terminals, and because
a transformer responds only to the presence of a differential signal, none of
the common-mode signal transfers to the secondary. In the example shown in
Figure 7–4, a common-mode noise signal appears on both sides of the power
line, in phase, and equal in amplitude, measuring 1 V peak, with respect to
ground (meters 1 and 2).

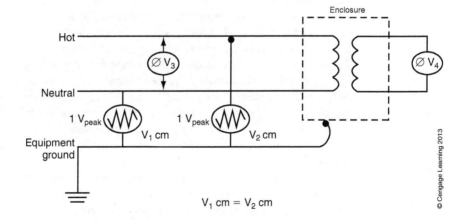

Figure 7–4
Common-mode
(cm) noise.

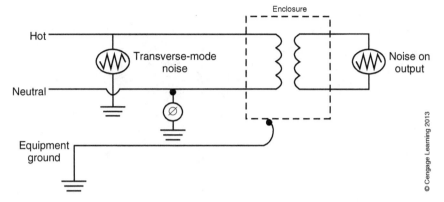

Figure 7–5
Transverse-mode
noise.

© Cengage Learning 2013

The voltage on the secondary reads zero because a differential quantity does not exist between the main input terminals, as shown on meter 3. As a result, transformers with isolated secondary windings are immune to common-mode noise.

Do all noise signals appear as common mode? The answer is no. Some noise appears as normal-mode noise, often called transverse-mode noise. Transverse-mode noise (Figure 7–5) occurs when noise is present only on one side of the power line, as measured to ground. In such cases, a differential noise measurement is measured across the hot and Neutral of the power line. Because transverse noise shows up as a differential quantity across the power line terminals of the primary, it passes through to the transformer secondary, superimposed on the source.

Transformers with isolated secondary coils are only effective at eliminating common-mode noise; they have no effect on normal or transverse noise. In such cases, additional shielding is needed to help reduce the risk for possible pickup and transmission of transverse noise. Noise superimposed on a signal often is difficult to remove and in some cases requires special filtering. Nevertheless, although filtering may work as a remedy to help reduce noise, it may also introduce distortions into the waveform as an undesired side effect because of the additional signal processing and circuitry. This is especially true for audio signals, where the least amount of filtering and processing typically results in the best and truest quality signal, more closely related to the original recording.

Autotransformer

An **autotransformer** is a single winding transformer with an adjustable center tap. Part of the coil acts as the primary input, whereas the section connected to the adjustable center tap provides the output and connection to the load. An autotransformer does not provide isolation, because the secondary is physically sharing the same circuit with the primary; they are in series, and share the same ground. Figure 7–6 shows an autotransformer.

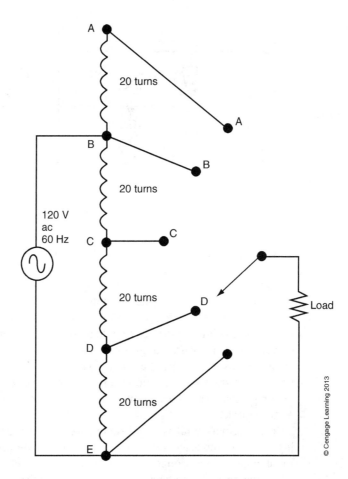

Figure 7–6
Autotransformer.

SEC 7.2 ALTERNATING CURRENT VOLTAGE REGULATORS

Alternating current (ac) voltage **regulators** are devices that make constant adjustments to the line voltage of a load to maintain a constant output voltage. To be effective, the line input must still stay within a reasonable range, not deviating too far; otherwise, correction by the control circuit is not possible. Three types of ac voltage regulators include *tap-changing regulators*, *buck-boost transformer regulators*, and *constant-voltage transformer regulators*.

Tap-Changing Regulators

Tap-changing regulators are devices that mechanically adjust the ac output by switching between multiple coil taps on an isolation or autotransformer coil. Figure 7–7 illustrates the concept. Depending on the degree of line input variations,

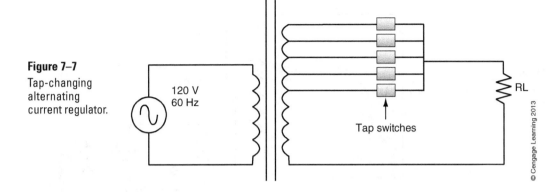

Figure 7–7
Tap-changing alternating current regulator.

most switching occurs automatically at the zero transition point of the current and voltage waveform to help minimize distortion. Taps usually are spaced at 4% to 10% increments, up or down, depending on the design of the regulator, to ensure that the output does not change in a sudden or erratic manner.

Buck-Boost Transformer Regulators

Buck-boost transformer regulators use a special type of transformer that allows inductive feedback to either increase (boost) or decrease (buck) the level of voltage on the output. The buck and boost feedback coil levels are controlled by a semiconductor control circuit that constantly monitors the output and then adjusts accordingly to maintain a constant output voltage. Extra filtering also is included to help reduce any distortions or harmonics in the waveform, ensuring that the output generates a pure sine wave. Figure 7–8 shows a simplified block diagram of a buck-boost transformer regulator. The benefit of buck-boost–type regulators is that they provide smooth and continuous outputs, even when connected to nonlinear-type loads such as computer systems. Normally, nonlinear loads can generate harmonics and distortions on the power line. This is caused by the sudden switching nature of the power supplies. The buck-boost regulator helps to smooth out and keep the power line clean. Combining a buck-boost regulator with an isolation transformer and outer shielding cover also helps to provide common-mode noise rejection and minimize the passage of transverse noise.

Figure 7–8
Buck-boost alternating current regulator.

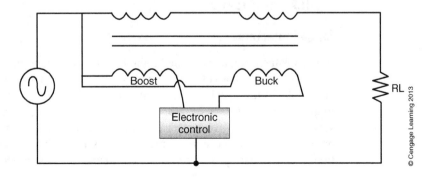

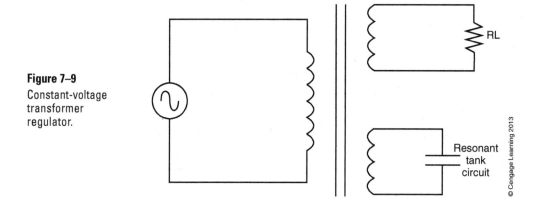

Figure 7–9
Constant-voltage transformer regulator.

© Cengage Learning 2013

Constant-Voltage Transformer Regulators

The **constant-voltage transformer regulator** (CVT; Figure 7–9) uses a saturating transformer and the coupling of a resonant tank circuit to achieve regulation. The resonant tank circuit is made from the internal inductance and capacitance of the transformer. The main concept is that the resonant circuit provides the additional output when the input decreases during voltage lulls; also, during voltage spikes or unexpected supply increases, the saturating core has the counteracting effect of reducing output voltages back to normal levels. Care must be given not to overload CVTs because excessive load currents can cause the tank circuit to go out of resonance. During start-up, excessive inrush currents also can cause severe core saturation and reduce outputs. In such cases, loads may need to be brought on-line more slowly to help minimize the effects of the sudden inrush. The benefit of CVT regulators is that they use no moving parts or active electronic components, which over time can wear out or fail. Once again, isolation transformers and shielding can be used in conjunction with the CVT regulator to help eliminate common-mode signals and minimize transverse noise at the output.

SEC 7.3   SURGE PROTECTORS, CONDITIONERS, AND FILTERS

Surge Protectors

Surge protectors are also called lightning arresters, transient suppressors, surge arresters, or transient voltage suppressors. The purpose and use of such devices is twofold: to limit peak voltage surges and to divert power surges and transients to ground. High-current surge and transient protectors often are used at the main service panel as the primary protector, whereas lower current protectors are used on branch circuits and subpanels as secondary protectors. A third type of protection, called tertiary protectors, is also available; tertiary protectors have smaller ratings and often are used on point-of-use branch receptacles or inside of power strips or equipment.

The primary protector is used to protect the main system from higher voltage lightning strikes or switching transients on the input side of the power line; whereas the secondary protectors are needed to help limit lower voltage surges generated farther up line within the system and caused by internal transients and power surges on the branch circuits. Tertiary protectors are used to protect against surges originating at a circuit receptacle. Primary and secondary protectors typically are made from carbon blocks, gas tubes, or sometimes solid-state devices. Tertiary devices usually are solid state. Figure 7–10 illustrates the placement of primary, secondary, and tertiary protectors.

Figure 7–10
Primary, secondary, and tertiary protectors.

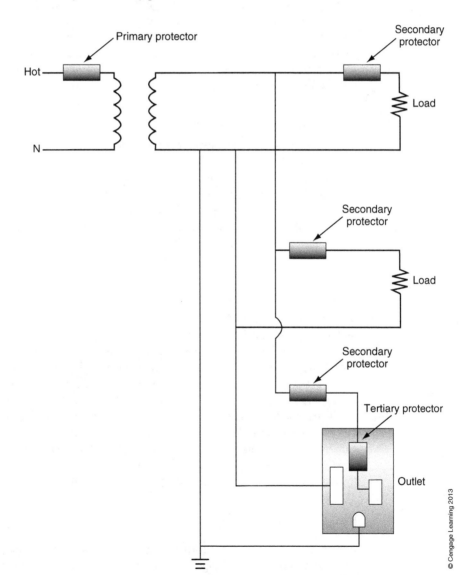

Lightning Arresters

Lightning arresters are primary protectors that are used to limit lightning and switching transients to safe levels. Lightning arresters appear as insulators or open circuits to all voltages below a specific trigger threshold. Once the threshold has been reached, the internal resistance of the device decreases significantly, providing surge currents as an alternate path to ground, safely bypassing the connecting loads. As line voltages return to their normal levels, dropping back below the trigger threshold, the device then turns off, preventing the flow of currents to ground. Such devices can be used repeatedly without risk for damage or need of replacement.

Carbon block arresters consist of blocks of carbon and an air gap. During high-voltage transients, typically in the range of 300 to 1000 V, currents arc across the air gap, conducting the surge currents to ground.

Gas tube arresters work in a similar manner to carbon block, except that currents arc across a gap within a gas tube. Gas tubes usually are more reliable and provide tighter tolerances for arc compared with voltages. The voltage breakdowns for gas tubes are also typically lower than carbon block.

Varistor-Type Lightning Arresters

Varistors are voltage-dependent resistors. They are solid-state devices that significantly decrease their internal resistance once a specific voltage threshold has been reached across the connecting terminals. Once activated, varistors do not short; instead, they clamp voltages to some maximum level while allowing surge currents to pass through the device. The most common type of varistor used in surge protection is the metal oxide varistor; in addition, there are other types of varistors also available that can provide high current protection in the range of 70,000 A for power applications. Varistors can be made to provide surge protection at all levels.

Combination Varistor/Spark Gap Suppressor

Varistors also can be made with internal spark gaps. The gap is used to shut off current flow through the device when transients are not present on the power line. During high surge conditions, the buildup and dissipation of excessive internal heat sometimes can damage the varistor. The addition of the spark gap protects the device from possible damage. When a transient appears on the line, an arc develops across the gap, forcing the varistor to turn on. Current remains flowing until the transient has passed, at which point the varistor returns to a high-resistance state, forcing the arc within the gap to self-extinguish. The device then resets to an off state and current stops flowing. Compared with a standard type of varistor, the combination varistor/spark gap suppressor only allows current to flow through the device in the presence of a transient, thus reducing the potential heat buildup and damage to the device.

Power Line Conditioners

Power line conditioners incorporate combinations of regulation, filtering, and surge protection within a single unit. Such units, however, are not able to provide protection against power line abnormalities such as variations in frequency or power blackouts. Most conditioners also provide a form of current limiting to limit the maximum load current deviations to a range of 150% to 200% at full voltage.

Harmonic and Noise Suppression Filters

Harmonic filters are used to reduce the presence of voltage distortions on the power line that often can affect the operation of sensitive electronic equipment. The combination of multiple harmonic frequencies and high-frequency noise on a power line can also be destructive and may ultimately cause conducting cables, motors, and transformers to overheat and burn up. To solve the problem, a combination of series and parallel-connected capacitors and inductors is used to help filter out the damaging effects of unwanted harmonics and high-frequency noise within the circuit. The phenomenon known as skin effect also helps to compound the problem. As described in Chapter 2, skin effect causes the resistance of conductors to increase significantly when transmitting high-frequency noise or harmonics. The increased resistance results in higher levels of voltage and heat dissipation on the conducting cables and less voltage for the load. Although less voltage at the load may appear to be a primary concern, a more serious problem is the potential threat of fire due to the increased levels of heat dissipation on the connecting cables, transformers, and motor windings. For this reason, power supplies and power lines must be properly maintained and filtered against the presence of harmonic and high-frequency noise; this is especially true for circuits carrying high levels of current flow. Figure 7–11 illustrates harmonic distortion compared with that of a normal sine wave.

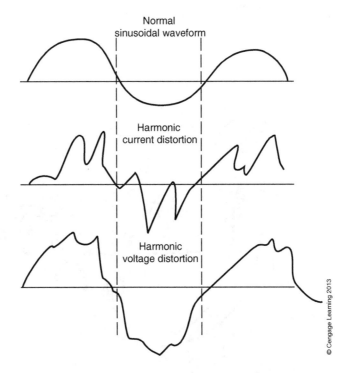

Figure 7–11
Harmonic voltage and current distortion.

© Cengage Learning 2013

The addition of low-pass, high-pass, band-pass, and band-rejection filters (see Chapter 5) also can be used in various combinations to help suppress power line noise, which often is a result of electromagnetic (EMI) or radiofrequency interference (RFI). Sensitive electronic equipment, such as medical equipment, also needs special isolation, filtering, and shielding from power line noise to prevent errors and device malfunctions.

SEC 7.4   GENERATORS

Generators come in two varieties: those driven by motors and those driven by engines. The basic parts of a working generator must include a magnet or a stationary field coil, referred to as a stator, and a secondary rotating coil called a rotor or armature (Figure 7–12).

When a stationary field coil is used, a direct current (dc) source provides the necessary current flow though the windings to energize the magnetic field. The rotation of the armature inside of the magnetic field develops the output voltage and current flow through the load. The speed of rotation determines the output frequency and the level of voltage. Power is delivered to the connecting load through slip rings that are mounted on the shaft of the armature. The electrical connection is made through carbon brushes that press up against the slip rings as they rotate. Depending on the type of generator, some designs are brushless.

Figure 7–12
Generator model.

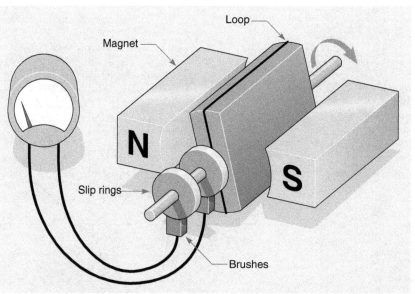

Generator

© Cengage Learning 2013

Not all generators develop ac outputs. Some generators produce dc. In such cases, a commutator is used instead of slip rings. The commutator rectifies the voltage and current alternations as the armature spins. Rectification converts ac into dc, by forcing the current to flow in only one direction, thereby preventing alternation. The principles of rectification are discussed further in Section 7.6.

The amount of dc flowing through the field coil usually determines the quantity of voltage generated at the output. By controlling the amount of dc in the stator, the output can be regulated. Regulation of the output often involves comparing the output voltage and frequency of the generator with internal references. A sensing circuit then develops a dc control voltage proportional to the amount of error at the output. The adjustment to the output then can be made instantly by feeding back the dc error voltage to the field coil, thereby forcing the generator into regulation. Figure 7–13 presents the block diagram of a regulated generator. In this case, the generator is engine driven, using a starting battery at the input and a power transfer switch at the output.

The traditional methods used to spin the armature of a backup generator involve either an ac or dc electric motor or a gas engine, whereas power companies often uses wind, water, or nuclear energy to spin the armatures of their generators. As stated earlier, a mechanical coupling links the rotation of the motor or engine shaft with the armature of the generator. The rotational speed of the shaft and the ratio of magnetic poles within the stators determine the frequency of the output waveform. The number of stator coils placed within the 360° rotation of the armature, as shown in Figure 7–14, determines the number of magnetic poles.

Alternating Current Motor Generator

The typical use of an ac, motor-driven generator is to provide a source of isolation from the main utility power line. The motor and generator are typically mounted

Figure 7–13
Regulated power generator.

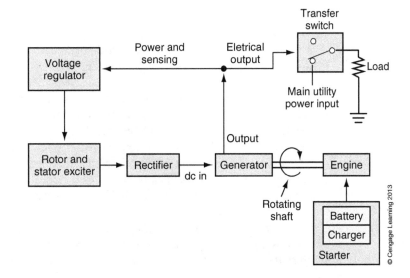

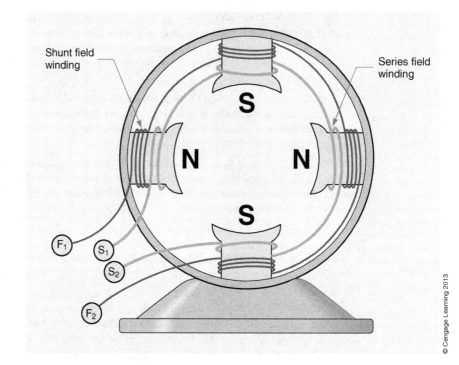

Figure 7–14
Generator field windings, two-pole.

© Cengage Learning 2013

on a common platform, with the rotating armatures of each connected through a mechanical linkage. The main utility power line often serves as the source of power for the motor. The rotation of the motor, in turn, spins the generator, because they share the same mechanical linkage. But although they are connected physically, they are not connected electrically. As a result, the output of the generator remains electrically isolated and floating, just as the secondary of an isolation transformer. *NEC Article 250.35* requires that permanently installed generators are grounded. *Article 250.34* states that portable and vehicle generators are not required to be grounded if certain conditions are met.

Motor-driven generators have the added advantage of being able to provide ac-to-ac conversion, or dc-to-ac conversion, depending on the type of motor at the front end of the system. Ac-to-ac conversion also can include a frequency conversion, such as 60 to 400 Hz. Higher frequencies increase the horsepower of a motor; they also provide cleaner, rectified dc outputs with lower ripple percentages, which often require less filtering for use by the load. Ripple is a small ac variation superimposed on the dc output, resulting from the rectification process. In most cases, it should be removed to prevent the transmission and generation of noise throughout the system. Figure 7–15 graphically illustrates the concept of ripple. Such noise may often be the source of error and malfunction when connecting sensitive circuits and devices.

Motor-driven generators can also help to protect loads from voltage sags, swells, or surges. The regulation of motor speed solves any power swells or surges, whereas the addition of a flywheel to the armature of the generator can help to

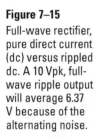

Figure 7–15
Full-wave rectifier, pure direct current (dc) versus rippled dc. A 10 Vpk, full-wave ripple output will average 6.37 V because of the alternating noise.

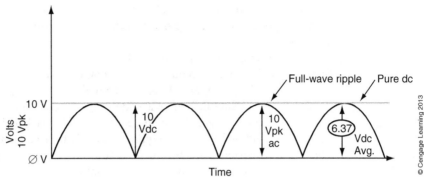

maintain a constant output voltage during power line lulls or sags. The flywheel on the armature provides the device with the needed rotational momentum and inertia to help keep the output voltage and frequency constant. As a result, the generator continues to turn, providing a steady output to the load, which helps to mask any momentary lulls from the main power utility.

Engine-Driven Alternating Current Generator

Engine-driven ac generators often are used as a source of backup power in cases where the main power line has failed. During prolonged power outages, such systems typically use automatic transfer switches to provide the needed isolation from the main electrical utility of the building to the backup generator. Ac generators are also called alternators because they convert a rotating mechanical energy into an alternating electrical energy. A typical engine-driven generator uses 12 or 24 V dc batteries for energizing the field coil and starting the system. Once the system is operational, the output then is rectified and fed back to the field coil as a self-generating source of dc. To help ensure future starting, the dc output also simultaneously recharges the battery through a charging circuit. See Figure 7–13 for an example of an engine-driven generator.

SEC 7.5   STATIC UNINTERRUPTIBLE POWER SUPPLY

Although primary protectors, regulators, and backup power often serve to protect most systems from momentary or even prolonged outages, more sensitive systems, such as computer or telecommunications equipment, need to have a more stable level of protection. Computer networking systems, in particular, can sustain permanent damage from a single power surge or lightning strike. The surge can travel instantly from an outside transformer through connecting wires, data and phone lines, and into computer devices and peripherals such as modems, router, switches, and network hubs. One direct hit can permanently damage motherboards, memory chips, and network interface cards (NICs). As a result, thousands of dollars of equipment could be lost in an instant. To help prevent such an occurrence, computer and telecommunications systems need a more robust system of protection. Static **uninterruptible power supplies (UPSs)** can provide such protection.

A UPS is an electronically controlled power system that provides ac regulation and filtering to loads requiring consistent, uninterrupted, steady-state power during a partial or even total failure of the main supply input. A variety of UPS systems are available that offer a wide range of configuration options, often based on the design and size of the system. Additional options often include power line conditioning, lightning protection, EMI protection, and redundancy. The size and capacity of a UPS can range from less than 100 W to as high as several megawatts. Single- or three-phase outputs also are available, with typical frequencies of 50, 60, or 400 Hz. The output waveform also can vary from that of a near-perfect sine wave to a square wave, depending on the design of the system.

A UPS system consists of a rechargeable battery, a rectifier and battery charger, a static inverter, and a power transfer switch. Two types of UPS systems exist: double- and single-conversion UPS systems.

Double-Conversion Uninterruptible Power Supply

Figure 7–16 provides a block diagram of the double-conversion UPS system. The system first converts the original ac into dc. As a result, all harmonic distortions, noise, power surges, and power lulls are removed from the source. The dc then drives the input of a static inverter. The inverter reconverts the rectified dc back into a pure, filtered, ac source for use by the connecting load. The **battery** provides a means of secondary power to the inverter during the momentary or long-term loss of primary ac at the input. In most cases, a zero time transfer between the rectifier and battery can be obtained because they are connected in parallel at the inverter. The inverter continues to function, uninterrupted, for a predetermined amount of time based on the capacity of the battery and power requirements of the connecting load.

The double-conversion UPS system is the electrical equivalent of a motor-driven generator. The advantages of using a double-conversion UPS system include excellent frequency stability, a high degree of isolation from the primary ac line, and quiet, low-decibel operation because there are no moving parts.

Low-power, double-conversion UPS systems providing less than 20 kW often can have the following disadvantages compared with higher power systems: They require larger dc supplies (typically 1.5 times larger than the maximum load

Figure 7–16

Double-conversion uninterruptible power supply.

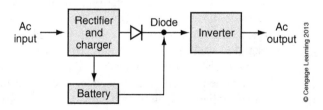

© Cengage Learning 2013

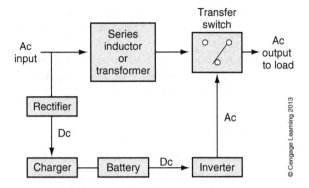

Figure 7–17

Single-conversion uninterruptible power supply.

rating of the UPS); they are lower in efficiency; they provide poor noise isolation between the line and load; and they dissipate more heat, which often can shorten the useful service life of the device.

Single-Conversion Uninterruptible Power Supply

The single-conversion UPS system uses a transfer-switching circuit for switching the output between the ac input and the output of the inverter. Figure 7–17 provides a block diagram of single-conversion UPS.

In such a system, the ac power is supplied to the load through a series inductor or an isolating transformer. The series inductor method helps to smooth out any power surges and power lulls on the input. In the single-conversion UPS system, the inverter output is used only as a backup when the main ac input fails. The battery charger maintains a full charge on the battery only while the ac input is present. During a power failure, the inverter is only able to supply a useful output for a predetermined amount of time based on the capacity of the battery and the power requirements of the connecting load.

SEC 7.6   DIRECT CURRENT POWER SUPPLIES

Although ac power is acceptable for devices such as lighting, heaters, fans, motors, pumps, and blowers, it is not a suitable source when powering sensitive electronic equipment. In most cases, dc power is more preferable and necessary. Computers, audio and video equipment, telecommunications equipment, and sensitive medical equipment are all examples of electronic circuitry that requires dc power to operate. In such cases, the dc power must also be free from noise, RFI, and EMI, or circuitry malfunctions and data errors may occur.

The downside to using dc power is that it does not pass through to the secondary of a transformer. Therefore, simple voltage or current corrections based on winding ratios are not possible. As a result, when using dc power, cable runs typically are limited to a maximum length based on the gauge of wire, to ensure the efficient transfer of power to the load with minimal loss. Small-gauge wire inherently has more line loss over long runs because of the increased resistance of the cable.

When transporting power over long distances, ac remains the best choice; however, eventually there must be a conversion to dc for use on various circuits and devices. A rectifier performs such a conversion.

Rectifiers

A rectifier converts ac into pulsating dc. Rectification occurs through a diode. A diode is a semiconductor device that allows current to flow in only one direction, acting much like an electronic switch. The circuit shown in Figure 7–18 is a half-wave rectifier.

The input supply is a 60 Hz ac sine wave. During the positive cycle, from 0 to 180°, the diode turns on, and current flows through the load. At this point, you may be wondering why the diode arrowhead points against the direction of current flow. The answer is that the arrow points in the direction of conventional current flow and not electron flow. It is simply a matter of orientation. Engineers traditionally think in terms of conventional current flow, which is why the schematic symbol is pointing in that direction. Regardless of which way you think, the current is still flowing in only one direction.

When electrons flow into the arrowhead, the diode forward biases and turns on. The forward voltage drop of a silicon diode is about 0.6 to 0.7 V (0.3 V for a germanium diode), depending on the amount of current flow. As the ac input reverses direction, from 180° to 360°, the diode reverse biases and shuts off. Load current stops flowing during reverse bias, as indicated by the flat line on the output waveform.

The output waveform is not considered to be pure, but rather is pulsating dc. Even though the current flows in only one direction, the periodic pulsations indicate that the output still has an ac irregularity. The pulsating dc output requires

Figure 7–18

Diode rectifier, half-wave output, pulsating dc.

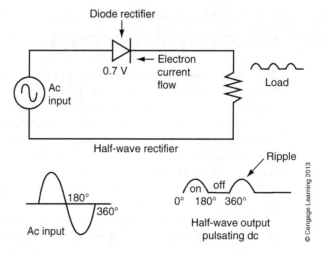

additional filtering and regulation to remove the rippling noise pulses before powering an actual load.

The output frequency of the ac ripple for a half-wave rectifier is equal to the input frequency, in this case, 60 Hz. A capacitor filter is used to help remove the majority of the 60 Hz ripple from the output of the rectifier. The capacitor acts as a low-pass filter, having a cutoff frequency less than 60 Hz, effectively removing most of the ac ripple from the load. At start-up, the capacitor charges to the peak of the ac input, minus the forward voltage drop on the diode; in most cases, the 0.7 V diode drop is small enough to ignore.

Voltage Regulators

Voltage regulators are much more complex than rectifiers, often involving transistors and active feedback circuits. For matters of simplicity, we discuss here a basic voltage regulator to illustrate the function and purpose of the circuit. A basic voltage regulator can be made from a single-series resistor and a Zener diode (Figure 7–19).

The Zener diode is a special type of diode that allows conduction during reverse bias, only at a particular Zener voltage. Once the Zener diode activates in the reverse bias mode, it maintains a constant output voltage; any attempted increase above the Zener voltage simply draws more current through the diode. The design of the diode determines the magnitude of the Zener voltage; commonly used values are 5, 12, and 24 V.

The series resistor is required as a dropping point for the excess circuit voltage, over and above the regulated output. It also serves to limit the total current flow to some maximum value. The level of voltage on the series resistor can be calculated by subtracting the output voltage of the regulator from the input voltage.

Capacitor Filter

A capacitor filter is used to filter out the ripple, or pulsating dc, from the output of a rectifier. During the positive cycle, the diode turns on, and the capacitor charges to the peak input voltage. Later in the cycle, when the diode turns off, the capacitor is forced to discharge through the load. In essence, the capacitor acts as a reserve energy tank, supplying voltage and current to the load when the diode is not conducting.

Figure 7–19
Simple Zener
voltage regulator.

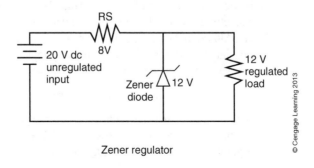

Zener regulator

© Cengage Learning 2013

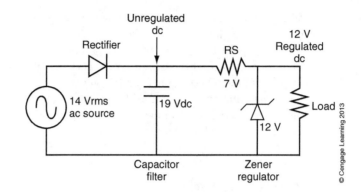

Figure 7–20

Completed power supply: rectifier, capacitor filter, and Zener regulator.

Although the capacitor works quite well to remove unwanted ripple from the output, it cannot maintain a constant voltage regulation. The peak capacitor voltage ultimately decreases as the demand for load current increases. A Zener regulator must be added to the output of the capacitor filter to achieve regulation. In essence, the voltage on the capacitor becomes the unregulated input, whereas the voltage of the Zener helps to maintain a constant regulated output voltage at the load. In Figure 7–20, the input capacitor filter measures 19 V dc, whereas the load sees 12 V because it is in parallel with the 12 V Zener diode. The difference of 7 V is dropped along the series resistor R_S.

R_S also limits the total current flow through the Zener and the load. Not using the series resistor causes the 12 V Zener to overload and burn up. Where a 5 V output is desired, the Zener must be exchanged for a 5 V Zener. The voltage of an individual Zener is constant and cannot be altered or changed.

Full-Wave and Bridge Rectifiers

The half-wave rectifier is the simplest of all rectifier designs, but it is also the most inefficient because it ignores the entire second cycle of incoming ac. A better choice would be to use a full-wave or bridge rectifier to gain back the unused portion of the input and increase efficiency. Figure 7–21 shows the design of both.

A full-wave or bridge rectifier allows current to flow continuously for the entire 360° input cycle. During the negative alternation of the input ac, the output is shown to flow in a positive direction. As a result, the output now appears to have two positive cycles for every one input cycle of ac; current also flows through the load more continuously and efficiently. As a result, the frequency output of a full-wave or bridge rectifier doubles because of the added output cycle, in this case, from 60 to 120 Hz. Although the ripple still exists, it is significantly lower than that of the half-wave design. Mathematically, the unfiltered ripple percentage is calculated by dividing the measured root-mean-square (rms) ripple at the output by the measured dc.

$$\text{Unfiltered ripple \%} = \text{rms ripple} \div \text{dc output}$$

For an unfiltered, half-wave rectifier, the ripple percentage measures 121%, which means that the output ripple is actually 21% larger than the intended dc; this is not good. The full-wave or bridge output is much better, measuring only 48%.

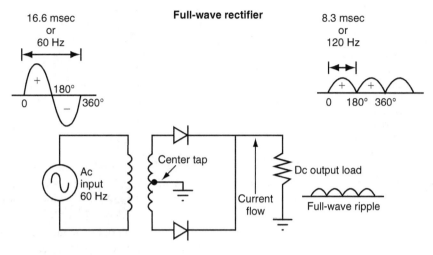

Figure 7–21

Full-wave center-tap rectifier and bridge rectifier.

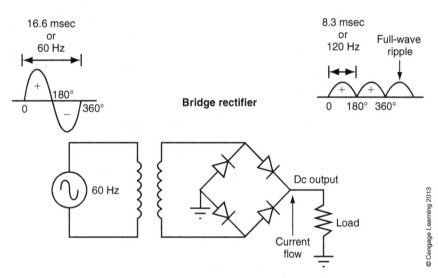

This means that the dc is more than double the ripple voltage, which is a step in the right direction. The main idea is to eliminate the ac ripple altogether. Pure dc has 0% ripple and on an oscilloscope appears as a flat line floating above the zero reference or ground. Overall, the full-wave or bridge rectifier provides a better output signal, double in frequency, and has a significantly lower ripple percentage, which ultimately requires less filtering to achieve pure dc.

Full-Wave versus Bridge Rectifiers

When comparing a full-wave with a bridge rectifier, the main difference is that a full-wave rectifier requires a center-tapped transformer to operate, whereas the bridge rectifier can rectify any ac signal, with or without a transformer.

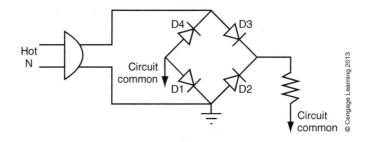

Figure 7–22

Bridge rectifier without input transformer. Never connect circuit common to ground, or D1 will short out. An input transformer is needed to provide isolation from the grounded power input.

In addition, special care must be taken with bridge rectifiers to ensure that the rectification of a hot power line does not occur because it could potentially introduce a direct short circuit across the rectifier and input power line if the load were grounded. Figure 7–22 illustrates the problem. Remember, the Neutral side of the ac power is already at ground potential. It would be a mistake to place an additional ground at the load because it would introduce a direct short across rectifier diode D1 and the input power line. A solution to the problem would be to use an isolation transformer. The isolation transformer can be connected between the hot line and the bridge rectifier input, or between the rectifier output and the connecting load. The load can only be grounded safely once isolation has been achieved. It is best practice to use a bridge rectifier only in conjunction with an isolation transformer.

Types of Direct Current Power Supplies

There are two types of dc power supply, linear and nonlinear. A linear power supply develops an output that is directly proportional to the input. Simply stated, if the input voltage doubles, then the output voltage doubles. Figure 7–23 presents a block diagram of a linear power supply.

The basic elements involve a rectifier, a filter, and a regulator. When using a nonlinear power supply, the output does not respond in direct proportion to the input source. As a result, a doubling of input voltage does not necessarily double the output; in fact, it may often increase significantly higher. Nonlinear supplies often are referred to as switch-mode power supplies because they typically involve the use of switching regulators. Nonlinear, switch-mode power supplies are far more efficient, smaller, and lighter weight than linear supplies. As a result, they often are found inside televisions, video equipment, computers, and communications

Figure 7–23

Linear power supply block diagram.

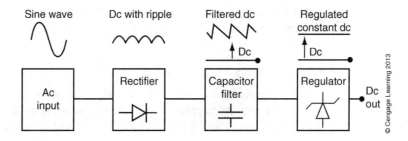

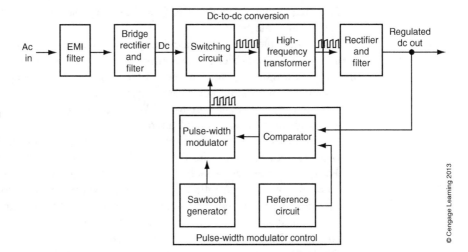

Figure 7–24

Block diagram of a typical switch mode power supply.

equipment. Figure 7–24 presents a block diagram of a basic, nonlinear, switch-mode power supply.

The circuit involves an EMI filter and a bridge rectifier, followed by a dc-to-dc converter, an output rectifier, and a filter. Notice how the regulated output feeds back to the dc-to-dc converter. Regulation is achieved by sampling the output and comparing it with a known internal reference voltage. The feedback provides instantaneous self-correction to sudden changes in the output.

The dc-to-dc converter receives the dc input from the bridge rectifier. The input dc energizes the switching circuit though the primary winding of a high-frequency transformer. A pulse-width modulator rapidly turns the current flow to the transformer on or off. The pulse rate determines the amount of inductive kick at the output as the high-frequency transformer switches on and off. The width of the pulse determines the amount of output from the high-frequency transformer. Although the input from the bridge rectifier may only be 5 to 12 V, the output often can be in the hundreds of volts or even 1000 V, depending on the design of the feedback circuit.

The final stage of the switch-mode power supply rectifies and filters the high-frequency output, converting it to dc. The feedback to the comparator circuit provides output regulation. The magnitude of the output is not directly dependent on the level of input at the bridge rectifier, but rather on the width of the pulses applied to the switching circuit. By varying the width of the pulse-width modulator and adjusting the comparator circuit, the output voltage can be adjusted to nearly any required value. As stated earlier, a doubling of the input voltage does not necessarily double the output voltage, which is why it is considered a nonlinear power supply.

Power Supply Selection

The selection of a power supply often depends on *Code* requirements, load requirements, input frequency, voltage, and in the case of a UPS, desired run time.

Code requirements involve listing and labeling, primary protectors, and the classification of the circuit.

Load requirements involve calculating the normal and critical power requirements of the system. Critical power involves a worst-case scenario to ensure that the supply can safely power the load. In most cases, a power supply is oversized by a minimum of 50%. In such a case, a system requiring 500 W of power needs a supply that can safely handle up to 750 W. Adding the extra capacity also ensures enough available power in case of future expansions.

The input frequency of most power supplies is either 60 or 50 Hz. Sixty hertz is the frequency standard for power in the United States/Canada, whereas 50 Hz is used more commonly in Europe and Asia. When building systems intended for overseas installation, the power supply specifications become a critical part of the design. Items such as input transformers and connecting hardware must be chosen carefully, depending on the frequency and voltage of operation. The type and style of electrical plugs and outlets also can vary greatly from country to country; wire color codes often differ as well. To ensure the proper connectivity and the functional use of an international system, it becomes necessary to first research the fine details and requirements of power, typically involving voltage, frequency, hardware, and color codes. Doing so also helps to ward off any possible confusion among technicians at various locations around the globe.

Power supply voltage in the United States often can include 120 V, 208 V, 208 V three-phase, 240 V, and 490 V three-phase. Other parts of the world often use a 230 V system, including 400 V and 400 V three-phase. All of these types of systems use different connecting hardware and color codes to help distinguish among them. The job of determining the proper connection and installation of power falls to that of an electrician. Do not ever attempt connection to high voltage, 120 V or greater, unless you are properly trained and qualified to do so. Otherwise, you might be putting your life or the life of another person at risk.

UPS run time involves the power capacity of a UPS during a main power outage. During a total power blackout, the run time of a UPS often can vary from 15 to 30 minutes, depending on the size of the supply. The intent of a UPS is not to provide unlimited power, but instead to provide power for only a short duration of time while backup generators come online. In cases where backup generators are not being used, the UPS provides the needed time to safely shut down the system, thus preventing damage or loss of data. In some instances, large systems can also be physically dangerous to technicians or operators if they are not shut down by a careful, step-by-step procedure. The run time of the UPS allows for such a shutdown.

Regulated versus Unregulated Supplies

Let us consider two power supplies: one 12 V, 500 mA, regulated; and another 12 V, 500 mA, unregulated. What is the main difference between the two? The graph in Figure 7–25 illustrates the difference.

Essentially, the 12 V regulated supply maintains a constant output voltage, up to 500 mA, as indicated by the flat line. At load currents greater than 500 mA,

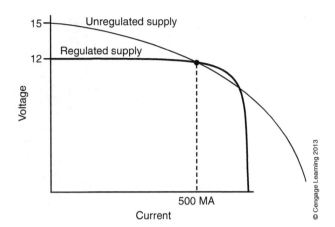

Figure 7–25
Unregulated
versus regulated
power supply.

the voltage decreases dramatically. The regulated supply only maintains output regulation provided the output current stays less than the 500 mA rating.

Let us now look at the graph of the unregulated supply. The unregulated supply never maintains a constant output voltage. In fact, every time the output current increases, the output voltage decreases. The 500 mA rating, in this case, is not a maximum; instead, it represents the output current that corresponds to 12 V. Simply put, the output voltage decreases for load currents greater than 500 mA and increases for currents less than 500 mA.

In summary, a regulated 12 V, 500 mA supply supplies a constant 12 V output up to 500 mA, whereas the unregulated 12 V, 500 mA supply only supplies a 12 V output at 500 mA. The difference may be significant because the unregulated supply drawing a load current of 10 mA outputs 20 V, whereas the regulated supply at 10 mA maintains a constant 12 V.

SEC 7.7   STORAGE BATTERIES

A battery converts chemical energy into electrical energy. A simple battery cell consists of two dissimilar metals separated by an electrolyte. The combination of dissimilar metals determines the amount of cell voltage. For instance, copper and aluminum produce one level of voltage, whereas nickel and zinc, or copper and silver, produce another. The electrolyte separating the two metals provides the pathway for current flow between the metal plates and the connecting load. By connecting individual cells in series, any desired level of output voltage may be achieved. Figure 7–26 illustrates the construction of a storage battery. *NEC 480.6* defines the insulation of storage batteries having connecting cells that do not exceed 250 V. *NEC 480.7* defines the insulation requirements for cells exceeding 250 V.

A simple voltaic cell (Figure 7–27) can be made by using a potato with aluminum wire and copper wire connected to a voltmeter. The acid within the potato

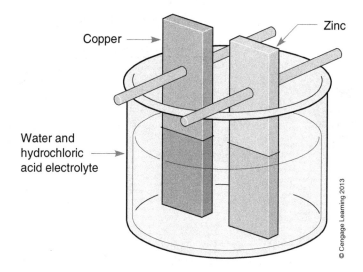

Figure 7–26
Copper-zinc cell.

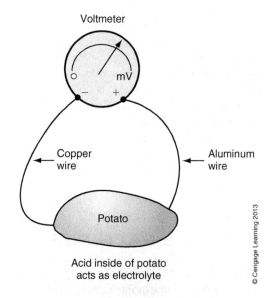

Figure 7–27
Potato power supply. The acid inside the potato acts as an electrolyte.

acts as the electrolyte between the aluminum and copper. As an experiment, one could change the combination of metal wires to see what happens to the level of output voltage. Different metal combinations produce different output voltages.

Besides voltage ratings, batteries are also rated for ampere-hours, indicating the typical output current one can expect from a cell over a specified period. For example, a battery rated for 20 ampere-hours can output 20 A for 1 hour, 10 A for 2 hours, or 5 A for 4 hours. Once the 20 A have been depleted, the cell is completely discharged. In most cases, rechargeable batteries are recharged using 1/10 the ampere-hour rating of the device. For the previous example, the charger would need to output 2 A for a period of 10 hours to replenish and restore the battery to full voltage and charge. The output voltage of the charger must also be slightly greater than the maximum battery voltage, under full charge, to achieve recharge. It is important not to use excessive charge voltage and charge currents because they can cause damage and possible destruction to the device. Batteries often heat up when charged. Improper charging may cause excessive heat buildup within the device, resulting in a possible explosion or fire. For this reason, always refer to manufacturer specifications for information on how to recharge a cell safely.

Wet- or Dry-Cell Electrolyte

The electrolyte of a battery can be of the wet- or dry-cell variety. Wet cells use a liquid electrolyte, whereas dry cells use an electrolyte paste or gel. The benefit of using dry cells is that they can be transported easily and placed physically in any orientation without the risk for acid spillage, and there is nothing to maintain. Wet cells, in contrast, tend to be cumbersome, difficult to move and transport, and must be maintained periodically, typically every 18 months, by testing the specific gravity of the acid-water electrolyte with a hydrometer. Specific gravity is a measurement that determines how much acid is contained within the water. Depending on the type of battery and the manufacturer, a typical specific gravity measurement can range from 1.215 to 1.28, with 1.0 being the specific gravity of pure water. The hydrometer is inserted into the well of the battery to measure the condition and charge quality of the electrolyte. A low specific gravity measurement may indicate that the battery is being insufficiently charged or that the battery may be nearing the end of its useful life span and must be replaced.

Primary versus Secondary Cells

Two types of battery cells exist: those considered primary and those considered secondary. **Primary cells** are batteries that are not rechargeable. Once the internal energy of the cell has been depleted, it must be thrown away. Primary cells include those of carbon-zinc, alkaline, button, or lithium-type cells.

Secondary cells are rechargeable. Often called storage batteries, they are used in a wide variety of applications, including automobiles, emergency lighting, fire alarm systems, UPSs, and generator starting systems. Secondary storage cells require the input of electrical energy to charge or activate the internal chemical reaction. Once the load has been connected, the stored energy within the cell then discharges through the load, reversing the initial charging process, thereby depleting and neutralizing the chemical reaction of the electrolyte. If a load is not connected, a charged battery can maintain full charge for only a finite period, depending on the type of battery. Two of the most commonly used types of storage batteries are those of lead-acid and nickel-cadmium (Ni-cad). Lead-acid batteries can maintain a no-load condition at full charge for a period of 3 to 6 months in a 77 °F environment. Ni-cad batteries can last for up to 2 years. The storage temperature greatly affects the shelf life. Cold temperatures typically increase the storage life of a cell by slowing down the internal chemical reaction.

Just as in primary cells, the voltage of a secondary storage battery is dependent on how many cells are series connected. Each cell of a lead-acid battery typically measures 2 V. Ni-cad cells are lower, measuring only 1.2 V. Therefore, a 6 V lead-acid battery requires only three series-connected cells, whereas Ni-cad batteries require five. As stated earlier, *NEC 480.6* is for cell groupings measuring less than 250 V, and *NEC 480.7* is for cell groupings measuring over 250 V.

Lead-Acid Batteries

The design of a lead-acid battery includes two plates, one of lead and the other of lead dioxide. Hardening agents such as calcium or antimony also are added to the lead because of the inherent softness of the metal. The plates are immersed within a bath of sulfuric acid and water, which serves as the electrolyte. Other designs use the sealed dry cell technique, replacing the liquid electrolyte with a sulfuric acid suspended in a gelatin. Often called gel cells, the sealed variety of lead-acid batteries have gained more popularity over the years because they can be placed in any position without the risk for spilling the acidic internal contents. Another change in the sealed design is that the lead plates often are replaced by lead and lead dioxide paste; one-way pressure valves, 2 to 6 psi, are also included to help relieve the buildup of gas pressure when charging. As stated earlier, the sealed, dry-cell batteries are virtually maintenance free and are easier to use and install.

Charging Cautions and Concerns

Overcharging a cell can cause possible damage or even permanent destruction to the device. Excessive current flow in the charge cycle may result in a lower than expected specific gravity, indicating that the acid content of the electrolyte has not been suitably replaced. As a result, the cell's output capacity is reduced, forcing the periodic recharge of the device to take place more often than typically planned. High-charge currents and mechanical shocks, such as dropping or bumping the battery, can also cause flakes to break away from the internal lead plates. As the flakes fall to the bottom of the cell through the liquid electrolyte, they eventually build up to the point of internally short-circuiting the lead plates along the bottom of the cell; once this occurs, the cell no longer operates. Flaking does not occur in gel cells or sealed dry cells because the plate material typically is made from a lead paste. Over time, a solid metal plate corrodes or deteriorates. The lead paste, in contrast, does not lend itself to becoming brittle or flaking, and because the electrolyte of a sealed dry cell is also of a gel or paste variety, the possibility of internal shorting is virtually nonexistent; this is just another benefit to using dry or sealed gel cells whenever possible.

Manufacturers often indicate the recommended operating temperature of a battery, the ampere-hour capacity, the number of possible lifetime charge cycles, the recommended charge voltage and charge current, the recommended charge time for 100% discharge, the type of electrolyte, the vibration resistance, and the vented gas combination indicating the type of acid vapor being released from the cell. It is important to know all of these specifications to properly maintain a battery and ensure a long and useful life span for the device. Not following recommended discharge, charge, and maintenance procedures ultimately destroys the reliability of the power source.

Nickel-Cadmium Batteries

Ni-cad batteries use a positive plate made of nickel hydroxide mixed with graphite and a negative plate of cadmium oxide. The graphite in the positive plate increases the conductivity of the nickel hydroxide. The electrolyte consists of potassium hydroxide, having a specific gravity of approximately 1.2. The benefit of using Ni-cad batteries is that they have a long shelf life and a long life span, typically allowing more than 2000 charge/discharge cycles. The initial cost of Ni-cad batteries is considerably greater, but they tend to pay for themselves over the life of the battery.

Figure 7–28, which compares Ni-cad batteries with lead-acid batteries, shows that Ni-cad batteries maintain their voltage better and do not experience the gradual drop off through discharge. The downside, however, is that a Ni-cad cell is much lower in voltage, typically around 1.2 V, and requires more cells to achieve the same result of a lead-acid battery. Carbon-zinc and alkaline primary cells are more in the range of 1.5 V, whereas secondary lead-acid cells are typically up in the range of 2 V.

Lastly, a Ni-cad battery must not be recharged too early in the discharge cycle. Instead, the cell should be allowed to discharge fully before recharging because it has the ability to build up an internal memory of the usage cycle. Ultimately, if a Ni-cad cell is used routinely at low current and not permitted to discharge to at least the 80% point, over time it can develop a new characteristic cure, one that

Figure 7–28
Lead-acid battery versus nickel-cadmium.

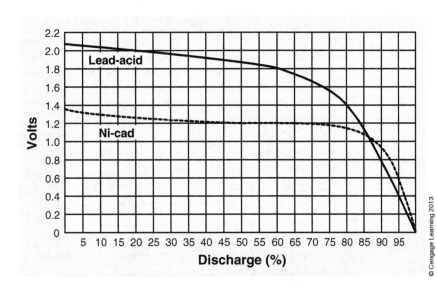

matches the usage cycle rather than the ideal maximum capacity of the device (see Figure 7–28). Lead-acid batteries, however, do not have this problem and may be recharged at any point along the discharge curve.

SEC 7.8   *NATIONAL ELECTRICAL CODE ARTICLE 700*

NEC Article 700 governs the requirements of emergency systems. Emergency systems include backup power sources and lighting intended for use in the event of a main power failure. Hospitals, fire alarm systems, critical computer systems, and manufacturing processes are all examples of systems requiring emergency backup power during the event of a main power failure. Any environment or process that has the potential to introduce a direct threat to the life and safety of a human being because of a sudden loss of power, needs to provide some form of secondary, emergency backup power and lighting to ensure the safety of all within the space. Emergency systems are legally required and classified as emergency systems by municipal, state, and federal codes. In certain cases, insurance companies may also require additional emergency systems over and above the required code. At the end of the day, the authorities having jurisdiction (AHJs) have the final word on the issue.

According to *NEC Section 700.3*, the AHJs require the periodic testing and maintenance of all emergency systems. Before using the emergency system for the first time, the AHJ must personally conduct or witness a system test and verify its functionality in writing. The AHJ must also approve the periodic schedule for testing and maintaining an emergency system to ensure that the system remains functional throughout the life of the system. Testing and maintenance also include the periodic maintenance of starting batteries for generators, a dated and signed written record or log of all test and maintenance procedures, and the means of testing all emergency lighting and power systems under maximum load.

In addition, *NEC Section 700.7* requires signs to be posted at the service entrance indicating the location and type of emergency power source. Individuals or service technicians must be made aware of backup power systems as a matter of safety; otherwise, someone turning off the main breaker may inadvertently assume that all power has been removed from the circuit. Such an assumption could ultimately be a life-threatening mistake, especially once the backup system energizes.

NEC 700.12 details the general requirements of emergency power sources. The following list outlines the basic requirements:

10-second activation: After a main power failure, the activation of an emergency system must not exceed 10 seconds.

Storage batteries and UPSs: Storage batteries and UPSs used to supply power to emergency systems must be able to supply maximum load capacity for a period of not less than 1.5 hours, with the load voltage not falling to less than 87.5% its normal value.

Figure 7–29
Unit equipment.

© Cengage Learning 2013

Engine-driven generator: When an engine-driven generator is used as a source of emergency backup power, the generator site location must provide a sufficient supply of fuel to operate the generator for up to 2 hours, at maximum load capacity.

Unit equipment: An example of unit equipment would be battery-powered emergency lighting (Figure 7–29). The purpose would be to provide illumination during a main power failure to provide safe assistance to individuals attempting to evacuate a space.

The unit equipment must include a rechargeable battery, a battery charger, one or more lamps mounted on the equipment, and a relaying device, allowing for the automatic illumination of the lamps during a main power failure. The batteries must be able to supply a total lamp load for a period of at least 1.5 hours, with the output voltage not decreasing to less than 87.5%, or the unit equipment must be able to supply and maintain illumination to not less than 60% of the initial emergency illumination for a period of at least 1.5 hours. Flexible power cords may be included on the unit equipment provided they do not exceed 3 ft (0.91 ft). The branch circuit supplying the unit equipment must not contain any local switches ahead of the unit equipment that could inadvertently disengage the system. Local switches may only control the normal lighting within an area or space, but they must not be able to control the unit equipment. Branch circuits feeding unit equipment also must be clearly marked at breaker panels indicating the type of emergency system and the location.

In addition, *NEC 700.10 (A)* states that all boxes and enclosures containing emergency systems must be clearly and permanently marked to indicate the presence of emergency circuits or systems. Wiring of emergency systems must also be kept entirely separate from all other system wiring. The design and location of emergency system wiring must also be able to withstand certain hazards, including flood, fire, ice, and vandalism.

Emergency systems may be installed only by electricians and then approved by the AHJ. This chapter has been included to provide the reader with a basic understanding of power supplies and emergency systems, because many types of low-voltage systems often require connection to secondary, backup power sources as a means of ensuring the safety, security, and reliability of a system for the protection of human life.

CHAPTER 7 FINAL QUESTIONS

1. What is the ratio between the primary and secondary of an isolation transformer?
2. The output of an isolation transformer is _____.
   a. floating
   b. grounded
   c. hot
   d. physically connected to the primary
3. A common-mode signal _____.
   a. is present on both sides of the power line, measuring equal in level to ground
   b. measures zero when measuring differentially between the power line terminals
   c. will increase the amount of total system noise on the output
   d. a and b
   e. a and c
4. What is the name given to the transformer that has an adjustable center tap?
5. The _____ transformer regulator uses inductive feedback to increase or decrease the levels of voltage on the output.
   a. buck-boost
   b. constant-voltage
   c. tap-changing
6. The purpose of a surge protector is to _____.
   a. limit voltage outputs to a maximum average value
   b. limit voltage outputs to a maximum peak value
   c. divert power surges and transients to ground
   d. a and c
   e. b and c

7. Compare the primary surge protector to a secondary surge protector.
   a. A primary surge protector is connected to the transformer primary, whereas a secondary surge protector is connected to a transformer secondary.
   b. A primary surge protector protects the main circuit panel from high-voltage surges, whereas the secondary surge protector is connected farther up line within the system on the branch circuit.
   c. The primary surge protector protects the main circuit panel, whereas the secondary surge protector is connected at equipment.
   d. The primary surge protector is connected at or on specific equipment at an installation, whereas the secondary protector is connected up-stream from equipment on the branch circuit.
8. The carbon-block lightning arrestor _____.
   a. operates at higher voltages than the gas tube arrestor
   b. operates at lower voltages than the gas tube arrestor
   c. is not as reliable or as accurate as the gas tube arrestor
   d. a and c
   e. b and c
9. Varistors are solid-state devices that significantly _____ their internal resistance when a voltage threshold has been reached.
   a. decrease
   b. increase
10. Once activated, a varistor _____ and allows surge currents to pass through the device.
    a. shorts
    b. opens
    c. clamps the output voltage to a maximum level
11. Power line conditioners are not able to protect against _____.
    a. variations in frequency
    b. a surge in current
    c. power blackouts
    d. a and c
    e. b and c
12. High-frequency noise or harmonic distortion on a power line _____.
    a. is not a concern, provided the ac power line is operating below 120 Hz
    b. can result in a fire due to skin effect and increased heat dissipation
    c. is only a concern on ac power lines operating above 1000 Hz
    d. can be minimized by connecting through an isolation transformer
13. There are two types of generators: motor driven and _____.

14. An ac power generator develops an output through the interaction of a rotating coil in a magnetic field. The magnetic field is generated in the _____.
    a. armature
    b. brushes
    c. rotor
    d. stator

15. What device changes ac into dc?
    a. Transformer
    b. Regulator
    c. Rectifier
    d. Varistor

16. What type of device converts dc into ac?
    a. Conditioner
    b. Rectifier
    c. Inverter
    d. Transformer

17. Which statement is true about a double-conversion UPS?
    a. The load is always connected to the output of the inverter.
    b. The load is connected to the output of the inverter only during a power failure.
    c. A backup battery is not required.
    d. The low-power systems are more efficient.

18. A full-wave rectifier provides current flow for _____ of the input cycle, and the ripple frequency is _____ the input frequency.
    a. 180°, equal to
    b. 360°, equal to
    c. 180°, double
    d. 360°, double

19. What type of diode is used as a voltage regulator?
    a. LED
    b. Rectifier
    c. Zener

20. What circuit component is used to filter out the ac ripple?
    a. Capacitor
    b. Diode
    c. Resistor
    d. Zener

21. Pure dc will have _____ ripple.

22. Power companies in the United States commonly use a frequency of _____Hz.

23. Explain the difference between a regulated 12 V, 500 mA power supply and an unregulated 12 V, 500 mA power supply.

24. The *NEC* groups storage batteries into two categories, those that measure less than _____ V and those that are over.

25. Which type of battery is rechargeable?
    a. Primary cell
    b. Secondary cell

26. A 15–ampere-hour Ni-cad battery can provide 1.5 A of current for _____ hours.

27. Rechargeable batteries are recharged using _____ the ampere-hour rating of the device.
    a. ¼
    b. ½
    c. ⅓
    d. ¹⁄₁₀

28. Which article of the *NEC* governs emergency systems and power supplies?

29. How much time does the *NEC* require storage batteries and emergency power supplies to maintain maximum load capacity?
    a. Not less than 30 minutes
    b. Not less than 60 minutes
    c. Not less than 90 minutes
    d. Not less than 120 minutes

30. According to the *NEC*, emergency systems are required to activate within _____ seconds of a power failure.
    a. 5
    b. 10
    c. 15
    d. 20

Chapter 8

Article 725 of the National Electrical Code, Classification of Circuits

Objectives

- Explain the differences among remote-control, signaling, and power-limited circuits.
- Define Class 1, 2, and 3 circuits.
- Identify Class 1, 2, and 3 circuit identifications and markings.
- Identify Class 1 power-limited circuits, including conductor size, use, and insulation rating, and list their power source requirements.
- Describe Class 1 circuit overcurrent protection requirements, and designate the location of such protection devices.
- Identify acceptable practices for placing conductors of different Class 1 circuits in the same cable, enclosure, or raceway.
- Identify Class 2 and 3 circuits, including conductor size, use, and insulation rating, and list their power source requirements.
- Explain wiring methods for Class 2 and 3 circuits for both supply-side and load-side applications.
- Explain reclassification guidelines of Class 2 and 3 circuits.
- Define the separation requirements between Class 1 circuits and Class 2 and 3 circuits.
- Outline the installation requirements for conductors of different Class 2 and 3 circuits in the same cable, enclosure, or raceway.
- Calculate the number and size of conductors in a raceway.

Chapter Outline

Sec 8.1 Types of Electrical Circuits

Sec 8.2 Article 725 of the National Electrical Code

Sec 8.3 Classification of Circuits and Class 1

Sec 8.4 Class 2 and 3 Circuits National Electrical Code Article 725.121

Sec 8.5 Power-Limited Tray Cable and Instrumentation Tray Cable

Sec 8.6 Reclassification of Class 2 and 3 Circuits, Markings, and Separation Requirements

Sec 8.7 Installation Requirements for Multiple Class 2 and 3 Circuits and Communications Circuits, National Electrical Code 725.139

Sec 8.8 Class 2 or Class 3 Circuit Conductors Extending beyond One Building

Sec 8.9 Support of Conductors and Cables

Sec 8.10 Calculate the Number and Size of Conductors in a Raceway

255

Key Terms

Class 1 circuit	inherently limited	power-limited tray cable (PLTC)
Class 2 circuit	instrumentation tray cable (ITC)	remote-control circuit
Class 3 circuit	overcurrent	signaling circuit
communication circuit	power-limited circuit	
conduit fill		

SEC 8.1   TYPES OF ELECTRICAL CIRCUITS

Electrical wiring in most modern office buildings, shopping malls, schools, colleges, hospitals, and industrial manufacturing plants may often include any or all of the following types of circuits and systems:

- Electric light and power circuits
- Motor control circuits
- Instrumentation signaling circuits
- Process control circuits
- Thermocouple circuits for measuring temperature
- Distributed control systems for controlling large-scale automation processes of complex building systems
- Programmable Logic Controllers to control and automate repetitive manufacturing processes
- Communication systems: audio, video, radio, television, closed circuit television, computer networking, and wireless

Except for the electric light and power circuits, all others are examples of specialized circuits, which as the *NEC* states, *This article covers remote control, signaling, and power-limited circuits that are not an integral part of a device or appliance* (see *NEC Section 725.1*). Wiring not considered an integral part of a device or appliance includes all circuit conductors that are connected and run externally to and from devices as required by the design of a system and the specifics of a location.

In comparison, integral circuit wiring typically is factory installed and not subject to change or alteration by the user. Integral circuit wiring is essentially the internal functional wiring of a device. In most cases, integral circuit wiring is concealed within a device, and thus is inaccessible to the user or system installer. An example would be the internal wiring of a washing machine. Except for the external power cord, there are no electrical connections to be made external to the device because all the control circuits and system wiring are self-contained within the unit. Also, integral wiring is outside of the AHJ's span of control and therefore is not subject to the AHJ's inspection privileges.

The connection of a coaxial cable from a parking lot video camera to a guard station video monitor would be an example of a **communication circuit**. The wiring of the circuit conductors between the camera and the monitor is not considered an integral part of the two devices because it must be applied externally and installed according to the unique physical parameters and restrictions of the parking lot and building. Also, the external cable does not actually make the camera or monitor function; it simply transfers a communication signal between the two devices.

Obviously, each installation is different according to the location, and in some cases, specialized hardware and wiring techniques may be required under special circumstances; special circumstances may include hazardous locations, as outlined in Chapter 5 of the *NEC*, or issues of life safety.

The specific definitions, classifications, and requirements of specialized circuits, other than electric light and power, that are not considered an integral part of a device or appliance are given in *Article 725* of the *NEC*.

SEC 8.2 ARTICLE 725 OF THE NATIONAL ELECTRICAL CODE

Article 725 of the *NEC* outlines the requirements of three general categories of circuits that are considered separate from those of electric light and power. They are remote-control, signaling, and power-limited circuits. Such circuits typically are designed and designated for a specific use and purpose. In addition, the requirements of their power supplies are often made more restrictive than those of electric light and power circuits. As an example, **power-limited circuits** typically operate from lower levels of voltage and current than the rest, and often implement some form of internal, active current limiting.

Notably, the wiring methods and installation requirements of electrical circuits, as presented in Chapters 1 through 4 of the *NEC*, still apply in all cases, except where they are amended by *Article 725*. *Article 725* does not replace Chapters 1 through 4 of the *NEC*, but instead only amends certain requirements, as determined by the type and classification of a circuit.

When designing and laying out an installation, remote-control circuits, signaling circuits, and power-limited circuits must each be further classified as either a **Class 1**, **2**, or **3 circuit**, depending on the voltage and current requirements of the connecting loads and the type of power supplies being used. Additional considerations also include the size and placement of circuit conductors, type of conductors, conduit fill, **overcurrent** protection, and the requirements of power supplies.

Let us first define the four types of circuits and then the specific ratings and variables of Class 1, 2, and 3 circuits.

Remote-Control Circuits

Remote-control circuits are circuits that control other circuits. Control voltages often can vary from as high as 600 V to as low as 5 V, depending on the type of system. As an example, motor control circuits typically use 120 V on starter coils, but many other types of control circuits operate from 24 V or less.

Low-voltage, low-current circuits commonly are used to control high-voltage, high-current systems. Coiled relays, transistors, and silicon-controlled rectifiers (SCRs) are often used to accomplish the task. The inputs to these devices typically require control voltages ranging from 5 to 24 V, with current levels in the milliamperes. Such minimal input specifications have the ability to control large output loads because of the electrical isolation and high impedance values measured between the inputs and outputs. Common uses for remote-control circuits are motor controls, elevators, conveyer systems, automated processes, and garage door openers.

Signaling Circuits

Signaling circuits are circuits that activate notification devices. Examples may include lights, doorbells, buzzers, sirens, annunciators, and alarm devices.

Power-Limited Circuits

Power-limited circuits are limited in output and capacity by the use of any or all of the following types of devices: overcurrent devices; overvoltage devices; and internal, active electronic circuitry. The power supplies of such circuits may be used on remote-control or signaling circuits, as well as in various other types of electronic circuits, including communications circuits, audio, video, computers, or any other type of specialty low-power application. *Chapter 9, Tables 11 (A)* and *(B)* of the *NEC* outline the limitations of Class 2 and 3 power sources for alternating current (ac) and direct current (dc) applications. In addition, *Chapter 9, Table 12 (A)* and *(B)* outline the power source limitations for ac and dc power-limited fire alarm (PLFA) circuits. These tables are discussed in more detail later in this chapter.

Communication Circuits

Communication circuits include telegraph and telephone, alarm systems, radio and television systems, community antenna television (CATV) and radio distribution, computer networks, and network-powered broadband systems. Although *Article 725* briefly discusses communication circuits, as related to the *Installation of Conductors of Different Circuits, NEC Article 725.139,* and the *Installation of Circuit Conductors Extending Beyond One Building, NEC Article 725.141*, the installation specifics of such systems are covered in *Articles 800, 810, 820,* and *830* of the *NEC*.

SEC 8.3   CLASSIFICATION OF CIRCUITS AND CLASS 1

Once a circuit has been designated as either remote-control, signaling, or power limited, it then must be further classified as a Class 1, 2, or 3 circuit for cabling and installation purposes. The circuit classification then defines the type of power supply to be used as well as the voltage, current, and insulation requirements of the conductors.

Class 1 Circuit

Class 1 remote-control and signaling circuits are circuits on the load side of a circuit breaker or fuse and are used to provide power to Class 1 circuits. Because these circuits connect directly to a panelboard, they are considered to be a shock hazard and danger to unknowing individuals. The installation of such circuits therefore must be performed by licensed electricians, in accordance with Chapters 1 though 4 of the *NEC*, as if they were electric light and power circuits. Licensed power-limited technicians and low-voltage installers are not allowed to service or install such circuits.

Two types of Class 1 circuits exist: Class 1 remote-control and signaling circuits, and Class 1 power-limited circuits.

Class 1 Remote-Control and Signaling Circuits

Class 1 remote-control and signaling circuits include all wiring connected between the load side of an overcurrent device and the connecting equipment. In most cases, overcurrent devices are either fuses or circuit breakers.

Class 1 remote-control and signaling circuits do not require power limiting and shall operate at levels up to but not exceeding 600 V. In the majority of cases, however, such circuits rarely exceed 120 V.

Because of potentially high levels of voltage and current, the wiring methods for all Class 1 circuits shall be installed in accordance with Chapter 3 of the *NEC*, just as electric light and power circuits. But be careful not to mistakenly jump to the next conclusion, as most people do; Class 1 circuits are not to be intermixed in the same cable, raceway, box, or enclosure with electric light and power circuits.

Two exceptions, however, do exist. Power conductors and Class 1 circuit conductors may be installed in the same cable, raceway, box, or enclosure when they are functionally associated, or when they are installed in factory or field-assembled control centers. The term *functionally associated* means that the power conductors and the Class 1 remote-control or signaling circuits are connecting to the same piece of equipment.

Factory or field-assembled control centers are considered a low risk for intermixing power conductors with those of Class 1, because they are typically in locations staffed by qualified personnel who are routinely servicing and maintaining the facility.

Separation of Unrelated Circuits

To prevent the intermixing of unrelated power conductors and Class 1 circuits, you must separate power conductors from those of Class 1 by a solid and firmly fixed barrier. For cables in cable trays, the barrier shall be made from a material compatible with the cable tray. As an alternative, when cables are in the same device box, outlet box, or cable tray, if a barrier is not used, either the power conductors *or* the Class 1 circuit conductors must be placed inside of a metal-enclosed cable as a means of providing the necessary isolation.

Two or more Class 1 circuits are allowed to occupy the same cable, cable tray, enclosure, or raceway, provided that all the conductors are insulated for the maximum voltage present.

In practice, it makes good sense to separate control circuit wiring from electric light and power conductors to help minimize the risk for unwanted electrostatic or electromagnetic noise interference on the lines. Metal conduits, although expensive and time consuming to install, act as a noise shield to high current–carrying conductors, helping to greatly reduce the effects of unwanted interference and noise spikes on sensitive circuits.

Class 1 Power-Limited Circuits

Class 1 power-limited circuits were designed and intended to handle the demands of higher power loads, where the use of Class 2 or 3 would prove to be insufficient or impractical. Although not as common as Class 1 remote-control and signaling circuits, Class 1 power-limited circuits are used in a variety of low-voltage applications, most commonly on the damper controls of environmental air systems, which are typically found in commercial and industrial building spaces.

Class 1 Power-Limited Circuit Specifications

Class 1 power-limited circuit specifications are as follows:

Class 1 power-limited circuits shall be supplied by a source having a limited output of 30 V (ac or dc) and 1000 volt-amperes (VA).

Class 1 circuits are not allowed in the same cable with communication circuits.

Class 1 transformers must comply with *Article 450* of the *NEC*.

Class 1 power-limited circuits should be supplied only through listed and labeled, Class 1, power-limited sources. *Section 725.41(A)(1)* and *(2)* outlines the specific details and specifications of these supplies.

Class 1 power supplies, not including transformers, shall have a maximum output of 2500 VA. The product of the maximum voltage and maximum current output shall not exceed 10,000 VA; these ratings are determined when overcurrent devices are bypassed.

Class 1 Overcurrent Protection Devices

Transformer supplies that feed Class 1 circuits must be protected by an overcurrent device rated to 167% of the volt-ampere rating of the source divided by the rated voltage. The overcurrent device can be built into the power supply, but it cannot be interchangeable with devices of a higher rating; interchangeable fuses are not allowed.

Class 1 overcurrent protection devices must be located at the point of supply, where the conductors receive their power.

Single-phase, single-voltage transformers may have their overcurrent devices placed on the primary side of the circuit, provided that the maximum rating does not exceed the value determined by multiplying the secondary load current by the secondary-to-primary voltage ratio of the transformer. Consult *NEC 450.3* for further details on transformer overcurrent protection.

Transformers with multiple secondary taps or windings must have their overcurrent devices placed on the secondary side of the circuit, at the point of supply. Although it is true that a device on the primary side would be able to protect the transformer from a total power overload, it would not be able to differentiate or protect individual secondary outputs to their required power-limited maximums.

Likewise, each output of a multioutput electronic power source must be protected from shorts and overloads individually, through separately connected overcurrent protection devices. Where a fuse or breaker on the input side would be able to protect the source from a total power overload, it would not be able to differentiate and protect the individual outputs to their required power-limited maximums.

Class 1 Conductor Ampacity Ratings

The load capacity for 18 AWG (American Wire Gauge) conductors shall not exceed 6 A16; 16 AWG conductors shall not exceed 8 A, as listed in *NEC 402.5*.

Ampacity levels for conductors larger than 16 AWG must be taken and calculated from *NEC 310.15* and *Tables 310.15(B)(16)* through *310.15(B)(21)*. These calculations should be left to licensed electricians and therefore are beyond the scope of this book.

Insulation Requirements

Insulation requirements for Class 1 conductors 18 and 16 AWG must not be less than 600 V. Conductors larger than 16 AWG should comply with *Article 310* of the *NEC*. All conductors must be listed for use in Class 1 circuits.

As stated earlier, all Class 1 circuit calculations and installations must be performed by licensed electricians. The Class 1 circuit ratings and regulations presented in this book have been included for general informational purposes only, as a comparison with those of Class 2 and 3 circuits. Figure 8–1 has been included for such a comparison.

SEC 8.4   CLASS 2 AND 3 CIRCUITS *NATIONAL ELECTRICAL CODE ARTICLE 725.121*

Class 2 Circuits

Class 2 circuits include all wiring connected between the load side of a listed and labeled Class 2 power source and the connecting load or equipment. Class 2 circuits are power limited and are considered safe from a shock hazard and fire initiation point of view.

Figure 8–1

Comparison of Class 1, 2, and 3 circuits.

Class of circuit	Insulation rating of conductors	Circuit voltage limit	Power limit	Load current maximum
Class 1 remote control and signaling	600 V Conductors larger than 16 AWG shall comply with *NEC Article 310*	600 V	No limit	See *NEC Table 402.5* 16 AWG not more than 8 A 18 AWG not more than 6 A The ampacity of conductors larger than 16 AWG shall be calculated according to *NEC 310.15* Flexible cords shall comply with *NEC Article 400*
Class 1 power-limited	Same as above	30 V	1000 VA	Same as above
Class 2	150 V	150 V See *Chapter 9, Table 11 (A)* and *(B)* of the *NEC*	100 VA Exceptions: Ac between 30 and 150 volts is limited to 5 mA × Vmax Dc between 60 and 150 volts is limited to 5 mA × Vmax See *Chapter 9, Table 11 (A)* and *(B)* of the *NEC*	See *Chapter 9, Table 11 (A)* and *(B)* of the *NEC*
Class 3	300 V	150 V See *Chapter 9, Table 11 (A)* and *(B)* of the *NEC*	100 VA	See *Chapter 9, Table 11 (A)* and *(B)* of the *NEC* Conductors shall not be smaller than 18 AWG

Class 3 Circuits

Class 3 circuits include all wiring connected between the load side of a listed and labeled Class 3 power source and the connecting load or equipment. Class 3 circuits also are power limited and are considered safe from a shock hazard and fire initiation point of view, but because they often operate at greater voltage and current levels, as compared with Class 2 circuits, they must therefore include additional safeguards to help protect individuals from possible electric shock.

Class 2 and 3 Power Sources

Class 2 and 3 circuits can derive their power from any of the following sources:

- A listed and labeled Class 2 or 3 transformer
- A listed and labeled Class 2 or 3 power supply
- Any alternative listed equipment marked and identified as a Class 2 or 3 power source

Alternative power sources may include computer circuit cards that supply Class 2 or 3 power to circuits, stored energy sources, batteries, and thermocouples.

Dry-cell batteries are considered to be **inherently limited** Class 2 power sources, provided that they are less than 30 V and have a maximum current capacity equal to or less than series-connected, #6 zinc cells.

A power source designed to be inherently limited is clamped internally and unable to deliver more than a specific amount of energy to a load. Any attempt to push an inherently limited source past its maximum limit causes it to either shut down or self-destruct in a safe manner. Both Class 2 and 3 power sources are available as either inherently limited or noninherently limited. Power sources that are inherently limited do not require overcurrent protection; to do so would be fine but redundant.

An example of an inherently limited power source would be an inherently limited Class 2 transformer. Such a supply often contains an internal fusible link, which is buried deep within the windings of the secondary. The link, however, is not user serviceable, thus forcing the operator or technician to replace the entire device if load currents were to increase above maximum levels. The actual replacement of the link would require a total rewinding of the transformer secondary, a job many people would find highly impractical and not easy to accomplish.

Chapter 9, Table 11 (A) and *(B)* of the *NEC* outline the specifics of ac and dc Class 2 and 3 power sources. Specifications for both inherently limited and noninherently limited sources are given in the table. In general, it should be noted that the power output maximum for almost all categories of Class 2 and 3 sources is 100 VA. The only exception would be that a type ac, Class 2 inherently limited source between 30 and 150 V must be limited to 5 mA; and a type dc, Class 2 inherently limited source between 60 and 150 V also must be limited to 5 mA.

Let us now look at the specifications for current limitation, I_{max}, and maximum current in *Chapter 9, Table 11 (A)* and *(B)*. At first glance, it may appear a bit confusing. You will notice that I_{max} in the 0 to 20 V inherently limited column, is listed as 8 A; but the maximum current specification listed at the bottom of the table states 5 A. To understand the difference, you must first read note number 1.

Note number 1 explains that the I_{max} specification is a measured current limitation when using a noncapacitive load with the current-limiting protection of the power source disabled or bypassed. The 8 A I_{max} rating therefore represents a worst-case scenario rating for the power source if the current-limiting protection ever fails or is bypassed. The 5 A rating is the maximum operating current of the power source under normal circuit conditions.

The power calculation for the above example then can be shown to be 20 V × 5 A = 100 VA; this is correct. (Remember, this is the 0–20 V column of the table.)

Also notice that VA_{max} for power sources not inherently limited is limited to 250 VA, noticeably greater than the 100 VA circuit limit. Once again, just as in the previous example for maximum current, note number 1 explains that the VA_{max} measurements are to be taken when the current-limiting protection has been bypassed or disabled.

Be careful not to confuse the current-limiting protection with the current-limiting impedance. Note number 1 also states that the current-limiting impedance is *not* to be bypassed. The current-limiting impedance of a device is a series-connected output impedance used to help match the supply to a desired load impedance, possibly 50, 75, or 600 Ω; impedance matching helps to achieve maximum power transfer from the source to the load, making the circuit more efficient. The current-limiting protection, in contrast, is used to limit load currents to some maximum level.

The Interconnection of Power Sources

As a final note, Class 2 and/or Class 3 power sources must never have their outputs connected in parallel across a single circuit or load, or interconnected in any manner as to boost the overall current output or capacity of the sources. Because the rated outputs of individual Class 2 and 3 power sources are required to be limited to 100 VA, for purposes of safety they must be kept separate. Connecting two Class 2 or two Class 3 power sources in parallel would ultimately increase the total output power capacity of the source above the 100 VA limit. The circuit then would have to be reclassified as Class 1 and approved by the authorities having jurisdiction.

Wiring Methods for Class 2 and 3 Circuits, Supply- and Load-Side Applications, *725.127* and *725.130*

The supply side of all Class 2 and 3 power sources must be wired in accordance with Chapters 1 through 4 of the *NEC*; overcurrent devices on the supply side must not be rated greater than 20 A. Remember too that supply-side connections are considered electric light and power circuits; for this reason, they must be installed by a licensed electrician. A power-limited technician can work only on the load side of a power-limited source or circuit, whereas the supply side is reserved for an electrician.

Wiring methods of Class 2 and 3 power sources on the load side of the circuit can be accomplished in one of two ways:

1. The circuits can be reclassified as Class 1 and installed in accordance with other Class 1 circuits. In such cases, the Class 2 and 3 circuit markings must be eliminated, and the derating factors of the conductors, as given in *310.15(B)(2)(a)*, shall not apply. Reclassification as a Class 1 circuit also requires the installation and future servicing to be conducted by a licensed electrician. To reclassify a Class 2 or Class 3 circuit ultimately would mean that the circuit is now off limits to power-limited technicians.

2. The Class 2 and 3 circuits can be installed under the requirements of *725.133* through *725.179*, which lay out the listing and marking of conductors, the

installation and separation requirements from **non–power-limited circuits**, and the application requirement of cables with regard to plenum, riser, general purpose, and dwelling, and their permitted substitutions.

Voltage and Insulation Requirements for Class 2 and 3 Circuits

Voltage and insulation requirements for Class 2 and 3 circuits are as follows:

Class 2 circuits shall be rated for not less than 150 V.

Class 3 circuits shall be rated for not less than 300 V.

Single-conductor Class 3 wires shall not be smaller than 18 AWG, and shall be listed and labeled as Class 3 (CL3), Class 3 plenum (CL3P), Class 3 riser (CL3R), or Class 3 residential (CL3X) cable.

Class 2 cables shall be listed and labeled as CL2, CL2P, CL2R, or CL2X.

The use and type of cable needed for an installation shall be determined by the location. Examples include general purpose location (CL2 or CL3), plenum space (CL2P or CL3P), riser (CL2R or CL3R), or dwelling (CL2X or CL3X). Be sure to refer back to Chapter 2 of this book for a more complete description of cable types and insulations.

Cable Substitutions

As an installer of electrical wiring, it is important to know which types of cable may be used for the various types of installations. In many cases, certain types of cable may be substituted for others, as permitted by location. The examples include plenum, riser, general purpose, or dwelling. *Table 725.154 (G)* of the *NEC* details the specific use, classification, reference number, and permitted substitutions of all types of Class 2, Class 3, and **power-limited tray cables (PLTCs)**.

Based on the permitted substitution chart, *Table 725.154 (G)* of the *NEC*, communications plenum (CMP) is shown to be suitable for use in any location. The logic then follows that plenum grade cable can be used as a substitute for riser cable, riser cable can be substituted for general-purpose cable, and general-purpose cables can be used in one- and two-family dwellings. The allowable substitutions drop down through the list but are not backward compatible.

SEC 8.5 POWER-LIMITED TRAY CABLE AND INSTRUMENTATION TRAY CABLE

PLTC is a special type of listed and labeled, nonmetallic-sheathed cabling that is intended for use in cable trays of factories or industrial establishments. *Table 725.154 (G)* of the *NEC* also states that PLTC may be used as an approved substitute for Class 2 or 3 wiring in general-purpose or dwelling locations. PLTC is rated for 300 V and can be purchased in sizes 22 AWG through 12 AWG.

For industrial or factory settings, another alternative to PLTC would be to use **instrumentation tray cable (ITC)** (see *Article 727* of the *NEC*). The main difference is that ITC can only be used and installed in industrial establishments that are maintained and supervised by qualified personnel. That being the case, a further benefit ITC has over PLTC is that it can be installed under the raised floors of control rooms or equipment rooms and without the use of cable trays or raceways. The only stipulation is that cables must be mounted securely and protected against physical damage. ITCs installed in open settings between cable trays and equipment racks, without the use of conduits or raceways, may extend to a maximum length of 50 ft (15.2 m); in such cases, the cables must be mounted and secured at intervals of not more than 6 ft (1.8 m).

Because ITC circuits are not under the regulations of *Article 725,* they are not considered to be Class 1, 2, or 3 circuits; instead, they are governed solely by *Article 727* of the *NEC*. For this reason, ITC shall not be used as a replacement or substitute for Class 2 or 3 wiring in a nonindustrial setting; only PLTC can be used as an approved substitute for Class 2 or 3 wiring in that case.

ITC specifications are limited to 150 V and 5 A for sizes 22 to 12 AWG. Size 22 AWG is limited to 150 V and 3 A. Although the specifications of ITC and Class 2 cable may look similar, they are not the same and should not be confused with each other, because ITC can only be installed in industrial settings.

Some manufacturers do, however, offer a brand of cable that can be used as PLTC or ITC, having a dual rating; in such cases, the cable is labeled appropriately for use as PLTC/ITC.

Neither PLTC nor ITC can be used on Class 1 circuits.

SEC 8.6 RECLASSIFICATION OF CLASS 2 AND 3 CIRCUITS, MARKINGS, AND SEPARATION REQUIREMENTS

Reclassification Guidelines of Class 2 and 3 Circuits

725.31(A) of the *NEC* states that remote-control circuits of safety-control equipment shall be reclassified as Class 1 if an equipment failure would introduce a direct fire or life hazard to the environment. Controls such as thermostats, heating and air-conditioning circuits, water temperature devices, or electrical household control devices are not to be included or regarded as safety-control equipment. The main designation would be the failure of the equipment to introduce a *direct* fire or life hazard.

One example of circuit reclassification may include the exhaust blower fans of a paint booth. The failure of the fans to remove paint fumes from the internal atmosphere of the booth would result in the accumulation of toxic vapors; such a situation would be considered a definite safety hazard from an explosive gas point of view, as well as life threatening to the operator. In such instances, the control circuits on the blower exhaust fans would have to be reclassified as Class 1 circuits.

NEC 725.31(B) also discusses the issue of possible damage to the remote-control circuits of safety equipment. In such cases, where the possibility exists for potential damage to control circuits of safety equipment, all conductors shall be installed inside of either rigid metal conduits, intermediate metal conduits, rigid nonmetallic conduits, electrical metallic tubing, type mineral-insulated (MI) cable, or type metal-clad (MC) cable, or be otherwise suitably protected from the environment and the elements. The control circuit conductors must be protected from possible abuse or damage to ensure that the functional use of safety equipment is not compromised or made inoperable.

Class 1, 2, and 3 Circuit Identification and Markings, *National Electrical Code 725.30, 725.154,* and *725.179*

Class 1, 2, and 3 circuits shall be identified at terminals and junctions in a manner that prevents the unintentional interference by service technicians. When servicing a system, there should be no confusion on the part of the technician as to which wires and cables are to be tested. The unwanted interference to other circuits can be avoided easily by simply marking and labeling conductors as to their type and proper designation.

Separation Requirements for Power-Limited Class 2 and 3 Circuits from Those of Class 1 and All Other Non–Power-Limited Circuit Conductors (See *National Electrical Code 725.133* and *725.136(A) through (I)*)

A common thread exists through all of *NEC 725.136 (A through I)*, the requirement that Class 2 and 3 circuits must be separated from electric light and power circuits, Class 1 circuits, non-PLFA, or medium-powered, network-powered broadband communication circuits.

The general requirement of *725.136(A)* states that Class 2 and 3 circuits shall not be placed inside of any cables, cable trays, compartments, enclosures, manholes, outlet boxes, device boxes, raceways, or any other similar fitting with Class 1 or any other non–power-limited circuit conductors.

Non–power-limited circuit conductors include electric light and power conductors, non-PLFA, or conductors of medium-powered, network-powered broadband communication circuits. However, provided that separation does exist between power-limited Class 2 and 3 circuits and those conductors of Class 1 and non–power-limited circuits, then certain allowances are permissible, as referenced by *725.136 B* through *I*.

There are a number of reasons why the separation ruling by the *NEC* is required, and also why it simply makes good sense. The reasons include protecting conductors of power-limited, Class 1, and non–power-limited circuits from internally shorting to each other; isolating circuits based on the rights and permissions of individuals to work on and maintain them; the safety and protection of individuals from their own technical ignorance; and also to electrically isolate power-limited circuits from possible electrostatic and electromagnetic interference.

First, if power-limited Class 2 or 3 circuits were to ever short out across Class 1 circuits, or to other non–power-limited circuits, the surrounding environment could easily develop into a potential shock or fire hazard. Remember also that the insulation requirements for Class 2 or 3 circuit conductors are significantly less than those of Class 1 or electric light and power circuits. Imagine what would happen if a shorted Class 2 circuit conductor, having a maximum insulation rating of 150 V, were to be exposed to the higher voltage and current specifications of a 600 V rated, electric light and power conductor. Ultimately, the Class 2 cable would not be able to hold back the potential threat, and over time the buildup of heat from the additional current flow resulting from the short would start to deteriorate the internal and external insulation and eventually burn the conductors. The separation of power-limited conductors from those of Class 1 and non–power-limited conductors is required to prevent such an occurrence.

A second reason for the need to separate power-limited circuits from those of Class 1 and non–power-limited circuits has more to do with the rights and restrictions of those individuals intending to work on and maintain such circuits. High-power circuits are intended to be installed and maintained by trained, licensed electricians. Master electricians are allowed to work on any type of circuit ranging from electric light and power to those of power limited and communications. Master electricians are also allowed to install any type of circuit inside of a hazardous location, as stipulated by *Article 500* of the *NEC*. Power-limited technicians, however, are far more restricted with regard to the type of work they can perform; they also are not allowed to work in hazardous locations. The permissible types of circuits that power-limited technicians can install include Class 2 and 3 circuits, audio circuits (as referenced by *Article 640* of the *NEC*), PLFA circuits (as referenced by *Article 760* of the *NEC*), fiber-optic circuits (as referenced by *Article 770* of the *NEC*), and communication circuits (as described in *Articles 800, 810, 820,* and *830* of the *NEC*).

For obvious reasons, power-limited circuits should be logically separated and isolated from those of Class 1 and non–power-limited circuits as a guarantee that the power-limited technicians do not accidentally interfere with conductors and circuits that they are restricted from working on. Separation also helps to reduce the risk for potential shorts on the lines of Class 1 or electric light and power circuits because of accidents, misunderstandings, or misconnections by service technicians.

From a safety point of view, Class 1 and non–power-limited circuit conductors should also be separated from those of power-limited circuits as a way to help protect unwitting individuals from a potential threat or shock hazard. Most individuals seem to believe that simply because they were able to install their own car stereo system, this somehow makes them knowledgeable enough to install all types of communications and electrical circuits. In certain situations, such beliefs can be dangerous. To help protect these individuals from themselves and their own ignorance, having power-limited circuits separated and installed inside of isolated compartments, and away from electric light and power circuits, is a wise and useful practice.

The last reason why circuit separation simply makes good sense is to help protect low-voltage conductors from the unwanted electrostatic or electromagnetic interference of high-power conductors. Often, low-voltage circuits with a high degree of sensitivity are unable to tolerate even small levels of electrical noise or interference on their lines; in some cases, noise levels as low as 10 mV often can generate unwanted intermittencies or circuit failures.

One example may be the need to separate audio conductors from those of electric light and power conductors. Grounded, unshielded audio lines running long distances ultimately pick up an annoying 60 Hz power hum as a result of their proximity to high-current power lines. The unwanted interference then is permanently imposed on the intended audio signal and amplified, to the great annoyance of listeners. To help prevent such unwanted interference, unbalanced audio conductors should be placed inside of insulated raceways, separate and away from high-power circuits and conductors.

The permitted subparts to *NEC 725.136* allow some deviation from subpart A, which insists on maintaining the absolute separation and isolation of Class 1 and non–power-limited circuit conductors from those of power-limited Class 2 and 3 circuits. The permissions, as stated in subparts B through I, all include some form of separation and isolation, either by barriers or isolated compartments or by placing raceways within enclosures. Let us now look at the specifics of subparts B through I.

725.136 (B), Separated by Barriers

Class 2 and 3 circuits may be installed with conductors of Class 1 and non–power-limited circuit conductors where they are separated by a barrier.

725.136 (C), Raceways within Enclosures

Inside of enclosures, Class 2 and 3 circuits may be installed inside of isolated raceways as a means of separating them from electric light and power, Class 1 circuits, and all other types of non–power-limited circuit conductors.

725.136 (D), Associated Systems within Enclosures

Class 2 and 3 circuits may be installed inside of compartments, enclosures, device boxes, or outlet boxes with those of electric light and power circuits, Class 1 circuits, or all other types of non–power-limited circuit conductors, provided that the circuits are functionally associated and that they enter the enclosure through separate openings.

Functionally associated implies that the non–power-limited conductors are providing the means of power to the equipment within the enclosure, and the Class 2 or 3 conductors are being used for remote-control or signaling purposes. In such a situation, the Class 1 or non–power-limited conductors must be routed and secured within the enclosure to maintain a $\frac{1}{4}$-in. separation from all power-limited conductors.

An alternative would be to install all circuits operating at less than 150 V to ground as Class 3 circuits, by using CL3, CL3R, or CL3P cabling. In such cases, the conductors extending beyond the jacketing must maintain a $\frac{1}{4}$-in. separation from all other conductors, or the Class 3 connections can be placed inside of a nonconductive barrier or nonconductive sleeve, such as flexible tubing.

A last alternative would be to install the Class 2 or 3 circuits as Class 1, in accordance with all the rules and regulations of *725.41* of the *NEC*.

725.136 (E), Enclosures with Single Openings

Class 2 and 3 circuit conductors can be included inside of enclosures with single openings with those of Class 1 circuits and non–power-limited circuit conductors, provided that they are functionally associated and that the Class 2 or 3 conductors are permanently separated from all others by a continuous, nonconductive type of insulator, such as flexible tubing. When entering the enclosure, the insulator is required to be placed over the outer jacketing of the power-limited cables as an added layer of protection against possible shorts and contact with the non–power-limited conductors.

Fire alarm sprinkler heads are a perfect example of this type of assembly; that is, where the high-power conductors are used to power the water flow valves, and the power-limited conductors are used to control the initiating relays and also to establish the supervisory circuit for the fire alarm control panel.

725.136 (F), Manholes

Inside of manholes, one of the following three possibilities exists when combining conductors and cabling from various types of circuits:

Either all of the electric light and power, Class 1, non-PLFA, and/or medium-powered, network-powered broadband communication circuits are to be installed inside of a metal enclosed cable or type underground feeder (UF) cable;

Or the Class 2 or 3 circuit conductors are installed inside of a firmly fixed, nonconductive barrier, such as flexible tubing, as a means to permanently separate the power-limited conductors from all other Class 1 and non–power-limited conductors;

Or the Class 2 or Class 3 circuit conductors are to be securely fastened to rack mounts, insulators, or any other approved supports meant to permanently separate them from all Class 1 and non–power-limited circuit conductors.

725.136 (G), Cable Trays

Power-limited circuits may be installed inside of cable trays with conductors of non–power-limited circuit conductors, provided that they are separated by a solid and firmly fixed barrier of a material compatible with the cable tray.

A second alternative would be to place the power-limited conductors inside of type MC cable (as described in *Article 330* of the *NEC*).

725.136 (H), Hoistways

Inside of hoistways, which were designed for the purposes of elevators or dumbwaiters, power-limited circuit conductors are to be installed inside of rigid metal conduit, rigid nonmetallic conduit, intermediate metal conduit, liquid-tight flexible nonmetallic conduit, or electrical metallic tubing, as a means of separating them from all Class 1 and non–power-limited circuit conductors.

725.136 (I), All Other Applications

In all other cases, Class 2 or 3 conductors must maintain a minimum separation of 2 in. from all Class 1 and non–power-limited circuits.

The 2-in. (50 mm) separation, however, is not required when the Class 1 and non–power-limited conductors are placed inside of isolated raceways, or are in metal-sheathed, metal-clad, nonmetallic-sheathed, or type UF, underground feeder cable.

The 2 in. (50 mm) separation also is not required if, instead, the Class 2 and 3 circuit conductors are placed inside of the isolated raceways, or are in metal-sheathed, metal-clad, nonmetallic-sheathed, or type UF, underground feeder cable.

A third and last possibility allows for the Class 2 and 3 circuit conductors to be separated from those of Class 1 and non–power-limited circuit conductors by placing them inside of a firmly fixed nonconductor, such as a porcelain tube or flexible tube.

In all the previous listed examples of *subparts A through I*, the basic rule is simple: non–power-limited conductors are required to be separated from power-limited conductors.

Electric light and power circuits, Class 1 circuits, non-PLFA, and medium-powered, network-powered broadband communication circuits are all considered to be non–power-limited circuits; and for the purpose of safety and fire prevention, they must maintain a fixed and continuous separation from all Class 2 and 3 circuit conductors. Other reasons for the separation may include:

- The prevention of possible short circuits between power-limited and non–power-limited conductors
- The isolating of circuits based on the rights and permissions of individuals to work on and maintain them
- As a means to help protect individuals from their own technical ignorance
- To electrically isolate power-limited circuits from electrostatic and electromagnetic interference

SEC 8.7   INSTALLATION REQUIREMENTS FOR MULTIPLE CLASS 2 AND 3 CIRCUITS AND COMMUNICATIONS CIRCUITS, *NATIONAL ELECTRICAL CODE 725.139*

Class 2 and 3 Circuits in the Same Cable, Enclosure, or Raceway

When intermixing multiple conductors of Class 2 and 3 circuits within the same cable, enclosure, or raceway, the following rules apply:

Two or more Class 2 circuits may share the same cable, enclosure, or raceway.

Two or more Class 3 circuits may share the same cable, enclosure, or raceway.

Class 2 circuit conductors may be included inside of the same cable, enclosure, or raceway with conductors of Class 3 circuits, provided that the Class 2 conductors are insulated to the minimum requirements of the Class 3 circuit, greater than 300 V.

Class 2 and 3 Conductors with Communications Circuits

Class 2 and 3 circuit conductors are permitted to be installed in multistranded cables together with communications circuits provided that they are reclassified as communications circuits and are installed in accordance with *Article 800* of the *NEC*. In such cases, the cabling must be a listed communication cable or multipurpose cable.

Composite Cables

Compared with the previous example, a composite cable is somewhat different; it is not the same as having multiple circuits running through a single multiconductor cable. Instead, cable vendors often engineer a composite cable, made up of individually listed Class 2 and 3 cables, combined with those of communications cables, all under a common outer jacket. The fire resistance of the entire composite cable then is tested and rated based on its own individual performance. In such cases, all conductors within the composite cable then are permitted to be reclassified as communications cables, and are installed in accordance with *Article 800* of the *NEC*.

Class 2 or 3 Cables and Other Circuits

Jacketed Class 2 or 3 circuit cables are allowed to be installed in the same enclosure or raceway with jacketed cables of the following:

- PLFA circuits
- Nonconductive and conductive optical fiber cables, in accordance with *Article 770*

- Communications circuits, in accordance with *Article 800*
- CATV and radio distribution systems, in accordance with *Article 820*
- Low-power, network-powered broadband communications, in accordance with *Article 830*

Class 2 or 3 Conductors or Cables and Audio Circuits

NEC 725.139(F) requires that audio circuit conductors, as described in *NEC 640.9(C)*, installed as Class 2 or 3 wiring from an amplifier, shall not be permitted to occupy the same cable or raceway with other Class 2 or 3 conductors or cables. Essentially, this ruling requires that audio cables must maintain a permanent separation from remote-control or signaling conductors, even though they may both be considered as Class 2 or 3.

SEC 8.8 ## CLASS 2 OR CLASS 3 CIRCUIT CONDUCTORS EXTENDING BEYOND ONE BUILDING

NEC 725.141 requires that Class 2 or Class 3 circuit conductors, extending beyond one building, in an exterior setting, shall be installed according to *Chapter 8* of the *NEC*. When cables are installed outside of buildings, they are susceptible to lightning strikes and therefore must be appropriately grounded and protected by primary protectors at the point of entrance. The other issue is the proximity of Class 2 and Class 3 circuit conductors to electric light or power conductors operating over 300 volts to ground. *Article 725* of the *NEC* references *800.44, 800.50, 800.53, 800.93, 800.100, and 800.170(A)* and *(B)* for all communication cables, and *820.44, 820.93,* and *800.100* for coaxial cables. The installation of exterior cables entering buildings is discussed in greater detail during the grounding and bonding discussion in Chapter 3 of this book.

SEC 8.9 ## SUPPORT OF CONDUCTORS AND CABLES

NEC 725.143 states that conductors of Class 2 or 3 circuits are not to be strapped, taped, or attached by any means to the exterior of any conduit or raceway as a means of support. The same applies to fire alarm cables (*760.143*), communication cables (*800.133(B)*), coaxial cables (*820.133 (B)*), and network cables (*830.133 (B)*). One permitted case, however, does exist, as referenced by *300.11(B)(2)*.

300.11(B)(2) allows for a Class 2 cable to be attached to the exterior of a conduit when the conduit internally contains functionally associated power supply conductors. This specific example exists in almost every house across the United States; the Class 2 thermostat control wires that are tied to the exterior of the main electrical power conduit of the furnace. In such cases, the internal conduit wires supply power to the furnace, whereas the external thermostat wires provide the Class 2 control circuit; both are functionally associated to the furnace.

Except for cables being installed for heating, ventilating, and air-conditioning systems, there are few additional cases, if any, where Class 2 conductors or cables are allowed to be attached to the exterior of a conduit or raceway. Notice also that Class 3 conductors or cables are never mentioned by *300.11(B)(2).* Even if they are functionally associated, Class 3 circuit conductors do not have the same options available to them as those of Class 2.

Not attaching cables to the exterior of any conduit or raceway is usually a wise and good practice. To do so would make future changes or remodeling efforts far more difficult to achieve, often requiring entire systems to be pulled out altogether, replaced, or rewired. In such instances, it makes life a lot easier to keep dissimilar systems isolated from each other and installed on separately supported mounting hardware.

As a related topic, *725.24,* entitled the *Mechanical Execution of Work,* states that circuits shall be installed in a neat and professional manner; cables installed on the outer surfaces of walls and ceilings shall be structurally supported so as not to subject them to possible damage or abuse from normal building use. Support hardware, such as hangers, straps, and staples, should also be designed and used appropriately, and not cause any undue stress or damage to cables.

An example of normal building use would be the opening and closing of a door. Obviously, cables should not be hanging down in such a manner so as to be trapped or pulled by the movement of a door throughout the course of a day.

A last topic related to the installation practices of cables involves access to electrical equipment behind panels. *725.21* states that the accumulation of wires and cables shall not prevent the removal of access panels, including suspended ceiling tiles.

When installing cables above suspended ceilings, cables must never be allowed to rest on the top of tiles, because the accumulation and weight of cables make it nearly impossible to remove tiles at a later date. Over time, the excessive weight of cable bundles may also cause tiles to warp or crack.

When installing cables above a suspended ceiling, cables also must never be tied to the suspended ceiling support wires. Instead, cables and conductors should be installed and supported by an approved means. Approved means indicates that the installer is using an appropriate mounting hardware acceptable to the electrical inspector. At the end of the day, conductors must be mounted up and out of the way of removable tiles or panels.

Regulation by the *NEC* regarding the support of cables and the mechanical execution of work simply makes good sense. In many instances, the regulations are intended to help installation technicians assume a reasonable amount of pride in their work. A job performed and installed in a neat, organized, and professional manner results in a safer environment and most likely will not develop into a potential safety or fire hazard.

SEC 8.10 CALCULATE THE NUMBER AND SIZE OF CONDUCTORS IN A RACEWAY

NEC 300.17 determines the purpose of **conduit fill**, concerning the number and size of conductors in a raceway. In addition, the *Informational Note* of *300.17* lists all the sections of the *NEC* that require conductors to be installed in accordance with *Chapter 9, Table 1* of the *NEC*. (An example problem is shown below.) When installing electrical wiring within a raceway, the number of conductors to be run should be limited to a certain percentage of fill. Limiting the number of conductors within a raceway allows for the easy addition and removal of future wires, to avoid damaging those already existing. In addition, not stuffing the raceway to the maximum fill allows room for the dissipation of heat by the conductors.

When determining the number of conductors permissible in a particular conduit, the starting point should be the type of raceway to be used. From there you need to identify the number of conductors that are placed to be inside the raceway, remembering to take into account the *NEC* fill restrictions; the pipe then must be sized accordingly. Let us take a closer look at the steps involved.

Example

What size flexible metal conduit (FMC) is required for the following conductors: 5—#14 THW; 6—#12 TW; 7—#10 THHN; and 8—#8 THWN? To solve this problem, you must first find the article specifically dedicated to FMC (see *Article 348*). As you read *Article 348,* be aware that there are some additional details you need to know. First, where do you find the number of conductors allowed in this type of conduit, and where do you find the percentage fill requirements for FMC?

Notice that *348.22* refers to *Chapter 9, Table 1* of the *NEC*. In looking at *Table 1,* you should determine that because there are more than two conductors running through the raceway, the allowable fill requirements should be 40%. We use this percentage later when trying to determine the right size pipe to use in the installation. Now, even though *Table 1* is quite small, we still cannot ignore the notes supplementing the table. *Table 1,* note 6, for example, tells exactly where to find the dimensions of different types and sizes of conductors; it also points to the appropriate table for sizing out conduit once the cable dimensions are known. To find the total area of all conductors, in square inches, we must first go to *NEC Chapter 9, Table 5.* First, locate the types and sizes of wire that run through the raceway. Then, multiply the area of each, in square inches, by the number of conductors. In this example, the total area in square inches, for each type of conductor is:

$$\#14 \text{ THW} = 0.0209 \times 5 \text{ CONDUCTORS} = 0.1045 \text{ in.}^2 (67.4 \text{ mm}^2)$$
$$\#12 \text{ TW} = 0.0181 \times 6 \text{ CONDUCTORS} = 0.1086 \text{ in.}^2 (70.1 \text{ mm}^2)$$
$$\#10 \text{ THHN} = 0.0211 \times 7 \text{ CONDUCTORS} = 0.1477 \text{ in.}^2 (95.3 \text{ mm}^2)$$
$$\#8 \text{ THWN} = 0.0366 \times 8 \text{ CONDUCTORS} = 0.2928 \text{ in.}^2 (188.9 \text{ mm}^2)$$

The totals for all conductors in the above problem are then added together to give a grand total of 0.6499 in.² (419.3 mm²).

Now, turn back to *NEC Chapter 9, Table 4,* and find Flexible Metal Conduit (see *Article 348*). Looking at the 40% fill column, because there are more than two conductors in this problem, compare the total calculated area of all conductors (0.6499 in.² [419.3 mm²]) with the trade size dimensions of various conduits listed in the table. Choose the conduit size that is large enough to fit all of the conductors. The answer should be a 1.5-in. (38.1 mm) FMC. Why? Because the conduit one size smaller measures only 0.511 in.² (330 mm²), and this would be too small to fit the required total of 0.6499 in.². As a result, the next size up, 0.743 in.², is the obvious choice.

In the previous example, none of the cables is a multiconductor, and all are listed in *Chapter 9, Table 5.* If multiconductor cables are used, refer to *Chapter 9, Table 1,* notes 5 and 9. Note 5 states that the actual dimension of the multiconductor cable can be used to calculate conduit fill. Note 9 also states that for multiconductor cables with an elliptical dimension, the widest or major diameter of the cable shall be used when calculating the cross-sectional area. All that is required, therefore, is to simply measure the widest outer diameter of the cable manually, and then calculate the cross-sectional area by using the formula for area: area $= \pi r^2$. Be sure to divide the diameter by 2 to obtain the radius; otherwise, the calculation will be off by a factor of 4.

The last point to make has to do with the 60% fill column in *NEC Chapter 9, Table 4.* When can you use 60% fill? The answer can be found in *NEC Chapter 9, Table 1,* note 4. Note 4 states that 60% fill can be used when a conduit connects between two boxes, cabinets, or enclosures, not exceeding 24 in. in length. This would be the only case for using 60% fill. In this special case, adjustment factors as given in *NEC 310.15(B)(2)(a)* need not apply.

CHAPTER 8 FINAL QUESTIONS

1. Give an example of a remote control circuit.
2. Give an example of a signaling circuit.
3. What defines a power-limited power supply and how is it different from a non–power-limited power supply?
4. Define the term *functionally associated.*
5. The voltage ratings of an 18 AWG, Class 1 conductor shall be _____.
6. The minimum voltage rating of PLTC shall be _____.
7. Class 1 power-limited circuits shall not exceed _____ V and _____ VA.
   a. 30, 1000
   b. 100, 1000
   c. 150, 1000
   d. 600, 1000
8. Which Chapter and Table of the *NEC* define the specifications for Class 2 and 3, ac and dc power supplies?

9. Single conductors of Class 3 circuits shall not be smaller than
   _____.
   a. 18 AWG
   b. 20 AWG
   c. 24 AWG
   d. 26 AWG

10. Class 2 cables shall have a voltage rating of not less than _____ V.
    a. 100
    b. 150
    c. 300
    d. 600

11. Class 3 cables shall have a voltage rating of not less than _____ V.

12. _____ conductors shall be permitted to be attached to the exterior
    of a functionally associated conduit.
    a. Class 1
    b. Class 2
    c. Class 3
    d. Class 2 or Class 3

13. Class 2 and Class 3 circuit conductors not installed in conduits shall be
    separated by not less than _____ from exposed conductors of any
    electric light, power, Class 1, non–power-limited fire alarm, or medium-
    power network broadband communication circuits.
    a. 2 in. (50.8 mm)
    b. 4 in. (101.6 mm)
    c. 12 in. (304.8 mm)
    d. 6 ft (1.8 m)

14. Class 2 and Class 3 conductors shall be permitted to be installed with
    conductors of Class 2 or Class 3 audio loudspeaker circuits.
    a. True
    b. False

15. What would be a permitted substitution for CL3 cable?
    a. CMG or CM
    b. PLTC
    c. CMR, CL3R or CMP, CL3P
    d. All of the above

16. Which of the following is not permitted as a power source for Class 2 or
    Class 3 circuits?
    a. Listed Class 2 or Class 3 transformer
    b. Field-constructed Class 2 or Class 3 power supply
    c. Other listed equipment marked to identify the Class 2 or Class 3
       power source
    d. Listed information technology equipment–limited
       power circuits

17. Class 2 conductors are installed in an enclosure that also contains electric power circuit conductors for the related equipment power.

    a. Acceptable

    b. Violation

18. What is CL3X or CL2X cable used for?

    a. Industrial applications

    b. Cable trays only

    c. One and two-family dwellings

    d. Risers

19. Conductors of one or more Class 2 conductors shall be permitted within the same cable, enclosure, or raceway with conductors of Class 3 circuits, provided

    a. the circuit voltages are less than 30 volts.

    b. that all circuits are functionally associated.

    c. that the insulation of the Class 2 circuit conductors is at least that required of Class 3 conductors.

    d. the job foreman has approved it.

20. What size electrical metallic tubing would you need to accommodate twenty 18 AWG XFF and eighteen 16 AWG XFF conductors?

Chapter 9

Fire Alarm Systems

Objectives

- Define fire alarm system.
- Understand the various types of fire alarm systems and their components.
- List and define the operation of various types of alarm-initiating devices.
- Identify the types of alarm-indicating appliances and explain their function in the fire alarm system.
- Understand the operation and use of manual fire alarm boxes, automatic fire detectors, and audible and visible signaling.
- Define the differences between Class A and B fire alarm circuits.
- Identify the electrical requirements of a power-limited and a non–power-limited fire alarm system.
- Practice safe wiring test guidelines.
- Understand basic design guidelines of a fire alarm system.

Chapter Outline

Sec 9.1 The Intended Purpose of Fire Alarm Systems

Sec 9.2 What Is a Fire Alarm System?

Sec 9.3 Control Units and Alarm-Initiating Devices

Sec 9.4 Notification Appliances

Sec 9.5 Electrical Requirements of Power-Limited and Non–Power-Limited Fire Alarm Systems, *National Electrical Code Article 760*

Sec 9.6 Wiring of Fire Alarm Systems

Sec 9.7 Troubleshooting

Key Terms

annunciator

annunciator panel

central station

circuit integrity (CI)

control unit

emergency telephone

end-of-line device

end-of-line relay

fire alarm circuit

fire command center

heat detector

initiating device circuit (IDC)

master box

mineral insulated (MI)

non–power-limited fire alarm circuit

notification appliance circuit (NAC)

photoelectric smoke detector

power-limited fire alarm circuit

proprietary station

protected premises

public fire alarm

public fire alarm reporting system

pull station

remote station

signaling line circuit (SLC)

smoke detector

strobe

supervisability

supervisory station

survivability

SEC 9.1 THE INTENDED PURPOSE OF FIRE ALARM SYSTEMS

Fire alarm systems are necessary for the protection and preservation of life, property, and the individual mission of an institution or business. First and foremost, fire alarm systems are installed for the preservation of life—all other issues remain secondary. The secondary goals and requirements of the fire alarm system are dependent on the type of property being protected and the overriding mission of the organization. As a result, fire alarm systems are tailor-made to better address the special needs of a business; for example, electrical and chemical plants, paper mills, hospitals, or large manufacturing facilities all have differing goals and critical processes that may require special engineering and handling to better accommodate and protect the people within the space, property and inventory, and the surrounding environment.

Property protection remains secondary to life safety, and it is often the focus of an institution needing to minimize damage to the internal contents of a building or structure. Examples may include a public library, museum, or storage facility. In such cases, the actual replacement and recovery of damaged or lost property would be nearly impossible to achieve, and in most cases the contents are irreplaceable. The obvious loss of such items, often priceless in nature, would be a great tragedy to educators and society as a whole. Fire alarm systems designed to protect valuable property are different in nature from those of life safety. Equipment and hardware vary, depending on the degree of protection needed and the type of institution.

The protection of business based on an operational mission is used by institutions requiring minimal interruption to their daily flow of operational services. A hospital, for example, would be required to continue operational services to customers, uninterrupted, because of the critical nature of their business. Energy plants, security control systems, and telecommunications centers are also examples of businesses that may be *mission* oriented when designing and engineering the fire alarm system for their facility. Organizations such as these cannot be made to completely shut down or evacuate, because their survival and presence are required to maintain the safety of others.

Fire alarm installations often conform to the minimum requirements of the National Fire Alarm Code® (National Fire Protection Agency document 72), and the *NEC* (NFPA document 70). NFPA 72 details all of the minimum recommended *Code* requirements associated with the design, installation, and testing of fire alarm systems, together with the related information on initiating and signaling devices, sensors, power supplies, communications, inspection, and system maintenance. Other authorities and code requirements pertaining to fire alarm systems may include:

NFPA 101, Life Safety Code®: Life Safety Code is concerned with the protection of human life. Life Safety Code defines early warning fire notification, allowing for the safe evacuation of individuals within a building or structure. From the life safety point of view, fire alarm systems must be continually tested and maintained at regularly scheduled intervals to ensure their reliability and ability to effectively warn individuals during an actual alarm.

International Building Code (IBC)/Municipal and Local Building Codes: International Building Code is regarded as the minimum code standard for building construction in the United States. Municipal and local building codes, however, may differ from place to place. In most instances, local municipalities adhere to the minimum *Code* requirements of the IBC, the *NEC* and the National Fire Alarm Code, but often there are additional *Code* requirements, often based on standards developed by standards agencies, such as Underwriters Laboratories (UL). The type of installation and the AHJs determine the correct use of *Code* and standards. Local authorities do have the right to alter and add to the basic *Code* guidelines, and do so in many situations.

Insurance Companies: Insurance companies may also require additional *Code* and safety requirements for installations, depending on the type of policy, risk analysis (to property and individuals), and market value of the potential loss.

At the end of the day, although the AHJs have the authority to interpret the *Code*, their rulings can always be appealed to the next level up in the chain of command. Final rulings would ultimately be decided by either the local municipal county attorney or, ultimately, at the state level.

NFPA 72 also covers the requirements of partial systems and nonrequired systems. A nonrequired system includes any additional design parameters or special performance criteria that may have been added to a system by a building owner or resident, over and above the minimum requirements mandated by building code, fire code, or electrical code. Ultimately, the AHJs must approve all parts of an installation and it shall be approved only after the AHJ has determined that the additional features are equivalent or superior to the minimum code requirements. NFPA 72, Section 1.5, Equivalency determines the requirements of nonrequired systems. In some cases, a system or system device found not meeting specific *Code* or standards requirements may need to be submitted to a testing laboratory, such as

UL, or a professional engineer for a determination of whether the device meets the initial intent of the *Code*. Only on approval by the testing laboratory or processional engineer will the system or system device be listed as suitable for use.

SEC 9.2   WHAT IS A FIRE ALARM SYSTEM?

A fire alarm system is a system that monitors and annunciates (announces) the status and conditions of a fire, to initiate an appropriate response to the warning. In most cases, the system notification not only alerts the occupants of the building and emergency response personnel but also provides the necessary control functions to utility and building services, such as elevators, heating and air-conditioning fans and dampers, electric door holders, and emergency lighting, just to name a few.

NFPA 72 Section 3.3.95, Definition: Fire Alarm–*A system or portion of a combination system consisting of components and circuits arranged to monitor and annunciate the status of fire alarm or supervisory signal-initiating devices and to initiate appropriate response to those signals.*

What sets fire alarm systems apart from other electrical systems is their ability to continually monitor and test for system readiness, together with the overall integrity of active devices and circuit functionality. Often referred to as **supervisability**, the fire alarm system contains a secondary supervisory circuit that acts as an active status loop, allowing the main fire alarm panel to continually monitor system devices for readiness, as well as the continual verification of circuit connectivity and integrity. As a result, a fire alarm system performs routine checks on all field wiring, making sure there are no open wires, shorts, or accidental grounds. The system also continually verifies the functionality of all active devices and components such as smoke alarms, strobes, or horns. A single detected failure alerts the supervisory staff by generating an audible and visible trouble alert signal within the control panel, supervisory device, or supervisory alarm center. The trouble alert indicates that the integrity of the system may have been compromised. An example of a trouble alert in a residential setting would be a low battery indicator on a smoke alarm. The alarm ultimately begins to beep and flash in response to low power, indicating that the device is not functioning properly. Appropriate action then must be taken when a trouble alert signal occurs.

In larger commercial systems, the fire alarm control panel indicates the status of the system and connecting devices in one of four ways: normal, trouble alert, alarm, or supervisory signal. The control panel indicates the status of all events through either a series of colored light-emitting diodes, an alphanumeric display, or a graphical display. In the event of a trouble or alarm indication, the visual display on the panel is also followed by a uniquely identifiable audible beep or alert tone.

A trouble alert is issued whenever a device is found to be inoperable, or if system wiring has been somehow compromised. In the event of such an occurrence, all system errors must be verified and repaired in a timely manner to ensure the functionality of the overall system and reliability of the connecting devices. The time to detect a system failure is not during an actual fire alarm. A supervisory signal from the panel indicates that an off-normal condition on a suppression system, such as a sprinkler system, wet/dry chemical suppression system, or gaseous suppression system, has occurred. In such cases, the supervisory signal indicates that water may be flowing or that pressure switches, valve tamper switches, temperature switches, or water level switches have changed state.

Types of Fire Alarm Control Panels

Fire alarm control panels are listed for specific purposes of use and given a UL rating based on the type of services they can perform. The types of ratings include:

- Service Type M, indicating the connection of manual devices, such as pull stations
- Service Type A, indicating the use of automated detectors, such as heat detectors, smoke detectors, and duct detectors
- Service Type SS, indicating the use of sprinkler supervisory devices, such as sprinkler valves, water level and temperature sensors, and dry pressure switches
- Service Type WF, indicating water flow devices or flow switches
- Service Type WS, indicating watchman supervisory devices, such as fire watchmen and guard tour check stations

For Service Type WS, in some large buildings, watchmen and guards often tour the facility at regular intervals. The guard-tour check stations, located throughout the **protected premises,** ultimately tie back to the main fire alarm control panel. Guards, who regularly tour the facility performing the supervisory functions within specified zones of the fire alarm system, manually activate tour results back to the main control panel, where they are recorded. The system automatically transmits start signals to the receiving locations, which then must be initiated by guards at the start of their tours. Delinquency signals are automatically transmitted by the system 15 minutes after a predetermined actuation time if the guard has failed to actuate the start of the tour. A finish signal then is transmitted within a predetermined time interval after the start of the guard's tour. All start, delinquency, and finish signals are recorded at the signal-receiving locations.

It should also be noted that fire alarm control panels listed for households or residential use are never to be used in commercial settings. The NFPA defines a household as a one- or two-family residential unit. Although apartment buildings and condominiums are considered to be commercial spaces, residential control panels can be used within the individual dwelling units; however, outside of the individual living units, all devices and control panels must be listed for use in commercial spaces. For more information on the subject, UL 864 discusses fire alarm control panels and UL 985 discusses household fire warning system units.

Types of Fire Alarm Systems

The various types of fire alarm systems and components as defined by the NFPA are listed in the following sections.

Single-Stage Fire Systems

A single-stage system is designed so that when activated, the alarm signal is immediately transmitted throughout the building to warn the occupants that a fire emergency exists.

Household Fire Alarm Systems (Residential Systems)

Household fire alarm systems may only be installed in dwelling units, including individual houses, apartment buildings, and hotel rooms. The system usually consists of a single **control unit** that acts as the master alarm system. The control unit provides input connections for initiating devices, such as smoke detectors and heat detectors, together with output connections to notification appliances, such as strobes and horns. The system may also be configured for single-station smoke alarms, providing single-room notification, or multistation smoke alarms for use in various locations around the living space. As an example, a multistation smoke alarm would announce an alarm in more than one room of the dwelling. In such a case, if the living room detector were activated, the alarm signals would activate in the living room and also in multiple locations around the dwelling, such as the bedrooms, kitchen, and family room.

Commercial Systems

Auxiliary Fire Alarm Systems. An auxiliary fire alarm system provides direct communication from a protected premise to the fire department over some type of local or public alarm-reporting system. Examples of reporting systems may include manual fire alarm pull stations located at various locations around a municipal area, local telephone lines, public radio transmitter systems, or even some form of designated hardwired circuit connecting back to the master fire alarm box of a town or municipality and the main fire station. In most cases, however, towns and municipalities are no longer providing local, manual pull stations because of the exorbitant maintenance costs such systems can incur and also because of the high degree of false alarms often attributed to them as a result of curious children or pranksters.

Cellular phones also are being used by some systems as a means of secondary backup to the hardwired telephone line in the event the main line were to ever become damaged or inoperable.

The Function of the Central Control Unit. Fire alarm systems operate from a central control unit. The central control unit, located within the protected premises, continually monitors circuit integrity, checks the status of initiating devices, controls the triggering of notification appliances, and also provides connection to local authorities or a public fire service communication center.

Figure 9–1
(A) Master control unit. (B) Auxiliary control unit.

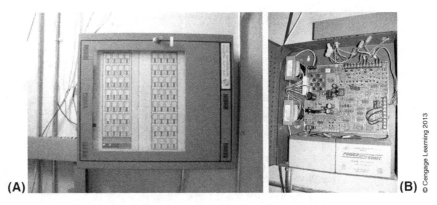

(A) (B) © Cengage Learning 2013

In large facilities, there often may be several auxiliary control units communicating back to a master control unit (Figure 9–1). In such cases, the master control unit communicates out to various zones and auxiliary control units within the building or structure. As shown in Figure 9–2, each zone may be monitored directly by the master control unit or indirectly through an auxiliary control unit. The auxiliary control ultimately communicates back to the master fire alarm control panel when an alarm is detected. The master control unit then reports all communications to a **supervisory station** manned by trained personnel or to a separate control center during the issuance of a trouble alert or an actual alarm. In many instances, a control center is servicing multiple clients and locations simultaneously and continuously, 24 hours a day.

When an alarm is detected, the central control unit activates all notification appliances circuits (NAC) such as horns, strobes, door releases, and ventilation damper controls, and simultaneously initiates communication of the event to the supervising authorities in a control room, or in some cases, to a **master box** located on the inside or outside of the protected premises. The supervising personnel, or master box, then notifies authorities of the apparent emergency by retransmitting the alarm to the public fire service communication center. On receipt of the alarm, authorities then will dispatch a fire crew or, in some cases, multiple fire brigades, to the appropriate location.

The report to the supervisory station or master box from the protected premise can be made over a series of public telephone line, through a coded wiring system, or over a coded public radio connection or cellular phone. In cases where the connection is coded, the coding is decoded at the supervisory station before retransmission to the local fire service center. The exact location of the fire alarm is apparent to the service center, as the decoded message arrives from the supervisor station or master box.

Internal circuitry within the incoming call center continually monitors the integrity of the system between the supervisory station or master box and the service center, providing immediate action and response to an alarm by supervising personnel once the call has been received.

Operating power from the protected premises to the supervisory station or master box is provided through the central control unit. Auxiliary battery power also is required to help protect the system and maintain communications during a main power failure.

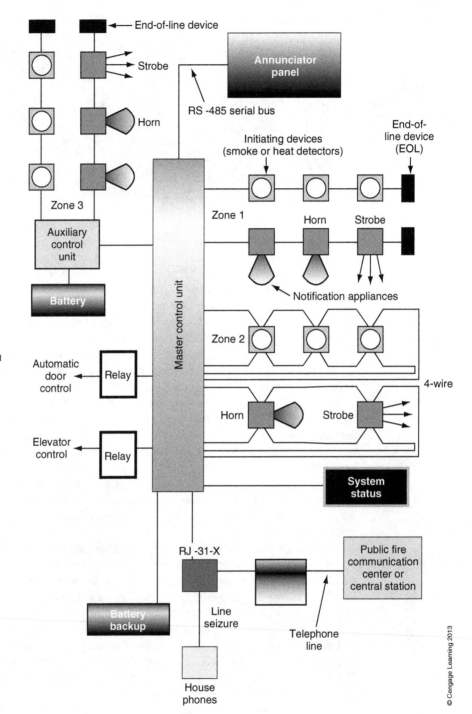

Figure 9–2
Block diagram of a fire alarm circuit.

© Cengage Learning 2013

Two-Stage Fire Systems. In a two-stage alarm system, a distinct alert signal first advises the supervisory staff of the fire emergency. The signal usually is coded and its meaning is apparent only to designated building staff members. The staff are expected to immediately investigate the source of the alert and, if a fire exists, to activate the alarm signal.

The alarm signal automatically activates after a predetermined period (usually 3 to 5 minutes) if the supervisory staff have not already activated or reset the system. If, however, after investigation it is determined that the alert is a false alarm, the staff can silence the coded alert signal and reset the system.

Proprietary Station Systems. **Proprietary station** systems are privately owned and operated by the property owners of the protected premises. The structure can vary from a single-building complex, such as a shopping mall or high-rise, to several buildings within a corporate or collegiate complex. Properties can be contiguous or noncontiguous, provided they are all communicating to a supervised central station, one that is properly staffed by qualified and trained personnel. The proprietary station must always have the ability to continually monitor and respond to all fire alarms, supervisory signals, and trouble signals originating from the protected premises, as needed, any time of day. Noncontiguous properties must use some sort of remote communication to the central station. A hotel chain would be an example of a noncontiguous property if the multiple sites around the country are all communicating to a central service center.

The control panel listed for use by a proprietary supervising station is identified by the letter *P* in the Type column of the UL Fire Protection Equipment Directory, followed by the words *Protected Premises*. The control panel at the monitoring facility is identified by the letter *P* and indicates that it is a *Receiving Unit* (see NFPA 72, 3.3.266.2, 26.4).

Central Station Service. **Central station** service fire alarm systems typically are owned and operated by private companies who are then responsible to monitor, manage, and report the occurrence of fire alarms and supervisory trouble signals from various clients and facilities within a municipal area. The central station service organization must be listed and approved by a recognized testing agency and also approved for use by the AHJs. Not all monitoring companies are listed and approved as central station service centers; the simple fact that they are a monitoring company does not also mean that they are listed as being a central station service.

The control panel for a central station service is identified by the letters *CS* in the Type column of the UL Fire Protection Equipment Directory. The panel at the protected premises may be the same as the panel in the control center.

The types of clients who would normally choose to purchase the services of a central station fire alarm system usually would be involved with more high-risk or high-valued products such as manufacturing facilities, chemical plants, or even hospitals and retirement homes, which often have large numbers of disabled or high-risk patients within their facility who need personal 24-hour care. The personnel working within a central station system therefore must be highly trained, on guard

24 hours a day, and able to respond to all system troubles, supervisory alarms, or fire alarm signals as they occur. Clients who use central station services are basically paying for extra security; they want the assurance that someone is always on call and continuously monitoring the status of their facility (see NFPA-72, 3.3.266.1, 26.3). Examples of central station systems include ADT and SIMPLEX.

Remote Station Fire Alarm Systems

Remote station fire alarm systems are used where a protected premises does not have access to a **public fire alarm** system or does not use a central station service or proprietary supervising station system. Remote stations are not certified systems, and companies installing them do not need to be UL certified. Often, landlords with building structures in isolated locations use remote systems as a way to link multiple structures to a single, centralized, manned control center. Remote stations transmit fire alarm signals by means of a private radio, dedicated phone circuit, one-way phone, or by any approved method deemed acceptable by the AHJs. Monitoring of supervisory or trouble alarm signals from a remote protected premise can be achieved through a central control unit, fire command center, or supervising station, provided they are continually staffed and attended by qualified response personnel (see NFPA 72, 3.3.266.3, 26.5).

The Fire Command Center (NFPA 72, 3.3.9.6)

The **fire command center** is an approved, designated area that typically is located at the entrance to a building, which the fire department can use as a central communication command post. In the event of an actual emergency, the fire command center is used by the fire department to direct and manage all activities related to building evacuation, automated building controls, emergency communications, and firefighting. Large building structures may have multiple command centers operating simultaneously, but only one can be designated as the main or central command center; multiple command centers must be able to communicate their operations and activities back to the central command center. In most cases, two-way telephones are installed between control centers, and in building stairwells, that allow firefighters the ability to communicate through dedicated, fire-protected lines. All stairwell construction and two-way telephone wiring must be rated and installed to accommodate 2 hours of flame resistance.

Notably, emergency telephones used by the fire department between fire command centers and building stairwells are not connected to or part of the main public telephone system. **Emergency telephones** are isolated, separate systems that provide two-way communications or party-line connections to multiple handsets throughout the building. Emergency telephones also do not need to be dialed; connection to the communication loop is made simply by lifting the handset. All parties can communicate simultaneously without the risk for external interference or line loss.

Where multiple command centers are used, the central command center must be plainly marked by a visible indicator that designates it as the central command center. The fire command center may also have access to and direct control of other building functions such as air dampers, water flow, elevators, and emergency communications, as permitted by the AHJs; the command center may also be linked to other security centers as permitted.

Remote Fire Alarm Annunciator Display Panel

Located within the fire command centers is a remote fire alarm **annunciator** display panel (Figure 9–3), which visually and audibly indicates the location of the fire by building, floor, or zone. The **annunciator panel** typically is separated into zones that correspond to the various sections of the building or floors of a protected structure. The specified zones visually and audibly indicate their alarm status on the front panel through either an alphanumeric display, printed output, colored light-emitting diodes, back-lit indicator labels, or graphically. Firefighters use the panel indicators to help secure the space, gauge the apparent level and scope of

Figure 9–3

Remote fire alarm annunciator display panel.

© Cengage Learning 2013

the emergency, and also estimate how fast the fire may be spreading to unaffected areas within the protected space.

The fire alarm annunciator panel receives its communication from the main fire alarm control panel, which may often be in a separate isolated room having also been built to accommodate a 2-hour fire rating. In most cases, the remote annunciator panel is in the entryway of the building, and the central control unit is in a separate protected location. The annunciator panel typically receives communication over an Electronic Industries Alliance/Telecommunications Industry Association (EIA/TIA)-RS-485 serial bus connection (Figure 9–4). The bus consists of a 2-wire, shielded twisted pair, 120 Ω, balanced (differential), high-speed serial connection; the data transmit over a bidirectional, half-duplex connection to a maximum of 32 transmitters or receivers.

The bus can functionally communicate up to 200 kbps over 1200 m or 10 Mbps over 50 m. In a multicommand center design, each command center annunciator panel would be communicating over the high-speed serial bus connection back to the main fire alarm central control panel or central control unit (see NFPA 72, 1–5.7).

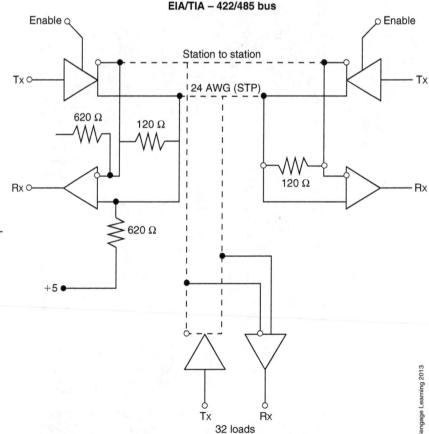

Figure 9–4

Electronic Industries Alliance/Telecommunications Industry Association (EIA/TIA) 422/485 bus.

© Cengage Learning 2013

SEC 9.3 CONTROL UNITS AND ALARM-INITIATING DEVICES

Central Control Unit

The central control unit is the main brain of the fire alarm system. It monitors all input sensors and initiating devices, such as smoke or heat detectors or radiant energy devices, and also controls alarm and signaling devices such as horns, strobes, ventilation dampers, elevator recall, and automatic doors. The control unit also performs supervisory services by continually monitoring the functionality of the system, connecting wires, power, and active devices. Any system failure triggers a trouble detector indicator on the panel, as well as an audible beep. The alert from the control unit must be in the form of a visual and audible response. Supervisory staff then must be on hand to troubleshoot the failure and also reset the system back to its normal operating mode (see NFPA 72, 10.11.3.(1) and 10.12.6 (1)).

Pull Stations (NFPA 72, 27.6)

Pull stations are examples of **public fire alarm reporting systems** and are manually activated by an individual during a fire emergency (Figure 9–5). The pull station is essentially nothing more than a shorting switch contact. When the contact closes, the supervised alarm notification circuit is shorted, causing the fire alarm control panel to activate all alarm notification devices, such as horns and strobes.

Pull stations must be easily recognizable, clearly visible for public access, and have operating instructions marked plainly on the exterior of the device. The pull station can be designed as either a single-action device, meaning it requires

Figure 9–5
Pull station.

© Cengage Learning 2013

a single lever pull to activate, or it can be a double-action device, which requires the user to first remove or break a protective cover before pulling the alarm-activating lever.

Pull stations must also be visually inspected semiannually and functionally tested annually. The AHJs may also require more frequent testing, as deemed necessary.

Heat Detectors

A **heat detector** is designed to detect abnormally high temperatures within a given space or area. The terms *heat* and *temperature* are not exactly the same. Heat represents a quantitative transfer of energy, measured in either Joules (J) or British thermal units (Btu). Temperature represents a quantity of heat within a given mass of material. Temperature is therefore intensity, and is measured on either the Celsius or Fahrenheit scale.

Heat flow or heat transfer can occur in one of two ways, as either convective or radiant heat. Convective heat transfers through currents or jets; examples include the heat flowing across a ceiling from the movement of hot gases. Radiant heat is absorbed into a detector from the available energy radiating out from a source, similar to a radiator heating the surrounding air within a living space. Convective heat typically provides the majority of heat flowing into a detector, whereas the amount of actual radiated heat transfer is much smaller by comparison.

Because of the differing methods of heat transfer, a heat detector may operate under any of the following three principals: as a fixed-temperature detector, rate compensation detector, or rate-of-rise detector. The detector can be designed in any one of the three manners or it can be all inclusive, depending on the manufacturer. As a result, there are many types of heat detectors available, all having slightly different functions and uses.

Types of Heat Detectors

Fixed-Temperature Detector. A fixed-temperature detector (Figure 9–6) responds when an internal detecting element reaches a predetermined set temperature point, similar to a thermostat.

Rate Compensation Detector. A rate compensation device responds when the surrounding air temperature reaches a predetermined temperature level, regardless of rate of rise. Rate compensation detectors primarily operate based on the expansion and contraction of two opposing metallic elements within a column of air. An outer cylindrical element is made to respond faster to changing temperature, whereas a secondary inner element takes more time to absorb the heat transfer. The detector only actuates at a predetermined temperature when both elements reach equilibrium. As a result, rapid temperature shifts affecting the outer element are not detected; the device only activates once the shielded inner element has had time to reach the predetermined trigger level.

Figure 9–6
Heat detector.

© Cengage Learning 2013

Rate-of-Rise Detector. A rate-of-rise detector responds when a shifting temperature varies excessively, beyond a predetermined value. Rate-of-rise detectors can either operate from an electrical conductivity point of view, as resistance changes are monitored over temperature, or they can be pneumatic in style, in which they respond to changing air pressure, as a chamber of air or line-type tubing is measured for atmospheric fluctuations.

Range of Heat Detection

The actual range of heat detection within a given space falls into one of two categories: spot-type or line-type detection. A spot-type detector concentrates its detecting element in one specific location. A line-type detector is a linear device that can sense temperature continuously along a pathway over an extended distance or range. Line-type sensors often look like a spider web of wires extending out over a given pathway, or may even be wrapped around a bundle of cable within a cable tray for extra protection.

The most commonly used temperature ratings for heat detectors may be 57°C (135°F), 88°C (190°F), or 93°C (200°F). Higher temperature ratings can be purchased, between 121°C and 302°C (250°F and 575°F), depending on the requirements and type of protected space.

As a general guideline, maximum allowable ceiling temperatures are typically 20° less than the temperature rating of the detector. The difference between the maximum ambient temperature of the ceiling and the rated temperature of the detector should be as small as possible to help reduce the response time of the device.

Heat detectors are not considered to be life-safety devices. In most cases, heat detectors are used as secondary property protection devices, because the presence of smoke will usually be detected prior to heat. It is also common to install heat detectors in unoccupied areas or in locations that are deemed unsuitable for smoke detectors.

Reset versus Nonrestorable Initiating Devices

All types of heat detectors, whether they be fixed, rate-of-rise, or rate compensation, fall into one of two general categories: those that can be reset and those that must be replaced. Detectors that can be reset are preferable because the periodic testing of the device is possible and easier to perform. Nonrestorable initiating devices, in contrast, self-destruct when activated and can be used only once.

Types of Heat Detector Technologies

There are many technologies available to detect the presence of heat. Manufacturers of heat detectors may include any of the methods in the following sections when designing and engineering a detector.

Expanding Bimetallic Components Bimetallic components are made from the bonding together of differing metal strips. Heat causes the two metals to bend, because the differing physical properties of each allow expansion at different rates. In most cases, the bending of the metal physically activates a mechanical switch, causing normally open contacts to close, thereby signaling the alarm circuit. Expanding bimetallic components reset to their normally open state as soon as the temperature drops below the device activation point.

Fusible Links, Eutectic Solders, and Eutectic Salts Eutectics refers to the thermal equilibrium of two elements as they transition from a solid to a liquid. A eutectic equilibrium results when two elements are soluble as a liquid but insoluble as a solid. When two elements are combined at optimum percentage ratios, the eutectic point represents the lowest possible melting point of the alloy. The transition between liquid and solid also becomes quite instant, allowing the mixture to bypass an unwanted pasty state often common in most other compounds. As a result, the percentage of each element used within the mixture becomes critical to achieve the optimum eutectic point and lowest possible melting point. Traditional lead–tin solder mixtures for example use a 60:40 percentage ratio. Any other value would cause the melting point temperature of the compound to be much greater, which would be a disadvantage, and it would also inhibit the solderability and performance of the junction, because of the return of the unwanted pasty state transition, ultimately making a smooth-flowing junction nearly impossible to achieve.

When the alloy reaches its rated temperature and liquefies, the device then is activated by a spring-loaded plunger, which either mechanically closes the contacts of a switch or forces a shorting bar across the terminals of the connecting wires. In either case, the system alarm activates. These types of detectors cannot be reset and must be replaced once activated.

Melting Insulators In most instances, a melting insulator results in a circuit short, as a spring-loaded plunger connects the two normally open contacts of a switch. Melting insulator sensors cannot be reset and must be replaced once activated.

Thermistors Thermistors are made from a combination of ceramic and metal-oxide materials. As temperatures increase, the overall electrical resistance of the device either increases or decreases, depending on the temperature coefficient. A device having a positive temperature coefficient exhibits an increased resistance with increasing temperature, and a device having a negative temperature coefficient exhibits a decrease in resistance as temperature increases. The rate of change over temperature is not linear, which means that the detector must be carefully engineered to work at appropriate temperature values. Devices using thermistor technology automatically reset once temperatures drop below calibrated trigger levels.

Temperature-Sensitive Semiconductors Diodes, transistors, and microchips are all example of semiconductors. In most cases, they are made from either silicon or germanium, are nonlinear in nature, and exhibit a negative temperature coefficient. Most temperature-sensitive semiconductors are nothing more than specially designed transistors that produce a changing output current based on the shifting temperature of the device. Devices using temperature-sensitive semiconductor technology automatically reset once temperatures drop below calibrated trigger levels.

Expanding Air Volume As air is heated, it expands. The calculation of air temperature can be determined by measuring pressure changes within a known volume of air. In most cases, the volume of expanding air, when heated, is made to activate a normally open switch contact within the device. The alarm circuit is activated once the increasing air pressure causes contacts to close. Expanding air volume type devices reset as air temperatures drop below calibrated trigger levels.

Expanding Liquid Volume The variations of pressure within a known volume of liquid can also be used to determine the temperature of a device. The device functions in a manner similar to the expanding air volume device, as listed earlier. An expanding liquid volume device resets as internal temperatures drop below the calibrated trigger level of the device.

Temperature-Sensitive Resistors All metals exhibit a changing electrical resistance over temperature. A PRT, or platinum-resistive thermometer, is an example of a temperature-sensitive resistor. A coil of platinum wire is wound to an exact known resistance value, typically 100 Ω at a 0°C (32°F) reference temperature. The measured resistance of the coil then increases as the applied temperature increases, and decreases as the applied temperature decreases. A PRT sensor can be accurate to within 1/100 of a degree, and it often is used to measure temperatures ranging between –200°C and 850°C (–328°F and 1562°F). A calibrated circuit internal to the device triggers an alarm as the temperature and internal resistance of the sensor rises above a required level. Devices using temperature-sensitive resistors reset as internal temperatures drop below the calibrated trigger level of the device.

Thermopiles Thermopiles are made from series-connected thermocouples. A thermocouple is created whenever two dissimilar metals come together at a common junction point. A temperature-dependent voltage is developed at the junction as a result of thermal gradients along the wires on either side of the junction. As temperature gradients shift, so too do measured voltages at the junction point of the thermocouple. So, then, what is a thermopile? A thermopile is a series-connected set of thermocouples, connected end to end as a way to help boost the overall output voltage of the junction and increase the sensitivity of the device. Modern-day thermopiles are constructed on a microchip-sized scale, having low mass and little surface area, thus allowing the response times of the device to increase dramatically as the transfer of thermal radiant energy moves through it. As the temperature and internal voltage of the thermopile increase above a required level, an internally calibrated circuit triggers an alarm. Devices using thermopiles reset as internal temperatures drop below the calibrated trigger level of the device.

Electrical Conductivity Heat Detectors An electrically conductive heat detector can be of a spot- or line-type; such detectors exhibit a changing electrical resistance over temperature and reset as temperatures drop below calibrated trigger levels.

Heat Detector Location

A spot-type detector must be located not less than 4 in. (100 mm) from the sidewall or on the sidewalls, between 4 and 12 in. (100 and 300 mm) from the ceiling (Figure 9–7). Special placement considerations must be taken into account when dealing with nonstandard-type ceilings, which are irregular, sloping, peaked, shed, or high (see NFPA 72, 17.6.3.4 and 17.6.3.5). Spacing of heat detectors also is determined by a full-scale fire test. Reductions to the spacing can be made based on the height of the ceiling, as given in NFPA 72, *Table 17.6.3.5.1*. Always refer to NFPA 72, 17.6.3 and the manufacturer's instructions when considering correct placement and location of heat detector devices.

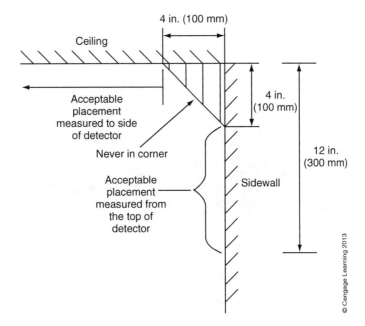

Figure 9–7
Acceptable placement of heat detector.

Smoke Detectors

A **smoke detector** is a device that detects visible or invisible particles of combustion (Figure 9–8A, B). Types of smoke detectors include ionization smoke detectors, photoelectric smoke detectors, air sampling-type, projected beam-type, and spot-type.

Figure 9–8
(A) Combination smoke and heat detector.

(Continued)

Figure 9–8
(Continued)
(B) Spot-type smoke detector.

© Cengage Learning 2013

Smoke detectors should not be confused with smoke alarms, which are often installed in residential settings as self-contained devices intended to notify the occupants of a single room or dwelling unit of a possible fire. Smoke alarms are not connected to a fire alarm control panel; instead, they are isolated devices, consisting of a smoke sensor and an audible alert circuit built into a single housing. They are typically powered by a 9 V battery, but sometimes are hard wired to the main building electrical circuitry.

Smoke detectors, in contrast, are smoke sensors; they terminate back at the central fire alarm control panel, or central control unit, and are intended to initiate an alarm for an entire building. Also, smoke detectors are considered input devices because they can only initiate, not announce, an alarm.

Types of Smoke Detectors

Ionization Smoke Detectors Ionization detectors use a radioactive material to ionize the air between charged electrodes. The conduction of the air changes as particles enter the ionized chamber. Ionization detectors are considered slightly more sensitive than other detectors because they are able to detect invisible particles, typically smaller than 1 μm. Ionization detectors are not as responsive to large-particle emissions, which often result from smoldering fires.

Photoelectric Smoke Detectors **Photoelectric smoke detectors** are meant to detect visible particles larger than 1 μm. The sensing circuitry typically consists of a light-emitting diode and a photosensitive receiving element. As emissions of smoke particles enter into the pathway between the light source and receiving element, light beams scatter, causing the signal intensity of the receiver to drop off sharply, indicating the presence of unwanted particles.

Projected Beam-Type Detector A projected beam-type detector uses the transmission of a photoelectric light beam to a detecting element some distance from the source. The projected beam-type detectors typically are used in large spaces, such as atriums and auditoriums, where high ceilings are commonplace. The beam usually spans the protected area and must always be installed according to manufacturer's instructions, based on certain range specifications and device limitations. As rising smoke begins to obscure and distort the transmission of the beam, the receiving detector, some distance away, automatically triggers an alarm, alerting the presence of a fire. For additional information on the requirements and spacing of projected beam-type smoke detectors, refer to various sections of NFPA 72, 17.7.3 and 17.7.3.7.

Air Sampling-Type Smoke Detectors, Duct Detectors (NFPA 72, 17.7.3.6) An air sampling-type detector uses piping or tubing networks to transport air from the protected area back to the detector element. The detector element is typically photoelectric in nature. As the air is forced through the tubular portals by way of an aspiration fan, the calibrated electronic circuitry analyzes the sample for the presence of smoke particles or emissions.

Air sampling-type detectors are typically installed inside air ducts and air passages or in locations where concerns of high airflow are evident (NFPA 72, 17.7.4 and 17.7.5), such as in computer rooms and underfloor and above-ceiling spaces, where environmental air normally travels. The primary purpose of duct detectors is to shut down air-duct fans when smoke is detected. Duct detectors are not to be used as a substitute for area smoke detectors.

To prevent false readings and ensure optimum use of the detector, air sampling-type devices are not to be placed in direct contact with any turbulent airstreams or supply registers and must also be far enough down line from duct bends or turns. Typical guidelines state that duct detectors should be placed 6 to 10 duct widths downstream from or before an air-duct restriction or bend. Duct detectors should also be mounted in locations before connection to a common air return or fresh air inlet. An access panel must also be provided for all detectors mounted inside of air ducts (see NFPA 72, 17.7.5.5.3).

Detectors must be able to withstand the flow of dust from dirty environments through some type of filtering mechanism or be electronically calibrated to discriminate particle sizes and types of emissions. Care must be given to the movement and location of dampers, making sure that their normal use does not block airflow to the detector.

Two types of air sampling-type smoke detectors are available: active and passive. Active air sampling devices use a negative pressure vacuum pump and blower assembly to help draw the air sample into the device. Passive detectors use the tubing and aspiration fan method for drawing airflow through the detector chamber. The active device would more likely be used in areas where normal airflow is quite low.

Air-sampling or projected beam-type smoke detectors shall be installed in accordance with manufacturer's published instructions. Refer to NFPA 72, 17.7.5.5 for more information on the location and installation of duct detectors.

Spot-Type Smoke Detectors Spot-type smoke detectors are engineered to detect the presence of smoke in a particular location. NFPA 72, 17.7.3.2 outlines the specifics and guidelines of spot-type smoke detectors.

Smoke Detector Location

Smoke detector locations and evaluations are based on specific guidelines, as described in NFPA 72, 17.7.3, and on engineering judgment.

Installation guidelines state that ceiling-mounted detectors should not be less than 4 in. (100 mm) from a sidewall to the edge of the detector, or if on the sidewall, placed between 4 and 12 in. (100 and 300 mm) down from the ceiling to the top edge of the detector. Detectors must also be mounted at least 3 ft (9.1 m) from air diffusers.

Spacing of detectors on flat ceilings can be up to 30 ft (91 m). In addition, all points on the ceiling shall have a detector within a distance equal to 70% the selected spacing. The distance from the corner of the room must be 70% the spacing rating or less. On peaked ceilings, detectors should be mounted within 3 ft (9.1 m) of the highest point. In all cases, be sure to follow manufacturer's instructions.

Radiant Energy Detectors (NFPA 72, 17.8)

Radiant energy detectors are used to sense the presence of flames, sparks, or embers. They are electronically calibrated to optically sense electromagnetic energy spectrums in the ultraviolet, visible, or infrared regions, depending on the type of detector being used and the associated wavelengths of the emission.

All burning elements have an individual signature related to their overall intensity level and measured wavelength. As an example, gasoline, wood, or carbon dioxide all emit energy in the ultraviolet, visible, and infrared regions, but to different degrees. The type of detector chosen therefore must match the specific emission signature of the combustible element to work effectively and not develop false triggers within the system. The choice of radiant energy device can only be decided once a determination has been made as to the material makeup of the combustible element. Refer to NFPA 72, 17.8.3 for information on spacing requirements when considering the use of radiant energy detectors.

Carbon Monoxide Detectors

Carbon monoxide (CO) is a by product of combustion. Furnaces, ovens, and open fires all can potentially produce large quantities of carbon monoxide.

UL standard 2034 references the acceptable standards for carbon monoxide detectors. Because carbon monoxide is colorless, odorless, and deadly, all residential installations must include carbon monoxide detectors, as well as smoke alarms. Furnaces and gas heaters can give off high quantities of carbon monoxide when they are not properly tuned and cleaned, but the quantity of smoke emissions from such appliances remains virtually nonexistent. In such cases, not having a carbon monoxide detector could be potentially deadly to the occupants of the residence because the smoke alarm remains unaware and unable to react to the problem.

Sprinkler Water Flow Alarm-Initiating Devices (NFPA 72, 17.12)

A sprinkler water flow alarm-initiating device outputs an alarm signal to the control panel within 90 seconds of a sustained water flow. Normally occurring pressure fluctuations along the system must be calibrated out to prevent false alarms. Trapped air and the use of excess pressure pumps, pressure drop alarm-initiating devices, and pressure regulators may generate longer response times within the system. For this reason, sprinkler systems and controls should be installed by trained professionals to be sure of their reliability. For more complete details on sprinkler systems, water sprinkler designers and installers must be familiar with NFPA 13, Standards for the Installation of Sprinkler Systems.

Automatic Extinguishing Systems (NFPA 72, 17.13)

Automatic extinguishing systems are meant to extinguish detected flames once their presence has been verified. In most cases, an automated control valve is activated, allowing for the mechanical release and flow of water, foam, carbon dioxide, or any required wet/dry chemical compound needed to extinguish the flames within the protected area. The release switches must be listed for use with the required system. Any initiating devices must also be listed for use to ensure that all critical functions operate appropriately within the required system.

Supervisory Signal-Initiating Devices (NFPA 72, 17.16)

Supervisory signal-initiating devices are of a type that continually monitor the status of the device and notify the central control unit when an automatic or manual change of status has occurred. Such devices are also considered tamper proof because any forced change or even sudden disconnection of the device is immediately detected by the central control unit. In such cases, an alarm or trouble signal is triggered by the system. Following is a brief description of the various types of control valves and devices that operate in a supervisory manner.

Control Valve Supervisory Signal-Initiating Device

A supervisory control valve contains electronic circuitry that sends out two distinct types of signals, one based on the normal position of the valve, whether it may be normally open or closed, and the other based on the off-normal position, meaning the valve has moved off the normal position and is now on the way to changing state. It also is important to keep in mind that valve logic is opposite electronic logic; a closed valve indicates the off position, whereas in electronics, a closed switch indicates the on position. If the valve is being used for water flow and it is closed, then water is not flowing.

The off-normal indicator is not a normal state, and it can only be triggered when valve movement or rotation is detected by the device. The movement may be automated, as in the event of an actual alarm, or it may indicate that someone may be manually tampering with the device. The off-normal supervisory signal triggers any time the valve is moved one-fifth the travel distance between normal and full span or at any time within the first two revolutions of the control wheel. The normal signal indicator is only restored once the valve has regained the normal position.

Pressure Supervisory Signal-Initiating Device

A pressure supervisory signal-initiating device outputs two distinct electronic signals, one based on the normal calibrated pressure setting and the other based on a changing or not-normal setting. The not-normal setting usually is triggered by a 10 psi increase or decrease in pressure.

Water-Level Supervisory Signal-Initiating Device

A water-level supervisory signal-initiating device outputs one of two electronic signals, one indicating a normal water level and the other indicating a shifting water level. For pressure tanks, supervisory signals must trigger the control panel any time a 3-in. (76 mm) level shift has been detected, up or down. For all other types of tanks, a 12-in. (300 mm) shift typically triggers the control panel that a change has occurred.

Water Temperature Supervisory Signal-Initiating Device

Water temperature devices are used to monitor and help prevent the possible freezing of water tanks. The device outputs one of two distinct electronic signals, one indicating a decrease in temperature to 40°F (4.4°C) and the other as temperatures are restored above 40°F (4.4°C).

Room Temperature Supervisory Signal-Initiating Device

Room temperature devices operate just as water temperature devices do. The device outputs one of two distinct electronic signals, one indicating a decrease in temperature to 40°F (4.4°C) and the other as temperatures are restored above 40°F (4.4°C). The only difference between the two devices is that the room temperature device is measuring air temperature and the water temperature device is measuring water temperature.

SEC 9.4   NOTIFICATION APPLIANCES

Notification appliances are considered output devices from the point of view of the control unit. During an alarm status, the control unit outputs a control signal activating all notification appliances within the system. Notification appliances are used to indicate the presence of an alarm. Devices usually are audible and visual in nature. In accordance with the Americans with Disabilities Act, all public spaces must now provide a combination of both audible and visual indicator devices to better accommodate individuals with special needs.

The number of devices on the system loop is critical and is determined by the manufacturer, because each device conducts a certain amount of minimal current during standby and maximum current during alarm activation. The standby current is used by the supervisory loop to verify device status and system readiness. Maximum current flows when the device goes into alarm. Manufacturers indicate maximum loop lengths and the number of active devices that can be connected safely in a circuit without adversely affecting the performance of the system.

Audible Alarm Indicator Devices

Once activated by the central control unit, **audible alarm indicator** devices output a uniquely identified audible alarm signal. All devices used within a single building should be of the same variety to avoid the confusion of differing signaling sources. The traditional methods of providing audible alarm indication may be either through a separate network of building loudspeakers or by bells, horns, or buzzers. Horns (Figure 9–9) are used in situations where high-power, high-decibel

Figure 9–9
Horn.

© Cengage Learning 2013

outputs are necessary for the coverage of large spaces or to help cut through a noisy environment. Examples of noisy environments might include manufacturing or factory settings where the normal audio levels of such environments measure as high as 80 dB or possibly more, depending on the type of machinery in operation or time of day.

Commercial audible devices are intended for use in public spaces, whereas residential audible devices are only intended for use in residential spaces. NFPA 72 requires that audible signals in public or residential spaces be 15 dBA greater than the ambient noise level of the environment, or 5 dBA greater than the maximum sound level for a period of 60 seconds, as measured 5 ft (1.5 m) from the floor. Total alarm outputs shall not exceed 110 dBA. Levels greater than 120 dBA can potentially cause permanent damage to hearing. For sleeping quarters, the requirements are the same, but the output level shall not be less than 75 dBA, as measured from the pillow level.

Distinctive Evacuation Signals (NFPA 72, 18.4.2)

Audible emergency evacuation signals shall be distinctive in nature, and are meant to precede any emergency voice paging or alarm communications. The intent of such signals is to alert the occupants of a building of an immediate evacuation or relocation to safer quarters within the structure. The standard fire alarm evacuation tones are temporal in nature and consist of the following order:

An ON phase lasting 0.5 seconds, followed by an OFF phase lasting 0.5 seconds.

The series then repeats for three complete cycles, followed by an OFF state lasting 1.5 seconds.

The entire sequence should be repeated for not less than 3 minutes, and all audible appliances should be synchronized within a notification zone.

Emergency Voice/Alarm Communications (NFPA 72, 18.4.1.5)

Emergency voice/alarm communications service provides voice instruction to the occupants of a building during an evacuation or relocation to safer quarters within the structure. The voice communication can be either automated or manual through an emergency telephone or microphone paging system. Loudspeaker systems used for background music and standard building paging are not acceptable for use unless they have been specifically designed to interface with the fire alarm system in accordance with NFPA 72, 24.4.2.28. Voice messages are not required to meet the audibility requirements of public or private audible devices, or audible devices in sleeping quarters, however they are required to meet intelligibility requirements of NFPA 72, 18.4.10.

Strobes (NFPA 72, 18.5 and 18.6)

Strobes are examples of visual alarm indicator devices (Figure 9–10). The central control unit outputs a control signal to activate strobes during an alarm status. Strobes visually flash synchronized indicator lights throughout the protected area,

Figure 9–10
Strobe.

© Cengage Learning 2013

similar to a police light, indicating the presence of an alarm. As stated earlier, a mixture of strobes and audible alarm indicator devices are now being required in all public spaces to accommodate visually or hearing impaired individuals. The requirements may be different for private or commercial spaces. Be sure to check with local and municipal building fire codes, because installation requirements may vary from town to town or city to city. In addition, areas that contain ambient noise levels greater than 105 dBA shall require visible notification appliances in accordance with NFPA 72, 18.5 and 18.6.

Emergency Telephones

Occupational Health and Safety Administration (OSHA) requirements state that employers must explain to employees the process and preferred method of reporting emergencies. When telephones are used as an emergency reporting system, a listing of emergency phone numbers must be posted at emergency stations or on employee bulletin boards for ease of use and availability.

All emergency messages must take priority over those considered nonemergency in nature. In addition, all live emergency voice messages shall take precedence over all other multichannel inputs, such as tape or digitally prerecorded evacuation messages. In most cases, electronic circuits or computer controls can be used to help automate and prioritize communications, thus blocking messages of lesser priority for those of higher priority when required. As an example, most commercial audio amplifiers provide precedence controls or offer special input modules, allowing the live, emergency channel inputs to have greater priority over all other secondary inputs. In such cases, the priority override controls are used to block or mute all nonpriority signals, allowing only the broadcast and transmission of emergency signals. In most cases, a uniquely identified alert tone, transmitting through a separate tone generator, shall precede the emergency voice message to get the attention of building occupants before the emergency page.

Additional Occupational Health and Safety Administration Requirements for Fire Alarm Systems

Some additional OSHA requirements regarding the installation and testing of fire alarm equipment are as follows:

1. All installed equipment must meet OSHA standards.

2. All installed equipment must be approved for the purpose for which it is intended.

3. The prompt restoration of all system settings back to normal operating function must occur after each alarm test. Any nonrestorable or nonre-settable devices destroyed during the test procedure must be replaced with normally functioning devices.

4. All systems and devices must be maintained as often as needed or recommended by the manufacturer or local fire codes to ensure the safety and reliability of the system.

5. All pneumatic devices installed after January 1, 1981, must be equipped with supervisory systems.

6. All maintenance, calibration, testing, and cleaning of fire alarm systems and devices must be performed by trained personnel.

7. Periodic cleaning for detectors that require it must be performed at regular periodic intervals.

8. Fire alarm equipment or devices installed in outdoor or potentially corrosive atmospheres must be protected by either a canopy, hood, or any other suitable means to ensure the safe function and reliability of the equipment.

9. All installed equipment must be protected from mechanical or physical impact that may possibly render the device inoperable.

10. Detectors must be supported by secure, independent means and not by attached wires or tubing.

11. Any automatic fire extinguishing equipment must be designed and tested to operate within a reasonable amount of time in response to an active alarm.

12. The number and spacing of fire alarm detectors must be based on field experience, design guidelines, tests, engineering surveys, manufacturer's recommendations, or a recognized testing laboratory listing.

13. All replaceable power supplies must be maintained as often as necessary to ensure the safe reliability of the system.

14. All manually operated devices must remain clear, conspicuous, unobstructed, and ready for use.

SEC 9.5 ELECTRICAL REQUIREMENTS OF POWER-LIMITED AND NON-POWER-LIMITED FIRE ALARM SYSTEMS, *NATIONAL ELECTRICAL CODE ARTICLE 760*

Access to Electrical Equipment behind Panels *(National Electrical Code Article 760.21)*

The unobstructed access to electrical equipment through service panels or suspended ceiling tiles shall be maintained at all times. The accumulation of conductors and cables shall not prevent the free access to, or the removal of, panels or tiles. To ensure that cables are not resting on access panels, they must be tied up and secured appropriately. The mounting hardware used in securing cables also must be approved for use and should not adversely affect the physical or electrical characteristics of the conductors. In addition, power-limited fire alarm (PLFA) cables or conductors must not be strapped, taped, or attached by any means to the exterior of any conduit or raceway as a means of support (see *NEC Article 760.143*).

Access to fire panels must be without incident, meaning that authorized personnel must have free, unobstructed access to the panel at all times.

Fire Alarm Circuit Identification *(National Electrical Code Article 760.30)*

Fire alarm circuits must be clearly labeled at every junction and terminal connection in a manner that prevents unintentional interference with the fire alarm system during servicing or testing. Service technicians need to know what they are looking at. The majority of problems can be solved quickly if the documentation of system wiring is up to date and labels are clearly marked and accessible.

Survivability (NFPA 72, 12.4, 23.10.2, and 24.4.2.6.2)

Survivability refers to the ability of a circuit to operate during an actual fire. Emergency voice evacuation equipment and notification appliances must be able to maintain their functionality during the process of an evacuation, allowing occupants the needed time to withdraw safely from the building or structure.

To ensure the survivability of the fire alarm circuit, cable assemblies must have a 2-hour rating. Only **mineral-insulated (MI)** cables and **circuit integrity (CI)** cables have the necessary 2-hour rating to ensure the functionality of the fire alarm circuit. To safeguard the system properly, all connections between the central control unit and the notification appliances must be protected against fire damage by using 2-hour rated cables, 2-hour rated shafts and enclosures, and 2-hour rated stairwells. Survivability requirements are required and must be applied to all non-voice systems used in the evacuation of occupants from a building, either by floor or by zone.

The fire alarm system should also be designed in such a way as to guarantee that a single fault or failed connection does not disable the rest of the building or any other **notification appliance circuits** connected down line from the actual failure. For example, the cabling for separate floors should run through separate risers as a way to help minimize the damage to multiple zones of a structure in the event of a catastrophic failure to a specific connection. The risk for total system failure would be too great if a circuit was allowed to feed multiple parts of a building structure through a single riser connection or cable (see NFPA 72, 12.2).

Mineral-Insulated Cable

MI cable (see *NEC Article 332, Part I*) is made of solid copper conductors insulated with magnesium oxide, covered over by a rigid copper sheath. The cable is rated for 2 hours of fire resistance, and it can be either single conductor or multiconductor, depending on the application. Previously, MI cables were widely used on emergency generators, fire pumps, low-voltage fire protection circuits, and emergency voice systems.

MI cable is not easy to install, and it requires special handling and training because of the rigid nature of the outer covering. Walls and shafts also need to include additional construction requirements to accommodate the installation. As a result, most current systems have opted to replace MI cable with the easier to use, flexible CI cable, which is now available through most cable manufacturers and electrical providers.

Circuit Integrity Cables

CI cables are made from solid, copper conductors, and a lightweight, fire-resistant insulation. The insulation remains flexible and easy to install at room temperatures, but when exposed to high temperatures, it changes state, becoming a 2-hour fire-resistant protective barrier to the central conductors.

CI cables can be easily pulled through raceways or conduits because of the soft nature of the outer jacketing material, and they require no special handling or training to install. The use of CI cables ensures that critical alarm circuits continue to operate longer under the adverse conditions of a fire.

Most plenum and high-temperature cables do not have the necessary CI rating, which guarantees their survivability over the required 2-hour time period needed to help evacuate occupants from a building or protected structure. CI cables are tested and listed by listing agencies, such as UL, and then labeled as CI, identifying them as suitable for use in fire alarm systems and fire-control applications. Applications of CI cables include firefighter elevator controls, smoke exhaustion fans, and fire alarm notification appliances. CI cables with low-capacitance specifications also are available, making the interconnection of computerized, high-speed data communication possible across the fire alarm network.

Non–Power-Limited Fire Alarm Circuits

Non–power-limited fire alarm (NPLFA) circuits operate at higher voltage and current levels and are rated as Class 1 circuits. The installation of such circuits must conform to the requirements of Chapters 1 through 4 of the *NEC*, as well as *NEC Article 760, Fire Alarm Systems.* NPLFA circuits must also be installed by licensed electricians. Power-limited technicians do not have the authority to install or service such systems. The information given in this section is provided for informational purposes only, as a comparison with the PLFA systems that are described later in this chapter.

Non–Power-Limited Fire Alarm, Circuit Power Source Requirements
(National Electrical Code Article 760.41)

The power source requirements for a non–power-limited fire alarm system shall not allow nominal output voltages greater than 600 V. An individual branch circuit is also required and no other loads shall share this circuit. The overcurrent protection device shall be permanently identified at the fire alarm control unit, and the circuit disconnect means shall have a red identification, accessible to only qualified personnel. The circuit shall be identified as "Fire Alarm Circuit." In addition, ground-fault and arc-fault circuit interrupters shall not be used on fire alarm circuits. All fire alarm circuits must be able to withstand a single ground fault and still operate.

Non–Power-Limited Fire Alarm Circuit Overcurrent Protection
(National Electrical Code Article 760.43)

Circuit overcurrent protection for conductors 14 AWG and larger must be provided in accordance with *NEC Article 310.15,* without applying the derating factors of *Article 310.15* to the ampacity calculation. Derating refers to the reduction of the total allowable ampacity of the cable, based on the number of conductors in a conduit or raceway and adjustment to the temperature rating of the insulation. Obviously, as more conductors are grouped together through a single opening or conduit, the total potential heat dissipation is increased, based on the internal heat dissipation of the individual conductors and the amount of specific heat radiating out from neighboring conductors; all conductors within proximity of a given measurement ultimately add to the total value. As a result, the internal and external heat dissipation of the conductors all combine to increase the total. For safety reasons, allowances therefore must be made by electricians to accommodate the grouping of multiple high-current conductors in a given raceway so as not to exceed the overall temperature rating of the insulation.

Overcurrent protection for NPLFA 18 AWG conductors shall not exceed 7 A, and those of 16 AWG shall not exceed 10 A.

Non–Power-Limited Fire Alarm Circuit Overcurrent Device Location
(National Electrical Code Article 760.45)

The location of overcurrent devices shall be placed at the point of the supply, where the conductors receive their power.

The following two exceptions may apply:

Exception 1: Overcurrent devices can be placed on the primary side of a single-phase, single-voltage transformer, provided that the protection is in accordance with *Article 450.3,* and that it does not exceed the value determined by multiplying the value of secondary current by the secondary-to-primary voltage ratio of the transformer.

Exception 2: For an electronic power source providing a single-output voltage, the rules are essentially the same. An overcurrent device can be placed on the input side of the supply, provided that the amperage rating is sized for the maximum current capacity of the non–power-limited conductors multiplied by the output-to-input voltage ratio of the power source.

The installation of electronic power sources having multiple output voltages is different. When multiple output voltages are available from a single electronic power source, the overcurrent protection must be placed at the point of supply to the connecting circuit conductors. Although an overcurrent device on the input side of the supply would ultimately protect the source from a total power overload, it would be unable to differentiate the level of individual currents at the various outputs. Therefore, to guarantee the protection of individual circuits, overcurrent devices must be placed at the point of supply, ensuring that they are individually protected to their maximum current ratings.

Non–power-Limited Fire Alarm Circuit Conductors
(National Electrical Code Article 760.49)

A. Only copper conductors can be used in fire alarm circuits. Wire sizes 18 or 16 AWG are permitted, provided that the load currents do not exceed 6 A for 18 AWG and 8 A for 16 AWG, as shown in *Table 402.5* of the *NEC*; the circuit wires must also be installed in approved raceways, enclosures, or listed cables. Cables larger than 16 AWG shall not exceed the load current ampacities of *NEC 310.15.*

B. Conductor insulation shall be suitable for 600 V. Conductors larger than 16 AWG shall comply with the wiring requirements of *NEC Article 310.*

C. Circuit wires can be of solid or stranded copper.

Multiconductor Non–power-Limited Fire Alarm Cables
(National Electrical Code Article 760.53)

Multiconductor cables shall be permitted for use on fire alarm circuits operating at 150 V or less. Multiconductor, NPLFA cables shall be installed in raceways, on exposed ceiling surfaces, sidewalls, or in concealed spaces. The cables must be protected to a height of 7 ft (2.1 m) by building construction, a metal raceway, or a rigid nonmetallic conduit. When passing through a wall or a floor, cables shall be installed in a metal raceway or rigid nonmetalic conduit, to a height of 7 ft (2.1 m). Cables installed in hoistways must be in a rigid metal conduit, a rigid nonmetallic conduit, an intermediate metal conduit, or electrical metallic tubing. A bushing shall be installed where cables emerge from a raceway. In addition, fire alarm cables in vertical raceways shall be supported according to *NEC 300.19* so that the weight of the cables does not cause undo stress or damage.

Non–Power-Limited Fire Alarm and Class 1 Conductors

NPLFA circuit conductors and Class 1 circuits shall be allowed to occupy the same cable, enclosure, or raceway, provided that all conductors are insulated and rated for the maximum voltage of any conductor within the enclosure or raceway.

Power supply and fire alarm circuit conductors shall also occupy the same cable, enclosure, or raceway, provided that the power supply and power cables are functionally associated with the fire alarm equipment (*NEC 760.48(B)*).

NEC 300.17 shall determine the number of NPLFA and Class 1 conductors in a raceway. All derating factors according to *NEC 310.15(B)(3)(a)* shall apply, provided that the conductors carry a continuous load current in excess of 10% the ampacity rating of each conductor (*NEC 760.51(A)*).

Allowable Numbers of Power Supply Conductors and Fire Alarm Circuit Conductors, *National Electrical Code Article 760.51*

The number of power supply conductors and fire alarm circuit conductors permitted to occupy the same raceway shall be determined according to *NEC 300.17*. All derating factors according to *NEC 310.15(B)(3)(a)* shall apply to all conductors where the fire alarm circuit conductors carry a continuous load current in excess of 10% the ampacity rating of each conductor and where the total number of conductors is more than three.

If the fire alarm conductors do not carry continuous load currents in excess of 10% the ampacity rating of each conductor, then the derating factors according to *NEC 310.15(B)(3)(a)* shall only apply to the power supply conductors, where the number of power supply conductors is greater than three (see *NEC 760.51(B)*).

As stated earlier, only licensed electricians can install or service NPLFA circuits; power-limited technicians are not authorized to install or service such circuits.

Types of Fire Alarm Cables Used on Power-Limited and Non–Power-Limited Circuits

NEC Article 760	Plenum	Riser	Commercial general purpose	Residential
Power-limited, fire-protective signaling cable	FPLP	FPLR	FPL	None
Non–power-limited, fire-protective signaling cable	NPLFP	NPLFR	NPLF	None

Type FPL and NPLF (fire power-limited and non–power-limited) cables are listed for general use only (not including ducts, plenums, or other space used for environmental air), and they shall be listed as being resistant to the spread of fire.

Type FPLP and NPLFP (fire power-limited and non–power-limited fire plenum) cables are "listed as being suitable for use in ducts and plenums and other space used for environmental air, and also shall be listed as having adequate fire-resistant and low smoke-producing characteristics." (Reprinted with permission from NFPA 70-2011). It is also worth noting that plenum cables are not to be installed inside of metal air ducts.

Type FPLR and NPLFR (power-limited and non–power-limited fire-protective riser) cables are listed as being suitable for use in risers or vertical shafts, and also shall be listed as having adequate fire-resistant and low smoke-producing characteristics, capable of preventing the spread of fire from floor to floor.

Power-Limited Fire Alarm Circuits

Power-limited fire alarm (PLFA) circuits shall be rated as Class 3 circuits. As a result, they may share the same cable, enclosure, or raceway with two or more PLFA circuits, low-powered network-powered broadband communication cables, or any other Class 3 circuits (see *NEC 760.139(A)*).

Class 2 circuit conductors may be installed with PLFA circuits provided that the Class 2 cables have been upgraded to the same minimum insulation rating as the PLFA and Class 3 circuit conductors; that is, not less than 300 V.

Power-limited circuit conductors must not be installed in the same cable, raceway, enclosure, or junction box with Class 1, electric light and power, non–power-limited fire alarm, and medium-power network broadband communications circuits (see *NEC 760.136(A)* through *(G)*). Because Class 1 and NPLFA circuits are rated for much higher voltage and current levels, they must remain physically separated from Class 2, Class 3, and PLFA circuits by a minimum distance of 2 in. (50 mm), or be permanently separated by a physical barrier.

Where separated by barriers, PLFA circuits shall be permitted to share the same raceway or enclosure with Class 1, electric light and power, non–power-limited fire alarm, and medium-power network broadband communications circuits. The insulation surrounding the PLFA conductors does not count as a separation barrier. To achieve proper separation, a physical barrier external to the construction of the cable must be placed between conductors. Special raceways are available that divide the internal traveling space of the conductors between permanently installed dividers or barriers. The barrier must also meet the 0.25-in. (6 mm) separation requirement of the *NEC* to keep the conductors of different classes isolated. Raceways within enclosures also are allowed as a means of separation and isolation.

Associated circuit conductors are permitted to share the same compartments, enclosures, device boxes, and outlet boxes, provided that the Class 1, electric light and power, NPLFA, and medium-power network broadband communication conductors are routed to maintain a minimum separation of 0.25 in. (6 mm) from the PLFA circuit cables. Associated circuitry implies that the Class 1 or power conductors are necessary to the functional operation of the power-limited circuitry; without the associated circuitry, the power-limited circuit would be inoperable.

Circuit conductors operating at 150 V or less may also be installed with PLFA cables, provided that they are associated systems and that they maintain a minimum separation of 0.25 in. (6 mm), either by a physical barrier, insulated sleeve, or appropriate mounting hardware.

Associated wiring entering enclosures with single openings shall allow conductors of Class 1, electric light and power, NPLFA, and medium-power network broadband communications conductors, provided that the PLFA conductors are separated from the other circuits by continuous and firmly fixed nonconductive flexible tubing. Because the cables must all enter the enclosure through a single opening, the addition of the flexible tubing supplies the necessary barrier separation needed to protect and isolate the different circuit classifications from each other.

In all other applications where PLFA and NPLFA circuits may come in contact with each other, a separation of 2 in. (50 mm) must always be maintained. The 2-in. (50 mm) separation also applies to all Class 1, electric light and power, and medium-power network broadband communications conductors. If not properly isolated from each other, higher voltage circuits could possibly induce damaging currents into the power-limited circuits, thereby interfering with normal circuit operations.

PLFA circuit conductors may also be re-rated and installed as non–power-limited circuit conductors as described previously or as stated by *NEC Article 760.46.* To do so would obviously add an extra expense to the installation process because the cabling would need to meet the higher 600 V rating of NPLF-, NPLFP-, or NPLFR-type conductors. It also would mean that a power-limited technician would be unable to install or service the circuits.

Audio System Circuits and Power-Limited Fire Alarm Circuits
(National Electrical Code Article 760.139 (D))

A new ruling in the 2005 edition of the *NEC* requires that audio circuit conductors, as described in *NEC 640.9(C),* installed as Class 2 or 3 wiring from an amplifier, shall not be permitted to occupy the same cable or raceway with other power-limited conductors or cables. Essentially, this ruling requires that audio cables must maintain a permanent separation from remote-control or signaling conductors and all other power-limited conductors, even though they may both be considered as Class 2 or 3.

Power Sources for Power-Limited Fire Alarm Circuits
(National Electrical Code Article 760.121)

Power sources for PLFA circuits are limited to a maximum output value of 100 VA. The supplies are internally limited by specially designed circuitry that makes the higher levels of output power attributed to NPLFA circuits impossible to achieve. Power sources for NPLFA circuits differ from those of PLFA in that they only limit output currents through some form of overcurrent protection; examples include fuses or circuit breakers. Power-limited supplies are protected by more complex internal circuitry that continually monitors output voltage and currents levels, as well as the total output power of a supply, to maintain a maximum rating of 100 VA. In addition, an individual branch circuit is also required, and no other loads shall share this circuit. The overcurrent protection device shall be permanently identified at the fire alarm control unit, and the circuit disconnect means shall have a red identification, accessible only to qualified personnel. The disconnecting means shall be identified as "Fire Alarm Circuit." Also, the branch circuit shall not be supplied through ground-fault and arc-fault circuit interrupters.

The types of power sources used in PLFA circuits may include any of the following:

- A listed PLFA or Class 3 transformer
- A listed PLFA or Class 3 power supply
- Any piece of listed equipment marked and identified as the PLFA source

Examples may include a listed fire alarm control panel, circuit card, or battery (stored energy device).

Conductor Size for Power-Limited Fire Alarm
(National Electrical Code Article 760.179)

Conductor size for PLFA must adhere to the following guidelines:

- Conductors shall be solid or stranded copper and with a voltage rating of not over 300 V.
- Multiconductor cable shall not be smaller than 26 AWG; 26 AWG cabling also requires specially approved connectors and terminators.
- Single-conductor cables shall not be smaller than 18 AWG.
- If used, coaxial cables are permitted to have a minimum of 30% conductivity along the central copper-covered steel conductor.
- All power-limited cables used for the survivability of critical circuits shall be CI rated.

In addition, *NEC Article 760.130(B)* states that power-limited circuit cables must be protected to a height of 7 ft (2.1 m) by the building construction or by a metal raceway or rigid nonmetal conduit. Where installed exposed, cables shall be securely fastened in an approved manner at intervals of not more than 18 in. (457.2 mm).

When passing through a wall or floor, cables shall be installed in a metal raceway or rigid nonmetalic conduit to a height of 7 ft (2.1 m).

Cables in hoistways must be installed in a rigid metal conduit, a rigid nonmetallic conduit, an intermediate metal conduit, or electrical metallic tubing.

In all cases, be sure to check with the manufacturer for any additional requirements before installation.

SEC 9.6   WIRING OF FIRE ALARM SYSTEMS

The wiring style of a fire alarm system may be designed around a 2-, 3-, or 4-wire format. Initiating devices, such as heat and smoke detectors, represent the input side of the alarm circuit, whereas notification appliances, including bells, horns, and strobes, represent the output side. Devices typically are grouped into zones, especially in large building structures or industrial complexes, to better assist support staff as to the most likely location of an alarm. Each zone represents an individual loop circuit within the fire alarm control panel. **End-of-line devices** consisting of resistors, relays, or diodes are then used on each loop circuit or zone to help supervise the presence of system activity.

Classification of Initiating Device Circuits

Initiating device circuits can be divided into the following classifications: Class A, style D or E, and Class B, style A, B, and C.

The Class B system connects through a 2-wire circuit (Figure 9–11) followed by an end-of-line device. The end-of-line device forces the total loop current to fall within a calibrated range, which allows the control panel to monitor circuit activity and provide supervisory functions. Not all systems use the same size end-of-line resistor. Some systems use a 4.7 kΩ resistor, whereas others use a 2.2 kΩ resistor. The manufacturer of the control panel specifies the correct size when installing the system.

The Class A system uses a 4-wire circuit (Figure 9–12); two wires representing the outgoing loop conductors and the other two representing the return. End-of-line devices are not used on 4-wire systems because the returning two wires are terminated back at the control panel, where the loop is monitored and supervised on the front end.

Style A is an outdated Class B wiring system that is no longer used in the United States. System alarms are triggered from a direct wire-to-wire short across the circuit conductors. Although style A was able to initiate a trouble response from a single open wire or a ground fault condition, it was unable to generate an alarm response from a ground fault, and therefore became outmoded and was dropped by manufacturers as a viable and reliable system.

Figure 9–11
Two-wire Class B circuit.

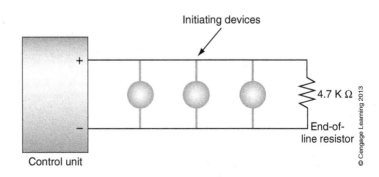

Figure 9–12
Four-wire Class
A circuit.

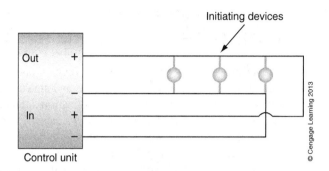

© Cengage Learning 2013

Style B is a Class B wiring system (Figure 9–13). Control panels using style B are required to receive an alarm from any device up to the point of a break or open in the initiating device circuit. System alarms are triggered from a direct wire-to-wire short across the circuit conductors. Trouble signals are initiated when a ground fault or open occurs at any point in the loop. The end-of-line device is used to supervise the presence of changing loop activity. An alarm can also be triggered in the presence of a single ground fault on the system; this is a direct fix to style A, which was unable to accommodate such a situation.

Style C, also a Class B wiring system (Figure 9–14), looks identical to style B, except that the initiating devices have a series-limiting resistor connected in line with the device terminals that typically is lower in value than the end-of-line

Figure 9–13
Style B, Class B
circuit.

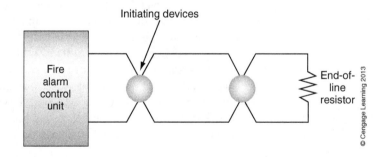

© Cengage Learning 2013

Figure 9–14
Style C, Class B
circuit.

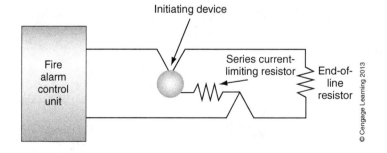

© Cengage Learning 2013

resistor. When the device initiates an alarm, the contacts close, placing the series-limiting resistor across the circuit. The level of loop current then increases, subsequently forcing the control panel into alarm. In this case, the alarm current is not a short-circuit current. For style C, the presence of a direct short across the system forces the control panel to issue a trouble alert, not an alarm, possibly indicating that one of the devices in the loop may be defective. A trouble signal is also generated during the presence of an open, short, or ground fault on the system conductors.

Style D is a Class A wiring system (Figure 9–15). An open or ground causes a trouble alert back at the control panel, whereas a short across the initiating loop conductors or return conductors causes the system to go into alarm. A system device also is able to activate an alarm during the presence of an open or broken loop; in such a case, the system acts as a 2-wire circuit, allowing all devices up to the point of the break the ability to still activate the system by placing a direct short across the initiating loop conductors. End-of-line devices are not needed for style D because the loop conductors return back to the control panel to provide the necessary circuit monitoring and supervision.

Style E is a Class A wiring system (Figure 9–16). It is simiilar to style D, except that the initiating devices have a series-limiting resistor connected in line with the

Figure 9–15
Style D, Class A circuit.

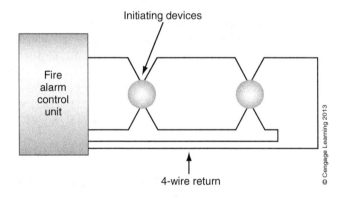

Figure 9–16
Style E, Class A circuit.

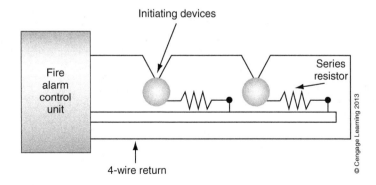

device terminals that typically is lower in value than the end-of-line resistor, just as seen in style C. When the device initiates an alarm, the contacts close, placing the series-limiting resistor across the circuit. The level of loop current then increases, subsequently forcing the control panel into alarm. In this case, the alarm current is not a short-circuit current. For style E, the presence of a direct short across the system forces the control panel to issue a trouble alert, not an alarm, possibly indicating that one of the devices in the loop may be defective. A trouble signal also is generated during the presence of an open, short, or ground fault on the system conductors. End-of-line devices are not needed for style E because the loop conductors return back to the control panel to provide the necessary circuit monitoring and supervision.

The manufacturer and the system classification determine the required specifications of fire alarm circuits. Parameters include the maximum number of devices that can be placed on the loop, the size of end-of-line devices, the size of series current-limiting resistors (as used in styles C and E), the maximum number of circuits or zones within a control panel, and the maximum number of buildings a monitoring station can monitor through **signaling line circuits (SLCs)** (see later in this section). Power supply requirements are also determined based on the size of a system and the number of zones and devices.

Classification of Notification Appliance Circuits

Notification appliance circuits operate devices such as bells, buzzers, horns and strobes. Notification circuits also are categorized by their classification and style of operation. Class A circuits are still 4-wire in nature, but Class B can be 4- or 2-wire, depending on the style. The style of the circuit ultimately determines the individual functionality of the circuit as related to ground faults, end-of-line devices, and opens. Class B has three styles associated with it: styles W, X, and Y. Class A, however, is associated with only style Z.

Style W

Style W is a Class B, 2-wire circuit operating with an end-of-line device such as a resistor or relay (Figure 9–17). All notification devices continue to operate up to

Figure 9–17
Style W and Y,
Class B circuit.

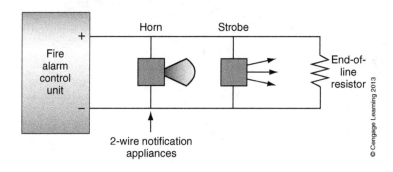

© Cengage Learning 2013

the point of a circuit fault or open. The occurrence of a ground fault may, however, disable the entire system.

Style X

Style X is a Class B, 4-wire circuit that can function under a single open but is unable to generate an alarm during the presence of a ground fault (Figure 9–18).

Style Y

Style Y is a Class B, 2-wire circuit that uses an end-of-line device (see Figure 9–17). Notification appliances continue to function up the point of a circuit fault or open. Ground faults in style Y are indicated separately from other faults by the control panel. A single ground fault does not disable the system.

Style Z

Style Z is a Class A, 4-wire circuit (see Figure 9–18). All notification appliances continue to operate under a single ground or open on the circuit. The control panel also has the ability to indicate a ground fault differently from other system faults as they may occur. The wiring of 4-wire notification circuits must also be polarized to ensure the appliances are being powered appropriately. The 4-wire appliance circuit travels out from the control panel, connecting all devices according to polarity. At the end of the line, the circuit then loops back to the control panel where conductors are once again terminated according to polarity.

During the presence of a single ground fault, the control panel has the ability to function as a grounded 4-wire loop. In the presence of a single open, the control panel also has the ability to drive the system from both ends, as two separate 2-wire circuits, up to the point of the break. All line faults trigger the control panel to issue a trouble alert. After the line fault has been repaired, a technician resets the system to resume normal operation as a 4-wire circuit.

End-of-Line Device

The control unit of a fire alarm system supervises a device circuit through the use of an end-of-line device. In most instances, the end-of-line device is

Figure 9–18
Style X, Class B
or Style Z, Class
A circuit.

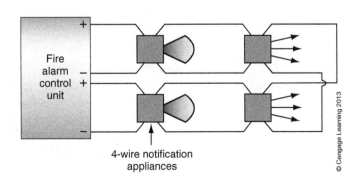

Fire
alarm
control
unit

4-wire notification
appliances

© Cengage Learning 2013

a resistor, measuring either 4.7 or 2.2 kΩ, depending on the calibration of the system and the manufacturer. The purpose of an end-of-line resistor is to allow a control unit to supervise a circuit against possible tampering or accidental shorts or opens. The end-of-line resistor adjusts the total loop current of a device circuit to a known fixed value. Any time a device changes states, such as from standby to alarm, or if the device loop were to encounter a sudden short or open, the fire alarm control panel immediately senses the occurrence of the event and issues a trouble alert. The control panel continuously monitors the total loop current of a fire alarm circuit as a means of supervision. The detection of any significant status change causes the control panel to trigger an immediate trouble alert or alarm signal. Most modern alarm control panels have the option of supervising any or all loops or zones with an end-of-line resistor. In some cases, end-of-line diodes also are used to help protect the circuit from accidental power reversal.

Door Releases

Door releases are either controlled by the central control unit or directly from the smoke detector (see NFPA 72, 17.7.5.6). Door releases are electromagnetic in nature. When the control panel issues an alarm, the circuit to the door release relay coils opens, causing the magnetic field to drop out and the doors to shut. During a fire, corridor and stairwell doors must be closed to help minimize the propagation of smoke and gases throughout the building.

End-of-Line Relay

An **end-of-line relay** is a device used to monitor the presence of building power in 4-wire systems. Referred to as a power supervision relay, this type of device is used to supervise power on 4-wire smoke detectors or the signaling power on addressable devices. To be effective, the end-of-line relay typically is installed within or near the last device on the loop. The occurrence of a power failure disengages the end-of-line relay, causing the switch contacts to reverse and thus opening the loop circuit. In such a case, the control panel would operate on backup battery power, and a trouble alert signal would be issued as the end-of-line relay disengages.

Multiplexed Systems

Multiplexed systems allow for multiple signals to be sent back and forth from alarm subpanels to the main fire alarm panel over a single communication line. The signals are uniquely encoded so that they are not confused in transition as they are sent or received between panels. Multiplexed systems also reduce the need for added control equipment and additional wiring, and they allow the entire system to operate from a single distributed power supply.

Addressable Systems

Addressable systems have the ability to identify and locate the origin of an alarm signal by an individual device. The exact location of an alarm can be instantly detected by the fire alarm control panel through a uniquely assigned device address. The addresses typically are assigned by selector switches on the device. One of the benefits of having an addressable system is that there is no longer a need to divide the devices into specified zones, because the fire alarm control panel has the ability to communicate directly with each individual device. The fire alarm control panel can then accomplish circuit supervision through a routine series of poles set up to communicate with devices one by one on a regular basis to verify the status and functionality of the overall system. Therefore, end-of-line devices are not needed in addressable systems because the communication is more comprehensive at the device level compared with traditional 2- or 4-wire systems.

Analog Addressable Systems

Analog addressable systems differ from standard addressable systems in that they are able to provide far more information to the control panel with regard to the specific details of an alarm. For instance, in an analog addressable system, the smoke detector not only sends an alarm notification to the control panel but also has the ability to pass on the specific data related to measured levels of smoke content or air contamination in the affected area. Because the data comparisons of each individual device can be stored and registered by the control panel over time, the control panel ultimately has the ability to issue a warning that a smoke detector may need servicing because of a buildup of dirt or dust on the detector elements. Detector sensitivity may deteriorate over a period of weeks or months as a result of the environmental conditions within the protected space. As an added level of protection against system failure, the analog addressable system issues trouble alerts based on self-evaluations of devices by control panels over a prolonged period. The measured levels of air quality are ultimately compared with sensitivity and calibration specifications by the control panel to make a determination on system reliability.

The control panel also has the ability to program and readjust device sensitivity as needed to maintain the recommended calibration requirements determined by the manufacturer. In instances where the measured outputs fall outside of acceptable calibrated norms and the control panel is unable to make programmed adjustments, the control panel then issues a trouble alert. In analog addressable systems, the trouble alert signals can be issued from five or more differing parameters, compared with the basic four indicators—normal, trouble, supervisory or alarm—on typical control panels. The specific nature of the failure, such as the detector is not answering a poll or the device is out of calibration, would also be indicated and recorded.

Manufacturers also specify and limit the number of devices on an analog addressable circuit and the overall distance requirements from the control panel to the farthest device in the loop. Addressable circuits operate at high speeds, and because high-speed data requires wide bandwidths to communicate, the circuit ultimately is limited by the total length and capacitance of the circuit conductors. As stated in Chapter 2, the level of capacitance on the line increases over the length of the cable, causing a reduction in bandwidth and signal attenuation. Manufacturers therefore limit the length of the conductors because most fire alarm systems are operating on two-conductor, twisted pair cables.

Signaling Line Circuits

Addressable systems use what are called SLCs for connecting and communicating between devices and the control panel. There are two common styles of SLCs: Class A or B.

Class B Signaling Line Circuits

Class B SLCs are 2-wire in nature and can be connected in any configuration; that is, they do not need to be connected in series from the control panel, as required by the more common systems, passing wires in and out of each device, eventually to an end-of-line resistor or relay. Because end-of-line devices are not needed in addressable systems, devices can be connected in any configuration, in series, or through individual parallel runs back to the main panel. From a supervisory point of view, any broken connection within the system results in a loss of communication to a device or series of devices. The control panel immediately senses the disappearance of devices through the failed attempts to communicate with each individual device. A trouble alert then sounds, indicating the apparent loss to the system.

Class A Signaling Line Circuits

Class A SLCs use a 4-wire connection, connecting each device, end to end, starting from the outbound loop of the control panel, and then returning on the back of the loop through an alternate route. Communication can take place either from the outbound or back of the loop. The benefit of Class A SLCs is that they have the ability to recognize breaks in the system and even detect the origin of the break by counting how many devices are on either side, counting first from the outbound side and then the back of the loop. Once the panel detects a break in the communication loop by losing continuity between the outbound and return side of the connection, it still has the ability to communicate with every device in the system by switching communication mode and communicating from either side of the break. Class A SLCs tend to be more expensive systems to install, but they provide more reliability and security than those of Class B.

Hybrid circuits do exist combining a mixture of Class A and B SLCs. An example would be a home run Class A SLC connecting multizone junction boxes. Each junction box can then be wired as a Class B SLC out to the individual devices of the zone.

Digital Alarm Communicating Systems and Telephone Line Seizure

When an alarm is activated, the fire alarm control panel communicates to a central station or fire alarm command center through a digital communicator, or digital alarm communicating system (DACS). The DACS is made up of two parts: the digital alarm communicating transmitter (DACT), which transmits signals out over a standard public phone line, and a digital alarm communicating receiver (DACR), which receives alarm signals at the central station or fire alarm command center. To ensure that the phone line is available for use, the DACT performs a procedure known as a *line seizure*. The line seizure gives the DACT priority access to the phone line, thus preventing it from being used simultaneously by the public during emergency transmissions. To accomplish a line seizure, the DACT activates an internal relay, switching the connection of the main phone line from the house phones to the DACT. The connection is accomplished through a special modular telephone connector known as an RJ31-X (Figure 9–19). The RJ31-X provides the necessary input and output pathways needed to toggle the DACT and house telephones to the main telephone line. In the process of seizing the line, the DACT subsequently disconnects all active calls on the house phones. The DACT maintains control of the telephone line for the duration of a transmission out to the fire command center. The call then terminates once the DACR has verified and acknowledged the transmission from the DACT. Public line usage transfers back to the house phones as soon as the DACT releases control and deenergizes the seizure relay.

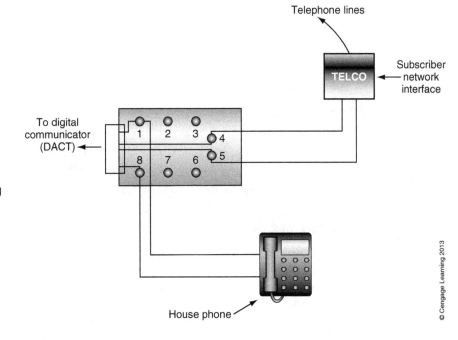

Figure 9–19
RJ31-X, line seizure connector. DACT, digital alarm communicating transmitter.

© Cengage Learning 2013

A DACT must be connected by two means of communications, with the main telephone line being the primary source. In some systems, cellular phones are now being used as a backup to the main phone line. In the event that main telephone service is ever disrupted, the DACT and house phones still continue to operate in an emergency situation.

Other communication examples now being used between fire control panels and central stations or fire alarm command centers may include the Internet, computer multiplexing, and long-range radio or satellite systems.

Computer multiplexing can involve either frequency division multiplexing, time division multiplexing, or wavelength division multiplexing. Regardless of how it is done, a central computer, located in a central station or fire alarm command center, can quickly control and communicate with multiple devices or locations over a single communication channel. The communication channel may consist of copper wire, fiber optic, or even wireless radio. Multiplexing allows for a variety of sites to be polled at regular intervals by the central computer, or each location may be given separate instructions as to when the desired information is required. The monitoring site determines the results of the data and whether a problem may exist, depending on the quality of the data or strength of the received signal.

SEC 9.7 TROUBLESHOOTING

When troubleshooting a system, always check for stray voltages, open or short circuits, and proper end-of-line resistance values.

Reduce Nuisance Alarms (False Alarms)

The most common cause of nuisance alarms in fire alarm systems is lack of routine maintenance. There are several steps you can take to ensure that the devices on the loop are working properly, such as these:

- Implementing a walk test every 6 months
- Scheduling a fire drill every 6 to 8 months to test the systems workability
- Cleaning devices at least once a year, which may include vacuuming dust from detectors and resetting or changing batteries
- Results from these tests should be maintained in a continuous logbook

Summary Statement

National or local building codes, fire codes, and electrical codes require the installation of fire alarm systems. The particular type of system and components to be used in a fire alarm system depend on the type of building occupancy, the relative size of a facility, and the number of occupants. To be effective, a high level of detection should be defined, and a fire alarm system must be customized to the building and the types of fire that could potentially develop. Life safety remains the primary concern of any fire alarm system, with property protection a secondary goal.

To ensure the long-term functional use of any system, be sure to practice safe wiring test guidelines. In addition, all conductors must be identified at every junction and terminal connection. Wires and cables also should be clearly marked and labeled for future use or possible renovation. Lastly, the location of devices and circuits must comply with governing codes and be installed in a neat and professional manner.

CHAPTER 9 FINAL QUESTIONS

1. List the primary and secondary goals of an effective fire alarm system.
2. What is the difference between a trouble alert signal and a supervisory signal?
3. Compare the difference between a single-stage fire alarm system and a two-stage system.
4. What type of fire alarm system is privately owned and operated by the property owners of the protected premises?
5. What is the main benefit of using a central station service fire alarm system?
6. When is a remote station fire alarm system used?
7. Where is the fire command center located, and what is the purpose of having such an area?
8. What is the purpose of the annunciator display panel?
9. Explain the difference between a single- and a double-action pull station.
10. Give an example of an initiating device.
11. A(n) _____ is used to provide live or prerecorded instructions to occupants during a fire.
12. What type of heat detector responds when an internal detecting element reaches a predetermined set temperature point?
13. The rated temperature of a heat detector is chosen to activate within _____ °F of the maximum ambient temperature of the ceiling.
14. Explain how a rate compensation detector initiates an alarm.
15. What type of heat detector responds when a shifting temperature varies excessively, beyond a predetermined value?
16. What is the main advantage of a resettable device over a nonrestorable initiating device?
17. A spot detector should be placed not less than _____ in. from the sidewall or on the sidewall between _____ and _____ in. from the ceiling.

18. On a flat ceiling, smoke detectors can be spaced _____ apart, and no point on the ceiling is more than _____ % of the selected spacing of the detector.

19. Which is more sensitive, an ionization type smoke detector or a photoelectric?

20. What is the primary purpose of a duct detector?

21. Can a duct detector be used as an area smoke detector?

22. What type of detector is used to sense the presence of flames, sparks, or embers?

23. Supervisory signal-initiating devices activate what type of equipment?

24. The Americans with Disabilities Act requires that notification appliances be _____.

25. The audio levels for audible notification appliances are measured _____ ft from _____.

26. Audible notification appliances must be _____ dBA above the average SPL or 5 dBA above the maximum SPL, lasting 1 minute or more.

27. In combination systems, fire alarm signals must be _____.

28. What is the purpose of an end-of-line resistor?

29. Do all fire alarm systems use the same value of end-of-line resistor?

30. What are the requirements for the main power connection to a fire alarm panel?

31. CI cables have a _____ hour fire rating.

32. A PLFA power supply is equivalent to a Class _____ power supply.

33. The fire alarm disconnect shall be clearly labeled "Fire Alarm" and colored _____, and the location of the disconnect shall be documented at the panel.

34. When passing through walls and floors, to a height of _____ ft, fire alarm cables shall be installed in _____.

35. The *NEC* requires that single conductors of PLFA cables not be smaller than _____ AWG.

36. What is the main difference between Class A and Class B initiation device wiring?

37. Who determines the maximum number of devices that can be placed on a supervised loop circuit?

38. What is one benefit of having an addressable fire alarm system?

39. How does a Class A SLC detect a break in a line connection between devices?

40. A DATC must be connected to _____ forms of communication, with the primary source being a _____.

Chapter 10

Fiber–Optic Cable and *National Electrical Code Article 770*

Objectives

- Give a brief description of light.
- Describe the construction of optical fiber cable.
- Explain the light-carrying capability of optical fiber cable.
- Describe how signal loss occurs in optical fiber cables.
- Compare the different varieties of optical fiber cables.
- Compare glass optical fiber with plastic optical fiber.
- Explain where optical fiber cables can be run with respect to Class 1, 2, and 3 conductors.
- List the different classifications of optical fiber cables described in the *National Electrical Code Article 770*.

Chapter Outline

Sec 10.1   Introduction

Sec 10.2   Basic Concepts of Light

Sec 10.3   Optical Fiber Cable

Sec 10.4   Applications

Sec 10.5   Varieties of Fiber

Sec 10.6   Types of Connectors

Sec 10.7   Classifications of Fiber and *National Electrical Code Article 770*

Key Terms

anaerobic

breakout cable

breakout kits

cladding

coating

coherent bundle

composite optical fiber cable

conductive optical fiber cable

core

critical angle

electromagnetic wave

graded index

Hotmelt®

index of refraction

light-emitting diode (LED)

loose-tube cable

multimode

nonconductive optical fiber cable

optical fiber cable

photoelectric sensor

randomized bundle

ribbon fiber-optic cable

simplex cable

single-mode

step index

tight-buffered cable

tightpack cable

transmission

zipcord

SEC 10.1   INTRODUCTION

In recent years, the trend has been to replace all copper communications cable with **optical fiber cable**. Unlike in Chapters 1 through 9, which deal with the passage of electrical signals and pulses through copper, fiber-optic cable communicates using light energy, not electrons. The advantages of using light can be summed up in five categories: speed, bandwidth, distance, resistance, and maintenance. You will see after completing this chapter that in all five categories optical fiber cable can simply outperform copper in every instance.

Before looking at the construction and statistical analysis of optical fiber cable, it is important to learn how fiber optics actually works. This chapter begins with the study of the basics of light.

SEC 10.2   BASIC CONCEPTS OF LIGHT

Light travels in the form of an **electromagnetic wave**; that is, the wave emits both electric and magnetic fields as it moves through space. The fields vibrate at right angles to the direction of movement and at right angles to each other (Figure 10–1). The amount of energy in a light wave is proportionally related to

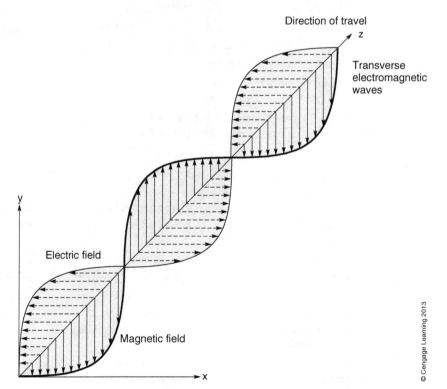

Figure 10–1
Transverse electromagnetic light waves.

© Cengage Learning 2013

its frequency: High-frequency light has high energy; low-frequency light has low energy. When traveling through a vacuum, light can move at speeds in excess of 186,000 miles/sec, or 300,000,000 m/sec. The speed of light is the fastest phenomenon known to humans. The concept of using light to carry signal transmissions over cables is ideal because it is the fastest available medium.

SEC 10.3   OPTICAL FIBER CABLE

Glass versus Plastic

Optical fiber cables typically are made of long, thin tubes of pure glass about the size of a human hair. The tubes are referred to as light pipes. Although plastic can be used in the construction of optical cables, glass tubes have proved to be far superior, but more expensive. One of the main disadvantages to using plastic is its inability to transmit across a broad light spectrum. Although both plastic and glass are unable to transmit wavelengths within the ultraviolet range, plastic also absorbs large portions of the infrared spectrum. Figure 10–2 compares the transmission efficiency of plastic and glass fiber optics. As shown, glass is clearly the better choice; it is far more reliable in its ability to transmit across the entire spectrum of visible and infrared wavelengths. To ensure maximum transmission when using plastic optical fiber (POF), you must limit signals to only those wavelengths within the visible light spectrum; additional light amplification may also be required. Plastic fibers are also less tolerant in extreme temperatures and are sensitive to many chemicals and solvents. One benefit to using plastic, however,

Figure 10–2
Spectral transmission efficiency in glass versus plastic fiber optics. LED, light-emitting diode.

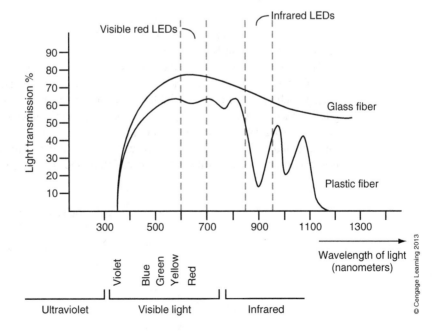

is its ability to survive under repeated flexing. Glass fibers, in contrast, are not so forgiving and may crack if the cable is bent beyond a recommended bend radius. Manufacturers specify the maximum bend radius of a product in their specification data sheets. As a general guideline, fiber optic cables that are not under tension should have a bend radius not more than 10 times the diameter of the cable.

Glass Fiber-Optic Construction

The construction of glass fiber-optic cable involves the bundling of hundreds of strands of cladded glass tubes (see the next section for a discussion of cladding), each strand having an individual diameter of about 0.002 in. (0.05 mm). Strands can be bundled in one of two ways, in either a **coherent bundle** or a **randomized bundle**. A coherent bundle means that the strands are carefully lined up end to end in such a way that an image can be transmitted through the cable and displayed in a viewable form at the opposite end. Coherent bundles are expensive to manufacture and often are used when a transmission requires a clear image at the output. When a clear image is not a priority, a less costly alternative would be to use randomized bundles. Randomized fiber-optic bundles are not organized in any special order; for this reason, the cables are easier to manufacture and are less expensive. Randomized cables are used in most cases where the application requires only the transmission of on/off light pulses. In such cases, the organization of individual strands within the bundle is not critical. POF cables differ in that they are typically made from a single strand of plastic, 0.01 to 0.06 in. (0.03 to 0.15 mm) in diameter.

Core, Cladding, and Coating

Figure 10–3 illustrates a closer look at a single glass tube. The illustration shows three separate parts: **core**, **cladding**, and a buffer **coating**, all of which play a crucial role in the **transmission** of signal.

The core diameter of glass fiber cables can measure from 9 to 200 μm, and the core diameter for plastic can measure up to 1 mm (0.04 in.). The glass or plastic core, which carries the pulses of light, is surrounded by a layer of cladding, a plastic spacer layer, a protective layer of Kevlar, and an outer sheath. The Kevlar is used to increase the tensile strength of the cable; it can be braided or applied longitudinally. The outer sheath or jacketing material varies in manner of protection and strength, depending on the installation and the environmental concerns

Figure 10–3
Optical fiber
construction.

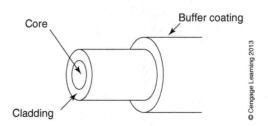

Figure 10–4

Fiber types and typical specifications.

Fiber type	Core/cladding Diameter (m)	Attenuation Coefficient (dBkm)			Bandwidth (MHz-km)
		850 nm	1300 nm	1550 nm	
Multimode/plastic	1 mm	(1 dB/m	@665 nm)		Low
Multimode/step index	200/240	6			50 @ 850 nm
Multimode/graded index	50/125	3	1		600 @ 1300 nm
	62.5/125	3	1		500 @ 1300 nm
	85/125	3	1		500 @ 1300 nm
	100/140	3	1		300 @ 1300 nm
Single mode	8-9/125		0.5	0.3	High

© Cengage Learning 2013

of the location. Examples may include outdoor, underground burial, plenum, or riser-type cable.

Two numbers typically associate the dimensions of an optical fiber cable. An example would be 62.5/125. The first number refers to the diameter of the core, and the second to the diameter of the core and cladding together, as measured in micrometers. Other sizes include 200/240, 50/125, 85/125, and 100/140. For plastic fiber-optic cable, the dimension typically is indicated by a single core/cladding measurement, often 1 mm (0.04 in.).

Figure 10–4 lists the core cladding diameters, as well as the typical bandwidths and attenuation losses, for the various types of fiber-optic cable. (The particulars of multimode and single-mode fibers are discussed later in this chapter.)

Transmission

Fiber-optic technology uses a laser or **light-emitting diode** (LED) to transmit data via light waves to a **photoelectric sensor** or receiver placed at the end of a cable. Figure 10–5 illustrates this concept.

Think of a piece of fiber as a long cardboard roll, which you would find at the center of a roll of wrapping paper. If you were to take a flashlight and shine it at one end of the roll, you would be able to see the light at the other end. Now, imagine the roll is bent favoring one direction; because light cannot bend around corners by itself, it would be hard to see the light on the other end of the tube, if at all. If we were to take a mirror and place it inside the tube at the point of the bend, then we would be able to reflect that light out the end of the roll. If we were to line the entire inside of the roll with mirrors, then we could achieve multiple bends within the roll with minimal loss. This is an explanation of how optical fiber cable

Figure 10–5

Optical fiber transmitter and receiver circuit.

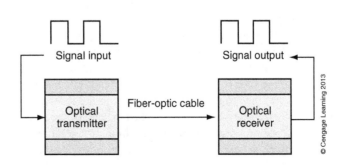

Signal input   Signal output

Optical transmitter   Fiber-optic cable   Optical receiver

© Cengage Learning 2013

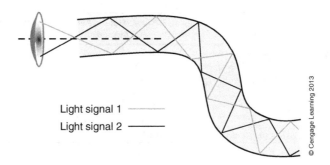

© Cengage Learning 2013

Figure 10–6
Two signals of light reflecting through an optical cable.

Light signal 1 ----------
Light signal 2 —————

works (Figure 10–6). The layer of cladding placed around the core acts as the reflecting mirror. In actuality, the cladding is made of a material having a different density from the core, to refract and reflect the light wave as it transmits though the cable. To understand the transmission in more detail, we need to discuss the index of refraction.

Index of Refraction

The **index of refraction** is a ratio comparing the speed of light in a vacuum with that of any other material or medium. In a vacuum, light travels at its maximum speed of 186 miles/sec. Through any other material, the speed of light slows down. For a vacuum, the index of refraction is equal to 1. As a result, all other materials or mediums exhibit a higher index of refraction because of the slower speed of travel. Water, for instance, exhibits an index of refraction approximately equal to 1.3, whereas glass and plastic are closer to that of 1.5. This means that the speed of light through water would be faster than it would be through glass or plastic.

So, what happens when light travels from plastic to water? The answer is that the light bends as it transitions between the boundaries of the two mediums. The sudden change in transmission speed, because of the differing index values, forces the light bend. As proof, have you ever placed a drinking straw into a glass of water and noticed how the straw appears to bend as it enters the water? The actual angle of apparent bend is directly related to how dissimilar the index of refraction levels are between the two mediums; in this case, the light first is traveling through air and then through water. Incidentally, if the two mediums were able to exhibit the same index of refraction, then the image of the straw would not appear to bend.

The actual construction of a fiber-optic cable makes use of this light-bending phenomenon, which is why the density of the core and cladding material are different. As a result, light traveling through the core eventually reaches the boundary of the core/cladding material and bend. The bending light then reflects back to the core, thus trapping the signal within the central portion of the cable. Ultimately, the traveling light wave reaches the end of the cable, as it bounces from one core/cladding boundary to another. In theory, total internal reflection can be achieved by choosing the correct refraction index for the core and the cladding material.

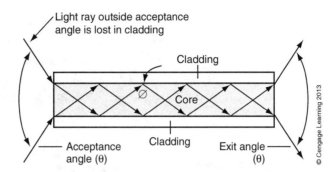

Figure 10–7
Critical angle.

Total internal reflection means that all of the light entering the core within a specific **critical angle** or aperture range remains trapped inside the cable, and losses are minimized. Light waves attempting to enter the core from outside the critical angle ultimately are rejected. On the specification data sheet for optical fiber, this critical angle is known as the numeric aperture. Figure 10–7 illustrates the concept of the critical angle.

Cable Designs

This section discusses three basic fiber-optic cable designs: loose-tube, tight-buffered, and ribbon fiber-optic cables. Each has a specific application according to the design specification of the cable (Figure 10–8).

   Loose-tube cable is used in outside plant applications; that is, the cable typically is exterior to the structure and is subject to environmental elements not normally seen on the interior of a building. In this design, color-coded plastic tubes house and protect the optical fibers. Because moisture is a concern in outdoor installations, this cable is designed with a gel filling compound to impede water penetration. To protect the fibers from damage during installation or environmental loading, a steel central member serves as an antibuckling element.

   A **tight-buffered cable** is more commonly used inside a building. It can come in two varieties, including single-fiber and multifiber tight-buffered designs. In the tight-buffered design, the buffering material is in direct contact with the fiber, making it less heat- and moisture-resistant than the loose-tube cable. The tight-buffered cable lacks the central steel member seen in the previous design. This design is most often used for intrabuilding, risers, general building, and plenum applications. More specifically, single-fiber tight-buffered cables are used as pigtails, patch cords, and jumpers, whereas multifiber tight-buffered cables are used primarily for alternative routing and handling flexibility for installations in tight areas.

   Ribbon fiber-optic cables are constructed by bonding up to 12 coated fibers into a flat, horizontal geometry. Multiple ribbons then can be stacked on top of each other within a single cable, which more effectively uses the size of a given space. A ribbon-style cable can pack more than 100 fibers into a 0.5-in. (12.7 mm) square, which is a more efficient use of the geometry of a small space and allows for higher density installations and lower costs from using less cable.

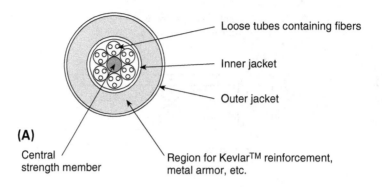

Loose tubes containing fibers

Inner jacket

Outer jacket

(A)

Central strength member

Region for Kevlar™ reinforcement, metal armor, etc.

Figure 10–8
(A) Loose-tube fiber-optic cable.
(B) Tight-buffered fiber-optic cable.
(C) Ribbon fiber-optic cable.

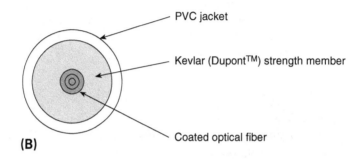

PVC jacket

Kevlar (Dupont™) strength member

Coated optical fiber

(B)

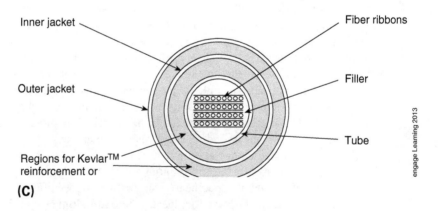

Inner jacket

Outer jacket

Regions for Kevlar™ reinforcement or

Fiber ribbons

Filler

Tube

(C)

engage Learning 2013

Cable Jacketing and Styles of Optical Fiber Cables

The statements in Chapter 2 about cable jacketing and types of insulation still apply to fiber-optic cables. When choosing the type of fiber-optic cable for a specific installation, just as with copper cables, the following factors must be taken into account:

- Insulation requirements (temperature and fire ratings)
- Environmental concerns (moisture, temperature, sunlight)

- Cable length
- Bandwidth
- Underwriters Laboratories and *National Electrical Code* (*NEC*) requirements
- Signal source power and receiver sensitivity
- Connectors and terminations
- Compatibility with additional systems

The physical construction of the optical cable can take on many forms, including:

- Simplex
- Zipcord
- Tightpack cable
- Breakout cable
- Loose-tube cables
- Hybrid or composite cables

Simplex fiber optic is a jacketed, single-fiber, tight-buffered cable that uses Kevlar for strength and reinforcement. **Simplex cables** typically are used as patch cords and in backplane applications (Figure 10–9). The backplane refers to the socketed interconnection of multiple cables and circuit cards for purposes of expansion.

Zipcord is a two-fiber version of simplex cable (Figure 10–10).

Figure 10–9
Simplex cable shown in cross section.

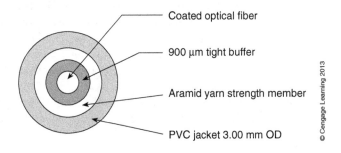

Figure 10–10
Zipcord cable shown in cross section.

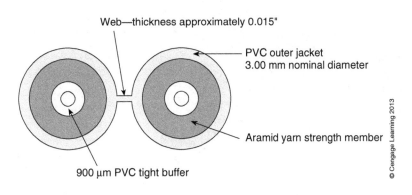

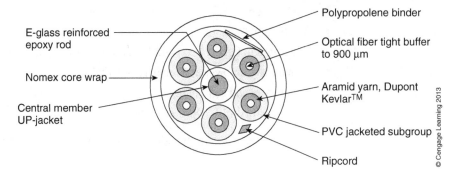

Figure 10–11
Tightpack cable shown in cross section.

Tightpack cables are used as distribution cables, containing several tight-buffered fibers bundled under a single jacket. Tightpack cables also use Kevlar for strength and reinforcement of the bundle. The individual fibers, however, are not reinforced separately and therefore must be terminated inside of patch panels or junction boxes. A typical use for tightpack cables is for a short, indoor, plenum, or riser installation (Figure 10–11).

Breakout cables combine multiple tightpack cables under a single jacket. For this reason, they are stronger, larger, and more expensive. Breakout cables typically are used for long conduit runs and riser or plenum applications (Figure 10–12).

Loose-tube cables contain several fibers under a single jacket. The cable is double-reinforced and insulated with extra layers of Kevlar and insulation jacketing, making it ideal for exterior use and underground burial. The cable is also strong enough to be strung from overhead supports (Figure 10–13).

Hybrid or composite cables come in one of two varieties: those that contain both optical fibers and copper conductors under a single jacket, and those that contain **multimode** and **single-mode** optical fiber under a single jacket. In either case, the different varieties are insulated separately, as required, because they are bundled together as a single cable. When producing such cables, a manufacturer must take into account the requirements and classifications of differing electrical circuits using copper conductors, as well as the insulation

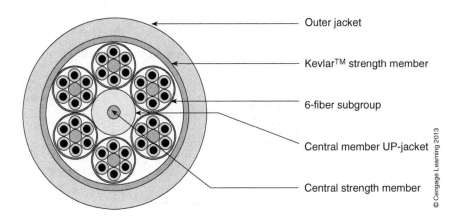

Figure 10–12
Breakout cable shown in cross section.

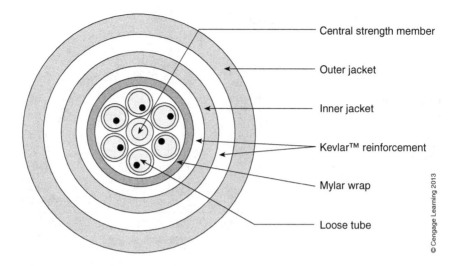

Figure 10–13
Loose-tube
cable shown in
cross section.

Central strength member

Outer jacket

Inner jacket

Kevlar™ reinforcement

Mylar wrap

Loose tube

© Cengage Learning 2013

barrier and separation requirements. Composite cables are commonly used for special applications where the need exists to transmit over multiple types of cable between systems.

Signal Loss

The transmission of light through fiber-optic cables is faster than copper conduction, offering transmission speeds greater than 2 Gbps, over long distances, without the use of repeater stations or amplifiers. See Figure 6-21 for a comparison of copper cables and those of multimode (MMF) and single-mode fiber (SMF). Compared with standard metal conductors, fiber-optic cables are not subject to electromagnetic interference, radiofrequency interference, ground loops, or lightning, thus making them virtually noise and error free. They also offer a high degree of security from communication taps and outside tampering.

Loss of signal in a fiber-optic cable can be the result of many factors. Two common losses include scattering and absorption. Scattering occurs when light waves bounce off individual atoms of glass, causing them to deflect outside of the critical angle. Light that scatters travels in multiple directions, often being absorbed into the cladding. In some cases, the scattering direction may cause light to travel backward toward the transmission source. Scattering is also an inverse function of wavelength. As wavelength increases, the degree of scattering goes down. The actual amount of reduction is proportional to the fourth power. As an example, a doubling of wavelength reduces the scattering by a factor of 16 ($2^4 = 16$). A tripling of wavelength reduces the scattering by a factor of 81 ($3^4 = 81$). It is therefore more advantageous to use long wavelength when long-distance transmission is required, to help reduce the amount of scattering and losses. Figure 10–14 compares attenuation losses caused by scattering and absorption, as related to specific wavelengths of light.

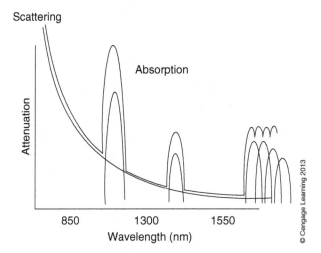

Figure 10–14
Fiber loss caused by scattering and absorption.

The glass tubes inside the cable also can cause signal losses, depending on their purity. For signal to travel through these tubes, the core must be made of ultra-pure low-loss glass. Any imperfection in the glass can cause degradation in the signal. Tight bending or kinking of the cable also limits the amount of light reaching the end point, resulting in attenuation of signal at the final destination. For this reason, bends along a fiber-optic cable should not be more than a 90-degree angle. In addition, splicing fiber-optic cable either mechanically or by fusing two ends together causes an average 0.2 dB loss of signal for every splice on the line, and a return loss of less than –50 dB. For connectors, the typical loss when mating connector pairs fall more in the range of 0.1 to 1 dB, with a return loss less than –30 dB.

SEC 10.4   APPLICATIONS

National Electrical Code 770.2 Optical Fiber Cables

Typical applications for optical fiber cable include communications, data transmission, instrumentation, process, and industrial control. Fiber-optic cables are lightweight and unable to spark, short-circuit, or ignite combustible materials, making them an outstanding option for hazardous locations where gaseous or chemically explosive environments may exist. The only downside to fiber-optic cable is that the material and installation costs are far more than those of copper conductors; but in most situations, the increased cost of metal conduits can be eliminated. Plastic cables, however, cannot be used in hazardous locations because of the off-gassing of plastic vapors. Especially in vacuum or sterile environments, the use of plastic cables is not an option.

SEC 10.5 VARIETIES OF FIBER

Fiber-optic lines come in the three basic varieties: SMF, MMF, and POF. They are generally used as trunk lines connecting major lines over relatively long distances.

SMF provides only one mode of transmission through a single strand of glass fiber, typically having a core diameter of 8.3 to 10 μm.

The small-diameter core can operate from a single light source (Figure 10–15), making it free from distortion which could result from overlapping light pulses; it also allows for greater transmission speed when compared with other types of fiber, with the least amount of signal attenuation. As a result, single-mode cable carries a higher bandwidth than multimode, but it requires a narrower spectral width light source to operate; for this reason, single mode typically uses laser light sources. Although transmission distances can be up to 50 times longer than those of multimode, the material and installation costs of SMF can run much higher. SMF typically is used by the communications and telephone industry because of its high performance over long runs, by community antenna television networks, and by high-speed digital networks operating in the gigabits.

MMF is made from glass fibers with diameters in the 50 to 100 μm range (the most common size is 62.5 μm). Although they are not as fast as single-mode cable, they can offer relatively high bandwidths at speeds of up to 200 Mbps over medium distances.

Multimode step index

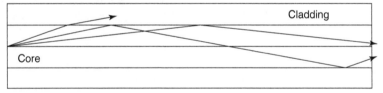

Figure 10–15
Three types of optical fiber.

Multimode graded index

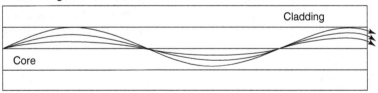

Single mode

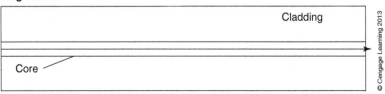

© Cengage Learning 2013

Multimode gets its name from the various light waves that are dispersed into numerous paths, or modes, as they travel through the cable's core, typically 850 or 1300 nm in diameter. The signal can therefore enter and exit the fiber at different angles. The main disadvantage of multimode is that over long cable runs (>3000 ft [914.4 m]), the multiple paths of light can cause signal distortion, resulting in unclear or incomplete data transmission at the receive end of the line. Also, signals traveling at higher modes can become attenuated as they travel down the fiber because of absorption of light into the cladding and also through scattering, as the light becomes bent at angles outside the critical angle.

There are two basic varieties of MMF: step index and graded index (see Figure 10–15). **Step index** MMF has a core composed of one type of glass. The light can travel through the fiber in one of two ways: either straight-line, along the central axis, or through a more indirect path, in modal form, reflecting back and forth along the core and cladding. Signals traveling along the more direct, straight-line axis transmit faster and achieve higher bandwidths; those traveling at higher modes have a longer distance and travel time, resulting in lower transmission bandwidths. Modal travel may also be subject to greater levels of signal dispersion and attenuation. For this reason, step index fiber typically is used over short, low-speed data links, in situations where higher bandwidths are not critical.

Graded index MMF comprises many different layers of glass, all chosen with a different index of refraction to produce an overall index profile approximating that of a parabola. Because light travels faster through a lower index of refraction, the light will travel faster along the outer edges of the cable and slower closest to the core. Light traveling at higher modes through graded index fiber therefore has an overall travel time similar to those of the straight-line, axial rays along the central core, thus allowing for minimal higher mode signal dispersion and higher bandwidths. Graded index fiber can provide nearly 100 times the bandwidth of step index fiber, and it is traditionally used in LED-driven data communications systems.

Multimode links typically use 850 to 1300 nm LED transmitters instead of lasers, because the larger core diameters of the fiber more readily accept the broader output patterns of the LED; also, because the LED is limited to speeds up to 200 Mbps, MMF remains the best choice.

POF cables are cheaper, thicker (typically 1 mm [0.04 in.] in diameter), and more limited in wavelength and bandwidth than those composed of glass fibers. For this reason, they are used only on specific applications where speed is not a necessity.

SEC 10.6   TYPES OF CONNECTORS

Fiber-optic connectors fall into one of two varieties: simplex or duplex. A simplex connector provides a single connection. A duplex connector contains two fibers within a single housing to more easily accommodate the connection of a transmit and receive signal at either end of the transmission line. Losses through a fiber-optic connector should not exceed 0.75 dB. Figure 10–16 illustrates a variety of possible simplex and duplex connectors. In addition, Figure 10–17 compares the use of these connectors within the data communications and telecommunications industries.

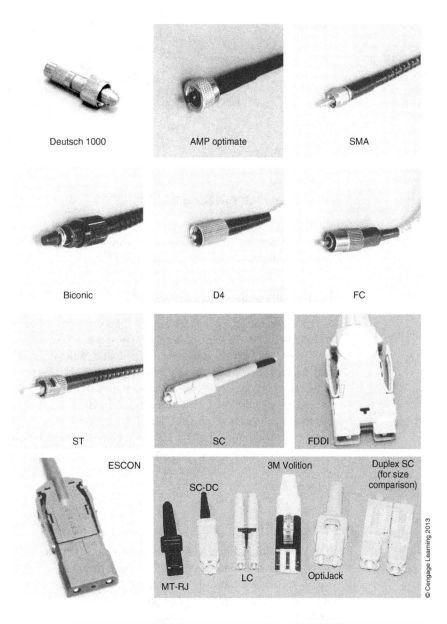

Deutsch 1000   AMP optimate   SMA

Biconic   D4   FC

Figure 10–16
Optical fiber connectors.

ST   SC   FDDI

ESCON   SC-DC   3M Volition   Duplex SC (for size comparison)

MT-RJ   LC   OptiJack

© Cengage Learning 2013

Data communications (mostly multimode)	Telecommunications (mostly single mode)
SMA (obsolete)	Biconic (obsolete)
ST (most widely used)	D4 (fading)
SC (for newer systems)	FC/PC (widely used)
FDDI (duplex)	SC (growing)
ESCON (duplex)	ST (single-mode version)
MT-RJ (new SFF duplex style)	LC (new SFF)
Volition (new SFF duplex style)	MU (SFF, outside United States)
Opti-Jack (new SFF duplex style)	

Figure 10–17
Popular connectors for data communications and telecommunications.

© Cengage Learning 2013

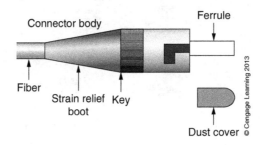

Figure 10–18
The components of a basic fiber connector

Small Form Factor Connectors (SFF)

Small form factor connectors, as listed in Figure 10–17, are a style of fiber-optic connectors that are small in size and low cost. Being smaller in size, SFF connectors offer a higher density of connections within a given space.

Components of the Basic Connector

A basic fiber connector is illustrated in Figure 10–18. While fiber connectors vary in size and shape, they are all basically composed of the following elements: strain relief boot, key, ferrule, dust cover. The fiber enters the strain relief boot from the back and passes through the key and into the ferrule. The strain relief boot acts as a protective barrier to the cable. The key may or may not be included, depending on the type of connector. The ferrule centers the fiber in the connector, and also provides mechanical protection. The dust cover protects the end of the fiber when it is not connected to a patch panel, and should always be used in such cases. In addition, multimode connectors should not be used on single mode fiber. The ferrule on a multimode connector has a larger diameter, and may cause misalignment when used on single mode fiber.

Breakout Kits

Prior to installing connectors on loose-tube cables, a breakout kit will be required to protect the individual strands of fiber once the outer jacketing has been removed. Breakout kits are loose-tube coverings that are placed over the bare fibers of loose-tube cables to separate and protect the individual strands while terminating connectors.

Basic Connector Installation Steps

The methods for installing fiber-optic connectors can vary. The most common methods include Hotmelt®, crimp, heat-cured epoxy, UV cured, and no-heat quick-cure epoxy or anaerobic. When installing fiber connectors, the basic steps include removing jacket and buffer, inserting fiber into ferrule, affixing fiber to connector, applying epoxy, crimping, scoring and removing excess fiber, air polishing, polishing, cleaning, and inspecting.

Methods of Connector Termination

Hotmelt® connectors are a 3M product. The connector is prefilled with an epoxy that must be heated in a Hotmelt® oven for about 2 minutes. Once the epoxy is softened by the heat, the unjacketed, unbuffered fiber is inserted into the connector. The connector is then air cooled prior to polishing.

Epoxy connectors use a syringe to insert the epoxy, and after the fiber is inserted, the connector is cured in an oven.

Crimp connectors can be found in two varieties. One type cleaves the fiber, and then the fiber is inserted into a pre-polished end prior to crimping. The second method requires the fiber to be scored and polished after crimping.

The UV method, similar to the epoxy method, requires a UV light to cure a UV-sensitive adhesive.

The anaerobic method is a nonheat curing method that uses a Loctite® 648 adhesive and a primer for curing.

Scoring and Polishing

Once the epoxy has cured, the protruding end of the fiber is scored, removed, and polished. Air polishing is used first to remove any sharp edges on the fiber. If canned air is used, it should be free of chlorofluorocarbon (CFC) to achieve the best quality. The final step requires the end of the connector to be inserted into a metal polishing puck and polished on differing grades of polishing paper. The polishing paper, like sandpaper, is available in coarse and fine grade. The puck is moved over the polishing paper in a figure 8 pattern. Periodically, the end of the connector should be inspected with a fiber microscope. Once the desired level of polishing has been achieved, the end of the connector is once again cleaned with air and a 99% isopropyl alcohol wipe, prior to attaching the dust cover. Fiber connectors should be cleaned every time they are tested or exposed to the environment to maintain maximum signal level.

Fiber-Optic Splices

There are two types of fiber-optic splices available, mechanical and fusion. Regardless of which method is used, the most critical step in preparing any fiber-optic splice is cleaving. Cleaving involves the precision cutting of a fiber-optic strand with a fiber cleaver. After the ends to be joined have been cleaved, the fibers are physically aligned in a mechanical splice block. The groove of the splice block is filled with an index matching gel. A light source and meter are then used to tune the mechanical alignment for maximum signal strength through the fiber. Once optimum alignment has been achieved, the splice block is crimped and protected from the elements with a splice enclosure. Some splice enclosures are designed for outdoor use, and some, for indoor. Be sure to always use appropriate hardware.

Fusion splicing involves the permanent joining of fibers by using a high-temperature electric arc. The temperature of the electric arc is near 2000°F (1093.3°C) in order to achieve the fusion or melting of fibers in the junction block. The splice is then protected with a splice enclosure once completed.

Fiber Color Codes

The individual fiber-optic fibers within a cable bundle are grouped in groups of 12. The basic color codes are as follows:

Basic 12 Colors

1–Blue

2–Orange

3–Green

4–Brown

5–Slate

6–White

7–Red

8–Black

9–Yellow

10–Violet

11–Rose

12–Aqua

Groups of 12 are then wrapped within color-coded tubes. For a 144-count fiber cable, the color codes are as follows:

Tube Colors (for a 144-count fiber cable)

Tube 1;	Fiber 1–12, Blue
Tube 2:	Fiber 13–24, Orange
Tube 3:	Fiber 25–36, Green
Tube 4:	Fiber 37–48, Brown
Tube 5:	Fiber 49–60, Slate
Tube 6:	Fiber 61–72, White
Tube 7:	Fiber 73–84, Red
Tube 8:	Fiber 85–96, Black
Tube 9:	Fiber 97–108, Yellow
Tube 10:	Fiber 109–120, Violet
Tube 11:	Fiber 121–132, Rose
Tube 12:	Fiber 133–144, Aqua

SEC 10.7   CLASSIFICATIONS OF FIBER AND NATIONAL ELECTRICAL CODE ARTICLE 770

National Electrical Code Section 770.2 Definitions

According to the *NEC*, optical fiber can be broken down into three classifications: nonconductive, conductive, and composite.

National Electrical Code Section 770.133, Installation of Optical Fibers and Electrical Conductors

Nonconductive Optical Fiber Cables

Nonconductive optical fiber cables are those that contain no metallic or conductive parts. Because of this, nonconductive optical fiber cables can be run in the same raceway or cable tray with non–power-limited, and Class 1 circuits. The primary reason for allowing nonconductive optical fiber cable and non–power-limited or Class 1 circuits in the same raceway or cable tray is that the fiber has no electrical characteristics, and it does not cause any signal attenuation, interference, or shorting to other circuits.

Nonconductive optical fiber cables cannot be run within the same enclosure or panel with non–power-limited or Class 1 circuits unless the systems are functionally associated, in which case, they are allowed. As for not allowing nonconductive optical fiber cables in the same enclosure with non–power-limited or Class 1 circuits, the primary reason is safety. As we know, Class 1 circuits can be supplied from power sources as high as 600 V. As a result, the installation and maintenance of such circuits can be dangerous and should only be performed by a licensed electrician. Having high-voltage, Class 1 circuits installed alongside optical fiber, inside of cabinets and enclosures, is simply not a good idea unless they are functionally associated circuits. However, nonconductive optical fiber cable can share the same raceway, cable tray, or cable with conductors of Class 2 or 3 circuits.

Conductive Optical Fiber Cables

Conductive optical fiber cables are made using a variety of metallic strength members, metallic vapor barriers, and metallic armor or sheathing. Although these metallic components are not meant to carry current, they are still conductive in nature. Because of the inherent conductive properties exhibited by these types of optical cable, they are not permitted to occupy the same raceway or cable tray with those of non–power-limited or Class 1 circuits, or any other type of non–power-limited circuit. Conductive optical fiber cable can, however, share the same raceway, cable tray, or cable with conductors of Class 2 or 3 circuits. As a final note, all metalic components of the cable must be grounded, as previously discussed in Chapter 3 of this book.

Composite Optical Fiber Cable

Composite optical fiber cables contain both optical fibers and current-carrying electrical conductors. In some cases, composite optical fiber cable may also contain noncurrent-carrying conductive members, such as metallic strength members and metallic vapor barriers. Composite optical fiber cables are classified as electrical cables and should be considered conductive. As long as the composite optical fiber cable contains current-carrying conductors associated with a non–power-limited or Class 1 circuit, they are allowed to be run in the same enclosure, cable tray, or raceway with Class 1 circuits. In addition, the manufacturer has engineered the composite cable for just such purposes, and provided the circuits are functionally associated, they are allowed to coexist within the same jacket cable.

National Electrical Code Section 770.179, Installation and Marking of Listed Optical Fiber Cables

Markings on optical fiber cables are listed in *Table 770.179* of the *NEC*.

National Electrical Code Section 770.154(b), Applications of Listed Optical Fiber Cables and Raceways

Plenum type nonconductive and conductive optical fiber cables such as OFNP (nonconductive optical fiber plenum cable) and OFCP (conductive optical fiber plenum cable) should be listed as such and should exhibit fire-resistant and low smoke-producing characteristics. Cables listed as OFNP or OFCP are suitable for installation in ducts, plenums, and environmental air spaces.

Nonconductive and conductive riser type cable should be listed as OFNR (nonconductive optical fiber riser cable) and OFCR (conductive optical fiber riser cable). These cables should be suitable for floor-to-floor runs and should also exhibit fire-resistant and low smoke-producing characteristics.

When choosing a general-purpose type optical fiber cable, types OFNG (nonconductive optical fiber general-purpose cable), OFCG (conductive optical fiber general-purpose cable), OFN (nonconductive optical fiber general-purpose cable), or OFC (conductive optical fiber general-purpose cable) should be used. These cables should be used in applications not including plenums or risers. General-purpose cables should exhibit a resistance to the spread of fire.

National Electrical Code Table 770.154(b), Cable Substitutions

Refer to *Table 770.154(b)* in the *NEC* for conductive and nonconductive optical fiber cable substitutions.

CHAPTER 10 FINAL QUESTIONS

1. What is the speed of light through a vacuum?
2. List the benefits of glass fiber over plastic fiber.
3. What are some of the disadvantages of using POF?
4. Which type of glass fiber-optic bundle is lined up from end to end to allow the transmission of an image at the other end of the cable?

5. When ordering optical fiber, what do the numbers 50/125 describe?

6. What is a typical core diameter for single-mode optical fiber cable?

   How does that differ from multimode optical fiber cable?

7. Explain the purpose of the cladding in the optical fiber.

8. What is the difference between grade index and step index MMF?

9. How does the application of loose-tube cable differ from that of breakout cable?

10. To reduce dispersion, multimode fiber-optic cable uses a _____ index profile to produce a net effect that has different modes of light arrive at the fiber end at the same time.
    a. buffered
    b. graded
    c. retracted
    d. step

11. Index of Refraction is the ratio of _____.
    a. the speed of light in a vacuum compared to a particular medium
    b. the speed of sound in a vacuum to a particular medium
    c. wavelengths in a vacuum to a particular medium

12. The buffer on a fiber-optic cable is _____.
    a. always the same size for all cables
    b. not required for single-mode cable
    c. removed when splicing and/or attaching connectors

13. Multimode fiber-optic systems use a(n) _____ as their light source.
    a. incandescent bulb
    b. neon bulb
    c. LED
    d. laser

14. What are the two varieties of hybrid or composite fiber-optic cables?

15. The most common core size for multimode fiber is _____.

16. _____ are used to protect the bare fibers that are removed from the tubes of loose-tube cables.
    a. Breakout kits
    b. Expansion tubes
    c. Innerducts
    d. Splice kits

17. The two main contributors to attenuation in fiber-optic cables are

    _____.

    a. absorption and scattering
    b. dopants and impurities
    c. fresnels and macrobends
    d. dopants and macrobends

18. The definition of critical angle or aperture range is _____.

    a. the largest angle that a light ray can enter a fiber and still propagate down the core
    b. the total amount of light reflected down the fiber core
    c. the fasted wavelength that light travels inside the fiber
    d. the inverse of the critical angle at the fiber's core

19. A good rule of thumb for cleaning connectors is that they be cleaned

    _____.

    a. weekly
    b. every time they are exposed to the environment
    c. annually
    d. biannually

20. The maximum loss through a fiber-optic connector should not exceed _____ per connector.

    a. 0.25 dB
    b. 0.5 dB
    c. 0.75 dB
    d. 1.0 dB

21. The most critical step in preparing for a fiber-optic splice is _____.

    a. polishing
    b. crimping
    c. cleaving
    d. cleaning

22. A general guideline for the bend radius of fiber-optic cable that is not under tension is _____ times its diameter.

    a. 10
    b. 20
    c. 30
    d. 40

23. What article of the *NEC* covers the installation of fiber-optic cables, and what three types of optical fiber cables does the *NEC* define?

24. Explain how conductive optical fiber cable can be run with respect to Class 1, 2, and 3 cables.

25. Which optical fiber cables can be substituted for type OFNG or OFCG cables?

Chapter 11

Telecommunications, and *National Electrical Code Article 800*

Objectives

- Define and describe the local loop of a telephone system.
- Describe and define the basic parts of a residential telephone system.
- Identify common types of signaling used by the telephone company.
- Define the basics of a PBX telephone system.
- Define and describe the various trunk lines associated with a PBX.
- Describe an electronic key system (EKS) and how it compares to a PBX.
- Identify the types of terminations and color codes for residential and commercial telephone wiring.
- Identify the requirements of *NEC Article 800*.

Chapter Outline

Sec 11.1 Telephone Basics

Sec 11.2 Residential Cabling

Sec 11.3 Commercial Systems

Sec 11.4 The Private Branch Exchange (PBX)

Sec 11.5 PBX Trunk Services

Sec 11.6 Electronic Key Systems (EKS)

Sec 11.7 Terminations and Color Codes

Sec 11.8 *National Electrical Code Article 800* Communication Circuits

Key Terms

area code

base unit

busy signal

central office (CO)

compensator

demarcation point

dial tone

dual-tone-multifrequency (DTMF)

duplex coil

earpiece

electronic key system (EKS)

exchange

handset

hook switch

horizontal wiring

intermediate distribution frame (IDF)

local loop	ring	TELCO
main distribution frame (MDF)	ring-back signal	tip
mouthpiece	ringer equivalence	trunk
private branch exchange (PBX)	number (REN)	

SEC 11.1   TELEPHONE BASICS

History of the Telephone

Alexander Graham Bell invented the telephone in 1875. Today's telephone was born out of the original telegraph, which used dots and dashes (Morse Code) to allow message to be sent, one at a time, over a single wire. It wasn't long before Bell took the process a step further by inventing the "multiple telegraph," which was made to send several different signals over one wire, provided the signals were transmitted at a different pitch. In experimenting with his recent invention, Bell realized that he could also hear the sound of a person's voice over the same wires. The original design worked on the principles of magnetic induction. By placing a thin, metal diaphragm in front of an iron core that was surrounded by a coil of wire, Bell discovered that he was able to induce voice signals. The sound waves hitting the core caused the magnetic field to fluctuate, inducing an electric current in the surrounding coil. The current was then transmitted over wires to a receiver coil, which induced a duplicate magnetic field in the receiver core, causing the receiver diaphragm to vibrate in unison. "Mr. Watson. Come here. I need you." were the first words transmitted in Bell's laboratory, and with that discovery, the birth of the modern telephone was born.

The Local Loop

The telephone company's **central office (CO)**, typically referred to as **TELCO**, is connected back to your house over a pair of twisted-pair wires called the **local loop**. These cables are the ones seen strung from telephone poles. The cables may also be buried underground to service customers. Originally, the local loop was designed for voice transmission using analog transmission technology. Today, because of the use of computer modems, the local loop uses both analog and digital transmission technology and is now sometimes referred to as the "subscriber loop." The local telephone utility is responsible for the installation and maintenance of the local loop wires leading up to any residence.

The Inside of the Telephone

The makeup of a basic telephone is relatively simple. This same technology has been used for many years and has stayed relatively unchanged. There are a few main parts that we address to aid in the understanding of the operation of the telephone itself.

The Mouthpiece

The **mouthpiece** contains a microphone that is used to generate a voice signal for transmission to the other end of the telephone line. When you speak into the microphone, the sound pressure created vibrates two thin metal plates filled with carbon granules. The resulting vibration changes the resistance (R) of the carbon inside the mouthpiece, and modulates the dc current (I) passing through the handset. Using Ohm's Law ($E = I \times R$) we can determine that the dc voltage across the microphone fluctuates with the changing pressure levels of your voice. This produces an ac voice signal riding on the dc current.

The Earpiece

The **earpiece** is made up of a small, inexpensive, 8-ohm speaker. When the dc current passes through the earpiece, it vibrates due to the ac voice signal produced by the mouthpiece. This vibration represents the voice signals sent by the person on the other side of the line over the local loop. The bandwidth of an analog telephone circuit ranges from 180 Hz to 3.2 KHz, just wide enough to cover the frequency response of the human voice.

The Hook Switch and Specifications of the Local Loop

The **hook switch** indicates whether the phone is "on hook" or "off hook." The switch can be easily identified as the plastic button depressed by the earpiece of a standard phone when it is on the hook. The switch is *open* when the handset of the telephone in on hook, indicating an open circuit, drawing no current. When the handset is taken off hook, the switch closes, and the loop circuit draws approximately 15–90 mA of dc current, depending on the total resistance of the local loop. The power source for the local loop ranges between –48 and –52 V dc. As the line begins to draw current, TELCO senses the connection and provides a **dial tone** indicating that the line is ready for use. The typical off hook resistance of a telephone is about 180 Ω, with an additional 200–400 Ω of series resistance occurring in the loop, over the connecting wires and through the telephone company switching circuits. The additional series resistance is necessary to prevent the customer from accidentally short-circuiting the phone line.

The Duplex Coil

The **duplex coil** is used to provide auditory feedback from the mouthpiece to the earpiece. It is important for people to hear themselves as they speak. This portion of the telephone controls the amount of voice signal feeding back to the earpiece of the handset so the speaker can hear a portion their sound. Not all of the voice signal is cycled through to the earpiece because a large portion is sent over the local loop to the other side of the line. Also, cycling the speaker's entire voice signal to the earpiece would create a large echo making is hard to talk into the telephone.

The Compensator

The **compensator** is the portion of the telephone circuit that compresses the dc current to control the loudness of the voice signal transmitting over the local loop.

In some cases, the loudness of a voice signal can vary to more than two times its original state. The compensator automatically adjusts the level of sound, so that listeners don't have to hold the receiver inches away from their ear to hear the speaker at a normal listening level.

Telephone Dialing

Originally when making a call, someone would pick up the telephone, give directions to the operator as to whom they would like to speak, and the operator would make the connection manually. Eventually, telephones were made that allowed direct dial, and operator assistance was no longer needed. The original direct dial phones used rotary dialers. This method was known as pulse dialing. When a number was dialed, the dc current on the line was pulsed the same number of times as the chosen digit. For example, if you were dialing the number 5, the dc current on the line was pulsed 5 times and recognized by the equipment at TELCO. The pulsing was then decoded and the correct connection was made once the pulsing stopped.

Today, because of the advancements in automated phone menus it is imperative that a phone utilize touch-tone technology. This technology uses **dual-tone multifrequency (DTMF)** to dial a number. When a button is pushed, a unique dial sound is created using a set of frequencies shown in Figure 11–1. The frequency is sent over the local loop and decoded by equipment at TELCO, which in turn connects you with the line you have dialed.

The Telephone Ringer

The telephone ringer, simply stated, is a device that alerts you of an incoming call. In the early days, the sound that phones used to make was produced by an electromechanical bell. There were two bells fitted with coils; when a ringing voltage was sent

Figure 11–1
Dual-tone multi-frequency chart.

	1209 Hz	1336 Hz	1477 Hz	1633 Hz
697 Hz	1	ABC 2	DEF 3	A
770 Hz	GHI 4	JKL 5	MNO 6	B
852 Hz	PRS 7	TUV 8	WXY 9	C
941 Hz	*	oper 0	#	D

© Cengage Learning 2013

from TELCO (approximately 90 V ac), a metal hammer placed between the two bells would move because of the magnetic field in the coil. The hammer would tap each bell alternately, creating a sound at a frequency of 20 Hz. This was done so as to be able to distinguish early telephone rings from other household bells such as the doorbell.

Today, most telephone ringing equipment is integrated circuit (IC) powered by the rectified ringing signal, and solid-state devices inside the ICs produce a digital ring tone.

The Ringer Equivalence Number

There is only a certain amount of current available on a given telephone line. Because it takes current to ring a telephone, if you keep adding telephones to your line, there comes a point where there is simply not enough current on the line to ring all the phones. In the United States, the telephone company only guarantees to ring up to three ringers. To be able to calculate whether you may have too many telephones connected to one line, you add up the **ringer equivalence number (REN)** of the phones. Every phone should have an FCC registration label that contains the REN. The format of the REN is standard across all phones. It contains NUMBER, then LETTER. The NUMBER is a decimal number representing how much power the ringer requires as compared to a standard phone. A standard phone is defined as a standard gong ringer as supplied in a telephone company standard desk telephone. The number 1.0 represents that the phone uses 100% of the power of a standard phone. A number of 3.2, for example, indicates that the telephone uses the power of 3.2 standard phones. The LETTER represents the frequency the telephone requires. The letter A signifies that a 20 Hz ring signal is required. The letter B signifies that the telephone rings at any frequency.

Signaling Tones

It is important that your telephone communicates back and forth with TELCO regarding the usage of the telephone line. This communication is called signaling. To be able to do this, the telephone company uses a series of tones for signaling states such as a busy line or a free line. When a customer is interested in placing a call, the phone is taken off the hook and current flows to the line. The telephone company acknowledges this action by sending a signal, called a dial tone, to the user. This tone, which is modulated at 350 Hz and is continuous, lets the caller know that the phone line is free and ready to be used. If a customer places a call to a line which is not being used, the caller hears a **ring-back signal**. The ring-back signal represents the sound of a bell ringing at 440 Hz. In the United States, the format for a typical ring-back signal consists of 6 seconds on, 2 seconds off. On the other hand if the line is being used, the caller is notified by a **busy signal**. The busy signal is sent at approximately 480 Hz and is not continuous, but rather is choppy. The typical busy signal is on for 0.5 seconds and off for 0.5 seconds.

Dialing

Every telephone company customer is assigned a specific telephone number that identifies the customer's line on the local loop. These numbers consist of 10 digits

Figure 11–2
Significance
of a telephone
number: area
code, exchange,
phone line.

Area code   Exchange   Phone line

© Cengage Learning 2013

and three identifiers. The first set of three numbers represents the **area code** you are dialing, which identifies the geographical region being called. The middle set of three numbers represents the **exchange**. The exchange is a switch that decodes incoming tones or pulses to identify the central office that provides service to the subscriber. The last four numbers represent the exact phone line.

Typically, if you are calling a number within the same area code as your own, you do not have to dial all 10 digits; just the 7-digit number is sufficient. Calls to other area codes usually incur an additional charge; however, most metropolitan areas cover several area codes that do not incur charges. Sometimes area codes cover such a large geographical area that an additional charge is incurred within the area code itself. This is seen in rural areas of the United States.

Each exchange can have a maximum of 10,000 phone lines because of the limitation of number combinations. This means that an area requiring 70,000 phone lines would need at least seven exchanges. The number of area codes has increased significantly over the past few years, due to the increased use of cellular phones and the high demand for multiple household connections. Landlines, traditionally used for telephones and fax machines, now primarily provide computer networks a connection to the Internet. And in many cases, individual households have given up their primary landline telephone in favor of a high-speed digital connection for their personal computer. Regardless of what the line is used for, an exchange number is still needed.

Cordless Telephones

Cordless telephones have been around since the early 1980s. These phones were originally primitive at best. Because of the original frequencies assigned to the phones, 1.7 MHz using AM modulation, and later 43–50 MHz for FM modulation, the sound quality was poor. The phones also picked up a lot of noise and a multitude of static. Channel allocation was also limited. The original cordless phones were equipped with only one channel, even though the telephone company was able to provide up to 10 channels for use. If the phone you bought had a conflict, you had to exchange your phone for one that ran at a different frequency.

The technology finally became viable when the FCC allocated the 900 MHz band in 1994. This increased the quality and range of signals significantly. Later, the frequency band was further increased to 1.9 GHz (originally developed in 1993, but not actually allocated until 2005) and then to 2.4 GHz, which was allocated in 1998. Later, in 2003, 5.8 GHz was allocated due to crowding of the 2.4 GHz band.

Today, cordless phones are commonplace in many homes and businesses. And being relatively inexpensive, they provide good coverage and clear digital sound quality compared to their predecessors. To understand the operation of a cordless phone, it is important to identify the two units that make up the phone.

The Base Unit

The **base unit** of the cordless phone is powered from an ac wall outlet. The base unit has circuitry so it can transmit and receive a signal. This signal is a frequency modulation (FM) radio signal that allows the base unit to communicate with the handset. When a call comes in, the base unit sends a signal to the handset. When the call is answered, a signal is sent back to the base station, notifying it to pass the voice signals to the handset. To place a call, the phone being taken "off hook" signals the base station to communicate a signal in the form of a dial tone if the line is ready.

The Handset

The **handset** is the mobile piece of the cordless phone. It does not have anything physically tying it down to a specific location, unlike the base unit. The handset also contains circuitry to transmit and receive radio signals. The handset contains a speaker, microphone, keypad, and rechargeable batteries. When a call comes in, the base unit signals the handset. The "talk" button on the handset has to be activated for the call to be answered. By doing this, the handset sends a signal back to the base unit, indicating voice signals should be sent. To place a call, the talk button is activated on the handset, and the base unit acknowledges that action by sending a dial tone. From there, the handset interacts with the base station by transmitting and receiving voice signals.

SEC 11.2   RESIDENTIAL CABLING

The wires in each twisted pair of a telephone cable are called the "**tip**" (green) and the "**ring**" (red). These terms date back to the time when telephone operators would "patch" calls manually at a switchboard. The actual names were derived from the electrical contacts on the original ¼-inch telephone plug, similar to that shown in Chapter 5, Figure 5–16.

The tip and the ring wires are twisted together in pairs so as to minimize any electrical interference from power lines on adjacent runs. Because of the twisting of the pairs, any induced voltage is canceled out.

Because there is more than one pair of wires per cable, we must address the assigned colors of the additional tip and ring pairs. It is assumed that the green and red wires are the primary, or first, pair. The subsequent color codes are designated by the chart in Figure 11–3.

On the primary pair, the tip (green) wire is connected to ground, and the ring (red) is connected to the –48 V dc battery at TELCO. When the phone is taken off

Tip wire	Ring wire
Green	Red
Black	Yellow
Blue	White
Secondary color with primary color stripes	Primary color with secondary color stripes

Primary colors: blue, orange, green, brown, and slate.

Secondary colors: white, red, black, yellow, and violet.

© Cengage Learning 2013

Figure 11–3

Tip and ring wiring chart.

hook, the tip and ring are connected together inside the phone, and current begins to flow. As soon as the central office (CO) detects the current flow, dial tone is sent, and the phone is ready to make a call.

Premise Wiring

Now that we are familiar with the local loop, we must talk about the wiring beyond the **demarcation point**, the junction of the local loop wiring and the premise wiring, often just called the "demarc" at a residence. As stated earlier, the telephone utility is responsible for the local loop wiring but is not responsible for the maintenance of the wiring inside a house, called the premise wiring. Premise wiring can be divided into two categories distinguishing its purpose: *residential* and *commercial*. Most telephone utilities service and repair premise wiring for an added monthly charge to your phone bill or by charging a fee for a repair call. The demarc is most commonly identified in residences by a 42A block (Figures 11–4 and 11–5) (see *NEC 800.90–800.93*).

This block may be located outside or in the basement and is connected to earth ground by an approved *NEC* means (as covered in Chapter 3 of this book). The telephone utility is responsible for providing a safety device, called a protector, just before the demarc. This protector is to prevent voltage from lightning or a fallen power line from getting on the telephone lines.

Residential Connections

From the demarc, all telephone sets (telsets) are wired in parallel. The telsets are connected to the premise wiring, using RJ-11 plugs and jacks. These are the connections you see every day when using or connecting a phone. The use of the RJ-11 plugs and jacks has been standard for many years. However, it is common in older buildings and residences to still see the screw-type block (42A block).

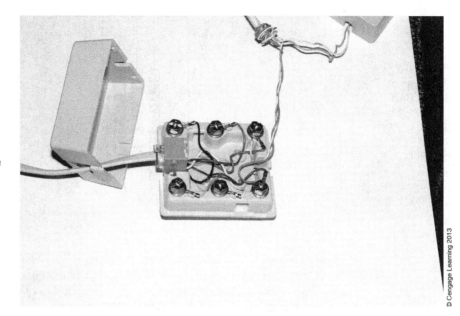

Figure 11–4
42A telephone connection.

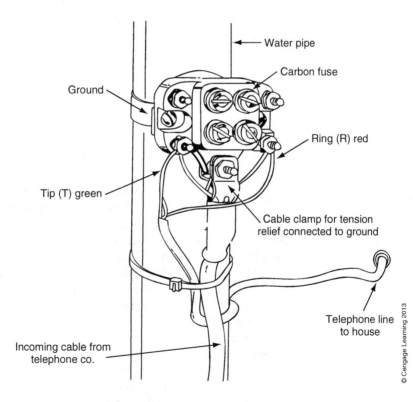

Figure 11–5
Residential 42A telephone connection.

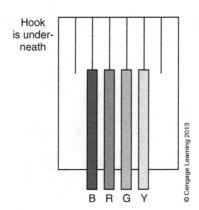

Figure 11–6
Universal Service Order Code (USOC) registered jack (RJ)-11 wiring standard.

Hook is under-neath

B R G Y

© Cengage Learning 2013

Figure 11–6 shows the USOC standard for wiring a RJ-11 plug.

Most telephone wires are one or more twisted pairs of copper wire. The most common type is the 4-strand (two-twisted pair). This consists of the red and green wires, which make a pair, and the yellow and black wires, which make the other pair. One pair is all that is needed to run one telephone line. The other pair of wires is often used to add a second line to the home.

The RJ-11 jacks either accept a four-conductor or a six-conductor type cable. This means they accommodate two or three pairs or two or three telephone lines per home. It is possible to split a six-conductor jack into two separate RJ11 jacks, each with one pair or one line.

SEC 11.3 COMMERCIAL SYSTEMS

An office building is a good example of where you would find commercial telephone wiring. A commercial business requires multiple phone lines, which in most cases connect through internal premise wiring to a **private branch exchange (PBX)**. A PBX is a privately owned telephone system within a business. The PBX can not only switch calls between users on local lines, but also allows a certain number of users to share the external phone lines from the CO. The main purpose of a PBX is cost saving. The business is charged a flat fee for the service as opposed to charging per line, per use. The PBX is operated and owned by the business, not the telephone company. The PBX is typically located in the basement or on the first floor of the establishment (Figure 11–7).

Commercial wiring commonly uses twisted-pair cable, like residential wiring. However, it is quite common to see commercial wiring utilizing coaxial cable, fiber-optic cable, or CAT5, depending on the type of circuit and speed of transmission.

Commercial wiring differs from that of residential with respect to the types of devices and the terminology. The starting point for a commercial telephone installation is the **main distribution frame (MDF)**. The MDF is the interface

Figure 11–7
Private branch
exchange (PBX).

© Cengage Learning 2013

between the outside world and the internal PBX. It serves two purposes: to provide protectors for incoming circuits against lightning and voltage surges and to cross-connect any outside lines with the inside lines of the building PBX. The MDF is usually located on the first floor or in the basement of the building and consists of two steel frames typically mounted on a wall of the telecommunications room housing the telephone equipment. All wire pairs terminating at the MDF are done so at type 66 punch-down blocks. Type 66 punch-down blocks contain V-shaped pieces of metal where the individual wires are forced, or "punched," down with a type 66 punch-down tool. Each contact can hold one or more wires. The benefit to using a punch-down block is that once a wire has been punched down, it can be easily removed and moved without having to run a new cable. Figure 11–8 shows multiple terminations in a 66 punch-down block for a typical MDF.

In tall buildings and high rises, where placing the MDF in the basement or first floor would cause cable runs in excess of thousands of feet, an **intermediate distribution frame (IDF)** is used. An IDF is a free-standing or wall-mounted

Figure 11–8
Type 66 punch-down blocks.

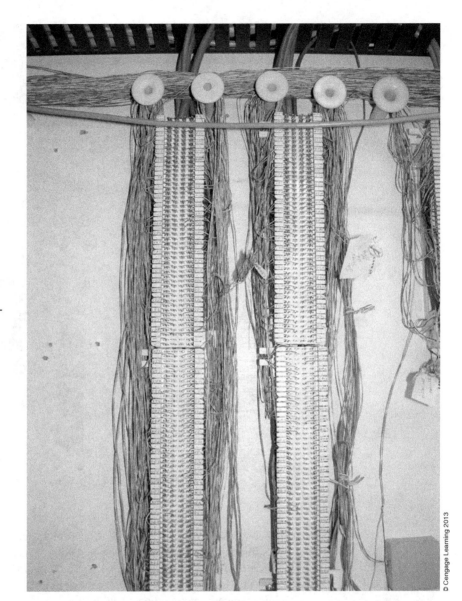

© Cengage Learning 2013

rack for managing and interconnecting the telecommunications cable between the telsets, fax machines, modems, or answering machines and a main distribution frame (MDF). IDFs can be located on every floor, every other floor, or where the designer deems necessary. In most cases, each floor has a designated telephone closet for just such purposes.

The wiring running from the MDF to the IDF, which completes a floor-to-floor run, is referred to as vertical wiring and utilizes a riser cable. The wiring running from the IDF to the user is referred to as **horizontal wiring** and utilizes a plenum or multipurpose cable.

SEC 11.4   THE PRIVATE BRANCH EXCHANGE (PBX)

There are three components that make up a private branch exchange (PBX). They are the common equipment, which includes the microprocessor or CPU and a switching matrix; the central office trunks, for connections outside the building, between the central office and PBX; and the station lines, connecting to the individual telephones throughout the business.

The CPU within the PBX runs all of the software necessary to manage calls between multiple intrabuilding stations or between inside station sets and outside callers, over the main trunk lines from the CO. The process is controlled digitally through a multiplexed switching matrix inside the PBX.

Station sets can be analog or digital; however, they cannot be intermixed and must connect to appropriate hardware within the PBX. Each station operates over either an RJ-11 or an RJ-45 connector, depending on equipment, and utilizes only one pair within the cable for communication. Digital station sets must never be connected to analog lines, and analog stations sets must never be connected to digital lines.

SEC 11.5   PBX TRUNK SERVICES

A **trunk** is similar but different from a telephone line. A telephone line connects between an individual phone and the CO switching equipment. A trunk connects switching circuits together, such as between the CO and a PBX or from one CO switch to another CO switch. The trunk service between a privately owned PBX and TELCO can happen in a variety of ways. The following sections describe and illustrate the possible options.

Ground Start

Ground start trunks are the most commonly used by PBX telephone systems. To begin a call, the PBX must seize the dial tone from the CO. This is accomplished by momentarily grounding the ring conductor of the loop circuit. Figure 11–9 illustrates the process. At the CO end of the circuit, the ring is connected to –48 V dc. As soon as the PBX attaches the ring to ground, the CO senses the flow of current and responds by grounding the tip conductor. The PBX then senses the flow of current along the tip and responds by removing the ring ground. The loop circuit is now connected, and dial tone is activated. To disconnect the circuit, the CO removes ground from the tip conductor, indicating to the PBX that the call has ended. The PBX then opens the loop circuit, and the call is released. Ground start circuits do not work if the polarity of the circuit is reversed.

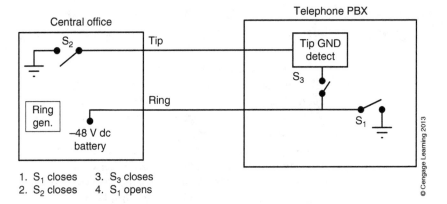

Figure 11–9
Ground start
trunk.

1. S_1 closes   3. S_3 closes
2. S_2 closes   4. S_1 opens

Loop Start

A loop start circuit is what residential telephone systems use, and it's the simplest configuration. Figure 11–10 illustrates the process. The trunk is idle until the station set goes off hook. An off-hook configuration connects the loop circuit, and current flows from ring to tip. The total circuit resistance is still about 600 Ω through the station set, the series resistance of the connecting wires, and the CO switching equipment, as previously discussed. In this configuration, the tip is always at ground, and the ring is supplied by –48 V dc at the CO. Once line seizure is recognized by the CO, due to the presence of current flow on the loop, dial tone is applied across the tip and ring. In this configuration, the CO is simply looking for the status of the hook switch, based on current flow in the local loop.

With loop start circuits, the ring conductor is not first required to initiate a grounding protocol on the customer side to initiate a call. The process is much more simplified.

The oversimplification of the loop start circuit can at times, however, be a disadvantage. Because grounding of the ring conductor on the customer side and the tip conductor on the CO side is not a requirement for making and releasing calls, there is no way of knowing when a call is about to occur until the phone actually starts to ring. In such cases, glare can occur. Glare is the simultaneous seizure of a circuit by two parties. Have you ever picked up a phone to dial a number,

Figure 11–10
Loop start trunk.

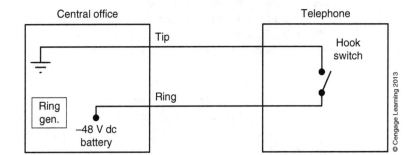

only to find someone else already on the line even though the phone never rang? The reason is because you picked up the phone just prior to the ring, and the circuit from the incoming call was already connected. This is an example of glare. With ground start circuits, this can never occur because of the grounding protocols required between the PBX and the CO to initiate and release calls. With ground start circuits, it is impossible for two customers to seize a line simultaneously.

Loop start circuits also have an issue with call disconnect. In some cases, when a caller hangs up and immediately goes off hook again to make another call, the CO may not have recognized the initial loss of current, and the original call is still connected. To release the call, the caller must hang up again and wait an appropriate amount of time for the CO to recognize that the circuit has opened. Once the CO releases the line, then the caller can begin another call, and dial tone is initiated.

Direct Inward Dialing (DID)

Direct inward dialing is a one-way trunk service that allows incoming calls from the outside to reach a specific station extension within the PBX, without requiring attendant assistance. There is no ringing voltage associated with DID. Instead, all that is required is a constant −48 V dc from the PBX to the CO. With DID, the CO provides the load on the line and closes the loop as calls arrive.

DID trunks do not have associated telephone numbers assigned to them. When a call is received, only the three or four digits of the subscriber's number are sent to the PBX. Internal software then routes calls directly to the desired station set.

With DID, the process is as follows. The PBX maintains a −48 V dc voltage on the tip and ring, looking back to the CO. When a call comes in, the CO places a load on the line. The resulting current flow causes the PBX to respond by momentarily reversing the voltage polarity on the tip and ring conductors, in acknowledgment that a call is being received. The CO then sends the extension digits to the PBX, which then routes the call to the desired station set. As soon as the station goes off hook, the PBX reverses the polarity so that the CO can start logging the call for billing purposes. Once the station goes on hook and the call is completed, the PBX then reverses the polarity back to the original state, and the CO removes the connecting load.

DID trunks are typically ordered in groups of 20, and standard practice is to have 10 telephones for every DID trunk. A one-to-one ratio is not required for DID.

E+M Tie Line

E+M tie lines are used to link multiple PBXs across a wide geographical network. In cases where a customer may have multiple PBXs located across the country, E+M tie lines are used to create a bidirectional trunk between locations. E+M uses a 2- or 4-wire configuration to transmit and receive. The E leads are used for incoming signals, and the M lead is used for outgoing signals. Communication protocols are then set up between the lines to determine when signals are to be transmitted and received. Three systems exist in E+M: immediate start, wink start, and dial tone start.

With immediate start, the PBX receiving the call answers as soon as the calling party initiates the call. In wink start, the system being called places a momentary signal reversal on the M lead (called a wink) to notify the sender that it is ready to receive signal data. The calling system then controls the transfer of information on the M lead while watching for return signals on the E lead. For dial tone start, the party on the far end signals that he or she is ready to receive data by sending a dial tone. As soon as the caller receives the dial tone, the transmission can begin. Depending on how the PBX has been programmed, the caller can still dial the desired extension prior to dial tone being received. In such a case, digits are stored by the calling PBX and sent as soon as dial tone is detected.

T1

A T1 line is a digital CO trunk that a customer leases. The T1 was created in 1960 as the first digital system supporting voice and data over a single connection. The T1 can provide up to 24 multiplexed voice and data channels at a rate of 1.544 Mbps. Digital sampling of the analog signal occurs 8000 times per second to generate an 8-bit word. The 24 channels are then placed into a 192-bit frame, and each frame is separated from the next by a single bit, making up a 193-bit block.

The T1 is commonly used to interconnect multiple PBXs or for customers that require high-speed data transmission over their computer network. A special router is required for T1 connections that utilize pulse code modulation (PCM) and time division multiplexing (TDM), as discussed in Chapter 13 of this book.

SEC 11.6   ELECTRONIC KEY SYSTEMS (EKS)

Electronic key systems are telephone systems for small businesses that allow them to share multiple lines from the CO without having to go through a PBX. The system is essentially a series of electronically controlled switches and relays that multiplex lines to specific station sets as needed. Customers can not only make calls between station sets locally but also share incoming lines from the CO. When using an EKS, each phone is equipped with a selection of LINE keys that the user must first select to be able to make an outside call. Indicators on the station set light up when a line is in use. With an EKS system, the user manually chooses an outside line; with a PBX, the line is chosen automatically. However, a PBX can be programmed to function like an EKS, through software programming, if desired.

A keyed system is not easy to set up, and because each manufacturer uses its own proprietary programming language, a technician intimate with the software is required to get the system up and running. The user can program very few features of an EKS system. The main system programmer must determine the design and all options, and any future changes must go through that individual. EKS systems are more expensive than analog systems, but for small operations they are less expensive than purchasing a PBX.

The EKS still operates over twisted pair to each station set, and some systems use multiple pairs for different functions within the system, depending on the manufacturer. The equipment from one EKS system is not interchangeable with other systems because they use different programming methods and do not digitally communicate in the same manner. Many systems are now fully digital and capable of supporting data and voice communications simultaneously. They can also access Internet Protocol (IP) telephony to networked systems in different locations, or Voice over IP (VoIP) to allow telephone calls over the Internet.

Each station set in an EKS system draws continuous current, transmitting digital data back and forth between the EKS and the handset, to determine the status of a pushbutton or station set feature. There are two types of keys in an EKS system, soft keys and hard keys. Soft keys are programmable at the station set, whereas hard keys are fixed and determined by the system. Disadvantages of using an EKS system are that during a power failure there is no dial tone or service, and moves, adds, and changes (MAC) can often be very expensive and difficult to accomplish.

SEC 11.7   TERMINATIONS AND COLOR CODES

RJ-21X

All trunk lines from the TELCO terminate at the demarcation point to an RJ-21X or to a binding post (Figure 11–11). The RJ-21X is a 50-pin connector that can deliver up to 25 individual lines from the CO. Connections are then made from the RJ-21X to a type 66 Style 50 pair block. Twenty-five pairs of the TELCO lines

Figure 11–11
RJ-21X
connector.

© Cengage Learning 2013

are terminated on one side of the block, and 25 pairs of the customer lines are terminated on the other side. The sides can then be connected through small, metal bridging clips, which connect the center two pins of the 66 block together. The clips make troubleshooting by technicians easy, because they can be pulled off at any time to break connections and help diagnose problems on a line.

110 Block

The 110 block is a different type of punch-down block for high-speed, structured wiring systems. In many cases, the 110 block (Figure 11–12) has replaced most 66 style blocks because it is CAT5 or CAT6 compliant and can support over 100 MHz of signaling. The 110 block still uses a punch-down tool, but it has a different cutting end. Most tools are now reversible, to be able to punch a 66 or a 110 block as needed. The 110 blocks can be purchased in 25-, 50-, 100-, or 300-pair sizes.

When installing 110 blocks, after the initial pairs have been seated and punched down, 3-, 4-, or 5-pair connecting blocks are installed on top of the wires to facilitate cross-connects or connection to patch cords.

Modular Patch Panels

Modular patch panels are used to terminate CAT3, 5, 5e, or 6 cables to modular 8-position jacks. The jacks allow connection to patch cables utilizing RJ-45 connectors.

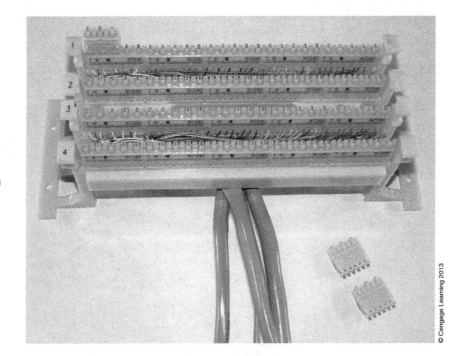

Figure 11–12
110 termination block.

© Cengage Learning 2013

Figure 11–13
Front of patch panel.

The front side of the patch panel provides connection to the RJ-45 (Figure 11–13), whereas the rear side utilizes the 110-type connecting block (Figure 11–14). Patch panels are mounted on a standard 19-inch rack and are available in groups of 24, 48, 72, or 96 output ports. The ports are arranged in groups of six. On the rear of the panel, each group of six ports is connected through six 4-pair 110 block connectors, segregated into two rows of three connectors.

The use of patch panels makes cross-connects and wiring changes easy to facilitate by simply moving patch cords and jumpers.

25 Twisted-Pair Color Codes

Twisted-pair telephone cables are available in pair counts of 25, 50, 75, 100, 150, 200, 300, and 600. Figure 11–15 shows a 100-pair cable. Cables having between 600 and 1800 pairs are also available, but only with a corrugated outer metallic

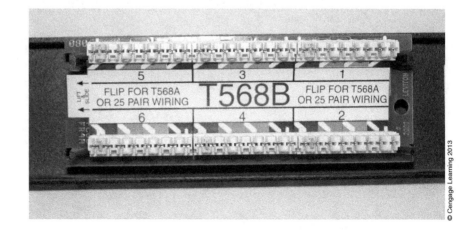

Figure 11–14
Rear of patch panel.

Figure 11–15
100-pair cable.

© Cengage Learning 2013

shield. Tables 11–1 and 11–2 define the colors for the tip and ring conductors and the color codes for a 25-pair cable.

When referring to the wire pairs within a cable, the colors are always listed in order of tip color first and then ring color. So, as an example, black/orange means black tip, and orange ring.

For cables having more than 25 pairs, the color codes repeat, but each bundle of 25 pairs is organized into binder groups. Each binder group is wrapped with a color-coded plastic binder string that is used to designate the binder number. Binder color codes are listed in Table 11–3.

Notice how the first five pairs use the white tip color, twisted with the five ring colors. Pairs 6–10 then use the next tip color, red, twisted with the five ring colors. The pattern repeats until all five tip colors are used with all five ring colors, for a total of 25 pairs.

Table 11–1 Tip/Ring Color Codes

Tip	Ring
White	Blue
Red	Orange
Black	Green
Yellow	Brown
Violet	Slate

© Cengage Learning 2013

Table 11–2 25-Pair Cable Color Codes

Pair 1	White/Blue Blue	Pair 14	Black/Brown Brown
Pair 2	White/Orange Orange	Pair 15	Black/Slate Slate
Pair 3	White/Green Green	Pair 16	Yellow/Blue Blue
Pair 4	White/Brown Brown	Pair 17	Yellow/Orange Orange
Pair 5	White/Slate Slate	Pair 18	Yellow/Green Green
Pair 6	Red/Blue Blue	Pair 19	Yellow/Brown Brown
Pair 7	Red/Orange Orange	Pair 20	Yellow/Slate Slate
Pair 8	Red/Green Green	Pair 21	Violet/Blue Blue
Pair 9	Red/Brown Brown	Pair 22	Violet/Orange Orange
Pair 10	Red/Slate Slate	Pair 23	Violet /Green Green
Pair 11	Black/Blue Blue	Pair 24	Violet/Brown Brown
Pair 12	Black/Orange Orange	Pair 25	Violet/Slate Slate
Pair 13	Black/Green Green		

© Cengage Learning 2013

Table 11–3 Binder Color Codes for 600-Pair Cable

Binder 1 Pairs 1–25	White/Blue	Binder 9 Pairs 201–225	Red/Brown
Binder 2 Pairs 26–50	White/Orange	Binder 10 Pairs 226–250	Red/Slate
Binder 3 Pairs 51–75	White/Green	Binder 11 Pairs 251–275	Black/Blue
Binder 4 Pairs 76–100	White/Brown	Binder 12 Pairs 276–300	Black/Orange
Binder 5 Pairs 101–125	White/Slate	Binder 13 Pairs 301–325	Black/Green
Binder 6 Pairs 126–150	Red/Blue	Binder 14 Pairs 326–350	Black/Brown
Binder 7 Pairs 151–175	Red/Orange	Binder 15 Pairs 351–375	Black/Slate
Binder 8 Pairs 176–200	Red/Green	Binder 16 Pairs 376–400	Yellow/Blue

(Continued)

© Cengage Learning 2013

Table 11–3 Binder Color Codes for 600-Pair Cable (Continued)

Binder 17	Yellow/	Binder 21	Violet/Blue
Pairs 401–425	Orange	Pairs 501–525	
Binder 18	Yellow/Green	Binder 22	Violet/Orange
Pairs 426–450		Pairs 526–550	
Binder 19	Yellow/Brown	Binder 23	Violet/Green
Pairs 451–475		Pairs 551–575	
Binder 20	Yellow/Slate	Binder 24	Violet/Brown
Pairs 476–500		Pairs 576–600	

© Cengage Learning 2013

SEC 11.8   *NATIONAL ELECTRICAL CODE ARTICLE 800* COMMUNICATION CIRCUITS

Article 800 covers communications circuits and equipment. Communication systems include voice, data, audio, video, interactive services, telegraph (except radio), and outside wiring for fire alarms and burglar alarms from the communication utility to the customer's communication equipment, up to and including terminal equipment such as a telephone, fax, or answering machine.

Article 800 is written as a self-contained document and is not subject to the requirements of *Chapters 1–7* of the *NEC* except where specifically mentioned.

The 2011 edition of the *NEC* has made a change to the previously used term *grounding conductor*. The grounding conductor has been replaced by either bonding conductor, or grounding electrode conductor (GEC). GEC is now used to indicate a direct connection from equipment to a grounding electrode. Bonding conductor is used to bond multiple electrodes or for bonding communication equipment to an intersystem bonding termination or a connection to building grounding means. Figure 11–16 illustrates the concept. Please refer to Chapter 3 of this book for the details of communication system grounding and bonding.

Wires and Cables Outside and Entering Buildings

Part II of *Article 800* of the *NEC* covers the installation of outside wires and cables entering buildings. In addition to communication cables, other types include remote control, signaling, and power-limited circuits (*Article 725*), fire alarm systems (*Article 760*), and fiber-optic cables (*Article 770*). For the installation of outside, overhead (aerial) cables, *NEC 800.44* requires the following:

- When attached to poles or in-span, communication cables shall run below electric light and power conductors.

- Communications cables shall not be attached to cross-arms that carry electric light and power conductors.

Figure 11–16
Grounding
electrode
conductor
(GEC) and
bonding
conductor.

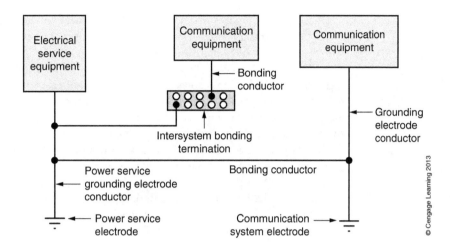

- Climbing space shall comply with *NEC 225.14(D)*.

- Supply service drops of 750 V or less, and running parallel to communication conductors, shall maintain a minimum separation of 12 in. (300 mm) at all points in the span, including attachment to buildings, provided that the nongrounded conductors are insulated and that a clearance of not less than 40 in. (1 m) is maintained between the two services at the pole.

- A vertical clearance of 8 ft (2.5 m) is required for all communication wires and cables from all points above a roof for which they pass. The 8-ft (2.5 m) clearance does not include auxiliary buildings, such as garages.

- Roof clearance can be reduced to 18 in. (450 mm) above an overhang portion of the roof, provided that not more than 4 ft (1.2 m) of the communication service-drop conductors pass above the overhang, and that they are terminated at or through an above-the-roof raceway or approved support.

- Roof clearance can be reduced to 3 ft (0.9 m), where the slope of the roof is not less than 4 in. (100 mm) in 12 in. (300 mm).

Underground communication wires and cables in raceways, handhole enclosures, and manholes shall be separated from electric light and power conductors, Class 1, or non–power-limited fire alarm circuit conductors by means of brick, concrete, or tile partitions, or by a suitable barrier.

NEC 800.50(B) states that communication wires and cable that require primary protectors shall be separated at least 4 in. (100 mm) from electric light or power conductors that are not in a raceway or cable. The 4-in. (100 mm) separation is not a requirement if the communication cables are permanently separated from conductors of other systems by a fixed nonconductor (in addition to the insulation on the wires) such as a porcelain tube or flexible tubing.

Communication wires that are installed near power conductors operating over 300 V to ground shall be separated from the building woodwork by a glass, porcelain, or other type of insulating support.

NEC 800.50(C) states that where primary protectors are installed inside of buildings, the communications cables and conductors shall enter the building through a noncombustible, nonabsorbent insulated bushing or through a metal raceway. The metal raceway is not required if the communication cables are already in a metallic sheathed cable or if they are entering the building through masonry.

Raceways or bushings shall slope upward from the outside, or drip loops shall be formed in the communication cables to prevent rainwater from entering the building.

Raceways shall be equipped with an approved service head, and any other metal raceways located ahead of the primary protector shall be grounded.

Where communication conductors for an entire street are run in an underground block distribution placed so as to be free from accidental contact with electric light and power circuits of over 300 V to ground, insulating supports shall not be required, and bushings shall not be required where the conductors enter the building.

Communication cables and conductors connected to a building shall maintain a distance of at least 6 ft (1.8 m) from lightning conductors (*NEC 800.53*).

Protection Devices (*National Electrical Code 800.90*)

A listed primary protector shall be provided on each circuit entering a building, aerial or underground, so as to protect them from exposure to lightning, or from accidental contact with conductors operating over 300 V to ground (Figure 11–17). A primary protector shall also be provided on each end of an interbuilding circuit.

A primary protector may not be required in large metropolitan areas where the buildings are close together and sufficiently high to intercept the lightning, or in areas where thunderstorms are fewer than five per year and earth resistivity is less than 100 ohm-meters.

Figure 11–17
Primary
protector.

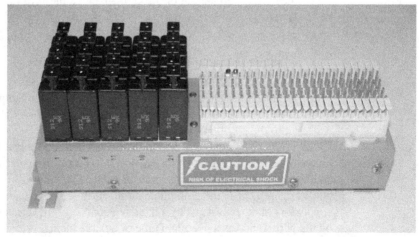

© Cengage Learning 2013

Primary protectors may also not be required on interbuilding runs shorter than 140 ft (42.7 m), where cables are directly buried or in underground conduit, provided the outer metallic sheath of the cable or the continuous metal conduit is connected at each end to the grounding electrode system of each building.

Primary protectors shall be either fuseless or fused and shall be located as close as practicable to the point of entrance to the building. Primary protectors shall not be installed in hazardous locations as defined in *NEC 500.5* or *505.5*, or in the vicinity of easily ignitable material.

Unlisted outside plant communications cables shall be permitted to be installed in building spaces other than risers, ducts, plenums, or other spaces used for environmental air, where the length of the cable from the point of entrance to the building to the primary protector does not exceed 50 ft (15.2 m). This, however, is not recommended if it is practicable to place the primary protector closer to the point of entrance. The main objective is to keep the lightning out of the building, and having the protector located as close as practicable to the point of entrance simply makes good sense.

Installation Methods within Buildings, *Article 800, Part V*

Communication system cabling for telephone or burglar alarms is type CMP, communications plenum; CMR, communications riser; CMG, communications general-purpose; CMX, communications—limited use; or CMUC, communications under carpet cable. The type of cable that you choose is dependent on the location in which you are wiring.

The listing requirements and cable markings for communications cables are given in *800.179* of the *NEC*. Type CMP cable should be used in plenum areas and be listed as being suitable for use in ducts and other air-handling spaces. Type CMR cable should be used in risers where a floor-to-floor penetration is required. Type CMG cable can be used in all other applications that are not risers or plenums. Type CMX cable should be used in residential installations only, and Type CMCU cable should be used when under carpet runs are necessary.

Table 800.154(b) provides the cable substitution chart for communications cables. The chart is a top-down substitution, starting with plenum; then moving to riser; to general purpose; and last, to dwelling. Cables must never be substituted in reverse order.

For the applications of listed communications wires, cables, and raceways in buildings, consult *NEC 800.154* and *Table 800.154(a)*. The 2011 edition of the *NEC* has added an extensive guide, cross-referenced with *Table 800.154(a)*, for choosing the right type of wire, cable, and raceway, based on location.

Location types include the following:

- In fabricated ducts
- In other spaces used for environmental air
- In risers
- Within building spaces other than air-handing spaces and risers

Within each of these locations, the categories are further broken down based on type of application, such as in vertical runs, in metal raceways, in fireproof shafts, in one- and two-family dwellings, in multifamily dwellings, in nonconcealed spaces, under carpets, in communication raceways (nonmetallic), in cable trays, in general spaces, and so on.

For a detailed breakdown of the installation of communications wires, cables, and raceways, see *NEC 800.113*. The details described in this section mirror the application details of *Table 800.154(a)* and provide all possible cable substitutions based on the locations listed above.

Cable Ratings

Communication cables shall have an insulation rating of not less than 300 V (*NEC 800.179*). The 300-V rating ensures that communications cables can accommodate the voltages ordinarily found on a telephone line (48 V dc plus a ringing voltage of up to 130 Vpk); it also allows communications cable to substitute for power-limited fire alarm cables or Class 3 cables, which are also rated at 300 V.

Separation of Circuits

Communications cables can be run in the same raceway or enclosure with Class 2 and Class 3 cables, coaxial cables, power-limited fire alarm cables, and fiber-optic cables, but cannot occupy the same raceway or enclosure with electric light and power conductors, Class 1 cables, non–power-limited fire alarm cables, or medium-power network broadband cables. To be able to do so, a permanent, nonconductive barrier is required to provide the necessary separation of low-voltage circuits from medium- or high-voltage circuits.

A minimum 2-in. (50 mm) separation is required between communication wires and cables from electric light and power conductors, Class 1 cables, non–power-limited fire alarm cables, or medium-power network broadband cables. An exception to this requirement would be if the communication cables are already separated by some type of suitable barrier, such as a raceway, or if they are enclosed within a metal-sheathed, metal-clad, nonmetallic-sheathed, Type armored cable (AC), or Type underground feeder (UF) cable. Another alternative would be to permanently separate the communication wires and cables by a continuous and firmly fixed nonconductor, in addition to the insulation on the wires, such as a porcelain tube or nonconductive flexible tube (*NEC 800.133*).

Also, any cables running together within the same raceway or enclosure shall be equally rated. This issue is most important when mixing Class 2 cables with those of Class 3 and communication conductors. Class 2 cables are rated for only 150 V (as stated by *NEC 725.179(G)*). To run with Class 3 or communications cables, Class 2 would first need to be upgraded to either a listed Class 3 or listed communications cabling to ensure an equal rating. A mismatch in cable ratings could pose a potential threat in a short-circuit situation. Ultimately, a chain is only as strong as the weakest link, and by mixing Class 2 cables with Class 3 cables and communications cables, the cables are only protected up to 150 V, and not the required rating of 300 V.

Communication wires and cables are not required to be installed in raceways, but when they are, the raceway shall be installed according to *Chapter 3* of the *NEC*. However, the raceway fill table requirements of *Chapters 3* and *9* of the *NEC* shall not apply to communication wires and cables (*NEC 800.110*). Although the *NEC* does not require raceway fill tables to be followed, it is still wise to do so, so as not to potentially damage the conductors or the insulation of the cables by having to pull on them too hard during the installation process.

Support of Conductors

Raceways shall be used for their intended purpose. Communication wires and cables shall not be strapped, taped, or attached by any means to the exterior of any raceway as a means of support. The only exception is the exterior mast supporting overhead (aerial) spans as they attach to the exterior of a building.

CHAPTER 11 FINAL QUESTIONS

1. The twisted-pair wiring from the CO of the telephone company, connecting to a dwelling or business, is referred to as _____.
2. The two conductors of a single telephone line are labeled _____ and _____.
3. Give the color codes for the primary pair of wires in a residential telephone connection.
4. What are the second three digits of a telephone number called?
5. What is the bandwidth of an analog telephone line?
6. Explain the term *demarc* and who is responsible for the wiring on either side of that point.
7. What is the standard connector used to terminate a telephone line to a telephone?
8. What type of termination block is traditionally used for the installation of telephone systems within large commercial and industrial buildings?
9. What type of termination block is used for high-speed data connections?
10. What is the purpose of the main distribution frame, and where is it typically located?
11. What is the typical off-hook voltage of a telset?
12. The telephone company provides an ac ringing voltage of _____ and _____ Hz.
13. If a telset has a REN of 1.5B, what does that mean?
14. Explain the difference between rotary dialing and touch-tone dialing.
15. What is a private branch exchange (PBX)?

16. How does an electronic key system (EKS) differ from a PBX?
17. What style of trunk line is most common between the CO and a PBX?
18. Which type of trunk line is similar to a standard residential phone line?
19. What are E+M tie lines used for?
20. Explain the purpose of a "wink"?
21. Which style of trunk line has the PBX providing voltage to the CO?
22. Twenty-five pairs of trunk lines from the CO are connected through a _____-type connector.
23. What is a binder group?
24. *Article 800* of the *NEC* has replaced the term *grounding conductor* with what two terms? How do they differ?
25. A vertical clearance of _____ is required for all communications wires and cables from all points above a roof over which they pass.
26. What is the purpose of a primary protector?
27. What is the voltage rating for a communication cable?
28. When Class 2 cabling is run with communication cabling, what needs to occur?
29. A minimum separation of _____ in. is required between interior communication wires and cables and those of electric light, power, and Class 1 conductors, provided that they are not already separated by a suitable barrier or running within a metallic raceway.
30. What type of listed cable can be used as a direct substitute for type CMG?

Chapter 12

Security and Access-Control System Basics

Objectives

- Explain the purpose of a security alarm system versus an access-control system.
- Identify the components of a security alarm system.
- Identify and describe the different types of sensors used in a security alarm system.
- Demonstrate the wiring of components to a security alarm system.
- Identify the components of an access-control system.
- Identify and describe the types of credentialing within an access-control system.
- Identify and describe the types of locking mechanisms within an access-control system.
- Demonstrate the wiring of components to an access-control system.
- Explain how an open source system differs from that of a traditional panel-based system.

Chapter Outline

Sec 12.1 Security Alarm Systems
Sec 12.2 Wiring a Security Alarm System
Sec 12.3 Access-Control Systems
Sec 12.4 Wiring an Access-Control System
Sec 12.5 Open Source Systems
Sec 12.6 Electrical Code

Key Terms

acoustic glassbreak sensor	maglock	radio frequency identification (RFID)
active infrared sensor	magstripe	
closed loop	microwave motion sensor	shock-type glassbreak sensor
electric strike	open loop	
fail-safe	panic button	ultrasonic motion sensor
fail-secure	passive infrared sensor	video motion detection
glassbreak sensor	proximity cards	Wiegand technology

SEC 12.1   SECURITY ALARM SYSTEMS

Purpose of Alarming

Ever since the events of September 11, 2001, the need for security and access-control systems to protect and notify individuals of a possible threat or takeover has become a priority. Although it may be true that since 9/11 the United States has not experienced anything nearly as extreme, even as threat levels have been reduced, detection and deterrence of a potential threat still remains the priority of any government, corporate, or private institution.

Over the past decade, the demand for security and access-control systems has surged, largely as a result of technological advancements which have significantly reduced cost and increased availability to the consumer. Today, such technologies are available not only to large governmental or corporate institutions but also to all homes and local businesses, no matter their size, and at a very affordable cost. In this chapter, we focus on the electronic devices and systems that provide such protection.

Security System Sensors

The complexity and design of any security alarm system primarily involve the type of sensing devices used to detect an intrusion. Different types of sensors are used to detect motion, sound, pressure, vibration, light, and particles in the air. Which type to use primarily depends on the specific layout of a location and the fine details of what is being secured.

There are many different types of sensors used for perimeter and interior protection. Some of these sensors may be interchangeable, but most are specified for either indoor or outdoor use. The goal of any sensor is simple: to indicate a change in state or condition. Depending on the type of sensor, a change in state typically results in either a shorted or open input on the system control unit. The control unit is the central brains of the security system. Its main function is to continually monitor the status of all sensor inputs and then sound an alarm when an intrusion has been detected.

Passive Infrared (PIR)

The **passive infrared sensor** detects motion by measuring the changes in temperature of a given space. The PIR sensor is available for indoor or outdoor use. Whether your goal is to detect an intruder in your backyard or to detect a safe being opened, the PIR is perfect. To trigger the device, the PIR must sense a change in heat. The PIR sensor monitors a given field of view; when the temperature of the field suddenly changes, the PIR reacts by sending a signal to the alarm panel, which in turn triggers an alarm. The PIR works well in environments where there are gradual temperature shifts, but sources of rapid heating and cooling should be avoided. Therefore, when installing PIR sensors, do not point or aim them in the direction of forced air ducts, fireplaces, space heaters, direct sunlight, or strong white lights, so as to prevent false alarms.

Active Infrared

Unlike the PIR, the **active infrared sensor** continually emits an infrared light beam by means of an IR-LED. The sensor measures reflected light beams as they bounce off objects and return. Any sudden changes in the reflection pattern results in a change of state for the sensor, and the output triggers an alarm. Active IR sensors are available for indoor or outdoor use.

Ultrasonic

Ultrasonic motion sensors use high-frequency sound waves to detect movement within a field of view. As a person or object moves past the sensor, sound waves bounce off and reflect back to a finely tuned receiver. An alarm is triggered if the frequencies of the returning sound waves are significantly altered. Only moving objects reflect an altered frequency, but static or nonmoving objects (table, file cabinet, desk, etc.) return the original. These types of sensors require careful planning to be effective. They typically are not used in outdoor applications where environmental effects such as wind may cause a false alarm.

Microwave/Radar

Microwave motion sensors operate in the same manner as ultrasonic, but at a higher frequency, up in the range of 10 GHz. These sensors use the principles of Doppler radar to detect motion and are highly sensitive, being able to pick up movement through glass. Therefore, the placement of such sensors must be carefully thought through to prevent false alarms due to possible movement on the opposite side of a window, for example. In such a case, the actual movement triggering the alarm would be from an occupant in an adjoining space, and not the secure space.

Dual-Technology Motion Sensors

Dual-technology motion sensors combine more than one technology in one enclosure to help eliminate false alarms. An example of a dual-technology sensor

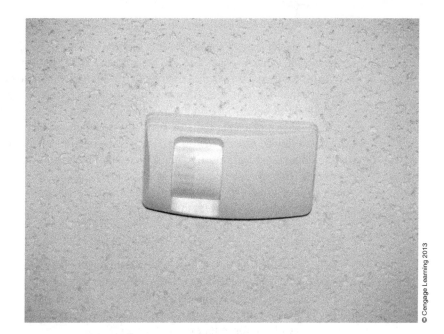

Figure 12–1
Motion
sensor.

may include PIR (Figure 12–1), microwave, and ultrasonic technologies for a more stable outcome. The sensor only alarms when two technologies have been activated. As discussed, each of these technologies uses a different means for determining when to trigger an alarm. The dual technology provides an added guarantee to ensure a more positive result, thereby lowering the risk of false alarms.

Acoustic or Shock-Type Glassbreak Sensors

Glassbreak sensors come in two varieties, acoustic or shock-type. These sensors are made for indoor use, for residential or commercial types of installations. The details for each are discussed below.

- **Acoustic glassbreak sensors** detect the unique frequency response of glass when it breaks. Regardless of the type of glass—plate, tempered, or safety—the sensor is usually able to decipher the difference and trigger the alarm. Installation and placement of the acoustic glassbreak sensor is typically on a wall within proximity of the window. Once installed, the sensor is tested for performance with a handheld, electronic glassbreak simulator. The simulator mimics the sound of many varieties of breaking glass to be able to activate the sensor and trigger an alarm. The sensor and tester are shown in Figure 12–2. When using these types of sensors, the windows should not be covered. To do so could potentially muffle the sound of the breaking glass, thereby making it difficult for the sensor to detect the intrusion.

- **Shock-type glassbreak sensors** are mounted directly on the window. They use a piezo electric transducer, which is tuned to the shock frequency of breaking glass. The nice thing about shock-type sensors is that they can be

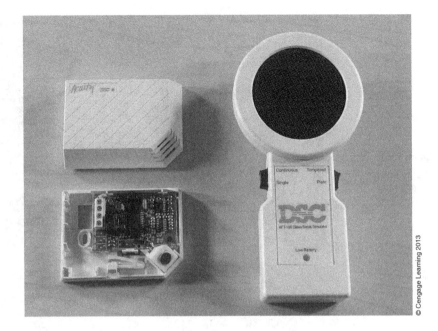

Figure 12–2
Acoustic glassbreak sensor and tester.

placed on any window, have no mechanical parts to wear out or fail, and can operate through heavy drapes or curtains. These types of glassbreak sensors are virtually failure proof.

Photoelectric Beams

Photoelectric beams consist of two parts: a transmitter and a receiver (Figure 12–3). The transmitter, on one side of the room, projects a pulsed infrared beam to a receiver on the other side. The control panel detects an alarm if the beam is broken or interrupted at the receiver. When using photoelectric beams, a direct line of sight is required so that the path from transmitter to receiver has no interference. Photoelectric beams are used in both indoor and outdoor applications.

Video Motion Detection

Video motion detection uses the changing field of view of a closed-circuit television (CCTV) camera to detect a security intrusion. Computer software analyzes the image and is able to detect unwanted movement. Once detected, the output of the CCTV system triggers an alarm relay that is tied in with the building security alarm. The benefit to using video motion detection is that a security guard is not needed to physically monitor a location in order to trigger an alarm. The system can be set up to record and/or notify someone if motion is detected. This not only saves space on the recorder or hard drive but also makes it possible for a guard to witness an act in progress. Usually, the motion triggers an audible alert signal at the control desk to turn attention to the correct video monitor. The sensitivity of the motion detection is set by the software driving it. Video detection is used on both indoor and outdoor cameras.

Figure 12–3
Photoelectric
beam sensor.

© Cengage Learning 2013

Magnetic Door/Window Contacts

A magnetic door/window contact is used to protect the perimeter of a facility and
to signal an alarm when a door or window is opened (Figure 12–4). Contacts can
be purchased that can be recessed into the door or window frame, thus hiding them
from view, or surface mounted on the exterior, depending on the level of security

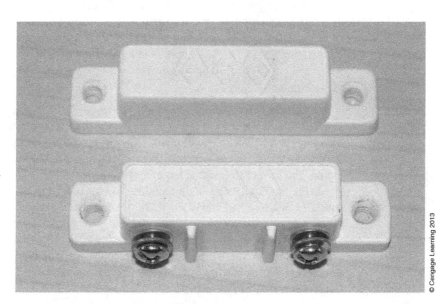

Figure 12–4
Magnetic Door/
Window Contact.

© Cengage Learning 2013

required. The type of door or window determines the style of contact. In addition, contacts are available for hot- or cold-temperature environments, aluminum doors, steel doors, sliding doors, high-security doors, freezers, and also a wide variety of windows. There are two parts to a magnetic contact: a switch that is mounted on the nonmoveable part of the door or window frame, and a magnet that is mounted on the door or window. When the magnet is within a couple of millimeters of the switch, the magnetic field forces the switch to change state, from either open to closed, or closed to open, depending on the design of the contact. Moving the magnetic field away from the switch causes it to revert back to its original position. When a change of state occurs, an alarm signal is sent from the control panel, indicating a security breach.

Normally Open (NO) or Closed-Loop Switch Most door or window contacts are "normally open," meaning the magnetic field of the magnet is used to close the switch. A NO contact is also referred to as a **closed-loop**, or Form A, switch, because a magnetic field is used to close the contact. In this scenario, a closed door places the switch in proximity to the magnet and the loop is closed.

Normally Closed (NC) or Open-Loop Switch The opposite configuration, a "normally closed" contact would open in the presence of a magnetic field. These types of switches are referred to as **open-loop**, or Form B, because a magnetic field is required to open the switch. An example for this type of switch would be to monitor the status of a door or window that needs to remain open.

Keypad/Control Panel

In an alarm system, the keypad/control panel is the interface between the user and the devices connected to the system (Figures 12–5 and 12–6). The keypad serves many functions, the most important being providing the ability to arm and disarm the alarm. Simply by typing a series of numbers into the keypad, a person may arm (turn on) or disarm (turn off) the alarm. The keypad typically is located at the main entrance or at an employee entrance where arming and disarming can be performed in approximately 30 sec. The time allowed is user specified and can usually be programmed to allow for up to a minute for both. Other features of the keypad allow the user to program certain devices and to enroll new users/codes into the system.

Display and Reporting Systems

Most residential or commercial security alarms have a central office to which they report to in the event of an alarm. The security system in the house or business connects back to the central office via a phone line; either wired or cellular backup is required. When a proprietary alarm is purchased, the central office charges the customer a monthly fee, over and above the installation cost, for the services of monitoring the location. As an example, in the case of a residential installation,

Figure 12–5
Alarm control panel.

© Cengage Learning 2013

Figure 12–6
Alarm keypad.

© Cengage Learning 2013

if an alarm is received by the central office, the attendant determines the type of alarm (break-in, fire, environmental, etc.) and either reports the alarm to the correct authorities or calls the owner of the residence to notify them of a possible intrusion.

Most large, commercial security systems are tied back to a designated computer that is capable of both programming the alarm system and displaying any actions that may have taken place. Usually, a series of guards are assigned to monitor the reporting station 24 hours a day, 7 days a week.

Alarm reporting can either be audible or silent. Depending on the situation, it might be more advantageous to use a silent alarm. For example, in a police interview room, if an officer is feeling uncomfortable or threatened by a detainee, it might be more suitable to alert backup, but not to let the detainee know of the call. This does not place the officer in any added danger. Another example is in the event of a bank robbery. A teller should be able to trigger a silent alarm by pressing a panic button. The **panic button** is typically placed in a hidden location so as not to draw attention to its being pressed, but well within reach of the teller's usual range of motion. Once triggered, the authorities and guards are instantly notified of the trouble, without drawing any unneeded attention to the event, thereby protecting the teller, coworkers, and customers from possible danger.

In other situations, audible reporting may be desired, typically in the form of a siren or bell. Compared to the silent alarm, audible alerts are used to scare off a potential intruder. In most cases, the audible alarm acts as a deterrent by forcing the intruder to leave, because attention is now being drawn to the area.

The monitoring and reporting computer should, together with reporting alarms, record the entire event, including the date, time, location, and any other user-specified options available. And in cases where the security system is tied in with surveillance cameras, the system should preserve all recorded images of the event for future viewing.

SEC 12.2   WIRING A SECURITY ALARM SYSTEM

The basic components of any security alarm system include door or window contacts, motion sensor, glassbreak sensor, bell or siren, and keypad. The next section details the wiring of these devices to the control panel.

Power

The control panel and devices of the security alarm are powered by a Class 2 transformer or Class 2 power supply. Battery backup is also required to guarantee circuit operability during a power failure. Do not connect the Class 2 transformer or power supply to a receptacle that is controlled by a switch. Also, the metal cabinet of the security panel shall be grounded to the building grounding system or to an earth ground connection, as previously described in Chapter 3 of this book. The output capacity of the power supply shall be rated sufficiently to be able to provide all the required power to the system. A good rule of thumb with power supplies is to load them not more than 70% of capacity; this means a 10 W transformer should not be required to power more than 7 W of load.

Battery backup uses either a sealed, rechargeable, lead-acid battery or gel-type battery, which is required to meet Underwriters Laboratories (UL) and Underwriters Laboratories of Canada (ULC) requirements for power standby. Residential/commercial installations require a minimum of 4-hour power standby time. Batteries should be checked periodically and replaced every 3–5 years. UL and ULC also stipulate that residential fire and healthcare installations require 24-hour power standby. ULC commercial burglary and fire-monitoring installations require 24-hour power standby. Just as with transformers and power supplies, when calculating the capacity in amperes per hour (Ahr) for a battery, it is good practice to load it not more than 70% the Ahr capacity of the cells to ensure that there is enough available power for the desired amount of time; this also helps to ensure a longer lifespan.

Required Wire

Type 18–22 AWG, twisted 2-wire, stranded cable is preferred when wiring a security system. Shielded wire is not required, and the maximum resistance of an individual wire run should stay below 100 Ω so as not to interfere with the supervised loop current.

Zone Wiring

Inputs for security systems are typically divided into zones. Depending on the make and model of your system, there are usually between six and eight onboard zones available for connection. If more zones are needed, additional modules can be ordered, and the system can be expanded.

The input circuit of a security system operates on a normally closed loop. A normal state is determined by the total resistance in the loop and the size of the end-of-line resistor. Most systems use either a 4700- or 5600-Ω resistor, depending on the manufacturer.

A loop resistance of 0 Ω (short) indicates a fault condition, and an infinite loop resistance (open or broken wire) indicates a tamper or alarm condition. The input configuration and system programming determine how the infinite loop is recorded by the alarm panel.

Device Configuration

As described earlier, door contacts can be either normally open (NO) or normally closed (NC), depending on how they have been designed by the manufacturer. A better practice, however, is to refer to them as open loop or closed loop, based on when the magnet is placed in proximity to the switch. When installing magnetic door contacts, closed-loop contacts are placed in series with the end-of-line (EOL) resistor attached to a zone, and open-loop contacts are placed in parallel. In the case of a closed-loop contact, an intrusion is detected when the contact opens and the EOL resistor becomes disconnected from the control panel. For an open-loop contact, the scenario is just the opposite. Because the contact is placed in parallel with the EOL resistor, a change of state causes the contact to short across the resistor and zone input, thereby forcing the panel into alarm.

Motion and glassbreak sensors can be wired as either NO or NC. The configuration is determined at power up, meaning that once power is attached to the device, the NO contact provides an open-loop configuration and the NC contact provides a closed-loop configuration. The output state of the device then changes when an intrusion has been detected. Wiring of a closed-loop or open-loop device is the same as that described above for magnetic door contacts.

Tamper

Motion and glassbreak sensors also have an optional tamper connection, which causes the device to go into alarm if someone tries to remove the cover to disable the circuit. The tamper circuit includes a series-connected switch that makes contact when the cover is attached to the device. In addition, the EOL resistor for the circuit is usually placed between internal jumpers within the device so that the device can monitor the loop current and tamper simultaneously. A tamper switch can also be wired into the door of the security alarm control panel to guard against anyone attempting to disable the system. A simple momentary single-pole, single-throw (SPST) switch works just fine for such protection.

End-of-Line Configurations

Inputs can be configured as either single end-of-line (SEOL) or double end-of-line (DEOL). A SEOL input circuit uses only one series-connected end-of-line resistor in the loop. In such a case, the normal state is equal to the end-of-line resistance. The system is programmed to recognize a minimal amount of wire resistance in the loop, typically kept below 100 Ω so as not to interfere with the supervised circuit. The DEOL circuit places one series-connected end-of-line resistor in the circuit for a normal state condition, and during an alarm places two series-connected end-of-line resistors into the circuit. Therefore, when the system detects double the loop resistance, a violation or alarm is indicated by the control panel. In the DEOL configuration, because double the resistance indicates an alarm, 0 Ω indicate a fault or short, and infinite resistance indicates a tamper condition.

Figures 12–7 and 12–8 illustrate all possible loop configurations for SEOL and DEOL circuits. As you can see, it is also possible to series-connect multiple devices and contacts. Devices can also be placed in parallel if they are using the NO configuration. In circuit drawings, NO appears as a capacitor symbol, and NC appears as a capacitor with a line through it. The circuit works fine provided that there is only one series-connected EOL resistor to indicate the normal state condition. Looking at Figure 12–8, you can see how the DEOL resistor is placed in parallel with a NC device. When the device goes into alarm, the NC contact switches to NO, and the loop resistance doubles to notify the panel that an intrusion has occurred.

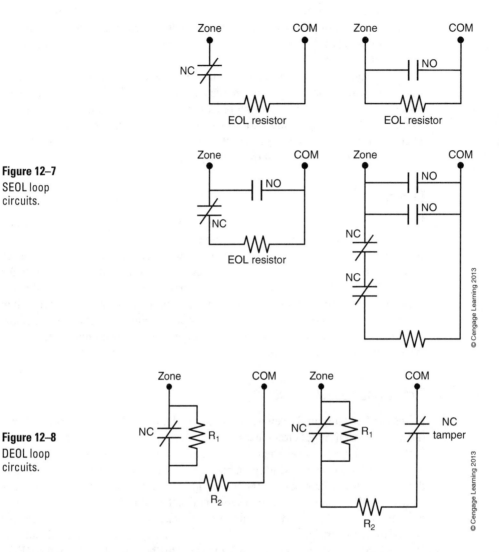

Figure 12–7
SEOL loop
circuits.

Figure 12–8
DEOL loop
circuits.

© Cengage Learning 2013

Bell Wiring

Depending on the manufacturer, bells or sirens can be wired to provide a constant
tone or a temporal, three-pattern signal, as required by NFPA 72. A jumper wire
within the device is used to switch the configuration. The bell is connected to the
output of the alarm system control panel. The panel typically provides 12 V dc to
the device during an alarm condition.

Telephone Connection

The control panel of the alarm system can be programmed to dial out to a central station or proprietary station through a telephone landline or over wireless
transceiver. The connection of the telephone line is made through a RJ-31X
connector, which is used by the panel to temporarily take control of the line

during an alarm condition. Figure 9–19, in Chapter 9, details the wiring of the RJ-31X. In a residential setting, for example, when the panel goes into alarm, a jumper across the RJ-31X connector is removed, temporarily disabling the house telephone; a preprogrammed call is then dialed out of the panel to the central station. Once completed, the panel goes off hook and replaces the jumper to the RJ-31X, restoring connection to the house phone.

Keypad

Depending on the type of system used, the number of system keypads is limited to a certain number. If more are required, expansion modules can be added to the system to accommodate additional keypads. Keypads are typically connected over four conductors; two are for power and two are for data. The maximum distance between the panel and the keypad is also determined by the manufacturer. It is typically based on the type of wire being used and the bandwidth of the data connection between the panel and the keypad. In most cases, keypads cannot be connected beyond 1000 ft of the panel without adding a subpanel to relay the signals.

The wiring diagram for a complete security alarm system is shown in Figure 12–9.

System Programming

System programming can be accomplished in a variety of ways. One method requires loading program codes into the system keypad. Programming options for the system can be accessed through the main system menu of the keypad after an administrator password has been entered to provide access.

A more user-friendly option is to connect a laptop computer or PC to the control panel through an RJ-45 network connection. Most manufacturers provide an Ethernet connection to the panel, either on the main system board or through an optional auxiliary module, which can be added if desired. Once connected, programming software must be downloaded from the manufacturer to enable communication between the PC and the security system control panel. The downloaded software makes the initial system programming and system management easier to facilitate; it also allows for the administrator to access the panel over the Internet and from any location within the building or structure, if desired.

SEC 12.3   ACCESS-CONTROL SYSTEMS

The Function of Access-Control Systems

The function of access control is to limit access to a desired location. To accomplish the task, you can use a series of strategies ranging from key cards and key fobs, to access codes, fingerprint identification, and retina scans. Such measures help to limit the traffic within a given area to only those individuals granted permission based on their credentials. Many types of devices can accomplish access control. A descriptive list is given in the following sections.

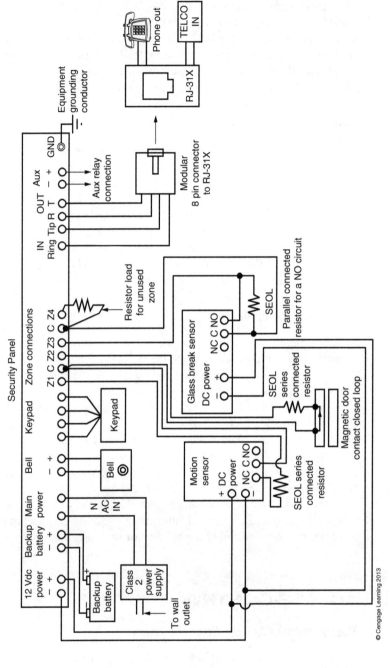

Figure 12-9
Complete security alarm system wiring.

© Cengage Learning 2013

Card Reader

A card reader can decipher, interpret, and understand the information on the access card. A reader detects the changes in a magnetic field within the access card. Most readers use one of the following three methods for reading the card:

- Swipe reader—The card is physically swiped through a long slot that allows passage on both ends.
- Insert reader—The card is inserted into a small receptacle that is just large enough to accommodate the card, just as it is at most automated teller machines (ATMs).
- Proximity card—The card is presented in front of the reader within a certain distance. Physical contact does not have to occur (Figure 12–10).

Card Entry

Card entry is possibly the most popular access-control measure used by companies to ensure only authorized entrance. An entry card, which resembles a credit card, is programmed into the access-control system to grant access to the user.

Figure 12–10
Proximity reader.

© Cengage Learning 2013

The access card typically contains printed information such as company name, user name, and a picture of the user. Several different card technologies are used, for example:

- **Wiegand technology**—Wiegand cards use short lengths of small-diameter, specially treated wire with distinctive magnetic properties. The wire, known as Wiegand Wire, is embedded into segments within the code strip of the card. When the card is presented to the changing magnetic field of a Wiegand reader, the different segments of embedded wire represent either a 1 or a 0 of a binary code. The digital format used by Wiegand is based around a 26-bit digital protocol. The code embeds a combination of user codes as well as facility data in order to provide individualized credentialing per user. If the binary code matches up with the code programmed into the access-control system, the user is granted access.

- **Magnetic stripe**—Magnetic stripe cards (often called **magstripe**) are typically identified by a black stripe on the back of the access card. The stripe itself is made up of tiny, iron-based magnetic particles in a plastic-like film. Each particle acts like a small bar magnet. Just as a bar magnet can be magnetized in either a north- or south-pole direction, so too can the particles on a magstripe card. When a card is "written to," the magnet particles embedded along the black stripe align to a desired code and are retained in magnetic memory. The reader must then recognize the unique north- or south-pole configurations along the magnetic stripe to gain access. Magstripe access cards have to be physically inserted or "swiped" to be read. To prevent accidental loss of data, care must be taken to keep magstripes away from strong magnetic fields that could potentially degauss the card and erase the magnetic alignment of the iron particles.

- **Proximity cards**—Proximity cards utilize contactless technology by communicating over radio frequency through a small, embedded antenna. These cards are also referred to as passive radio frequency identification (RFID) tags. The antenna is made from a tightly wound wire, attached to a digitally encoded, resonating circuit within the card. The unique card data from the resonating circuit transmits over the antenna between the card and the card reader when they are placed in close proximity; actual physical contact is not required. Proximity cards can be broken down into three general categories: immediate proximity, close proximity, and vicinity cards. Immediate proximity cards must pass within less than 1 mm (0.04 in.) of the reader. Close proximity cards can be read up to 10 cm (3.9 in.) from the reader. Vicinity cards can function in a range from 30 to 70 cm (11.8 to 27.6 in.) from the reader. A more detailed description of RFID is given in the next section.

Radio Frequency Identification (RFID)

Radio frequency identification (RFID) technology communicates over radio waves between an electronic tag and a reader. The tag, also known as a label, is composed of an embedded, digital radio circuit and an antenna. The reader, referred to as an interrogator, provides a radio frequency carrier signal, necessary to activate the tag. Once activated, the digital information embedded within the tag superimposes itself on the RF carrier signal and reflects back over the antenna to the reader as a modulated response. What makes RFID superior to other types of systems is that RFID does not need to have a visual line of sight between the tag and the reader, such as with bar code. With RFID, the tag can be concealed or may even be inside of boxes, cartons, or containers, and can still be read. The other interesting point is that RFID readers can read hundreds of tags simultaneously; passing the tags over the reader one at a time is not necessary.

RFID tags come in three varieties: passive, active, and battery-assisted passive (BAP).

- Passive RFID tags do not require a battery or power supply to operate them, and they are activated by the radio signal transmitting from the reader. As a result, the range of a passive RFID tag is not as far, requiring the tag to be placed in close proximity to the reader.

- Active RFID tags use an onboard battery to power up the internal circuit. The circuit is then able to continuously broadcast a beacon signal to the reader. Having an internal source of power also increases the broadcast range of active tags, making it possible to read them over a distance of many meters.

- Battery-assisted passive (BAP) tags use a much smaller battery than those of active tags. The battery is also not operating continuously, and the circuit is activated only in the presence of an RFID reader. As you would expect, the broadcast range of a BAP tag is wider than that of passive, but not as far as that of active.

RFID technology is not only being used by access-control systems but is also currently being used by manufacturers for inventory control and product shipping, in grocery stores, casinos, public transit, toll roads, and parking garages, and also for pet identification, asset management, and retail sales, just to name a few other applications. The technology continues to grow as more and more uses for it are discovered.

Biometrics

Biometric technologies are defined as "automated methods of identifying or authenticating the identity of a living person based on a physical or behavioral characteristic" ("Voice Security Systems: Biometric Technology," Voice Security Systems Inc., **www.voice-security.com/Biomet.html** [accessed June 2005].) For example, there are biometric readers that scan a fingerprint, and some may scan the retina of your eye, whereas others may memorize the geometry of your hand to grant access. These types of entry systems are most equated with futuristic movies that go to the extreme in utilizing the technology, but biometrics is a technology currently being used by many companies as a high-security access-control credentialing device.

Door Hardware

Granting access to a user through a specified door requires more than just the use of a card and a reader. The locking mechanism in the door must work hand in hand with the reader to release the door so it can be opened. Examples of locking mechanisms are magnetic locks and electric strikes.

- Magnetic locks—Magnetic locks, or **maglocks**, are locking devices that use an electromagnet and an armature plate to secure a door. There are no moving parts, and the device is relatively simple to install. When powered, the electromagnet attracts the armature plate, and the door is held secure. A maglock is shown in Figure 12–11. The holding force of a maglock ranges between 300 and 1200 lbf (1300–5300 N), depending on the manufacturer. Power to the lock requires dc, typically 12 to 24 V. Ac power sources do not work on maglocks unless they are first converted to dc through a rectifier diode. All magnetic locks are **fail-safe**. This means that they need a constant source of power to remain locked. If power is removed, the lock opens. In some cases, this may be the desired application, but if power fails and the company has a room that contains secret or

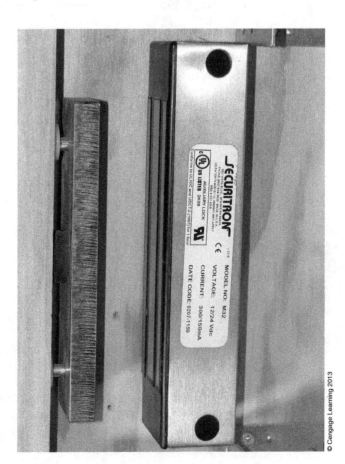

Figure 12–11
Maglock.

© Cengage Learning 2013

sensitive documents protected by a maglock, the door is no longer secure and anyone can enter the room. In this case, it may be better to use an **electric strike**.

- Electric strikes—The design of an electric strike is different from a maglock in that it uses a pivoting ramp plate to allow or deny access through the door. The device is mounted in the door frame where the door latch normally resides. When activated in fail-safe mode, an electronic solenoid within the strike latches the pivot plate, making it impossible to swing freely or move, and thereby securing access through the door. Once deactivated, the internal solenoid releases the pivot plate, allowing it to swing clear as the door is pushed open. An electric strike is shown in Figure 12–12. Electric strikes typically operate from 12 V or 24 V, but higher voltage strikes are available. The strike may run on ac, dc, or both, depending on the make and model of the strike. Compared to maglocks, electric strikes can be installed as fail-safe, or **fail-secure**. With fail-secure, power is required to unlock the strike, meaning that the lock stays locked during a power failure. This may be a benefit, as described in the above situation, but it may also prove to be a drawback. If a fail-secure lock is

Figure 12–12
Electric Strike.

© Cengage Learning 2013

used in a corridor where people are expected to evacuate during a fire alarm, the locked door may slow down the evacuation process and, in turn, endanger the occupants. Electric strikes may also be purchased with internal electronic buzzers that alert an individual on the other side of the door that the door is about to open.

Request to Exit (RTE)

A request to exit (RTE) is provided by a momentary normally open switch installed on the secured side of a door to allow people the ability to exit the space. By pushing the RTE switch, the panel is notified that an individual is requesting the door to be opened. In response to the request, the panel momentarily unlocks the door, allowing the individual time to exit. The door then is resecured after a predetermined amount of time that has been programmed into the system software. Reentrance to the space is only then allowed if a valid access card is presented to the reader on the unsecure side of the door.

SEC 12.4   WIRING AN ACCESS-CONTROL SYSTEM

The basic components to be wired to the control panel of an access-control system include the card reader, electric strike or maglock, RTE switch, door contacts, and an alarm siren to indicate a possible security breach. Instead of zone, an access-control panel separates the inputs and outputs by door. Terminal blocks are located on the main board of the control panel to designate the inputs and outputs for each door within the system. Input connections are used for card readers, door contacts, and RTE switches. Output connections are used for locks and alarm outputs. The next section details the basic connections to the panel.

Power

The connection of power to the control panel for an access-control system is different from that of a security alarm. With a security alarm, the main system power is usually large enough to power up all the devices and sensors within the system; additional power supplies can always be added if the current draw starts to exceed capacity of the main supply. For an access-control system, two or more system power supplies are typically required, one for the control panel and another for the locks. Depending on the number of locks, multiple supplies may be required. Maglocks especially require more power than any other device within the system, and for this reason they are usually separate.

Power to the panel typically requires a Class 2 transformer or Class 2 power supply, 12–24 V, ac or dc, depending on the manufacturer. Battery backup is also required to guarantee circuit operability during a power failure. Do not connect the Class 2 transformer or power supply to a receptacle that is controlled by a switch. Also, the metal cabinet of the card-access panel shall be grounded to the building grounding system or to an earth ground connection, as previously described in Chapter 3 of this book.

Just as with security systems, battery backup uses either a sealed, rechargeable lead-acid battery or gel-type battery. Be sure to calculate the Ahr capacity so that the required load is not more than 70% of the cells' capacity.

One important note on access-control power supplies: If you are planning to provide backup power to the door locks, be sure to provide sufficient backup power to the control panel, or you will not be able to get into the secured space during a power failure. Consider the situation where a lock is provided backup power, but the panel is not. In such a case, the lock remains locked, but the panel is inoperable and unable to read your card to provide access. Therefore, it is important to provide sufficient backup to all the system components to prevent a possible lockout.

Required Wire

Locks, RTE switches, door contacts, and alarm sirens are wired using Type 18–22 AWG, twisted 2-wire, stranded cable. The maximum resistance of an individual wire run should stay below 100 Ω.

Reader connections use a shielded, seven-conductor, stranded cable, 20–24 AWG.

Reader Connections

Proximity card readers require six conductors: reader antenna, beeper, power, ground, green LED, and red LED. The connection to the panel is shown in Figure 12–13.

Wiegand readers require seven conductors: Wiegand data0, Wiegand data1, beeper, power, ground, green LED, and red LED. The connection of a Wiegand reader is shown in Figure 12–14.

Swipe card readers usually only require four conductors, two for power and two for data.

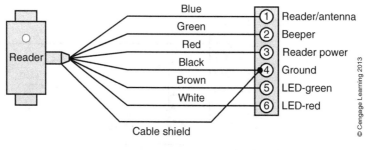

Figure 12–13
Proximity card reader wiring.

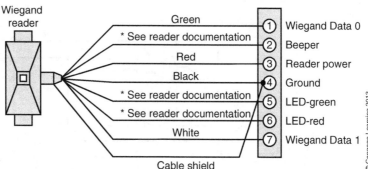

Figure 12–14
Wiegand reader wiring.

Connecting Maglocks or Electric Strikes to the Panel

A lock connects to the panel output through a single-pole, double-throw (SPDT), Form C relay. The connection is made to either the NO or NC contacts, depending on the style of lock. A fail-safe lock connects to the relay between the NC and common contacts. The NC connection provides continuous power to the lock. The relay then switches output states for either a signal from a RTE switch or a valid input from a card reader or biometric. A fail-secure lock connects between the NO contact and common, because the fail-secure only requires power to unlock the lock.

Transient Protection

Diodes, transorbs, or transient protectors are required on all switched circuits to prevent inductive kick when locks or relays shut down. Not using transient protection on locks or relay circuits opens the system up to possible failure. Inductive kick can generate thousands of volts when a device shuts down. The kick is developed from the collapsing magnetic field. The speed of the collapse determines the peak of the voltage kick. To prevent the kick, transient protectors are placed in parallel with the switched circuit. The transient protector provides a short-circuit pathway for the reverse current; otherwise the infinite resistance of the open circuit could potentially develop thousands of volts, and threaten the panel.

Figure 12–15 illustrates how locks are connected to the panel, including the placement of transient protection within the circuit.

Door Status Circuit

The door status circuit for access control is the same as the door contact circuit in a security alarm panel. A closed-loop, magnetic door contact is wired to the NC input

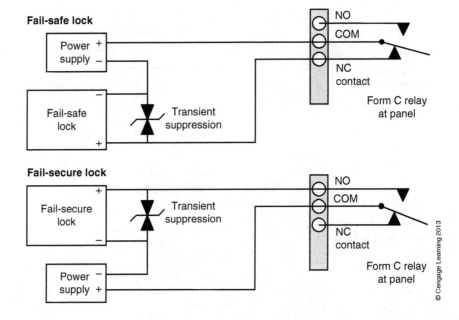

Figure 12–15
Wiring of locks to access-control panel.

© Cengage Learning 2013

for a desired door. The panel monitors the status of the input to ensure that the door remains closed. If the door were to remain open past the allotted time for a given access, the system would go into alarm until the contact loop was once again closed.

Request-to-Exit (RTE) Circuit

The RTE circuit uses a NO, momentary, single-pole, single-throw (SPST) switch to notify the panel of a request to exit at a desired door. The RTE switch connects to the NO input of the control panel for a desired door. The control panel monitors the input status of the switch and responds to a change of state when the switch is pressed.

Alarm Relay

The alarm siren connects to the alarm output relay of the panel. The siren requires a separate power supply that is wired through the NO output contacts. When a door does not close within the required amount of time, the system switches on the alarm relay to activate the circuit.

Figure 12–16 illustrates the connection of the door status circuit, RTE, and the alarm relay circuit.

System Programming

Access-control panels require a laptop computer or PC to program the system. Connection to the panel is usually made through a serial RS-232 connection or

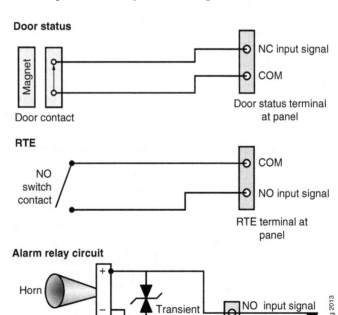

Figure 12–16
Door status, RTE, and alarm relay circuits.

© Cengage Learning 2013

through an RJ-45 network connection. Software must be downloaded from the manufacturer and loaded on the computer to be able to program the system. The benefit to using a network connection is that the system can be monitored and maintained from anywhere in the building or from a remote site over the Internet.

SEC 12.5 OPEN SOURCE SYSTEMS

Today, although connection of devices to system control panels may still exist, a more modern solution connects all system devices through the main building computer network, over (PoE) switches. A central computer is then used to program, monitor, and control the system. The various devices, whether they are door readers, locks, motion sensors, or cameras, are controlled through open source software. The benefit to using an open source system is that the configuration is completely customer controlled, and devices can be added or subtracted as needed without having to buy additional control panels or manufacturer-specific hardware. Hardware from virtually any manufacturer can be intermixed with others and controlled through device drivers. Device drivers are programs that run within programs to allow one type of product to communicate with another.

Now imagine that within your existing system, you want to install a new product that hasn't even been invented yet. Instead of having to purchase a new control panel and change out all of your system hardware to be compatible, the solution instead is to connect the device, when it becomes available, and simply download a device driver to be able to communicate with it. Software developers for open source systems are continually updating device drivers for new products as they are released to the public. So with an open source system, you never need to throw away your old hardware for new; instead, you are able to continually add on to your existing system or, if desired, upgrade old devices for new ones, without having to start from scratch. This is clearly a more economical way to grow and expand into the future.

Connection of Network-Based Devices to an Open Source System

Connection of network-based devices to an open source system involves a network of POE switches to various parts of the building. Devices are then wired locally to a control module located in the ceiling above the device. A card reader, for example, connects, as previously described, through six or seven shielded conductors from the reader to the above-ceiling control unit. The control unit is then connected to the building network over CAT5 or CAT6 cables and powered by a POE switch. If power capacity is required over and above that provided by the POE switch, an auxiliary supply can always be added to pick up the extra load. Power to a maglock, for instance, may push the switch beyond capacity. The addition of an auxiliary supply is then required to power the lock, and the control unit still gets its power over the Ethernet cable.

Figure 12–17 illustrates how a network-based security and access-control system are connected.

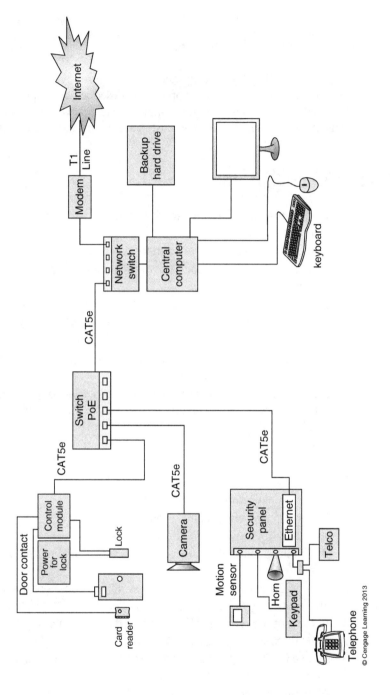

Figure 12–17
Open source
system
wiring.

© Cengage Learning 2013

SEC 12.6 ELECTRICAL CODE

Wiring and cabling for security and access-control systems shall be installed according to *NEC Article 800*, as previously described in Chapter 11 of this book. The installation of Class 2 power supplies shall follow the installation requirements of *NEC Article 725*, as previously covered in Chapter 8 of this book.

CHAPTER 12 FINAL QUESTIONS

1. What is the purpose of a security alarm as compared to an access-control system?
2. How does a passive infrared sensor differ from an active infrared sensor?
3. What is the main difference between an ultrasonic and a microwave motion sensor?
4. What are the two types of glassbreak sensors, and how are they different?
5. Which type of motion detector requires a line-of-sight pathway between transmitter and receiver?
6. A closed-loop magnetic door contact requires the use of a magnet to _____ the contact.
7. Give an example of why you would use an open-loop magnetic door contact. Draw how it would be connected to the control panel using an SEOL resistor.
8. How is connection to a central station monitoring facility made from the security system control panel at the protected premises?
9. The use of a _____ triggers a silent alarm.
10. Why is a silent alarm sometimes more desirable?
11. Power supplies should be loaded not more than _____ % of capacity.
12. The input circuits for security systems operate on normally _____ loops.
13. For a SEOL circuit, when a door or window is opened, the input loop to the control panel _____ (opens, closes, or shorts).
14. To prevent the removal of a device cover or the opening of the control panel door, a _____ circuit is required.
15. Draw a DEOL loop circuit for a zone that has a single, closed-loop magnetic door contact installed.
16. Wire run resistance for a zone should not be greater than _____ Ω.
17. What type of technology does a proximity card use?
18. Access cards come in three varieties; they are magstripe, _____, and _____.

19. Give an example of a type of biometric input for an access-control system.
20. Maglocks are fail-_____.
21. A fail-secure lock is connected to the normally _____ relay output at the control panel.
22. What type of switch is used for an RTE, and where is it typically installed?
23. How does an access-control panel monitor that a door has been propped open?
24. An installer provided a power supply with backup power to a door lock but failed to provide backup to the control panel. Why is this problematic?
25. Which devices in an access-control system require them to be installed using shielded wire, locks, or readers?
26. List the six connections to be made when wiring a proximity card reader.
27. Can a Wiegand card be used on a proximity card reader?
28. Transient protectors are used on _____ and _____ circuits to prevent inductive kick from damaging the control panel.
29. When installing access-control systems, which devices require separate power supplies?
30. Security alarms and access-control systems are installed according to which two articles of the *NEC*?

Chapter 13

Wireless Communications and *National Electrical Code Article 810*

Objectives

- Describe the fundamentals of wireless communications.
- Identify the basic components of a wireless system and explain the function of each.
- Identify different types of wireless antennas and explain the purpose and function of each.
- Describe and explain the principles of cellular communications.
- Describe and explain the principles of satellite communications.
- Identify the basic components of satellite communications.
- Explain the operating principals of wireless computer networks.
- Describe and explain the requirements for radio and television equipment installation, as required by *NEC Article 810*.

Chapter Outline

Sec 13.1 A Brief History of Wireless Communications

Sec 13.2 The Fundamentals of Wireless Communications

Sec 13.3 Antennas

Sec 13.4 Cellular Phone Communication

Sec 13.5 Satellite Communications

Sec 13.6 Wireless Computer Networks

Sec 13.7 Cable Television System

Sec 13.8 *National Electrical Code* Requirements for Radio and Television, *Article 810*

Key Terms

Advanced Mobile Phone Service (AMPS)

amplitude modulation (AM)

analog-to-digital (A/D) converter

antenna

Bluetooth

carrier frequency

channel

digital-to-analog (D/A) converter

downlink

Federal Communications
   Commission (FCC)

frequency division multiple
   access (FDMA)

frequency modulation (FM)

frequency shift
   keying (FSK)

full-duplex

geosynchronous satellite (GEO)

global system for mobile (GSM)

half-duplex

low-earth-orbit satellite (LEO)

low-noise block (LNB) converter

medium-earth-orbit satellite
   (MEO)

modulation

multiplexing

personal communication
   system (PCS)

phase modulation (PM)

picocells

polarization

pulse modulation

radiofrequency
   interference (RFI)

repeater

roaming

satellite

time division multiple
   access (TDMA)

transceiver

transmitter

transponder

unidirectional

uplink

wireless communication

SEC 13.1   A BRIEF HISTORY OF WIRELESS COMMUNICATIONS

Wireless communication involves the sending of signals by air instead of through a hardwired connection or cable. Scientists originally conceived of the concepts of wireless transmission and reception in the 1800s, but they were not put into practical use until 1901, when Guglielmo Marconi successfully engineered the first transatlantic communication by radio waves. Before that, all electronic communication involved the sending of signals over wires.

Samuel B. Morse started the first wired telegraph system in 1837, which issued a series of spark gap dots and dashes in coded format known as Morse code. The transmission channel for Morse code consisted of a network of telegraph wires mounted on poles and a switched battery.

Voice communication was not far behind. By 1876, Alexander Graham Bell invented the first hardwired, cabled telephone system. At the time, switchboard operators switched all channels of communication manually. Various parties on either end of the line were connected or disconnected as needed by way of patch panels and jumper cables. In many cases, multiple parties shared the same communication channel as a party line because of the limited number of available channels.

Although the theoretical framework for radio communications was developed by James Clerk Maxwell in 1865, it was first verified experimentally in 1887 by Rudolph Hertz. Working radio prototypes were realistically designed and put to use after Lee De Forest invented the vacuum tube amplifier in 1906. Incidentally, the predecessor to the vacuum tube was the invention of the lightbulb by Thomas Edison in 1879. All of these men, knowingly or unknowingly, added an individual piece to the puzzle, and the new technology of radio was born.

Military leaders of World War I were also quick to realize the benefits of the new technology, as they could now send messages instantly to the front lines, no longer needing to dispatch couriers through dangerous territory. It was not long after that the public use of radio communication was ultimately realized and made available to everyone around the globe. Isolated locations, islands, and ships at sea could now benefit from the new technology, providing for the first time an efficient and simple means of communicating to remote places.

By the late 1920s, radio broadcasting in the United States became a vehicle for entertainment. Entertainment, however, was never the priority or intention of radio. To this day, the **Federal Communications Commission (FCC)**, which oversees the regulations and standards of radio communications in the United States, considers the priority of radio communications to be that of a public emergency broadcast system. All other reasons to broadcast over wireless communications remain a secondary priority. Emergency broadcasting aside, since the days of the 1920s, wireless communication has grown into a much larger public service than anyone could have ever imagined.

Although the original infrastructure of wireless was designed primarily for the sending and receiving of analog voice signals, today it has grown into a far more complex system, supplying an endless stream of audio, video, and data signals between industry and business, colleges, schools, and homes. Some current examples may include power and utility companies communicating by wireless to electric, water, and gas metering systems, and home appliance repair technicians using wireless to access the availability of parts and copies of service manuals. In addition, auto dealers are now offering services such as OnStar, which provides a means of two-way, wireless communication between drivers and an OnStar service representative, 24 hours a day, 7 days a week. Through global positioning satellite (GPS) mapping, OnStar can not only locate the exact position of a vehicle but also can unlock doors remotely and notify the authorities of a potential emergency.

Moreover, in the 1920s who could have imagined the use of present-day cellular phones and paging systems, which now are a built-in necessity of modern life? Parents, children, business, and commerce all partake in and use such systems on a daily basis. As technologies continue to expand and integrate, the possibilities often appear endless as systems consolidate, offering a wider range of creative solutions to the problems of everyday people and business.

The use of wireless communication, however, has some main disadvantages compared with hardwired systems: Available frequencies are a limited commodity, bandwidth often is in short supply, and security can also be at risk. To help solve these problems, designers continue to think up new ways of stacking and combining multiple signals on a single-frequency carrier as a means of conserving space and increasing broadcast speeds. In the old days, party-line telephone service was required because current multiplexing techniques had not yet been made available. Issues of capacity, privacy, and security have always been the highest priority of designers and engineers of communications systems. To help solve such problems, the newer digital technologies offer scrambled and encrypted signals to prevent information from getting into the wrong hands. The recipient of such signals then must use special decryption software to retrieve and interpret the encoded communication.

At any given time, there are only a finite number of available frequencies that can be authorized for use. In the United States, the FCC oversees such authorizations. The FCC not only approves and monitors broadcast frequencies, but it also oversees the protection and security of the entire system. Station pirating and frequency jamming are considered federal offenses and are aggressively prosecuted by the FCC when encountered.

The FCC also provides the licensing and approval of wireless circuits and devices to ensure that their emission levels are safe and that they do not pose a threat to users or other electronic devices and systems through unwanted **radio-frequency interference (RFI)**. Airlines and hospitals, in particular, often require cell phones to be turned off to help avoid any inadvertent or accidental interference with other sensitive electronic equipment, typically providing the critical functions of navigation or life support. In some cases, it is always safer to err on the side of caution, because a malfunctioning wireless circuit could potentially do more damage than good. All wireless circuits have the potential to interfere with the functions of other electronic circuits and devices, however minimally. For this reason, manufacturers often build into their electronic products a level of protective shielding and a grounding scheme to ensure the safe, functional use of devices. The first step to troubleshooting an intermittent circuit suspected of receiving RFI is to check all equipment grounds and shielding to make sure that they have not been removed or tampered with. In addition, all radiating or sensitive equipment should be installed according to manufacturer's specifications to ensure proper operation and to reduce the risk for potential interference.

SEC 13.2 THE FUNDAMENTALS OF WIRELESS COMMUNICATIONS

A basic wireless system consists of a signal source, a transmitter, a receiver, and two antennas. The job of a **transmitter** is to send a signal source over a communication channel to the receiver. In telecommunications, the communication **channel** is the connecting wire or fiber optic between the caller and the recipient. In radio communications, the channel is the actual airspace between the transmitting **antenna** and the receiving antenna (Figure 13–1).

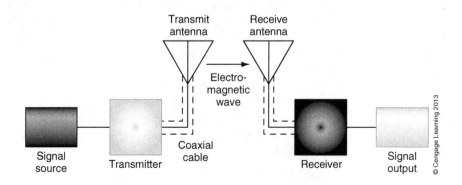

Figure 13–1
Block diagram of a wireless transmission model.

© Cengage Learning 2013

A high-frequency electromagnetic sine wave, called the **carrier frequency**, carries the signal as it radiates across the channel. The receiver then is able to receive the communication, provided that it is tuned to the same frequency as the carrier traveling through the channel. Voice, data, or video signals are superimposed on the carrier, through a process known as **modulation**, as a means of sending the signals out over the radiating antenna. The radiating signal then travels out over the airspace, at nearly the speed of light, to the receiver antenna, some distance away from the transmitter. The strength of a received signal is dependent on the wattage level of the transmitter output and the distance to the receiver.

A wireless system is **unidirectional** in nature; that is, signals can only travel out from the transmitter to the receiver; a receiver does not transmit, and a transmitter does not receive. To accomplish a two-way link requires an additional system installed in the reverse direction. In such a case, the receiving end needs a separate transmitter to communicate back to the originator, and the originator needs a receiver to receive a reply. Such a device does exist. It is called a **transceiver**. A transceiver combines the circuitry of a transmitter and receiver all in one box. Cellular phones are examples of transceivers. That cellular phones can transmit and receive signals simultaneously also makes them **full-duplex**. To accomplish a full-duplex link, the transceiver must transmit on a separate carrier frequency from the receiver; otherwise, the transceiver feeds back on itself during a transmission (Figure 13–2).

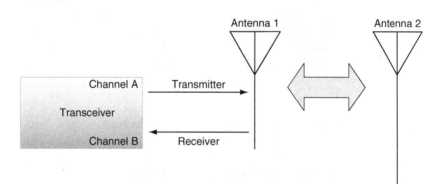

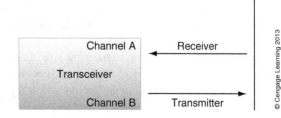

Figure 13–2
Full-duplex
communication.

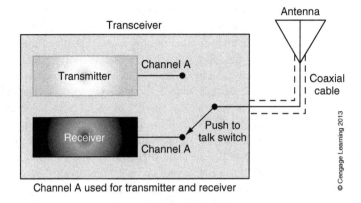

Channel A used for transmitter and receiver

Figure 13–3
Half-duplex
communication.

An alternative to full-duplex is a **half-duplex** system, where the transmitter and receiver are able to share the same frequency carrier; but in such cases, the receiver is disabled when the transmitter is operating and vice versa. Communication protocols also must be established where using a half-duplex system to help minimize confusion or interference between listeners on both ends of the link (Figure 13–3). Obviously, if each party is simultaneously transmitting a signal, then no one is listening; alternately, any attempt to simultaneously receive on both ends of the link means that no one is actually transmitting; thus the need for protocols. A CB radio is an example of a half-duplex connection. The microphone on a CB radio has a push-to-talk button. To transmit, the user must first press the button to activate the transmitter and disable the receiver. Receiver operations are then reestablished only after the push-to-talk button has been released.

Modulation

Modulation is a process through which a desired information signal is superimposed onto a carrier frequency. Various signals such as audio, video, or data can be transmitted over a carrier frequency, in an analog or digital format. The type of information signal and the format usually dictate the style of modulation to be used. Typical forms of modulation may include these:

- **Amplitude modulation (AM)**
- **Frequency modulation (FM)**
- **Frequency shift keying (FSK)**
- **Phase modulation (PM)**
- **Pulse modulation**

Other forms of modulation do exist, but they are often variations of the general methods listed above.

To be able to transmit an intended signal, the operating frequency of the radio station must be tuned to a specific carrier frequency. The intended signal can be voice, data, or video. As an example, station 92.5 FM corresponds to a carrier

Figure 13–4
Amplitude
modulation.

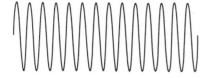

Unmodulated
carrier frequency

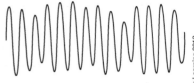

Amplitude-modulated
carrier frequency

frequency of 92.5 MHz. In the absence of a transmitting signal, the carrier is a continuous and constant sine wave. The process of modulation is what makes the carrier shift or vary in direct relation to the variance of the intended signal. The type of modulation determines how the intended signal is superimposed on the carrier and how the shift occurs. The following sections list the details of the more common modulation methods.

Amplitude Modulation

AM involves varying the amplitude of a carrier frequency (vertical deflection of the sine wave) at a rate equal to the frequency of the superimposed information signal. Figure 13–4 shows a carrier frequency modulated by a 1 Khz tone. In the absence of a modulating signal, the carrier maintains a constant vertical peak-to-peak level as a continuous, undistorted sine wave.

Frequency Modulation

FM involves varying the carrier frequency at a rate equal to the superimposed information signal (Figure 13–5). Where a 92.5 MHz carrier is FM modulated by a 1 kHz audio tone, the carrier frequency shifts 1 kHz above and 1 kHz below 92.5 MHz (92.499–92.501 MHz). The absence of a modulating signal causes the carrier to remain at a constant 92.5 MHz.

Frequency Shift Keying

FSK is a type of FM that causes the carrier frequency to vary with respect to two separate tones, to represent the highs and lows of a binary number. For example, the two tones may be 1 and 10 kHz. The 1 kHz tone can represent a binary 0, whereas the 10 kHz can represent a binary 1. By alternating the sequence of the modulating tones, the carrier can then be made to transmit a direct digital code.

Figure 13–5
Frequency
modulation.

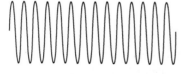

Unmodulated

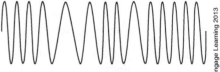

Frequency modulated

Phase Modulation

PM is similar to FM except that the phase of the carrier signal is made to shift at a rate equal to the superimposed information signal. PM often is used in digital communication circuits.

Pulse Modulation

Pulse modulation involves varying the amplitude, duration, or time interval between pulses to represent a formatted coded signal. The pulses are then superimposed on the carrier frequency for transmission.

Frequency Notation	
Hertz (Hz)	Cycles per second
Kilohertz (kHz)	Thousands of cycles per second
Megahertz (MHz)	Millions of cycles per second
Gigahertz (GHz)	Billions of cycles per second

© Cengage Learning 2013

All usable communication frequencies fall into specific bands, which are frequency ranges; audio is the lowest, representing the only frequency band the human ear can detect.

Frequency Examples and Standard Industry Bands	
Audio	20–20,000 Hz
Voice telephone	300–3400 Hz
AM radio	530–1600 kHz
FM radio	88–108 MHz
HF (high frequency)	3–30 MHz
VHF (very high frequency) broadcast television	30–300 MHz
UHF broadcast television (ultra high frequency)	300–3000 MHz
Mobile cellular phones	800–900 MHz
PCS cellular	1.85–1.975 GHz
Infrared	300,000–428,000 GHz
Satellite band frequencies:	
L band	1–2 GHz
S band	2–4 GHz
C band	4–8 GHz
X band	8–12 GHz
Ku band	12–18 GHz
K band	18–27 GHz
Ka band	27–40 GHz
Millimeter band	40–300 GHz
Submillimeter band	$\geq$ 300 GHz

© Cengage Learning 2013

Microwave and **satellite** signals are considered any frequency greater than 1 GHz.

Analog versus Digital

Analog signals are those that vary continuously, for each instance along a waveform, representing a value or signal level based on the output of a given device known as a transducer. Sine waves are examples of analog signals.

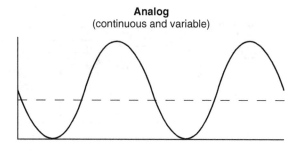

Analog
(continuous and variable)

Figure 13–6
Analog versus
digital.

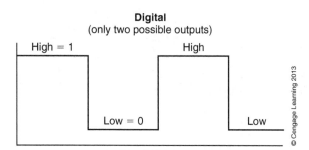

Digital
(only two possible outputs)

High = 1   High

Low = 0   Low

© Cengage Learning 2013

A transducer converts a mechanical property into an electrical signal. An example of a transducer is a microphone. The variations of sound pressure applied to the diaphragm of the microphone causes the electrical output to vary accordingly. Such a device generates a direct analog signal, which continuously changes with respect to the applied input. Ultimately, an infinite number of possible output variations can exist, based on the level of the applied input.

Digital signals, unlike those of analog, have only two possible output variations: high or low. A square wave is an example of a digital signal. The high state of the square wave may represent a logic 1, or on; the low state may represent a logic 0, or off. By using square waves, electronic signals can be binary coded, based on the transitions of a streaming series of rectangular pulses. Figure 13–6 compares the differences between an analog and a digital signal.

Analog-to-Digital Converters

Analog-to-digital (A/D) converters are used to convert real-world analog inputs, such as the audio pressure variations of a microphone or the temperature variations of a thermal resistor, into a digital format. Because computer- or microprocessor-based systems can only communicate with and interpret binary signals, all analog inputs must first be converted into some form of digital code for processing (Figure 13–7).

In most cases, computer outputs must be decoded back through a **digital-to-analog (D/A) converter** to make connectivity to analog devices possible. Output devices such as amplifiers, speakers, and video monitors often require analog input signals to operate. Current audio and video systems, however, are now

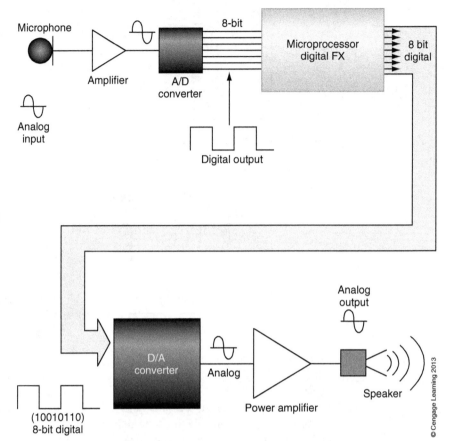

Figure 13–7
Complete circuit involving analog-to-digital (A/D) conversion and digital-to-analog (D/A) conversion.

providing input ports for both analog and digital signals, depending on the type of source and connectivity needed. In cases requiring human interaction, digital signals ultimately must be converted back to an analog state. As of yet, the human ear is incapable of decoding the pure binary data streams of digital audio format; therefore, until evolution moves mankind forward, analog is still the preferred choice of any listener.

Multiplexing

Multiplexing involves the mixing of several communication channels from multiple sources into a single, large-capacity transmission channel. On the transmission end of the communication link, signal sources are processed through a device known as a multiplexer. The receive side of the link then reverse processes the multiplexed signal through a demultiplexer, providing the necessary separation to individual channels and signal sources at the destination point.

SEC 13.3 ANTENNAS

The antenna provides the means of either radiating the modulated carrier frequency out into the surrounding environment or collecting radiated radiofrequency (RF) energy as it bombards the antenna of a distant receiver.

An antenna can be a simple loop of wire, a multielemental device, or a parabolic dish, depending on the type of transmission, wavelength of the carrier signal, and bandwidth of the tunable receiver. For optimal transmission and reception, antennas on each end of the communication link should be mounted well above trees and buildings, or above any obstruction. Natural barriers such as hills and valleys must also be taken into account, which is why most antenna systems are mounted on high towers.

Antennas are designed for a specific bandwidth of use, which is why radio antennas differ from those of television and satellite antennas. The wavelength of the carrier frequency and operating power at the transmitter determine the optimum length and size of the antenna elements (Figure 13–8). Most antennas are

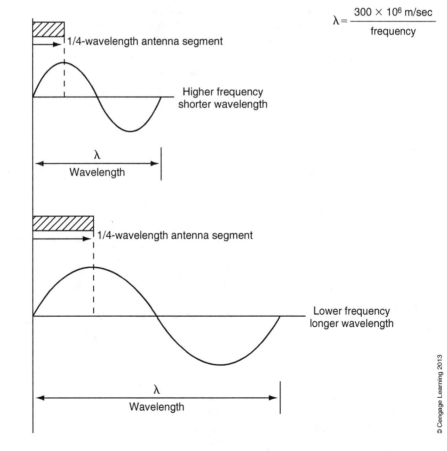

$$\lambda = \frac{300 \times 10^6 \text{ m/sec}}{\text{frequency}}$$

1/4-wavelength antenna segment

Higher frequency shorter wavelength

λ
Wavelength

Figure 13–8
Frequency versus wavelength.

1/4-wavelength antenna segment

Lower frequency longer wavelength

λ
Wavelength

© Cengage Learning 2013

sized at one-fourth the wavelength of the carrier frequency. A quarter-wavelength antenna segment produces maximum resonance at a specific frequency of operation, just as the tuning pipe of a pipe organ is carefully adjusted for length to recreate the exact frequency of a musical note. Lower carrier frequencies have longer wavelengths, requiring the antenna loops or elements to be long. In comparison, high carrier frequencies have much shorter wavelengths, thus allowing antenna lengths to be quite small.

The following formula is used to calculate the wavelength of a traveling waveform:

$$\lambda = V_p \div f_{carrier}$$

where λ represents the wavelength of one signal cycle; V_p is the velocity of propagation, which is nearly 300×10^6 m/sec, depending on the medium of travel, such as air or water; and $f_{carrier}$ is the carrier frequency in Hertz.

The length of a quarter-wavelength AM antenna tuned to 830 kHz is:

$$300 \times 10^6 \text{ m/sec} \div 830 \text{ kHz} = 361.45 \text{ m}$$

$$361.45 \text{ m} \div 4 = 90.36 \text{ m, or } 296.46 \text{ ft}$$

AM antennas are long, which also explains why most AM reception is lost when traveling in a car through a tunnel, simply because the wavelength of the propagating RF waveform is unable to fit through the required tunnel entrance. Most modern buildings, built around a grounded, metal grid-style framing structure, also prevent AM signal reception because the RF wavelengths are much too large to pass between the spaces of the support beam matrix (Figure 13–9). Building grounds also neutralize the signals as they hit the frame of the structure. A solution to this problem is to place a receiving antenna outside of the structure. The signal then could be brought into the building through the antenna lead-in wires. Antenna lead-in wires typically are 50 or 75 Ω coaxial cable, or 300 Ω balanced twin-lead, depending on the type of system and signal.

Example

Calculate the length of a quarter-wavelength FM antenna tuned to 104 Mhz.

$$300 \times 10^6 \text{ m/sec} \div 104 \text{ MHz} = 2.88 \text{ m, or } 1185.85 \text{ ft}$$

$$2.88 \text{ m} \div 4 = 0.721 \text{ m, or } 2.37 \text{ ft}$$

FM antennas are relatively short, which also explains why FM reception often is easy to receive; as a result, FM usually transmits through most small spaces and tunnels with ease.

In some cases, however, buildings are shielded because of the metal ductwork of the heating and ventilation system, or even by the sheet metal siding and flooring of the framing structure. Depending on the direction and polarity of the traveling waveform, some building structures simply act as mechanical filters, often preventing access to specific bands of frequencies, and in most cases deflecting or

Figure 13–9
The transmission of short and long wavelengths versus the size of the entrance point.

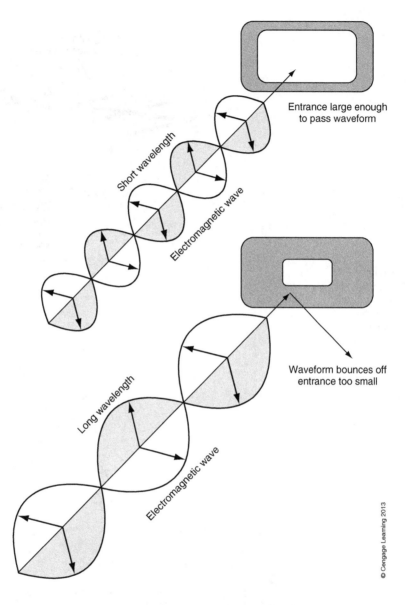

grounding out desired signals (Figure 13–10). Again, the only solution to such a problem is to install an exterior antenna to the building structure. The actual signal can then be brought into the building over the antenna lead-in wires. Once the signal is traveling through a cable, it can be sent just about anywhere.

Types of Antennas

Antennas come in all shapes and sizes, depending on the desired carrier frequency, operational bandwidth, and directionality of signal. Some antennas are tuned to

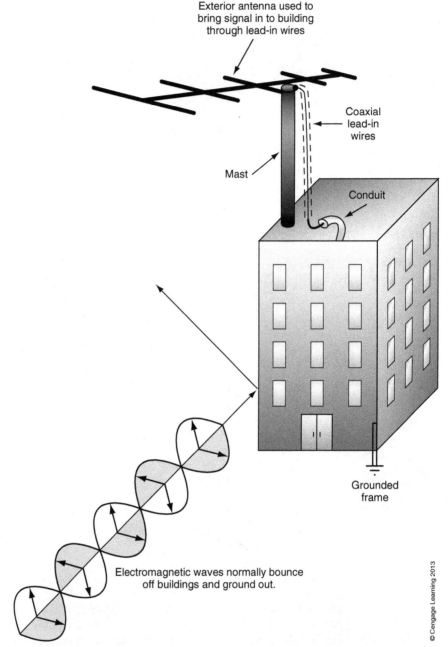

Exterior antenna used to
bring signal in to building
through lead-in wires

Coaxial
lead-in
wires

Mast

Conduit

Grounded
frame

Figure 13–10
Electromagnetic
waves normally
bounce off
building struc-
tures. Exterior
antenna brings
signal into build-
ing through
rooftop lead-
in wires.

Electromagnetic waves normally bounce
off buildings and ground out.

© Cengage Learning 2013

one specific frequency, whereas others are able to input or output a wider band of
multiple frequencies. Directionality can also be built into an antenna, as some are
built to be highly directional and only transmit or receive along narrow pathways,
whereas others are built to be omnidirectional. The radiation footprints of antennas are plotted and graphed as polar patterns, similar in appearance to the pickup

Figure 13–11
One-quarter
wavelength
monopole
antenna.

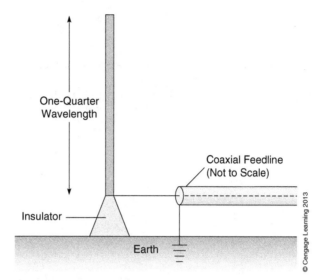

patterns of audio microphones (see Chapter 5 in this book). The shape of the polar pattern determines the directionality of the antenna.

Following are examples of commonly used antennas.

Quarter-Wave Monopole

Quarter-wave monopoles are whip antennas that consist of a quarter-wavelength segment of metallic element, vertically mounted and connected at the base to the antenna lead-in wires (Figure 13–11). A matching device sometimes is needed where a half-wave dipole is connected in series with a quarter-wave monopole.

The matching device typically consists of a coil and capacitor that often can be adjusted to achieve optimum tuning for the characteristic impedance of the system and cable. In many instances, antennas are not perfectly tuned and therefore must be adjusted for optimal performance and minimum signal loss, often caused by standing waves traveling on the connecting cable. Many antennas have matching devices connected to their lead-in wires.

Dipole Antennas

Dipole antennas are half-wave in length, consisting of two quarter-wave segments. A variation on the dipole is the folded dipole that often is used on FM radio receivers. Dipoles also are used on short-wave and amateur radio systems (Figure 13–12).

Single-Channel Yagi Antennas

Single-channel yagi antennas are multielement, high-gain, directional antennas, typically used for television reception and the UHF frequency bands. Figure 13–13 shows a yagi antenna.

Dipole antennas

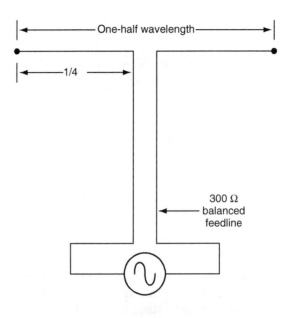

Figure 13–12
Dipole antenna
versus folded
dipole antenna.

Folded dipole

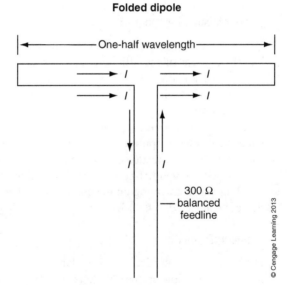

© Cengage Learning 2013

Broadband Antennas

Broadband antennas are often V-shaped, having multiple elements arranged by increasing size. The different sized elements make it possible for the antenna to pull in a wider range of frequencies. Broadband antennas are also directional in nature, and they must be mounted properly and pointed in the direction of the transmitter to be able to receive the strongest signal (Figure 13–14).

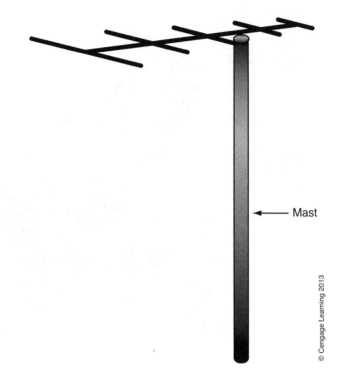

Figure 13–13
Yagi antenna.

Mast

© Cengage Learning 2013

Figure 13–14
Broadband
antenna.

© Cengage Learning 2013

Figure 13-15
Parabolic dish antenna.

© Cengage Learning 2013

Parabolic Dish Antennas

Parabolic dish antennas are used for microwave and satellite transmission and reception. The parabolic shape of the dish helps to concentrate a larger amount of low-intensity microwave signals to the antenna element, located at the focal point of the dish (Figure 13-15).

The antenna element of a receiving dish antenna is a waveguide called a **low-noise block (LNB) converter** that is used to receive the traveling microwave signal (Figure 13-16). The LNB converter also provides signal preamplification before sending it through a 75 Ω coaxial cable to the receiver. Waveguides are often rectangular, and they perform impedance matching from the dish to the input of the preamplifier.

Power to the LNB preamplifier is supplied through the central conductor of a coaxial cable. A dc voltage, generated at the receiver, travels with the ac microwave signal as a composite waveform along the central conductor. Care, however, must be taken where splitting signals because of the presence of dc on the line. As an example, a signal may often be split to a signal strength meter connected in parallel with the input of the satellite receiver. But remember, the dc voltage supplied by the receiver is intended for the LNB, not the meter. The signal strength meter does not need or expect to see the additional dc voltage on the line. In such cases, a dc block must be inserted in series with the splitter to prevent the unwanted transfer of dc voltage to other circuits. The dc block is simply a series capacitor and a 75 Ω matching device that is necessary to maintain the characteristic impedance of the connection.

LNB

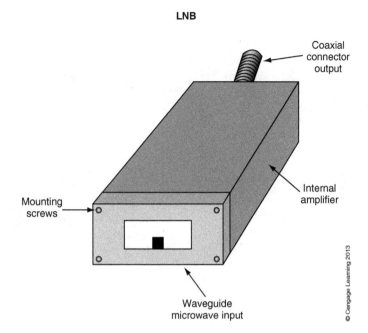

Figure 13–16
Low-noise block.

Helical Antennas

Helical antennas use a coil-shaped element placed lengthwise along the center of a reflecting surface. The reflecting surface can be made to be circular, rectangular, conical, or horn-shaped, depending on the requirements and directional pattern of the antenna. Helical antennas are used in satellite communications (Figure 13–17).

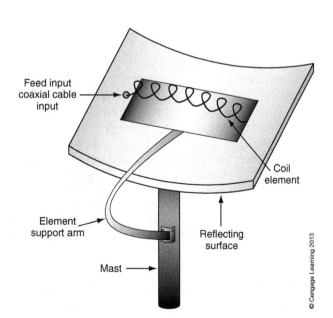

Figure 13–17
Helical antenna.

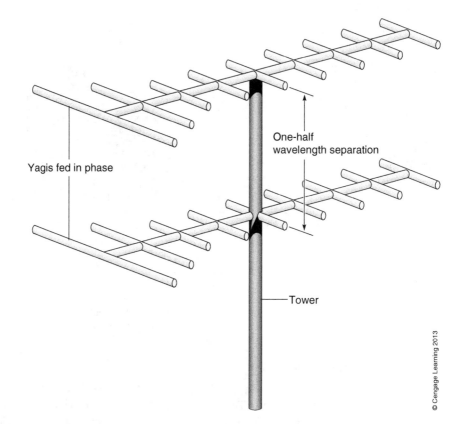

Figure 13–18
Phase-array
antennas.
Stacked yagi
antennas.

Yagis fed in phase

One-half
wavelength separation

Tower

© Cengage Learning 2013

Phase-Array Antennas

Phase-array antennas use multiple banks of antenna elements, arranged in a one- or two-dimensional pattern. The purpose of phase-array antennas is to increase the radiation pattern of a transmission. By using phase arrays, a designer can develop any shape antenna pattern desired and also substantially increase the overall gain of the transmission by multiplying and concentrating the signal intensities of various antenna elements into a desired location (Figure 13–18). The two-dimensional pattern also allows for the simultaneous transmission or reception of vertical and horizontal signals.

Antenna support structures must be grounded. The grounding requirements and other *NEC* requirements for antenna systems are discussed at the end of this chapter.

SEC 13.4 CELLULAR PHONE COMMUNICATION

Analog radiotelephone service in the United States was established in the early 1960s. The original system used transceivers mounted in moving vehicles and a series of repeater stations widely spaced over a given coverage area

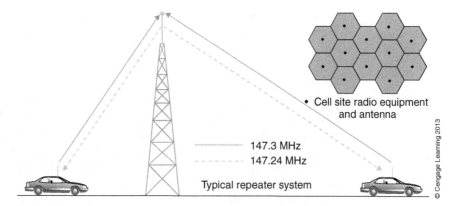

Figure 13–19
Cellular phone repeater systems and cell grid.

• Cell site radio equipment and antenna

——————— 147.3 MHz
– – – – – – 147.24 MHz

Typical repeater system

© Cengage Learning 2013

(Figure 13–19). Each repeater station was responsible for transmitting within a specified zone, now referred to as a cell. In addition, to achieve the best signal strength and coverage, service providers mounted antennas on top of high towers; today we call them cell towers.

Repeater

A **repeater** provides the function of passing data or radio transmissions from one system or antenna cell to another. Because the maximum transmission range of most radiotelephones is quite short due to limited output power, it becomes necessary to extend their total broadcast distance by bouncing signals beyond the local coverage area. In addition, repeaters link the overlapping cells of individual transmission zones to achieve a wider coverage area. Repeaters also link land-line phones to the wireless system, allowing access to the wireless system from external base stations or the local telephone company. The primary function of a repeater station is to receive signals and then pass them on, unaltered, to the next destination, acting much like an electronic data relay.

Mobile Radiotelephone History

The mobile radiotelephone service of the 1960s was originally intended for use on moving vehicles, with taxicabs, ambulance services, and dispatchers being the primary users. The hardware of such systems was somewhat bulky, high in wattage, and often required large power supplies to function. The original mobile radiotelephone systems were not designed for individuals to carry them around as **personal communication systems (PCSs)**. By the mid-1980s, however, the system was upgraded into that of a true cellular radio system, more closely resembling the systems currently in use. In 1983, Chicago became the first city in the United States to use such a system, which was known as the Improved Mobile Telephone Service (IMTS). Although the new system offered more channels and availability to consumers, it was still heavily dependent on assigning individual carrier frequencies to users. Modern techniques of frequency or time domain multiplexing were still not yet available within the IMTS systems; as a result, customers experienced signal loss and interference when traveling between adjacent cells attempting to assign the same frequency to different transceivers.

Roaming

The act of traveling and communicating between adjacent cells is known as **roaming**. Service providers add on additional charges when customers place calls outside of a service provider's home area. In addition, cell phones do not communicate directly between each other. Even if they are within the same room, the service provider must still assign frequencies controlled by the repeater and cell tower. Although some companies are now offering a talk-back feature, allowing two phones to beep each other and communicate without actually making a dialed call, they are still not communicating directly, for example, as walkie-talkies do. Instead, the repeater issues a single-carrier frequency, and the mobile units communicate through the cell tower in a half-duplex manner rather than over a conventional full-duplex link.

Advanced Mobile Phone Service

By the early 1990s, the next generation of cellular telephone came into being. The upgrade, known as the **Advanced Mobile Phone Service (AMPS)**, which is still in use today, offered three basic power classes of analog service (with respect to a half-wave dipole antenna):

Class 1 (mobile): 4 W ERP (effective radiated power)

Class 2 (transportable): 1.6 W ERP

Class 3 (portable): 600 mW ERP

In comparison, the cellular radiotelephone systems of the 1960s used high-power transceivers of 30 W or more, because cell towers were widely spaced and the need to transmit over larger areas was a necessity. With the advent of the AMPS system, cell towers were placed at closer intervals and wattage levels were reduced, allowing for smaller designs and safer emission levels. Finally, a portable phone having a maximum output of 600 mW allowed for designs small enough to be personally carried or worn on the hip. Advancements in computing also allowed for faster switching and the ability to increase the number of cell towers and repeaters in the system.

It did not take long, however, for customers to prefer the newer competing digital services being offered, and later digital PCSs, compared with AMPS, ultimately provided far more options, including access to e-mail, Internet connectivity, and text messaging. Let us now look at a comparison of the AMPS and the digital PCS.

Technical Description of Advanced Mobile Phone Service (AMPS)

The North American AMPS allocates 395 duplex analog voice channels for use in a single cell. The channels operate in a frequency band between 824 and 894 MHz. Each channel provides a pair of carrier frequencies, referred to as the forward channels and the reverse channels. The forward channels transmit from the base station to mobile phones, and the reverse channels from mobile phones to the base station. Computerized switching provides automatic repeater control, allowing for instant frequency and channel changes, as needed, based on which cells are providing the highest transmission signal and which frequencies are available for use. The use of dynamic frequency allocation solved many of the inherent problems of the older style mobile radiotelephone system. Dynamic frequency allocation allows the provider to allocate frequency changes instantly, as needed, to help solve issues of interference and busy or unavailable frequency carriers, as well as poor signal response; these are just some of the major advantages to the AMPS system over the previous mobile telephone designs.

In addition, the AMPS system provides space for 21 separate digital control channels, which are used to allocate frequencies, initiate calls, manage traffic flow, and also perform many of the administrative functions for the service provider. Control channels transmit intermittently, interspersing the digital control content throughout the analog audio signal, in 100-millisecond bursts, which is much faster than the human ear can detect. The audio, however, is perceived as one continuous transmission, even though it is actually being chopped up between the occasional bursts of control channel data. The recipient remains completely unaware of the interruptions.

When the AMPS system was originally designed, it was set up to promote market competition. For this reason, half the available channels of a cell are referred to as A carriers, and the others are B carriers. The A carriers are owned by wireless service providers, and the B carriers are owned by wireline or landline service providers. Figure 13–20 shows the various frequency breakdowns.

The AMPS system uses analog FM modulation, allowing for a maximum 12 kHz bandwidth and a channel spacing of 30 kHz. The 12 kHz bandwidth provides a decent enough frequency response for the audio, and the 30 kHz channel spacing ensures that adjacent channels are not riding on top of each other or interfering. In reality, channels typically are not allocated any closer than seven channel spaces or 210 kHz, which is much wider than the 30 kHz guard band. Transmitter frequencies in adjoining cells typically are separated by at least 60 kHz to help prevent

Figure 13–20
North American cellular radio frequencies.

Base Frequencies (forward channels)	Mobile Frequencies (reverse channels)	Type of Channel	Carrier
869.040 to 879.360	824.040 to 834.360	Voice	A
879.390 to 879.990	834.390 to 834.990	Control	A
880.020 to 880.620	835.020 to 835.620	Control	B
880.650 to 889.980	835.650 to 844.980	Voice	B
890.010 to 891.480	845.010 to 846.480	Voice	A*
891.510 to 893.970	846.510 to 848.970	Voice	B*

Table denotes transmit carrier frequencies. Mobile transmits 45 MHz below base.
A = nonwireline carrier (RCC)   B = wireline carrier (TELCO)
* * = frequencies added in 1986*

© Cengage Learning 2013

additional interference. The modulation technique that controls the allocation of channels to subscribers is **frequency division multiple access (FDMA)**; it allows customers to be reassigned frequencies as needed due to problems of interference and availability.

The AMPS system has one main disadvantage compared to the newer digital systems: Security and privacy are virtually nonexistent. Because transmissions can be picked up and received by just about any FM receiver or scanner tuned to the correct channel frequency, eavesdroppers or radio hobbyists can easily breach any or all transmissions. In 1988, in an attempt to curb the problem, the sale of unauthorized frequency scanners to consumers was outlawed in the United States, although such devices are still widely available in most other countries around the world.

Issues of security and privacy were obviously a primary concern to engineers and designers when they set out to develop the newer generations of digital cellular and PCSs. As a result, such systems are now coded, involving a complex series of mathematical algorithms known as encryption. The use of encryption makes the unauthorized decoding and reproduction of cell phone transmissions nearly impossible to achieve.

Technical Description of Personal Communication Systems Cellular Systems

Digital PCSs represent the second generation of cellular telephone (2G), with analog AMPS being the first generation (1G). By the early 1990s, the 800 MHz frequency band had reached its limit, offering no additional room for expansion. As customers continued to subscribe daily to an already overburdened AMPS system, the digital PCS was ultimately born out of necessity. In addition, PCS also promised to solve many of the security problems and switching issues of AMPS.

As a result, the PCS established six new frequency bands between 1850 and 1980 MHz (Figure 13–21).

Higher transmission frequencies meant that portable antennas were smaller. The total number of available cells also increased, which ultimately reduced the total coverage area of a single cell. Smaller cells allowed for even lower power levels because of the reduced transmission range and the ability to control a greater

Figure 13–21

Personal communication systems frequency bands.

Allocation	Base Transmit (Forward Channel or Downlink)*	Mobile Transmit (Reverse Channel or Uplink)*
A	1850 to 1865	1930 to 1945
B	1870 to 1885	1950 to 1965
C	1895 to 1910	1975 to 1990
D	1865 to 1870	1945 to 1950
E	1885 to 1890	1965 to 1970
F	1890 to 1895	1970 to 1975

Frequencies are in MHz.

© Cengage Learning 2013

volume of traffic by placing fewer customers within any given cell at any one time. Portable handheld phones were the obvious intent of the new PCS. PCS was also designed to be all digital, providing customers with a cleaner transmission signal and the availability of far more options and service from the provider. The only disadvantage, however, is that North America currently provides service to three separate digital PCSs, all of which are currently incompatible with each other. One standardized system is still a future dream.

Additional benefits to digital PCS include the ability to send and receive text messages, e-mail, and pictures; limited Web browsing is also available, depending on what features the service provider is willing to support.

The electronic benefits of digital transmission include the use of less bandwidth, more voice channels in a given spectrum, security through digital encryption, error correction, less susceptibility to noise and dropout, and the use of time and code division multiplex schemes, which have proved to be more flexible and easier to switch than those of analog AMPS.

Time Division Multiple Access, Global System for Mobile, and Code Division Multiple Access

The three competing digital PCS systems include IS-136 **time division multiple access (TDMA)**, **global system for mobile (GSM)**, and IS-95 code division multiple access (CDMA). Each system uses a different form of modulation, and in all respects they are quite different from each other.

Time Division Multiple Access

TDMA involves the dividing of channels into time slots rather than assigning a subscriber an individual frequency, as is commonly done with FDMA. Analog signals are first converted into digital and then assigned time slots among multiple subscribers. The main difference is that subscribers in the TDMA system share pieces of a communication channel with other subscribers, but they do not have access to the entire channel, as is allowed in FDMA. Digital audio signals are organized into data packets and then assigned to an allocated time slot within the complete data stream. Control data packets also are included in the mix, making the use of extra control data channels unnecessary, which is why the entire system can function comfortably within the six allocated frequency bands, compared with that of AMPS, which requires the use of 21 separate control channels to manage and administer the system.

Let us now look at how the time slot concept works. Imagine that a train is passing a station, and every third car has your name on it. The other cars represent messages for other individuals and the system administrator. Inside of your car is part of a message. The next train to enter the station brings the next piece of the message, and once again, you have to go to car number 3 to retrieve it. Because PCSs transmit in the 1900 MHz frequency band, signals rates are much faster, which allows for the transfer of multiple conversations in a compartmentalized fashion. Ultimately, TDMA can accommodate more subscribers more efficiently than AMPS, greatly increasing the overall capacity of the entire cellular system.

The modulation technique used on the carrier frequency for a TDMA transmission is a form of phase-shift keying known as quadrature phase-shift keying (QPSK). The system also still uses FDMA to allocate a carrier frequency for use, but now multiple customers are riding on the frequency as opposed to a single user.

Global System for Mobile

GSM is a PCS used in Europe, most of Asia, and parts of North America. GSM is a type of TDMA, but it uses a Gaussian minimum-shift keying method of modulation. Such a method reduces the overall bandwidth between the two shift-key modulation frequencies (see Modulation section earlier in this chapter), making the system far more economical than those employing standard FSK methods. The Gaussian method allows for eight time slots, whereas the TDMA system allows only three; and because transmitters operate only within their allocated time slots, they are far more efficient and use less power.

Code Division Multiple Access

CDMA uniquely codes digital transmissions by using a technique known as direct-sequence spread spectrum (DSSS); such a method of data coding is extremely secure. DSSS in relation to wireless networks is discussed later in this chapter. After DSSS coding, the digital data packets are interleaved over time into a format similar to TDMA and then modulated out onto the carrier using QPSK. CDMA also used FDMA to allocate a carrier frequency from the multiple available channels within the 1900 MHz frequency band.

The entire process resembles the following sequence:

- Analog signals are digitized and sometimes coded or encrypted.
- Digital bits are then coded for transmission through a technique known as DSSS.
- Coded packets are then interleaved into time domains, similar to the process of TDMA.
- A carrier frequency is then allocated using FDMA.
- Data is modulated onto the carrier, using QPSK.

CDMA can accommodate up to 20 times more users than FDMA, and it is far more secure than TDMA because of the DSSS coding. Figure 13–22 presents a graphic comparison of the three multiplexing techniques—FDMA, TDMA, and CDMA.

The Third Generation of Wireless (3G)

The third generation of wireless (3G) was agreed on in 1999; it became known as the IMT-2000 standards (International Mobile Telecommunication). The specifications included three CDMA standards, one TDMA standard, and one FDMA standard. In 1992, the World Administrative Radio Conference added additional frequency bands between 1885 and 2025 MHz and also 2110 and 2200 MHz, to help support the changes and the need for additional capacity.

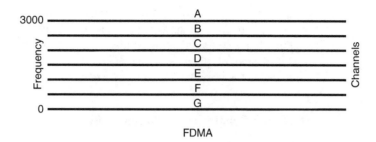

Figure 13–22

Frequency division multiple access (FDMA) versus time division multiple access (TDMA) versus code division multiple access (CDMA).

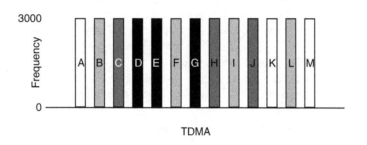

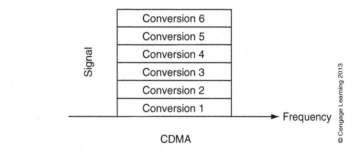

The three new CDMA standards—CDMA2000, W-CDMA, and TD-SCDMA—include greater capacity, adaptability, and standardization, ultimately making PCS more integrated and seamless.

CDMA2000 and W-CDMA both increase the bandwidth from 1.25 MHz to 5 MHz, and the data rate from 9.6 kbps to a maximum of 2 Mbps, making the system much faster, wider, and more secure. The difference between the two, however, is that CDMA2000 provides multicarrier service, up to a maximum of three, and W-CDMA provides only a single carrier.

TD-SCDMA differs in that it uses a technique known as time-division duplexing (TDD), which allocates the same RF channel for both the forward and reverse directions but uses different time slots for the digital packets.

The third generation of wireless includes a wide-band TDMA called UWC-136 (Universal Wireless Communications). UWC-136 includes a high-speed component that provides packet-data services for multimedia support at speeds of up to 384 kbps for moving vehicles and up to 2 Mbps for stationary reception or walking. Voice communication on this system is high-fidelity quality, comparable to that of landline phone systems. 3G also offers increased network security as compared to the previous 2G mobile system.

The Fourth Generation of Wireless (4G)

In 2008, the International Telecommunications Union responsible for radio communications (ITU-R) specified the IMT-Advanced requirements for 4G communications. Peak speeds for 4G were set at 100 Mbps for high mobility communications, and 1 Gbps for stationary or low-mobility communications.

4G communications provides comprehensive and secure mobile access to all-IP based wireless modems, smartphones, and other mobile devices. 4G systems are not only faster than 3G but can also provide voice communications and Internet access simultaneously, essentially providing all the functionality of a mobile laptop.

The method of transmission for 4G replaces the CDMA spread spectrum technology of 3G with orthogonal frequency-division multiple access (OFDMA) and other frequency-domain equalization schemes (see *802.11* in this chapter) combined with multiple-in multiple-out (MIMO), multiple antennas, dynamic channel allocation, and channel-dependent scheduling.

Picocells

Cellular reception is often unavailable inside of building structures or through underground concourses because of the electromagnetic shielding and polarization of signals, often resulting from design techniques and the type of building materials used during construction. A possible remedy to the problem is to install a picocell to help provide additional signal strength to those hard-to-reach radio-free dead zones or holes within building spaces. Essentially, **picocells** are small-range cellular repeater systems intended for indoor use that are installed to help bridge the gap between the outdoor radio signal and the shielded building. Elements of a picocell include an outdoor antenna, multiple indoor antennas, and a central repeater through which all signals pass. Picocell systems are often quite expensive to purchase and install, depending on the number of channels needed and the desired signal throughput; but in many cases, there simply are no other alternatives.

SEC 13.5   SATELLITE COMMUNICATIONS

Television stations first used satellites in the 1970s, primarily for the transmission of broadcast programming between network affiliates. It was common for a New York station to send programming to a California affiliate through a satellite feed early in the day. The affiliate station then rebroadcast the program later in the evening to accommodate the 3-hour time difference in the Pacific time zone. Today, satellite feeds between networks and news affiliates are still a common practice.

By the mid to late 1970s, cable companies also started using satellite as a means of receiving direct programming from various networks around the country. The local cable providers then modulated the individual satellite feeds on to new channel designations within their internal network for rebroadcast to customers over a multiplexed wideband coaxial connection. As an example, the local cable network might assign a satellite feed from broadcast channel 7 in Buffalo, New York, to channel 35, whereas channel 36 might be pulling in a feed from Atlanta, Georgia. In this way, cable companies were able to provide customers with dozens of channels and programming options instead of having to rely only on the standard three local networks: ABC, NBC, and CBS.

Satellite broadcasting made multichannel programming possible. Eventually, cable subscribers began to cut out the middleman by receiving direct signal feeds to their own personal satellite systems. And it was not long before business and industry began to use satellites for many industrial applications, including personal teleconferencing and data exchange.

Components of Satellite Communication

The components of satellite communication include a satellite, an **uplink** antenna, a **downlink** antenna, a feed horn, and a satellite receiver or transmitter (Figure 13–23). Three basic types of satellites exist with different functions and uses; they include **geosynchronous satellites (GEOs), low-earth-orbit satellites (LEOs)**, and **medium-earth-orbit satellites (MEOs)**. Let us first look at the basic operations of GEO satellites, which are more commonly used by television broadcast and cable companies (see later for a discussion of LEO and MEO systems).

Geosynchronous Satellite

A GEO is a multichannel, multifrequency transponder that rotates 35,880 km (22,300 miles) above the Earth in a geosynchronous orbit. The distance from the Earth to the geosynchronous orbit is the exact location where the centrifugal forces pushing the satellite out of orbit and the gravitational forces pulling the satellite down toward Earth are equal and balanced. The achievement of a geosynchronous orbit usually guarantees that the satellite remains in a fixed elevation above Earth, while appearing to be stationary from a ground perspective because of the speed of orbital rotation (Figure 13–24).

Transponder

A **transponder** is a repeater that responds to incoming transmissions of narrow beam, radiated microwave. The following sequence of events illustrates the process of communicating over a satellite transponder.

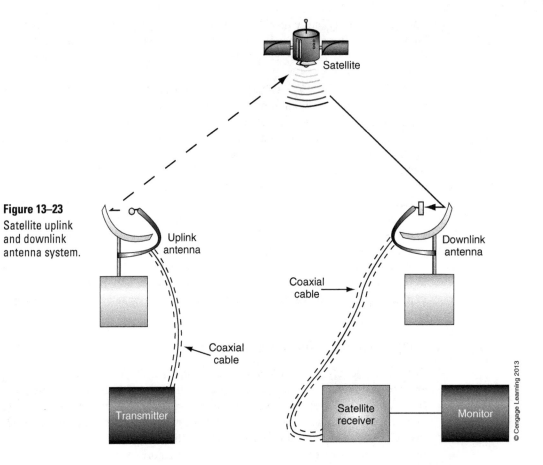

Figure 13–23
Satellite uplink and downlink antenna system.

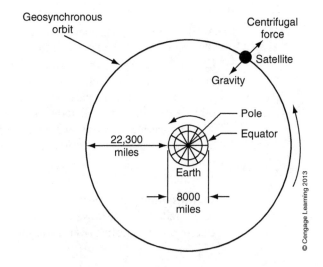

Figure 13–24
Geosynchronous orbit.

- An Earth station antenna transmits, or uplinks, microwave signals to an orbiting satellite.
- The transponder receives the signal on a specified uplink frequency channel.
- The transponder amplifies the received signal.
- The amplified signal is frequency shifted to an alternate channel to allow duplex communication.
- The transponder then retransmits a downlink back to Earth on the new channel frequency.

Types of Uplink/Downlink Configurations

Three general configurations of uplink and downlink ultimately determine the pattern or footprint of the returning microwave radiation. The footprint refers to the coverage area of a microwave signal as it returns to Earth from the transponder. The three configurations include point-to-point, point-to-multipoint, and multipoint-to-point.

Point-to-Point. Commercial communication carriers use a point-to-point configuration as a means of sending signals between two locations, from a single Earth station to another single Earth station. The use of narrowly focused, spot-type beams makes point-to-point an exclusive communication link. Such a connection is not, however, recommended where broadcasting to the general public because large areas of the country remain outside of the narrowly shaped footprint.

Point-to-Multipoint. Broadcast television, satellite, cable television, and radio networks often use a point-to-multipoint configuration as a means of providing a wide-beam downlink to multiple access points across the country. An example may include station TBS in Atlanta, which provides an uplink of their programming to a satellite transponder that then beams the signal down over a broad coverage area of the United States. Multiple affiliate cable systems around the country can then receive the TBS signal from just about any location, because the size of the footprint often is large enough to accommodate the reception.

Multipoint-to-Point. Multipoint-to-point involves the use of very small aperture terminal (VSAT) systems (small, portable, low-power satellite systems), typically used by mobile or fixed station sales personnel who are often spread out around the country and regularly in need of a communication link back to the home office. Multipoint-to-point also allows users of satellite-link telephone systems to connect to a common service provider from just about anywhere on the planet. Often, such systems use LEO or MEO satellites to interconnect their communication link.

Orbital Slots

An orbital slot defines the geosynchronous position a satellite might occupy in the sky for transmission to and from an Earth station antenna. From an Earth station antenna point of view, all satellites in geosynchronous orbit (GEO-type satellites) appear to arc across the sky in fixed free-floating positions. The longitudinal coordinate of an individual orbital slot defines the position of a specific satellite. See Figure 13–26 for the current listing of satellites and longitudinal coordinates of their orbital slots within the Western hemisphere.

So how do you get authorization to launch and orbit a satellite, and who chooses the orbital slot? This question actually has two parts. First, an international agreement between countries must define the ownership of a given arc in space. Once ownership of longitudinal coordinates has been decided, a regulatory body within each country is then responsible for licensing and assigning orbital slots within a given space. Orbital slots are a finite commodity that governing bodies must regulate and maintain to guarantee that satellites stay a required safe distance from each other and do not interfere with or adversely affect neighboring transmissions.

The FCC is the regulatory body of the United States that oversees the licensing of orbital slots and the frequency allocation of satellite transponders. The FCC also defines specific classifications for GEO, LEO, and MEO satellites, which often are based on the type of use and technical characteristics.

For purposes of discussion, let us look at the classifications for GEO television satellites and see how they affect issues of power, orbital slots, and frequency allocation.

For GEO satellite television reception, two classifications exist, those defined as fixed satellite service (FSS), often used by various broadcast and communications companies around the globe, and direct broadcast service (DBS), which provide broadcast television reception directly to homes using television receive only (TVRO) satellite systems. Examples of TVRO include Direct TV and Dish Network.

As the name implies, TVRO systems do not provide transmission or uplink capabilities. In cases where the customer may need to communicate back to the service provider, as is customary when ordering pay-per-view movie channels, for example, landline telephone connections provide the necessary communication link, over standard data modems built into the back of the receiver (Figure 13–25). The service provider authenticates the order by first determining the electronic serial number of the TVRO receiver. The satellite then downlinks the decryption codes to the waiting receiver, but only after the service provider has verified the order. The entire process takes a matter of seconds.

Signal Strength, Dish Size, and the Spacing of Orbital Slots

The total output power for a given satellite transponder determines the necessary spacing of orbital slots and the minimum size of an Earth station antenna. If orbital slots are too close, interference between neighboring satellites transmitting on the same frequency may occur. DBS orbits, for example, are more widely spaced because they are required to output greater power levels for reception on small TVRO dishes, typically 18 in. in diameter. The size of the dish directly affects the level of measured signal input. As a result, DBS systems output greater power levels to keep dish sizes small. In comparison, C- and Ku-band satellites, classified as type FSS, typically output lower power levels, allowing for more closely spaced orbital slots. For these systems, dish sizes must be considerably larger to ensure that signal-to-noise ratios are high enough; otherwise picture quality may suffer. Dish sizes of 8 to 12 ft (2.4 to 3.7 m) in diameter are common for C- and Ku-band systems.

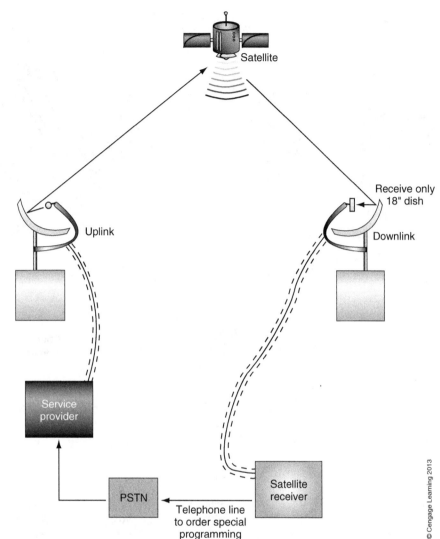

Figure 13–25
Television receive only satellite system. PSTN, Public Switched Telephone Network.

The orbital slots for GEO satellites classified for FSS include 62° to 103° and 120° to 146° west longitude for C-band satellites, and 62° to 105° and 120° to 136° west longitude for Ku-band satellites. Digital DBS has space allocated at 61.5° west longitude for the eastern United States; 101°, 110°, and 119° west longitude for the entire United States; and 148°, 157°, and 175° west longitude for the western United States. Figure 13–26 shows the current positions of various C, Ku, and DBS satellites.

Satellite Frequencies, Channels, and Types of Polarity

The transponders on each satellite accommodate a specific range of uplink and downlink frequencies, depending on classification and purpose of use. Figure 13–27 shows some of the typical allocations for a C-band satellite.

Figure 13–26
North and South American orbital slots.

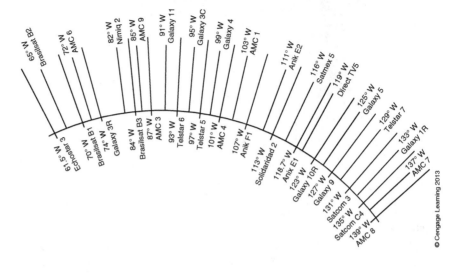

Figure 13–27
C-band transponder frequencies.

Uplink frequency (MHz)	Downlink frequency (MHz)	Polarization	Channel number	Alternate numbering system
5945	3720	H(V)	1	1A
5965	3740	V(H)	2	1B
5985	3760	H(V)	3	2A
6005	3780	V(H)	4	2B
6025	3800	H(V)	5	3A
6045	3820	V(H)	6	3B
6065	3840	H(V)	7	4A
6085	3860	V(H)	8	4B
6105	3880	H(V)	9	5A
6125	3900	V(H)	10	5B
6145	3920	H(V)	11	6A
6165	3940	V(H)	12	6B
6185	3960	H(V)	13	7A
6205	3980	V(H)	14	7B
6225	4000	H(V)	15	8A
6245	4020	V(H)	16	8B
6265	4040	H(V)	17	9A
6285	4060	V(H)	18	9B
6305	4080	H(V)	19	10A
6325	4100	V(H)	20	10B
6345	4120	H(V)	21	11A
6365	4140	V(H)	22	11B
6385	4160	H(V)	23	12A
6405	4180	V(H)	24	12B

Notice that the uplink frequencies are nearly 2 GHz away from those of down-link. The frequency separation is required to ensure that interference is not possi-ble. The separate frequency bands for uplink and downlink also provide the means of full-duplex communication.

A typical C-band satellite includes 24 transponders, all on separate channel fre-quencies within the 4 to 6 GHz band, depending on whether they are providing uplink or downlink communications. Each transponder has a 36 MHz bandwidth with a 4 MHz guard band. The guard band is necessary to separate neighboring channels from possible interference. Transponders can receive and transmit any type of signal, digital or analog; and because the transponder is nothing more than a repeater station, it retransmits, in exact form, any signal it receives. In essence, a satellite is simply a relay station that bounces signals from the uplink to the downlink as requested.

Polarization

A technique known as signal polarization is used to help minimize interference between adjacent transponders. **Polarization** can take the form of either a linear vertical or a linear horizontal orientation. The rotational angle of the pickup ele-ment inside of the antenna feed horn determines the polarity of the uplink or down-link. In most cases, antennas have dual elements installed, one being horizontal and the other vertical, making polarity selection a simple matter of switching between the two, as needed. As long as the polarity of the antenna element matches the polarity of the radiated microwave signal, communication between devices is pos-sible. Mismatched polarities, however, cause confusion and loss of signal.

DBS systems often use an alternative method of polarity referred to as cir-cular polarization. Circular polarization involves the simultaneous injection of a signal to the vertical and horizontal elements of a transmitting antenna, with one of the injected signals shifted 90° relative to the other. As a result, signals radiating out from the antenna begin to spin in either a left- or right-hand orienta-tion. Circular polarization does not then require a pickup element to be oriented in an exact position, as in the case of horizontal or vertical polarity, because the spinning waveform provides an equal level of reception at all angles. The main benefits to using circular polarization have to do with the conservation of space and added system capacity. Because some transponders allow for the simultaneous transmission of left- and right-hand polarizations, then theoretically, the number of available channels a satellite may support doubles. The downside of circular polarization, however, is that transmissions lose up to half their original intensity, or −3dB, where detected at the antenna feed horn. This is just another reason why DBS transmissions require greater power levels and wider orbital slots to accom-modate the eventual losses; remember too that the system also needs to support the signal power requirements for small-dish antennas to ensure decent enough signal-to-noise ratios.

Areas of Service

The area of service provided by a satellite greatly depends on the shape of the microwave beam as it leaves the transponder. Some transponders send out wide beam patterns that cover vast areas of the United States, whereas others may only send narrowly focused spot-type beams that tend to hit specific locations. The following list defines some common beam shapes and coverage possibilities:

- **Spot:** Spot beams cover a limited area, providing point-to-point connections between private parties. Television broadcasting rarely uses spot beams because they are too limited in their coverage area.
- **National:** National service areas require the use of wider beams to cover a more significant portion of the country. Television or radio providers use such beams for multipoint broadcasts.
- **Regional:** Regional involves beams that cover groups of entire countries, all-inclusive; an example is Europe or Central America.
- **Global:** A global beam covers the entire viewable area of the Earth, as seen from a transmitting satellite. Companies involved in international communications provide global coverage.

Direct communication between satellites is also possible where linking long-distance signals. In special cases where an Earth station uplink does not have a direct line-of-sight to a specific satellite, then a neighboring satellite may be used to direct relay a signal to an intended target. A single uplink and downlink connection between two Earth stations' antennas then must pass through a minimum of two satellites to accomplish the transmission (Figure 13–28).

The Satellite Dish and Alignment

A satellite dish is a microwave antenna. It provides the necessary means to uplink or downlink a signal to and from a satellite from an Earth-based station. The dish consists of a parabolic- or quasi-parabolic–shaped reflector and a feed horn. The antenna element, located inside the feed horn, can either transmit or receive microwave signals depending on the type of system. The feed horn structurally mounts within the focal point of the parabola, where the greatest concentration of microwaves exists as they reflect off the dish.

An uplink transmits a microwave signal from the feed horn to the satellite by bouncing the transmission off the reflecting dish and up to the receiver portion of

Figure 13–28
Crosslink satellite transmission, using two or more satellites.

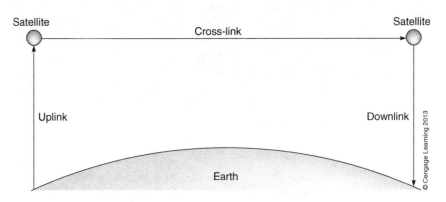

© Cengage Learning 2013

the transponder; in such cases, the dish acts as a reflecting mirror when transmitting the microwave signals. A downlink performs the function in reverse by collecting and concentrating microwave signals from the dish to the feed horn located at the focal point of the parabola.

Antenna alignment is critical to the quality of reception. The installer of a satellite dish must be sure to mount it vertically plumb and at the correct angle to the horizon. The technician must also adjust the dish azimuth (side-to-side positioning) to ensure that it is on target with the desired satellite and receiving the strongest possible signal. An installation technician is able to locate the optimum position for azimuth and angle adjustment by using a signal strength meter. A signal strength meter makes system fine-tuning possible by allowing the technician to see signal response in relation to the mechanical adjustments made to the dish; a peak signal indicates optimum positioning. If a signal strength meter is not available for use, the satellite receiver usually has a digital one located within the setup menu of the device. The downside to using it, however, is that a television monitor is needed when viewing it. In most cases, this could be a bit problematic because dishes are not often within viewing range of the monitor, making critical adjustments nearly an impossible task to achieve. The portable variety is far easier to use, allowing technicians to make the critical mechanical adjustments more precisely.

The line-of-sight from the feed horn to the satellite in the sky should also be clear and free from obstruction. Any physical barrier placed between the satellite and the antenna might adversely affect or block signals. Trees blowing in the wind often can reduce signal intensities if they are obstructing the pathway between the dish and the satellite. Small antennas and low-power systems are especially susceptible to interference, VSAT being one possible example.

In a downlink system, the LNB converter is the pickup element located inside the feed horn of the antenna. A 75 Ω coaxial cable is typically the connection between the LNB and the satellite receiver. The cable is wire tied to the support structure of the dish, if not otherwise placed inside of a metal raceway or conduit. For antennas mounted on rotational motors, or rotators, which allow the dish to be moved to various longitudinal positions across the geosynchronous arc, it is important to make sure that the connecting cable is not too tight and also that it has enough slack to move freely with the motion of the antenna. Specifications for coaxial cable should also match the specifications of the LNB, as determined by the manufacturer, or signal quality may be degraded. Characteristic impedance, frequency response, and signal attenuation are all cable specifications (see Chapter 2) that can make or break a system. Do not ever randomly choose coaxial cable based on appearance; to do so would most likely put system performance at risk. When in doubt, always consult with the manufacturer for a recommendation of cable type and optimal length to ensure the best performance of the system.

Other connecting cables to the satellite dish include power for the main dish rotational motor and power for the skew and polarity servomotors, which are used where fine-tuning and setting the rotational polarity of the LNB inside the feed horn.

At the head end of a receiving system, microwave signals reflect off the dish and onto the LNB for preprocessing and signal amplification. The LNB then down-converts the incoming high-frequency microwave to a lower frequency, somewhere in the range of 950 to 1450 MHz, for transmission through the coaxial cable to the satellite receiver.

The central conductor of the coaxial cable provides power to the LNB from the satellite receiver. Be sure to take special care where attempting to split signals off from an LNB to various other devices, such as spectrum analyzers or signal strength meters, because the unexpected presence of additional dc voltage on the line could potentially damage unsuspecting circuits. To prevent such an occurrence, be sure to add an in-line dc block to safely contain and isolate the LNB power from other attached devices. The splitting of satellite signals to multiple receivers, however, is not an option, because polarity adjustments made to the orientation of the LNB where selecting a channel do not allow the simultaneous viewing of neighboring channels, or those having opposite polarity. Where connecting multiple receivers to the same antenna, the feed horn must contain additional elements, one for each receiver. A separate coaxial cable connects to each element and then back to an individual receiver to ensure that channels are isolated and that polarity mismatches do not occur.

Satellite Receivers

Satellite receivers are similar to radios in that their main purpose is to tune in specific transponder frequencies. The receiver also acts as a television receiver by processing the incoming video and audio signals for line output connection to a monitor and audio amplifier; an RF output modulated on channel 3 or 4; or even optical and digital outputs, compatible with the newer digital high-definition television (HDTV) or digital audio formats.

In addition, the satellite receiver also performs the basic functions of channel ordering and decryption, which are necessary for the authorized viewing of scrambled or blocked analog or digital programs. As stated earlier, decryption codes download to the receiver from the transponder only after the service provider has verified the order and the electronic identification numbers of the receiver over a standard telephone modem connection. The satellite receiver includes an RJ-11 modem connector on the back of the device for connection to the telephone line.

Receiver identification codes are also changed routinely by the service provider to make sure individuals are not stealing service. The user identification card placed inside the front or back slot of the receiver acts as a programmable memory that can be changed as often as needed by the service provider to ensure that only valid receivers are online. The system occasionally checks receivers for authorized use during paid programming; otherwise, the service provider may turn off the system through a remote downlink command.

Propagation Times

The use of satellite communications ultimately causes a signal delay because the transmission must travel up to the orbiting satellite and then back down the Earth station receiving antenna. The following equation calculates the round trip propagation time of the signal:

$$t = d/v$$

t represents the time of the round trip propagation in seconds; d is equal to the round trip propagation distance in meters, from the uplink antenna to the satellite, and then back to the downlink antenna; and v is the velocity of propagation for the microwave signal. The v value is a constant, equal to 300×10^6 m/sec, which is the approximate velocity of propagation for microwave signals traveling through the air.

Example. How long does it take a signal to travel 50,000 km from an uplink antenna, and then 80,000 km to the downlink antenna?

$$t = (130 \times 10^6 \text{ m})/(300 \times 10^6 \text{ m/sec})$$

$$t = 0.433 \text{ second}$$

In this example, the delay is quite noticeable because the human ear can easily detect a 0.43-second pause (Figure 13–29).

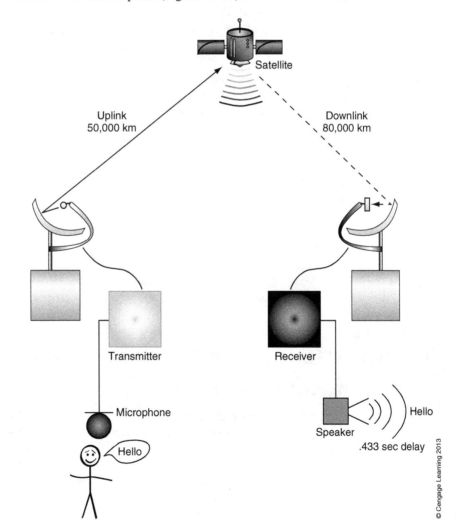

Figure 13–29
Satellite
propagation
delay.

© Cengage Learning 2013

Low-Earth-Orbit Satellites

LEO satellites are not stationary. Instead, they are moving continuously around the north and south geographical poles of the Earth, in a low orbit often at an altitude of 300 to 1500 km (186–932 miles). The benefits of using LEO include lower power and shorter signal delays, making them ideal for voice communications and cellular phone usage. A LEO satellite connection can achieve a real-time communication link from any point on the globe, with virtually no signal delay, provided that a line of sight exists between the Earth station antenna and the LEO. The LEO then relays a received communication directly to another LEO or to an available service provider, depending on which is closer. LEO satellites work much like cellular phones in the manner in which they transmit and receive signals. A third benefit to using LEO is that they are readily accessible in out-of-the-way locations such as deserts or onboard ships at sea, where standard cellular phone systems are not available.

As already stated, signal delays for LEO satellites are virtually nonexistent. Assuming that a LEO has a 1700-mile round trip between uplink and downlink, the propagation delay for the signal is about 1.4 milliseconds. In comparison, a GEO satellite, having a minimum propagation distance of 44,500 miles, round trip, depending on the number of hops and how close the uplink and downlink are from the transponder, typically sees signal delays of 1 to 2 seconds. For this reason, LEO satellites are obviously best suited for duplex voice communication.

Medium-Earth-Orbit Satellites

MEO satellites operate somewhere between LEO and GEO satellites, typically at altitudes of 8000 to 20,000 km (4970–12,427 miles). As is expected, because of the higher orbit and larger orbital circumference, MEO satellites require more devices in orbit than do LEO satellites to accomplish a smooth communication link between parties. Signals transmitting on MEO satellites may require more hops to achieve their final destination. Power requirements for MEO satellite transmissions are also higher because of the higher altitude, forcing hardware to be larger and less portable in comparison with the much faster and smaller, low-power LEO systems (Figure 13–30).

SEC 13.6   WIRELESS COMPUTER NETWORKS

Wireless Local Area Networks

Many businesses and organizations are now installing wireless local area networks (LANs) as a low-cost alternative to expanding their existing fiber or copper wire networks. For many years, companies were manufacturing proprietary wireless products that operated in an unlicensed frequency band of 900 MHz. Data rates for such systems were often 1 to 2 Mbps. By the early 1990s, additional wireless products were being produced, operating at much faster rates in an unlicensed band 83 MHz wide, up around 2.4 GHz. The wireless industry referred to it as the

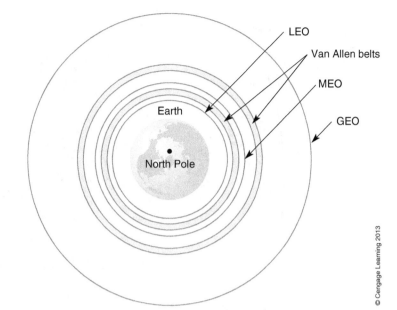

Figure 13–30
Geosynchronous satellites (GEO) versus medium-earth-orbit satellites (MEO) versus low-earth-orbit satellites (LEO) orbits (not drawn to scale).

industrial, scientific, and medical (ISM) band. Standardization, however, did not exist, and unfortunately it was often impossible for devices of competing manufacturers to intercommunicate as a unified system. By 1998, the Institute of Electrical and Electronics Engineers (IEEE) approved the *802.11* standards that finally put a framework around wireless LANs and defined the communication environment so that the chaotic mess of dissimilar manufacturing products could begin to speak the same language.

In addition, the need to communicate over wireless to peripheral devices began to show great promise. As a result, many portable, low-power, wireless products were being produced that could not only communicate with each other but also share files over short distances with local networks and personal computers. The wireless standard named **Bluetooth** provided such a connection for short-range communications, up to 10 m. Examples of Bluetooth products now include personal digital assistants (PDAs), cell phones, storage devices, digital cameras, and printers. This is obviously not a complete list, but it illustrates the range of possibilities. The next section discusses the basic concepts of wireless LANs, *802.11*, and Bluetooth technology.

The Components of Wireless Networking

Wireless networking allows computers and peripheral devices to communicate within a LAN or a WAN through an RF link. The basic elements of a wireless network include an access point, an antenna, and a network interface card (NIC). The access point provides a hardwired connection to the wired network while also acting as a hub and transmitter for the wireless network (Figure 13–31). The antenna transmits and receives the wireless signals within a given coverage

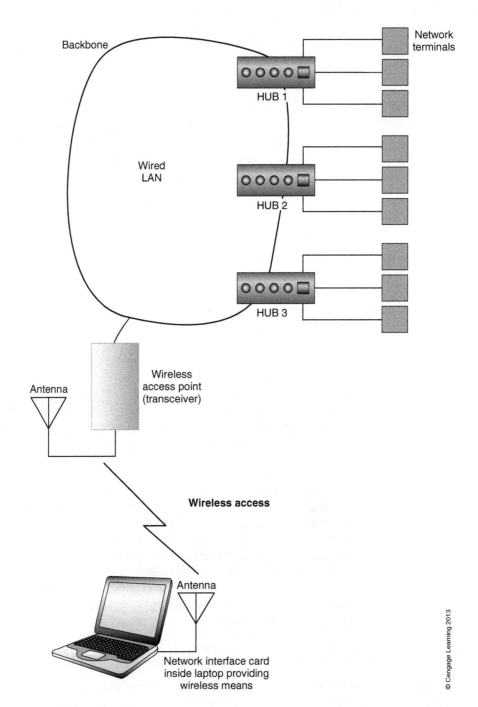

Figure 13–31
Wireless networks.

© Cengage Learning 2013

area. Some buildings or structures may often require the installation of multiple antennas to cover hard-to-reach spaces or radio-free dead zones. The NIC provides the same functionality of an Ethernet card, except through wireless means, by using either RF or infrared transceivers. Although RF can pass though walls and permeate most materials and building structures, infrared communications can only transmit or receive in either a direct line of sight or indirectly by bouncing off walls and ceilings. The user must be aware of the inherent speed, range, and frequency limitations of each type of product to be able to interface them effectively with existing equipment.

In 1998, the IEEE accepted the specifications and standards for wireless networks as defined by *802.11*. Four specification levels exist within the *802.11* family; they are *802.11, 802.11a, 802.11b,* and *802.11g.*

802.11 (Wi-Fi)

802.11 became the initial standard for wireless communication over a LAN, providing 1 to 2 Mbps transmission at 2.4 GHz. The regulation allows for either a frequency hopping spread spectrum (FHSS) or a DSSS transmission, to help conserve bandwidth, reduce the risk for interference from electromagnetic noise and neighboring frequencies, and increase the security and privacy of the signal. The *802.11* standard also specifies a carrier-sense multiple-access with collision avoidance (CSMA/CA) protocol, similar to that of wired Ethernets (see Chapter 6 in this book), to help control the flow of data and the number of collisions. The only difference is that the wireless version of CSMA/CA operates in a half-duplex mode, which is unable to detect collisions once the transmission has begun. The next two sections give a brief description of FHSS and DSSS.

Frequency Hopping Spread Spectrum FHSS transmits on multiple channels over a wide band of frequencies. As the name implies, a signal hops from one frequency to another, rapidly, not staying on any one channel for more than a few milliseconds at any one time. Although the hops give the appearance of being random, they are instead a carefully timed and preprogrammed sequence of events that the receiver must then tune to, in exact order, to intercept and reconstruct any meaningful intelligence from the transmission. If a technician views on a spectrum analyzer the entire band of frequencies within a hop sequence, the FHSS transmission appears as simple background noise because of the scattering and randomizing of the signal. An analogy is taking 10 individual 5 × 7 photographs, cutting them each into 1000 squares, shaking all of the squares in a bag, and then reassembling them randomly into one large rectangle. After having done that, now try to find picture number 1. Attempting to view an FHSS transmission on a spectrum analyzer is like trying to locate picture number 1. As a result, FHSS transmissions are more private and secure. They also offer more transmission capacity over a given bandwidth because issues of interference are virtually nonexistent to neighboring channels that are unable to recognize a frequency hop as anything other than simple noise. Where using frequency hopping, the modulation scheme is typically FSK (see "Modulation"). To help ensure that interference is not possible, the FCC regulates the sequences of hops, the number of hops, and the hop rates of FHSS transmissions. The requirements allow for a minimum of 75 hops per transmission and a hop rate of not less than 2.5 hops per second, with an output power ranging from 100 mW to a maximum of 1 W. The hop rate refers to the time between hops. The *802.11* standard typically uses 79 separate channels, 1 MHz apart, visited at a rate of 10 per second; this is slightly more than the FCC requirements.

Direct-Sequence Spread Spectrum DSSS is a method of coding digital bit streams before modulating them out over a wireless carrier. DSSS maps each data bit into a higher frequency code, or pseudorandom noise pattern known as a chip, or chipping code. The chipping code appears to be random 1s and 0s but is instead a carefully constructed repeating pattern, unique to a specific transmission. The chipping rate typically is 10 times faster than the rate of the actual data bits; this specification is referred to as the spread ratio. Because spread ratios are high, bandwidths for DSSS are often wide, typically in the range of 5 MHz, requiring the use of wideband receivers that can continuously monitor a broad range of frequencies. Whereas the FCC requires a spread ratio of 10 or more, the *802.11* standard requires a spread ratio of 11 for wireless networks. On the receive side of the communication link, the receiver must know the pattern of the chipping code to be able to extract the data bits. The process also requires a technique known as autocorrelation, which involves multiplying the incoming signal by a secondary signal generated from the chipping code. The level of math involved to perform such calculations is well beyond the scope of this book. A simple analogy to how the chipping code works is to have a room crowded with individuals all speaking different languages. The languages you do not understand appear as noise, whereas the language you do understand is clearly recognizable. Theoretically, you should be able to identify the person speaking your language even if the room contains a thousand unrecognizable conversations.

Compared with FHSS, DSSS offers a higher degree of privacy and security and is relatively free of interference from other signals. An individual attempting to intercept such signals must have an intimate knowledge of the codes and of computer programming to be able to accomplish such a task. PCS cellular phone systems operating on CDMA often use DSSS for the added security. DSSS also provides additional capacity to the bandwidth, allowing more users to share the entire frequency band, rather than simply providing individual customers access to only small, allocated time slots.

Another consideration when comparing FHSS with DSSS is that DSSS tends to offer a greater range of transmission, typically 100 to 200 m. System performance, however, often has more to do with the type of environment and the level of signal interference than it does with the distance of the transmission. The various types of objects within a given space affect the quality of a transmission, depending on the varying degrees of conductivity they may exhibit. As a result, cluttered environments, having many reflective surface areas, tend to favor FHSS over DSSS.

Interference from other sources of electromagnetic radiation may also degrade or possibly block the transmission of signals. Microwave ovens, for example, tend to operate in the 2.4 GHz range. Were it not for the use of spread-spectrum techniques such as FHSS or DSSS, wireless networks and microwaves would not be able to coexist within the same environment. This does not mean, however, that

FHSS and DSSS are able to prevent interference from all types of high-frequency electromagnetic waves. Eventually, a high-powered, high-frequency waveform is able to swamp out a 1 W wireless communication by degrading the signal-to-noise ratio to a point were reception is simply not possible.

Eventually, the IEEE added extensions to the *802.11* standards to accommodate faster transmission speeds and higher frequencies. The extensions are known as *802.11a, 802.11b,* and *802.11g.*

802.11a

802.11a is an extension of *802.11* that applies to higher speed LANs. *802.11a* supports data rates up to 54 Mbps at 5 GHz. Transmissions, however, do not involve FHSS or DSSS, but instead use an orthogonal frequency division multiplexing (OFDM) encoding scheme. Orthogonal simply means that transmitters are never at the same place at the same time, which otherwise causes interference.

802.11b

802.11b is an extension of the *802.11* standard that the IEEE ratified in 1999 for wireless LANS, allowing them a level of functionality comparable with the Ethernet. Also known as Wi-Fi, or *802.11* High Rate, the new *802.11b* can provide a transmission rate of up to 11 Mbps at 2.4 GHz. Fallback transmission speeds of 5.5, 2, and 1 Mbps are also available to be compatible with earlier systems. *802.11b* only uses DSSS as its method of transmission encoding.

802.11g

802.11g is an extension of *802.11* that applies to wireless LANs having minimum data rates of 20 Mbps, operating at 2.4 GHz. To be able to obtain faster computer speeds, *802.11g* uses an OFDM-encoding scheme, just as *802.11a* does. *802.11g* also can operate at data rates of 11 Mbps to be backward compatible with *802.11b*.

Figure 13–32 outlines the differences within the *802.11* family of wireless network standards.

802.11n

802.11n is an amended improvement to the *802.11* standard that adds multiple-input multiple-output antennas (MIMO). MIMO operates on both the 2.4 GHz and 5 GHz frequency bands, providing up to four transmission streams, using an OFDM method of modulation.

Bluetooth

Bluetooth is a wireless communication specification designed through a joint venture of Ericsson, IBM, Intel, Nokia, and Toshiba. The Bluetooth standard provides a short-range communication link between peripheral devices and a computer, typically 10 cm to 10 m apart. If need be, the transmission range could be increased up to 100 m through the use of additional RF amplification. To do so ultimately

	Maximum data rate*	Frequency	Transmission methods
802.11 300 ft maximum	2 Mbps	2.4 GHz	Frequency hopping spread spectrum or Direct sequence spread spectrum
802.11a 150 ft maximum	54 Mbps	5 GHz	Orthogonal frequency division multiplexing
802.11b 300 ft maximum	11 Mbps	2.4 GHz	Direct-sequence spread spectrum
802.11g 300 ft maximum	20 Mbps	2.4 GHz	Orthogonal frequency division multiplexing
Bluetooth version 1 33 ft maximum	Full-duplex 432 kb/s ↔ or 721 kb/s → 57.6 kb/s ←	2.45 GHz	Frequency hopping spread spectrum

Figure 13–32
802.11 and Bluetooth specifications.

* Data transmission rates depend on the distance between devices. Transmission speeds decrease as devices are spaced farther apart.

© Cengage Learning 2013

undermines the purpose of the standard because Bluetooth was originally designed as a communication link between small, inexpensive devices. Any additional amplification requires larger power supplies and bulkier devices. Notebook computers, PDAs, cellular phones, peripheral devices, modems, and loudspeakers are all examples of products now using Bluetooth technology. As one might expect, all devices containing Bluetooth can mutually communicate.

Bluetooth is similar to *802.11* in that it operates at 2.4 GHz, using a frequency hop spread spectrum of 79 channels, each 1 MHz apart, modulated by a two-level method of FSK. Bluetooth also allows for both audio and data transmissions, making it somewhat different from wireless Ethernet systems. The system can accommodate up to three audio streams, coded at 64 kbps operating in full-duplex mode. Data streams can flow bidirectional at 432 kbps; or in one direction at 721 kbps and the other at 57.6 kbps. The second data method is more useful where performing asymmetrical transfers between systems, where the slower 57.6 kbps side of the stream is mainly used to communicate system acknowledgments rather than actual data.

Bluetooth supports full-duplex communication through a procedure called TDD, which is similar to the procedure used in TDMA and GSM cellular phone systems. Bluetooth also can be used to set up wireless LANs; although the data speeds are not as fast as *802.11*, the cost is considerably lower than wireless Ethernet, only tens of dollars per Bluetooth device as opposed to hundreds of dollars for wireless Ethernet.

A Bluetooth network is called a piconet. Piconets can operate from two to eight nodes, or as bridges between other piconets. A master–slave relationship must exist between the various nodes to achieve communication. The master node controls the signal transfer, whereas the slave nodes must respond to the various commands. In a Bluetooth network, no single node acts as a constant master. Instead, the node initiating the communication temporarily becomes the master, forcing all others to act as slaves. The job of the master node is to control the frequency hops and the transfer of data between Bluetooth devices. The master node then relinquishes control of the network on completion of the data transfer.

Bluetooth Version 2 In 2004, an improved version of Bluetooth, version 2, was released, having a faster rate of data transfer. Version 2 is backward compatible with version 1 and can transmit up to 2.1 Mbps. Soon after, in 2007, version 2.1 was released, which not only helped to reduce the power consumption but also added extended inquiry response (EIR), which provides more information data during an inquiry to allow better filtering of devices prior to connection.

Bluetooth Version 3 +HS In 2009, version 3 of Bluetooth was released, which offered a high-speed communication of up to 24 Mbits/sec over Wi-Fi. Although the Bluetooth link is not actually transmitting at such speeds, it works in combination with *802.11* to achieve such rates. The initial network connection and profile configurations are still transmitted over the Bluetooth connection, but the larger quantities of data are now handled by the Wi-Fi. Therefore, a co-connection with the wireless Internet and Bluetooth helps to facilitate the desired high-speed data transfer rates.

Other Devices

802.11 and Bluetooth are just some of the current standards for wireless communication across a network or between peripheral devices and a main computer. Many more standards and devices do exist, however, that are similar in nature to what has already been discussed in this chapter. Other products such as wireless bridges and wireless modems may operate at 900 MHz or 2.4 GHz. The transmission range and speed of these devices may not always be sufficient where interfacing them with current system hardware. In addition, coding schemes and communication protocols may be different, depending on the type of product, which means that the user typically needs to install additional software and drivers before using them. Always be sure to check the manufacturer's specifications where attempting to integrate dissimilar devices to be sure of system compatibility.

Infrared Devices

As an alternative to RF, networking and peripheral devices can also transmit over infrared light waves. Although speeds are a bit slower, typically in the range of 4 Mbps or less, infrared does offer a nice low-cost alternative to RF. The only disadvantage to using short-range infrared is that although it has the ability to bounce and reflect off surfaces, it cannot pass through walls, floors, or ceilings, or to other parts of a building structure. As a result, the infrared transmitter and receiver must be within the same room, and in most cases a direct line-of-sight connection is optimal for achieving maximum transmission speeds. Any obstruction blocking the pathway between the transmitter and the receiver often causes a communication failure, especially if an alternate path for the signal is not available. Signals bouncing off walls and ceilings may eventually make their way to the receiver, but only after a short delay. Direct line-of-sight communications is therefore a more optimal and faster choice where using infrared. For more information, see Chapter 12 in this book for a discussion on the basics of infrared technology and sensors.

SEC 13.7 CABLE TELEVISION SYSTEM

Figure 13–33 illustrates a typical cable television system. There are three basic parts to a cable television system: the head end, the distributions trunk and feeders, and the subscriber drops.

Head End

The head end is where the cable provider receives its main programming. The satellite is used to pull in national networks and movie channels, such as TBS, CNN, MTV, HBO, Showtime, and so forth. The antenna and microwave inputs are used to pull in the local networks' broadcasts as well as any other community television feeds, including local high schools or broadcasts from the local government. Special programming such as pay-per-view or special events programming can be inserted into the system by satellite, or it can be played internally and placed within the system by the cable provider.

As stated earlier in this chapter, individual channels received at the head end of the system are shifted to different carrier frequencies and multiplexed out onto the provider's trunk line. The final distributed output from the service provider is then a multifrequency, multichannel broadband signal. In most cases, the subscriber

Figure 13–33
Basic cable television system.

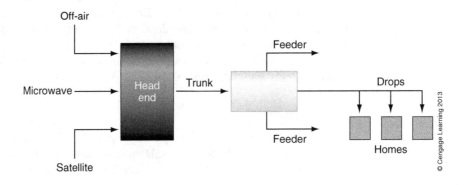

can directly tune such channels through a cable-ready television or DVR tuner. Provided that an individual channel does not require descrambling, direct tuning is possible. Where channel descrambling is required, the subscriber has to obtain a descrambling box from the service provider to be able to receive special pay-per-view programming or subscribed movie channels. The descrambling box also provides the function of demodulating the individual channels to be viewed on broadcast channels 3 or 4. In such cases, the subscriber leaves the television tuned to channel 3 or 4 and then simply tunes the descrambling box to view selected channels.

Program Jamming or Scrambling

Program jamming or scrambling can occur at the head end of the system or on the individual drop lines at the location of the subscriber. Injected scrambling at the location of the subscriber is known as program interdiction. Where head end scrambling is used, the subscriber must connect a cable descrambling box to view special programming because all special programming channels are universally jammed across the entire cable network. Where program interdiction is used, all channels are universally broadcast from the head end of the system unjammed, and the service provider controls the individual scrambling and unscrambling of services through a series of control signals to the selected service drops. The main benefit to using program interdiction is that, in most cases, the subscriber does not need to connect the descrambling box to view special events programming or subscribed movie channels.

Distributions Trunk and Feeders

The distribution trunks and feeders distribute the cable signal from the head end of the system to the subscriber. Although coaxial cable is often the primary choice of trunk and feeder distribution, more and more cable companies are switching over to fiber-optic cable, especially for distribution of the newer high-definition, digital video formats. Although fiber-optic cable provides the best signal quality, having the least amount of loss and signal distortion as compared with that of coaxial cable or twisted pair, it is still far too expensive for general distribution and use. In most cases, systems are still using traditional connections through coaxial cable, which means that signals through the distribution system must be routinely amplified and filtered before reaching the customer. Figure 13–34 presents a block diagram of a typical distribution system.

To help reduce the amount of signal loss within the system, amplifiers are usually placed at intervals of several thousand feet along the distributed trunk and feeder lines. Power for the amplification system typically is injected over the central conductor of the coaxial cable, often at levels of 60 V ac. Series-connected filters are also used to prevent the high-frequency television signal from shorting to the power supply. A single power supply often powers an individual segment of the distribution system. Power blocks are therefore required to prevent the cross-connection and interconnection of multiple supplies.

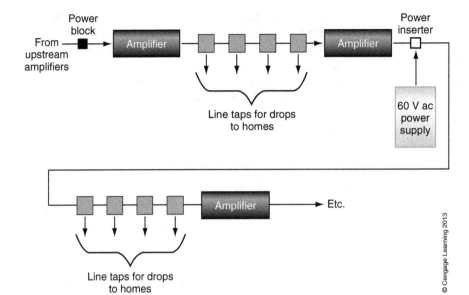

Figure 13–34
Cable distribu-
tion system.

SEC 13.8 NATIONAL ELECTRICAL CODE REQUIREMENTS FOR RADIO AND TELEVISION, *ARTICLE 810*

The scope of *Article 810* of the *NEC* covers antenna systems for radio and televi-sion receiving equipment, amateur radio transmitting and receiving equipment, and certain features of transmitter safety.

NEC 810.12 states that antennas shall be securely supported. Antenna or lead-in wires are not permitted to be installed on the electric service mast or on poles containing electric light or power conductors. Antennas must also not be attached to poles carrying trolley wires of 250 V or more between the conductors. All insulators used in supporting the antenna must be of sufficient mechanical strength to safely support the conductors. Lead-in wires must be securely attached to the antenna, and made from a high-strength, corrosion-resistant material, as stated in *NEC 810.11*.

The installation of outdoor antenna lead-in conductors installed between the antenna and a building structure must not cross over any open conductors of elec-tric light or power circuits. Where proximity cannot be avoided, then a 2-ft (0.6 m) minimum separation must be maintained between antenna lead-in conductors and all other power conductors measuring 250 V or less (see *NEC 810.13*). The instal-lation of antenna lead-in conductors should also not cross under open electric light or power conductors. For issues of safety, always try to avoid a close proximity among antennas, antenna conductors, and all types of power conductors.

The masts and metal support structures of antennas must be grounded to ensure the safety of individuals and equipment. *NEC 810.21(A)–(K)* covers antenna grounding. The bonding conductor or grounding electrode conductor shall not be smaller than 10 AWG copper, 8 AWG aluminum, or 17 AWG copper clad steel or bronze.

Bonding jumpers between radio and television equipment grounding electrodes and the power grounding electrode system at the building or structure shall not be smaller than 6 AWG copper. It often becomes necessary to bond the grounding electrodes of the radio and television equipment to the service-entrance grounding conductor of the building to eliminate ground loops, noise, or stray voltages that often can interfere with and disrupt the system.

Grounding to metal underground gas pipes is not permitted. Aluminum or copper-clad aluminum shall not be used as grounding conductors where direct contact to masonry or the earth may result or where they may be subject to corrosive conditions. In most instances, it is usually not a good practice to use aluminum wire because of issues of dissimilar metals, galvanic reaction, and possible corrosion. Bonding conductors or grounding electrode conductors (GEC) should always be made from a corrosion-resistant material, such as copper or copper-clad steel or bronze.

The bonding conductor or GEC may be run inside or outside of a building. Where they are run inside of metallic raceways, both sides of the raceway must be bonded to the bonding conductor or GEC, or to the same terminal or electrode to which they are attached.

The minimum size of outdoor antenna conductors for receiving stations can be found in *NEC Table 810.16(A)*. The required gauge of antenna conductors is based on the type of conductor material (such as aluminum alloy vs. copper-clad steel) and the maximum open span length of conductors (ranges include <35, 35–150, and >150 ft [<11, 10.7–45, and >45 m]).

Lead-in conductor sizes shall be of a size equal in strength (or greater) to those of the antenna conductors; that is, if antenna conductors are required to be made from a minimum of 20 AWG copper-clad steel, then the lead-in wires must be of a tensile strength equal to or greater than 20 AWG (see *NEC 810.17*).

Antenna lead-in conductors shall maintain a minimum clearance of 2 ft (0.6m) from all electric light and service conductors 250 volts or less, where proximity to such conductors cannot be avoided. A 10-ft (3.0 m) separation must be maintained between conductors greater than 250 V.

For circuits not more than 150 V, a minimum 4-in. (100 mm) clearance is allowed, provided that all conductors are securely supported by mounting hardware to ensure a permanent separation.

A 6-ft (1.8 m) separation must be maintained between antenna lead-in wires and lightning rod systems.

Underground conductors must maintain a 12-in. (300 mm) clearance from conductors of any electric light or power circuits or Class 1 circuits.

Indoor antennas and lead-in wires must maintain a 2-in. (50 mm) separation from conductors of other wiring systems, unless, of course, the other conductors are separated by a metal raceway or armored cable, or by a firmly fixed nonconductor, such as flexible or porcelain tubing.

Inside of boxes or enclosures, antennas and lead-in wires are permitted with conductors of other wiring systems, provided that they are separated by an effective permanent barrier.

Each lead-in conductor must be provided with a listed antenna discharge unit, except where the lead-in conductors are enclosed in a continuous metallic shield that has been permanently and effectively grounded, or is itself protected by an antenna discharge unit. *NEC Article 810.20(B)* states that discharge units shall not be located near hazardous locations, as classified in *Article 500* of the *NEC*. The discharge unit should be placed as close as practicable to where the conductors enter the building.

Discharge units are used to help drain excess static from the antenna and lead-in conductors.

Amateur Transmitting and Receiving Stations

For amateur transmitting and receiving stations, the size of antenna conductors and lead-in conductors shall not be smaller than the sizes listed in *Table 810.52* of the *NEC*. Wire gauges are determined based on the type of conductor material used and the maximum open-span lengths. Span lengths fall into one of two categories: less than or more than 150 ft (45 m).

Lead-in conductors for transmitting stations must be mounted at least 3 in. (75 mm) clear of the surface of the building, on nonabsorbent insulating supports, such as treated pins or brackets equipped with insulators. The 3-in. (75 mm) separation is not required where the conductors are enclosed inside of a continuous metallic shield that is permanently and effectively grounded. A coaxial cable is an example of a grounded shielded cable. In such cases, the grounded shield may also be used as a conductor.

Transmitter conductors shall also be located or installed in a manner to make accidental contact difficult.

Entrance to a Building

Except where the lead-in conductors are protected by a continuous metallic shield that has been permanently and effectively grounded, the following methods of building entrance are allowed:

- Through a rigid, noncombustible, nonabsorbent insulated tubing or bushing
- Through an opening that provides the entrance conductors a firmly secured, 2-in. (50 mm) minimum clearance
- Though a drilled window pane

Each lead-in conductor of a transmitting station shall also be provided with an antenna discharge unit to help drain off any stray static charge from the system. Installation of the discharge unit is not necessary where a continuous metallic shield, which is permanently and effectively grounded, protects the lead-in conductors or where the antenna itself has been permanently and effectively grounded.

The transmitting station shall be grounded in accordance with *NEC 810.21*. The protective grounding conductor must not be smaller than 10 AWG copper, bronze, or copper-clad steel. The operating grounding conductor shall not be smaller than 14 AWG copper or equivalent.

Interior Transmitting Stations

All conductors of transmitting stations must maintain a 4-in. (100 mm) clearance from all interior conductors of electric light, power, or signaling circuits except where all other conductors are separated by a raceway or firmly fixed nonconductor or as provided for by *Article 640, Audio Signal Processing, Amplification and Reproduction Equipment.*

Transmitters must be installed inside grounded metal enclosures. All external handles and controls must be grounded. Doors and access panels shall be provided with interlocks that disconnect all voltages greater than 350 V when opened.

Coaxial cable shall be permitted to supply low-energy power though the center conductor if the voltage is not more than 60 V and if the current is supplied from a transformer or an energy-limited device.

CHAPTER 13 FINAL QUESTIONS

1. List the components of a basic wireless system.
2. What organization oversees the allocation of all communication frequencies in the United States?
3. In telecommunications, the *channel* is the connecting wire or fiber between the caller and the recipient; in radio communications, the *channel* is _____.
4. What device combines the circuitry of a transmitter and receiver in one box?
5. Explain the difference between half-duplex and full-duplex.
6. A desired information signal is superimposed on a carrier frequency through the process of _____.
7. Compare a digital signal to an analog signal.
8. A _____ converts a mechanical property into an electrical signal.
9. The practice of mixing several communication channels from multiple sources into a single transmission channel is called _____.
10. Radio antennas are tuned to resonate at _____ the wavelength of the carrier frequency.
11. A GEO satellite rotates at _____ miles above the earth in a geosynchronous orbit.
12. Which type of satellite can operate at lower power levels while providing shorter signal delays: GEO, MEO, or LEO?
13. Which type of satellite configuration provides exclusive communications between two locations?
14. Can multiple satellite receivers be connected to a single LNB? Explain.
15. Explain line of sight and how it relates to the installation of a satellite system.

16. A _____ is used to retransmit signals between cell towers.

17. Which type of transmission divides channels into time slots?

18. Which type of transmission allocates and reassigns carrier frequencies to customers as needed?

19. Which type of transmission uniquely codes digital transmissions through DSSS?

20. Which wireless transmission standard provides the greatest distances and highest speeds, Bluetooth or Wi-Fi?

21. If you are intending to mount your equipment in a closed cabinet or closet, is RF or infrared the preferred choice of communication? Explain.

22. To remove the dc voltage from the central conductor of a coaxial cable, you need to install a _____.

23. In a cable distribution system, how far apart are line amplifiers typically placed?

24. Which type of program scrambling is controlled by the service provider through a series of control signals to the selected service drops?

25. For a typical cable distribution system, how does a feeder line amplifier receive its power?

26. In a typical cable distribution system, what do series-connected power filters prevent?

27. Which article of the *NEC* covers radio and televisions antennas?

28. According to the *NEC*, the minimum size grounding electrode conductor for an antenna system should be _____ AWG copper.

29. According to the *NEC*, antenna lead-in conductors must be separated from lightning rod systems by a minimum distance of _____ ft.

30. According to the *NEC*, indoor antenna lead-in conductors for receiver stations shall maintain a minimum _____-in. clearance from conductors of all other wiring systems.

Chapter 14

Closed Circuit Television Camera Systems

Objectives

- Identify the purpose of a closed circuit television (CCTV) system.
- List the components of a CCTV system.
- Determine what type of camera to use in a specific environment.
- Demonstrate how to match a lens to a camera.
- Calculate field of view and distance of a particular camera.
- Determine the correct transmission link for a CCTV system.
- Describe the difference between analog and digital recording devices.

Chapter Outline

Sec 14.1   Introduction
Sec 14.2   The Purpose of Closed Circuit Television
Sec 14.3   Closed Circuit Television Components
Sec 14.4   Closed Circuit Television System Specifications
Sec 14.5   Transmission Link
Sec 14.6   Viewing Monitors and Video Formats
Sec 14.7   Recording Devices
Sec 14.8   *National Electrical Code Article 820* Requirements

Key Terms

camera

cathode ray tube (CRT)

charge-coupled device (CCD)

closed circuit television (CCTV)

coaxial cable

depth of field

digital video recorder (DVR)

fiber-optic cable

field of view (FOV)

focal length (FL)

foot-candle

HDMI

lens

lux

monitor

multiplexer

power over Ethernet (PoE)

resolution

sensitivity

shutter speed

time-lapse video recorder

transmission link

varifocal

SEC 14.1 INTRODUCTION

Closed circuit television (CCTV) cameras continue to be one of the fastest growing technologies in the area of security and surveillance. Compared to that of a standard television system, a CCTV system differs in that it is not publicly transmitted. Instead, CCTV uses wired or wireless transmissions from the video camera to a private monitoring system and some type of recording device. With the rapid advancement of semiconductor technologies and the development of high definition (HD) television, it should be no surprise to learn that many security systems are now enhanced with the ability to take high definition (HD) photographs or video of a situation in progress. This is especially true in legal proceedings where eyewitness testimony can be replaced by a 10-megapixel digital recording, the quality of which can easily display the fine details for facial recognition software, license plate number retrieval, and a whole host of other information in the field of view.

Different security systems require distinct video capabilities. Whether it is in a casino setting where thousands of blackjack tables are being watched at one time, or a single camera planted in a retail shop to help identify shoplifters, the equipment must match the application. Because there are so many technical aspects to a CCTV system, we focus on the system components and how they work together to create a clear, usable image.

SEC 14.2 THE PURPOSE OF CLOSED CIRCUIT TELEVISION

The retail industry has been using CCTV for years to catch shoplifters in the act to help improve the likelihood of a criminal being convicted in a court of law. However, CCTV also has made a transition into the residential market as well, with cameras allowing homeowners some peace of mind in protecting personal property.

Where a CCTV camera is used, a second layer of protection is provided. Not only can the camera detect an intrusion but also the incident can be viewed later by security and law enforcement officials to obtain meaningful data concerning the actual happenings of an event.

SEC 14.3 CLOSED CIRCUIT TELEVISION COMPONENTS

Every CCTV system possesses five main pieces of equipment: a camera, lens, transmission link (wiring), monitor, and recording device. The many options available within these five categories vary greatly, as well as how they interconnect. The most important detail is first defining the environment and location of the camera. Is the camera going to be openly visible or hidden, ceiling mounted or wall mounted, indoor or outdoor, auto adjusted or with fixed settings? Once the fine details have been determined, then the pieces can be gathered, assembled, and installed to be an effective enhancement to any security system. The first step is to define the camera and the various types available.

Closed Circuit Television Cameras

A CCTV **camera** collects reflected images from objects in the environment and then converts them into electronic signals. The purpose of the camera is to capture a visual image of an event. After doing so, the camera converts the image into a usable analog or digital video signal. The electronic video signal then is sent to a viewing **monitor,** recording device, or central computer by using any one of several transmission links: coaxial cable, optical fiber, digital HDMI cable, or a network category 5(CAT5) or category 6(CAT6) cable.

Analog Tube Cameras

Although analog cameras are quickly being replaced by the newer digital, high-definition varieties, for the purposes of historical reference the following list displays the types and variety of analog cameras that have been used in the past and in some cases are still being used today.

Tube cameras convert a video image into an electronic signal by using a video vacuum tube or pickup tube. The pickup tube is a type of cathode ray tube (CRT), such as was used in older style television sets. A light-sensitive photocathode is mounted on one end of the tube. In simple terms, the camera lens focuses an image onto the photocathode. The photocathode then emits an electron beam of varying intensity, directly proportional to the detected level of light. A scanning circuit scans the photocathode and literally draws the image, line by line, to create an analog electrical equivalent of the focused image. The video output from the analog camera typically measures 1 $V_{peak-peak}$.

Tube cameras had certain limitations. Primarily, they could be easily damaged or destroyed if pointed directly at the sun or at intense light sources. In addition, they were also susceptible to image burn-in, especially when the image was static and nonchanging for long periods of time, such as a camera pointing at the entrance of a store front, continually, for 24 hours a day. Image burn-in cannot be fixed, and ultimately the camera tube must be replaced. The various types of tube cameras include the following.

Standard Vidicon A Vidicon camera was commonly used in environments having bright, consistent light levels. The best location was indoors, in well-lit rooms, where the lights were on the majority of the time. One benefit of using standard Vidicon cameras was that they were the most inexpensive. The downside was that they had a useful life span of only about 2 to 3 years.

Ultricon and Newvicon Ultricon and Newvicon tube cameras had a higher sensitivity to light, which made them ideal for use in low-light environments. This type of camera was also equipped with an auto-iris lens, which allowed it to automatically adjust to different levels of light. An ideal application for Ultricon or Newvicon cameras was an indoor situation that relied heavily on natural light or in situations where the amount of light in the room was frequently changing.

Semiconductor Intensified Tube The semiconductor intensified tube (SIT) was a very sensitive camera, originally designed to operate in extreme, low-light conditions, for example, nighttime. For this reason, the SIT was often used in parking lots, storage areas, and rooms with no windows. It was a good choice for a security camera, especially in cases where lights were low or possibly inoperable. One trade-off, however, was that it often had a grainier picture quality in low-light situations.

Semiconductor-Based Cameras

Charge-Coupled Device Cameras (CCD)

CCD cameras (Figure 14–1) use a semiconductor target instead of a vacuum tube to record an image. The photo sensor target is divided into a grid structure of capacitive elements called pixels. When light strikes the chip, it generates an electrical charge. The charge is stored as a voltage level within each pixel. An electronic circuit then converts the voltages of each pixel into an analog or digital format to be transmitted at the output. A CCD camera is designed to suitably record images in environments where the light level is low and variable.

Some of the advantages to using a CCD camera include these:

- A CCD camera can provide a full-resolution picture requiring less light than a tube camera.
- Generally, a CCD camera is physically smaller than a tube camera, and therefore can be used in tight or covert applications.
- The price of a CCD camera has decreased significantly over the years and is now very affordable.

There are two types of CCD cameras, interline transfer and frame transfer. A third type of semiconductor camera exists, which uses a complementary metal-oxide semiconductor or CMOS-based sensor. We will first look at the CMOS-based sensor and then compare the two varieties of CCD.

Complementary Metal-Oxide Semiconductor-Based Cameras

The sensor in a complementary metal-oxide semiconductor (CMOS)-based camera differs from CCD in that it does not store charges in capacitive pixels,

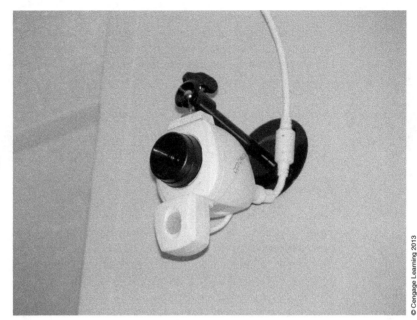

Figure 14–1
Charge-coupled device type, closed circuit television camera.

© Cengage Learning 2013

but instead uses an active array of transistor switches. CMOS sensors also process data frame by frame as opposed to line by line. Although this method does produce a high-quality image, the downside is that CMOS sensors tend to be much less sensitive than CCD and are therefore not typically used in low-light environments. CMOS sensors also do not perform well in infrared light. In the right environment, however, a CMOS-based sensor can provide a decent, high-quality image, and at a much lower cost than CCD. In the past few years, dramatic improvements have been made in CMOS technology, and it has found increasing use and popularity.

Interline Transfer Charge-Coupled Device Cameras

The imaging chip used in an interline transfer CCD is much more sensitive than that of a CMOS-based camera. Knowing this, the interline transfer CCD can perform over a wide range of low- to full-light conditions, with a clear and trusted picture. This type of camera also responds better to infrared light than the CMOS-based camera. Therefore, it works well in normal indoor light, outdoor daylight, and even during the nighttime hours.

Frame Transfer Charge-Coupled Device Cameras

A frame transfer CCD camera uses a multilayer chip that is composed of two elements, a photo-plane and a memory section. The frame is first exposed in the photo-plane and then quickly moved to the memory section. As the next image frame is being exposed, the memory section processes the previous frame, line by line. The advantage to using this method is that all the pixels in a frame are exposed at the same time instead of one line at a time, and therefore the sensor preforms better on faster moving images. This also allows the frame transfer CCD camera to perform even better in low- to full-light conditions.

SEC 14.4 CLOSED CIRCUIT TELEVISION SYSTEM SPECIFICATIONS

Camera Sensitivity

Sensitivity refers to the minimum amount of light that the camera requires to produce a full video image. Indoor applications pose few design problems; therefore, the sensitivity of the camera typically is not important. Outside, however, the amount of light in a particular area determines the type of camera that should be used. In determining the amount of light, a designer must not only take into account any light sources but must also consider the reflected light from objects in the area such as parking walls, buildings, or parking lot surfaces. Be careful to place your camera and lights in a manner that supports and benefits the existing lighting of the space.

Before picking the right camera for the job, it is important to identify how much light is in the given area. To do this, a light meter calibrated in **foot-candles** (fc) or **lux** (lx) may be used. A foot-candle is defined as the amount of illumination produced by a standard candle at a distance of 1 ft (0.3 m). Sometimes manufacturers specify light calibration in lux, so it is important to know that lux is a unit of illumination equal to 1 lumen/m^2 or 0.0929 ft-candle. For practical application related to outdoor and indoor illumination, refer to the tables listed in Figures 14–2 and 14–3.

Figure 14–2
Outdoor illumination table.

Direct sunlight	10,000 to 13,000 fc
Full daylight	1000 to 2000 fc
Overcast day	100 fc
Full moon	0.01 fc
Overcast night	0.00001 fc

© Cengage Learning 2013

Figure 14–3
Indoor illumination table.

Bank lobbies	20 fc
General work area	30 fc
Retail stores	50 fc
Hospital operating rooms	1800 fc

© Cengage Learning 2013

Black asphalt	5%
Gravel surface	13%
Trees on grass	20%
New concrete	40%
New snow	85%

Figure 14–4
Surface
reflectance
percentages.

© Cengage Learning 2013

Surface Reflectance

For a correct exposure, you also have to determine whether any reflective light is being introduced indirectly into the camera. For practical applications, Figure 14–4 provides some examples of surface reflectance.

The formula for determining camera sensitivity is as follows:

$$\text{Illumination} \times \text{Reflectance} = \text{Light at the camera}$$

Example

How much light is available in a parking lot with new concrete on a bright and sunny afternoon?

Answer

$$\text{Approximately } 12{,}000 \text{ fc} \times 0.4 = 4800 \text{ fc}$$

Converting Foot-Candles to Lux

As mentioned earlier, illumination and camera sensitivity often are listed in manufacturers' specification sheets in foot-candles or lux. For this reason, it is necessary to convert foot-candles to lux and vice versa. The relation between these two terms is as follows:

$$0.0929 \text{ fc} = 1 \text{ lx}$$

Example

In the above example, we determined that the available light is 4800 fc. Express this value in lux.

Answer

$$0.0929 \times 4800 = 445.92 \text{ lx}$$

There also are times when it is necessary to convert lux into foot-candles.

Example

The illumination level measures 1000 lx; express this reading in foot-candles.

Answer

$$1000/0.0929 = 10{,}764.26 \text{ fc}$$

Camera Resolution

Image **resolution** refers to the quality of image or the ability to see fine details in an image. It is also the measure of a camera's ability to distinguish between objects in close proximity to each other. When deciding on the best image resolution, a good rule of thumb is the higher the image resolution, the better the image. When a camera produces a picture, sequential scanning lines form the image. The number of scanning lines usually determines the quality of the image. For example, a picture that uses 200 lines is of poorer quality than a picture using 1200 lines. Because of the difficulty in determining the number of scanning lines, the National Television Standards Committee (NTSC) has developed an alternate system designed around aspect ratios, expressing the size of an image in width to height. For example, a typical standard television has an aspect ratio of 4:3, whereas the new high-definition television standards use a wider ratio of 16:9.

Resolution, however, is often one of those overrated issues. The first thing to know is that under normal circumstances, your indoor images do not necessarily have to be produced with a high-resolution camera. Because interior cameras tend to be much closer to their subjects, often in full-light environments, issues of clarity and fine detail typically are not a problem. Under such circumstances, a lower resolution camera works just fine. A high-resolution camera, however, has to be used in a more low-light environment; for example, in a parking lot, where the recording of such a large space requires the need to see fine detail more clearly. From a camera sensor point of view, determining resolution in an HD environment is more about pixels per ft within the field of view. A good rule of thumb is 50 pixels per ft, daytime; 80 pixels per ft, nighttime; and 125 pixels per ft, if facial recognition is required. Figure 14-5 displays a range of camera sensors and how the aspect ratios and total number of megapixels are inter-related. Therefore, if you are intending to record a space 8 ft wide, and you need to determine facial recognition clearly, then at a minimum, the sensor target will need to have a width of 8×125, or 1,000 pixels. In such cases, a 1.3 megapixel camera will be fine.

Figure 14–5

Digital camera sensor aspect ratio versus megapixels.

Width	Height	Aspect Ratio	Actual pixel count	Megapixels
640	480	4:3	307,200	0.3
1280	960	4:3	1,228,800	1.3
2048	1536	4:3	3,145,728	3.0
2272	1704	4:3	3,871,488	4.0
2560	1920	4:3	4,915,200	5.0
3008	2000	3:2	6,016,000	6.0
3072	2304	4:3	7,077,888	7.0
3264	2448	4:3	7,990,272	8.0
3648	2736	4:3	9,980,928	10.0
4000	3000	4:3	12,000,000	12.0

If the field of view were to expand to 25 ft, then you would be more inclined to use the 8 megapixel cameras to be able to capture the same level of detail. Keep in mind, however, that as you increase the megapixels, you may also need to increase the transmission bandwidth and data storage capacity, or the system can become slugish, bog down, or even crash. When designing a network based CCTV system, the amount of available bandwidth becomes the primary issue.

Now, although you may presently have the most expensive, highest resolution camera available, you still may not have a clear picture. The reason is that the resolution of a video image is only as good as the weakest link in the system. If a camera produces 1600 lines of resolution, but the recorder only has a playback resolution of 300 lines, then 300 lines is all that results. In this case, you may have wasted your money, unless, of course, you want to upgrade the rest of your system to better accommodate the specifications of the high-resolution camera.

In addition to resolution, the camera itself is somewhat useless, unless you know how to fit and use a proper len. The **lens** of the camera is a transparent optical device, used to converge or diverge transmitted light to form images. Where using CCTV as part of a security system, you want to ensure that your surveillance subject is large enough, in focus, and in the **field of view (FOV)** to be identified on the monitor. Choosing the correct lens contributes to the clarity of a picture.

Because lenses are often bought and sold independently of the camera itself, it is considered to be a peripheral of the camera rather than an integral part.

Camera and Lens Format

The format of a camera is defined as the maximum usable image created by the lens. Lens format is actually determined by the size of the opening in a camera where the lens attaches to the body of the camera. For perfect results, a lens must match the format of the camera. This means that if a camera has a 1/3-in. format, then the picture generated by the lens cannot be larger than 1/3-in., or it does not entirely fit on the viewing screen. Notice from Figures 14–6 and 14–7 that where a camera format does not match the lens format, parts of the picture overshoot the edges of the viewing monitor and thus are not displayed in their entirety.

Figure 14–6

Lens format matches camera format.

1/3-in. lens used with a 1/3-in. camera

© Cengage Learning 2013

Figure 14–7

Lens format is larger than camera format.

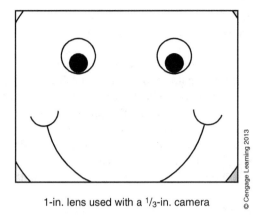

1-in. lens used with a $1/3$-in. camera

© Cengage Learning 2013

Lenses and cameras are commonly available in 1/4-, 1/3-, 1/2-, 2/3-, and 1-in. formats. The camera format, measured in inches, represents the diagonal measurement across the photosensitive target. But camera formats and lenses can also be expressed in millimeters, depending on the manufacturer; this means that the user has to be able to convert between the two to determine compatibility.

Types of Lenses

Choosing the correct lens for a particular camera usually depends on two factors: how far you need to see and the amount of available light. There are several choices available when it comes to choosing a camera lens. Picking the lens that best suits the application is the key to designing a good CCTV system.

Fixed-Focus Lens

CCTV systems more commonly use fixed-focus lenses than any other type. Fixed-focus lenses are not adjustable. They are set by the manufacturer to a fixed focal length and fixed aperture, and provide a range of focus that is sufficiently deep for most applications. These lenses are used in environments where the light stays relatively constant and the subject is stationary; for example, a camera pointed at an interior door. In this situation, the light remains essentially constant, and the subject is not moving anywhere, especially out of the camera's FOV.

Fixed-focus lenses are sometimes called fixed-iris lenses. The iris of the camera is the small opening through which light passes. The measurement of the size of the iris is called an *f*-stop (see the next section for more details). In basic terms, during full-light conditions, the iris is almost completely shut down, whereas in low-light conditions, the iris is nearly fully open. When using a fixed iris lens, the size of the iris is set by the manufacturer, and it is not adjustable. This is why, for this type of lens, you must have an environment where the level of light remains somewhat constant; otherwise, image quality changes throughout the day.

Varifocal Lens

The **varifocal** lens or auto-iris lens is used in situations where variable focal lengths (wide angle to telephoto) and the ability to adjust focus is a requirement. It is also a good choice of lens where lighting conditions can vary from full light to low light, from minute to minute. This lens is more expensive than the fixed-focus lens, but some argue that the extra money is worth the flexibility of using this lens in any location.

Varifocal lenses can also automatically adjust the focal length and diameter of the iris by using a small forward and reversible motor. This type of lens is ideal for indoor applications where lights are being turned on or off throughout the day or where the natural light conditions in a room normally fluctuate. Another application for a varifocal lens is an outdoor setting where both daytime and nighttime viewing is essential.

Choosing a Proper Lens

There are two important factors to consider when choosing a proper lens for a particular situation; these factors are **focal length (FL)** and FOV. FL determines the magnification of an object. FOV is the actual area that the camera is expected to cover, from a horizontal and vertical point of view. Typical lens FLs are 3.5, 6.0, 12, 35, 50, and 75 mm. You can figure out the format of the camera and the FL of the lens by looking at the manufacturer's specifications. Incidentally, the human eye has a focal length of about 17 mm (0.7 in.), and can be compared to a fixed focal length, auto iris lens. An auto-iris lens adjusts to changes in light automatically. It is relatively easy to determine the actual FOV from the chosen combination of camera format and lens FL.

Focal Length

FL represents a measurement internal to the camera. It defines the distance from the focal plane of the camera (θ) to the lens, where the lens is focused on infinity. The focal plane (θ) is the vertical plane inside the camera that coincides with the internal focal point of the lens, or it can be thought of as the physical point inside the camera where the photosensitive target resides. Figure 14–8 illustrates FL and

Figure 14–8
Focal length and focal plane.

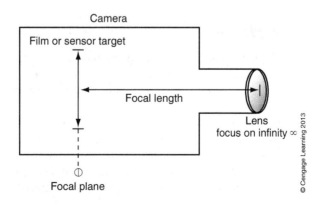

focal plane. Film cameras physically place the film within the focal plane; digital cameras place the sensor target within the focal plane.

F-stop

An *f*-stop mathematically represents the ratio of the FL to the diameter of the iris.

$$F\text{-stop} = \text{Focal length} \div \text{Iris diameter}$$

Common *f*-stop values are 1.4, 2.8, 4, 5.6, 8, 11, 16, and 22. A small *f*-stop number indicates a large iris opening and the availability of more light for the camera. Conversely, a large *f*-stop number indicates a small iris opening, indicating that little light passes through to the camera. In addition, with each successive increment of the *f*-stop dial, for example, from 8 to 11, the amount of available light is cut in half. Moving in the other direction, as *f*-stop values are decreased, for example from 8 to 5.6, the amount of available light doubles with each successive decrement.

Shutter Speed

The **shutter speed** of a camera can also be used to adjust for changes in light level. Although fast-moving shutters can capture high-speed images, they can also be used to cut down on the level of light if it's too bright. And conversely, slowing down the shutter will add more light to the camera, although the image may start to blur if the shutter speed is too slow. Shutter speeds can be auto-adjusted or manually adjusted, depending on the type of camera. Typical shutter speeds will range from 1/60 of a second to over 1/1000 of a second. Video cameras will shoot at 30 frames per second in most cases, but can also be adjustable to 24 or 25 frames per second (fps), equivalent to the rates of motion-picture film cameras.

Shutter speed also has a relationship to *f*-stop. A doubling of shutter speed will have the effect of increasing one *f*-stop, even though the position of the iris has not been changed. Cutting the shutter speed in half will conversely decrease the exposure one *f*-stop. So, in situations where there is not enough light and the lens iris is open 100%, the only way to increase the exposure is to decrease the shutter speed. Just keep in mind that slowing the shutter speed too much may cause blurring of image.

Depth of Field

Depth of field relates to the measured distance of focus in front of and behind the main subject of an image. As an example, when the camera is focused on a spot in the viewing frame, there will be a measured distance in front of and behind the spot that will remain in focus. The area that remains in focus is called the depth of field. Depth of field is affected by *f*-stop. As *f*-stop increases, less light enters the camera, and the light beams become more narrowly focused, this causes the picture to be painted with smaller dots of light. In such a situation, the depth of field increases. Going the other direction, as the *f*-stop decreases, the lens aperture widens and the rays of light are made broader, causing the depth of field to become shallower. As a rule, increasing the *f*-stop increases the depth of field and deepens the focus.

Putting it all together, it is easy to see that *f*-stop, shutter speed, and the level of light will not only affect the exposure, but also the focus—how sharp or blurry the subject matter is. It is important to know how all of these parameters work in connection with each other to create a useful image.

To calculate how far a camera must be placed from an object, define the following variables:

Distance (D): Distance is an external measurement representing how far a camera is from the focused subject matter. Distance can be measured in either meters or feet. The calculated examples in this chapter use feet.

Width of the camera imager in millimeters (W): This is the horizontal dimension of the camera format measured in millimeters. In general, the following conversions apply:

$$2/3\text{-in. camera} = 8.8 \text{ mm width}$$
$$1/2\text{-in. camera} = 6.4 \text{ mm width}$$
$$1/3\text{-in. camera} = 4.8 \text{ mm width}$$
$$1/4\text{-in. camera} = 3.2 \text{ mm width}$$

Millimeter widths vary among manufacturers. For this reason, it is important to check the specifications of a particular camera to verify correct width dimensions.

Horizontal field of view (HFOV): HFOV determines exactly what the object in view looks like along the horizontal plane of the viewing monitor, from left to right. Figure 14–9 illustrates the typical functions of a camera, including HFOV, distance from the camera, FL, and camera format.

The formulas shown below mathematically explain how these variables are related:

$$HFOV = W \times (D \div FL), \text{ or}$$

$$D = HFOV \times (FL \div W)$$

The HFOV equation typically is used in the second form because the correct distance from the object to the camera often is unknown. (The vertical field of view [VFOV] also can be calculated simply by multiplying the dimensions of the HFOV by 0.75; this works provided the aspect ratio is 4:3.) Knowing these variables and formulas assures proper camera placement. Let us take a closer look.

Example

A camera has to record a view that includes a double door measuring 6.5 ft horizontally. In addition, 1 ft on either side of the door also must be viewable, bringing the total horizontal view to 8.5 ft. The camera format is 1/3-in. (4.8 mm), and

Figure 14–9
Horizontal field
of view. CCD,
charge-coupled
discharge.

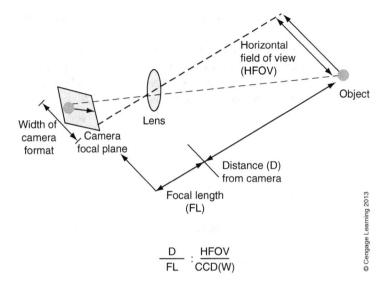

$$\frac{D}{FL} : \frac{HFOV}{CCD(W)}$$

© Cengage Learning 2013

the FL of the lens is 12 mm. How far out from the door does the camera have to be for there to be a viewable picture on the monitor? Calculate the distance as follows:

$$D = HFOV \times (FL \div W)$$

$$D = 8.5 \text{ ft} \times (12 \text{ mm} \div 4.8 \text{ mm}) = 21.25 \text{ ft}$$

Based on the chosen camera/lens combination, the placement of the camera has to be approximately 21 ft from the subject matter for ideal viewing. If the camera is placed any closer, the image appears too large and thus might not fit within the horizontal viewing limits of the monitor. In such a case, possible suspicious activity around the door frame might not be viewable. The other extreme is to place the camera too far from the door, which makes the subject matter too small; also, the presence of suspicious activity might be impossible to see with any degree of detail.

SEC 14.5 TRANSMISSION LINK

Every camera in a CCTV system is wired back to a central station, where viewing takes place. The path between the camera and this station is called the **transmission link**. Options for transmission links for CCTV systems may include coaxial cable, fiber-optic cable, or network CAT5 cable.

Coaxial Cable

Coaxial cable, typically referred to as "coax," is the most commonly used link in CCTV systems, mainly due to its low cost and reliability. As explained earlier in

Chapter 2, coax has a center conductor surrounded by dielectric insulating material, which in turn is covered by a braided shield to help reduce the level of interference on the transmission line. RG/59 is the most common type of coax used in CCTV systems; its electrical properties are ideal for video transmission.

The only drawback in using coax is the potential for ground loops. Ground loops occur where there is more than one ground along the path of a video signal. In most cases, the shield on a coaxial cable is connected to only one ground point to help eliminate possible ground loops.

Fiber-Optic Cable

A more reliable, yet more expensive option for CCTV transmission is to use **fiber-optic cable** (see Chapter 10). Optical fiber provides more bandwidth for video transmission, allows for longer distances between the camera and the viewing station, and is not susceptible to electrical interference of any kind.

Network Category 5 Cable

Many new CCTV camera systems are now being controlled directly through a computer network. CAT5 cable often is used in such cases not only to transfer video signals along the transmission link, but also to operate the automated digital camera controls. Control signals may include network addressing as well as some basic camera functions, including telephoto or pan-tilt-zoom (PTZ) controls, or remote camera positioning. The benefit to using a networked camera system is that, in most cases, additional cabling infrastructure does not have to be installed within the structure of a building. Installations are simplified because camera systems can be connected directly to preexisting computer networks with little to no need of running additional cable. Powering the camera is also simplified by connecting the CAT5 through a Power-over-Ethernet (PoE) switch.

Power over Ethernet (PoE)

Power over Ethernet (PoE) is used in situations where a standard USB cable is not practical. Like, USB, PoE provides power and a communication link to connecting devices. The benefit of PoE to CCTV is that PoE provides system power to the camera without the need for additional cabling or hardware. In 2009, the IEEE 802.3af (2003) standard was revised to the new 802.3at standard for PoE. Device power was increased from 12.95 to 25.5 W, with dc voltages ranging from 50 to 57 V at a maximum current of 600 mA per mode and a maximum cable resistance of 12.5 ohms. Two modes of PoE exist, Mode A and Mode B. Mode A delivers phantom power on the data pairs of 100BASE-TX or 10BASE-T. Mode B delivers power on the spare pairs. Internal current protection is also provided. If the power-sourcing equipment detects that the cable resistance is too high or too low, power to the connecting cable will be removed. In addition, the device must continually use between 5 and 10 mA for at least 60 ms with no more than 400 ms of down time, or power to the device will be turned off. The increase to the 802.3 at standard has only helped to broaden the horizon of PoE. As a result, in the last few years manufacturers have brought to market many more devices that take advantage of the new PoE standard, which has greatly simplified the installation, connection, and communication of all types of network devices.

SEC 14.6 VIEWING MONITORS AND VIDEO FORMATS

The video signal of a camera is sent over a transmission link to a viewing monitor, and in some cases to a peripheral recording device or central server. A monitor can be set up to display the pictures from a single camera or the pictures from multiple cameras within the CCTV system.

The size of a monitor can also vary, depending on the type of system, and the output can be a color or black-and-white image, as needed. A monitor is not a television; it does not have a built-in tuner and therefore cannot tune in UHF or VHF frequency bands, nor does it need to have this type of capability. The function of the monitor is to simply display the video picture as seen by the camera. For this reason, the signal being transmitted by the camera is transmitted either through a single-connection, composite video signal; component video, which involves a five-connection cable of red, green, blue, chrominance, and luminance signals, referred to as RGB; an HDMI digital connection; or by optical fiber.

CCTV monitors do not have the life span of conventional television sets. Because they are designed to operate 24 hours a day, 7 days a week, a monitor may only last about 5 to 7 years before it must be replaced.

Video Formats

Analog Video Formats

NTSC In 1953, the television industry in the United States adopted the NTSC color television transmission standard. The new standard replaced the oldest existing standard, the 1941 NTSC B&W standard, and was backward compatible. Known as the NTSC-M standard, the new standard consisted of 525 lines, scanned horizontally, with an aspect ratio of 4:3 and a video frame rate of 30 frames per second. Maximum channel bandwidth was set at 6 MHz, and each frame was divided into two interlaced fields, comprised of 262.5 odd lines, then even lines, with a refresh rate of 60 Hz. These numbers refer to the resolution of the monitor and the speed of the moving image; the more scanning lines on the viewing monitor, the clearer the picture.

Incidentally, there is a mismatch between film movie cameras and video. Film cameras do not move at 30 fps, but at a frame rate of 24 fps in the United States and 25 fps in Europe. The 30 fps rate for video was chosen because the standard power frequency in the United States operates at 60 Hz, mathematically double the 30 Hz frame rate. For this reason, it was more convenient to set the video frame rate at 30 fps rather than 24 fps.

PAL The Phase Alternating Lines (PAL) standard is not used in the United States, but it is found in many other parts of the world. PAL uses 625 horizontal lines and a 50 Hz vertical refresh rate. There are many different types; most common are PAL B, G, and H, and less common are PAL D, I, K, N, and M. All of these types are generally not compatible with each other.

SECAM The final standard is known as *Système Électronique pour Couleur Avec Mémoire* (SECAM). This standard also provides 625 horizontal lines and a 50 Hz vertical refresh rate. There are also different types of SECAM standards. They all use different video bandwidth and audio carrier specifications. For example, there are types B and D, typically used for VHF; types G, H, and K are used for UHF; and types I, N, M, K1, and L are used for both VHF and UHF. The different types of SECAM formats are generally not compatible.

Digital Television Standards

ATSC In the early 1990s, a group of electronics and telecommunications companies came together to design the high-definition television standard (HDTV) for the United States. They were named the Advanced Television Systems Committee (ATSC). The ATSC Standard A/53 was originally adopted by the FCC in 1995, but was later revised in 2009 to the ATSC Standard A/72. ATSC Standard A/72 introduced and defined the H.264/AVC video-encoding standard for high-definition television. Included in this television standard was Dolby Digital 5.1 surround sound, as well as many elements of MPEG video coding, and an 8-level vestigial sideband modulation (8VSB) scheme for terrestrial carriers to incorporate the mixing of the video and audio data on the same frequency carrier. The 8VSB was designed around the same 6 MHz bandwidth as the NTSC standard, but in many cases satellite and cable companies use alternate modulation schemes because they do not follow the same signal-to-noise ratios or compressing standards as local broadcasters.

There are three commonly used resolutions for ATSC Standard A/72: 480 scan lines, 720 scan lines, or 1080 scan lines. The frames can be interlaced or progressive: interlaced divides the frame into two fields, odd or even, as described in the NTSC format, and progressive simply scans the entire frame on the monitor all at once. Frame rates in ATSC are adjustable between 23.976, 24, 25, 29.97, 30, 59.94, and 60 per second, to accommodate the conversion from movie film to video in the NTSC, PAL, or SECAM formats. Figure 14–10 illustrates the options within the standard.

Multiplexers

Where it is necessary to view a large number of cameras, it is not practical to have a monitor for every camera. For example, large casinos may have hundreds of cameras, yet only have a limited amount of viewing monitors; this is done through

Figure 14–10
ATSC Standard
A/72.

Resolution		Aspect Ratio	Scanning
Vertical	**Horizontal**		
1080	1920	16:9	Progressive or Interlaced
	1440	16:9	Progressive or Interlaced
720	1280	16:9	Progressive
480	720	4:3 or 16:9	Progressive or Interlaced
	704	4:3 or 16:9	Progressive or Interlaced
	640	4:3	Progressive or Interlaced
	544	4:3	Progressive or Interlaced
	528	4:3	Progressive or Interlaced
	352	4:3	Progressive or Interlaced
240	352	4:3	Progressive
120	176	4:3	Progressive

© Cengage Learning 2013

multiplexing. A **multiplexer** takes the images from multiple cameras (4, 8, 16, 32, etc.) and displays them on one monitor in a pattern chosen by the user. If this is done, the monitor must be sized to the number of cameras. For example, displaying 4 cameras on a 17-in. monitor is acceptable; however, displaying 32 cameras on a 17-in. monitor means that individual viewing windows for each camera display might be too small and hard to view. In this case, a 25- to 27-in. monitor, or larger, might be appropriate.

SEC 14.7   RECORDING DEVICES

Most CCTV systems include a recording device to ensure that a particular event is captured in some permanent form, often to be used as evidence in a legal case. Whatever the reason, recording can be achieved through a variety of methods.

Time-Lapse Recorders

In the most basic, low-cost, older-style camera systems, the camera or multiplexer was connected directly to a **time-lapse video recorder**. This recorder is a lot like a typical videocassette recorder (VCR) using a tape to record events. However, a time-lapse recorder is unique in that it can take hundreds of hours of viewing and put it on a 3-hour tape. It does this by skipping frames to save space. Viewing a recorded image from a time-lapse recorder is different from viewing a real-time image. Because frames have been left out, the picture tends to look jerky, and some event details may be missed because they happened on a particular frame that was left out. Time-lapse recorders may often be connected to a motion sensor. In such cases, the recordings only take place when a movement in the viewing frame has been detected.

Digital Video Recorders

Today, computer technology has paved the way for digital recording devices. **Digital video recorders (DVRs)** use a hard drive–type storage device for recording images. This allows much more space for saving footage without eliminating frames. Therefore, the recorded picture being watched is in real time.

The DVR looks similar to a typical VCR except that there is no tape. It is capable of high-resolution picture playback and also gives the user the ability to almost instantly access a particular event, whereas with the time-lapse recorder, the tape must be searched in a linear analog fashion that often can be time consuming. It is also possible to view recorded images remotely through a secure connection using a DVR.

Figure 14–11 illustrates the block diagram of a typical CCTV camera system that includes a connection to an alarm system.

The layers include lighting, cameras, transmission layer, video processing, switcher/multiplexer, monitors, video recorders, digital controller, and connection to an alarm system.

The transmission layer involves the method of connecting the cameras to the systems, whether through coax, optical fiber, or a CAT5 connection.

The video-processing layer is used in situations where video amplification is needed, depending on the distance of the cameras to the switcher/multiplexer.

The switcher/multiplexer is the central hub through which all video signals pass for connection to monitors or recording devices. The switcher typically is programmed to sample each camera for the presence of motion. The video recorder then is activated, depending on the level of activity coming from either camera.

The monitors often are mounted at a guard station to be used by security personnel for the continuous viewing of cameras A and B.

Figure 14–11
Block diagram of a typical video system.

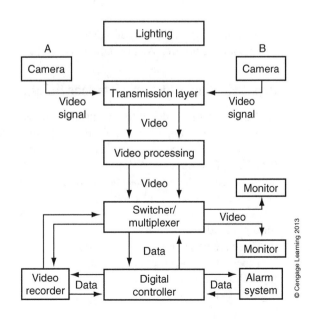

The digital controller, in most cases, is a central computer that drives the system. The controller also notifies the alarm system if there appears to be an occurrence of a recordable incident.

Open Source Software

To simplify the system even more, the modern CCTV systems now uses a centralized computer and open source software, which incorporates all the functionality of video processing through digital control, as illustrated in Figure 14–11, as well as giving the user the ability to add on modular security and card access capabilities, as discussed in Chapter 12. Recordings, in such cases, can be stored indefinitely in high-speed digital memory arrays or RAIDs, which can be instantly accessed, nonsequentially. RAIDs also help protect from loss of data through redundancy. The benefit to using open source software is that devices from multiple manufacturers and systems can be interconnected easily through software drivers into a modular, user-friendly package, designed and controlled by the consumer, not the manufacturer. Open source systems are not limited to the finite basic functions of a specific device or controller because they can always be updated and modified by installing a software revision or new device driver, as needed, without having to completely change out all system hardware for every new advancement in technology.

Connecting Analog Hardware to a Digital System

Analog cameras and peripherals can still be used in digital systems provided that they are linked through a digital video server. The digital video server includes an analog-to-digital converter as well as an IP transmission link to be able to communicate over the computer network. Input connections to digital video servers are typically BNC coax, as illustrated in the next section, and use an 8-pin modular, RJ-45 connector on the output. Also available are wireless, digital video servers that communicate over Wi-Fi. Figure 14–12 illustrates a typical IP camera system.

High-Definition Multimedia Interface (HDMI)

HDMI is an audio/video interface used for transmitting digital data among compatible devices, computers, and monitors. Prior to HDMI, video was transmitted over coaxial cable, composite video (RCA connector), S-Video, component video (RGB), D-Terminal, or VGA for connection to PCs. HDMI supports eight channels of audio, compressed or uncompressed, and any uncompressed format of TV or PC video. HDMI is also electrically compatible with digital visual interface (DVI) where a DVI-to-HDMI adaptor is used. DVI was an earlier style connector for digital video and is still found on some equipment.

There are five connector types for HDMI.

Type A is a 19-pin connector, approximately 14 × 4.5 mm, and is used on most HD monitors and DVD players. A Type A HDMI connector is shown in Figure 14–13.

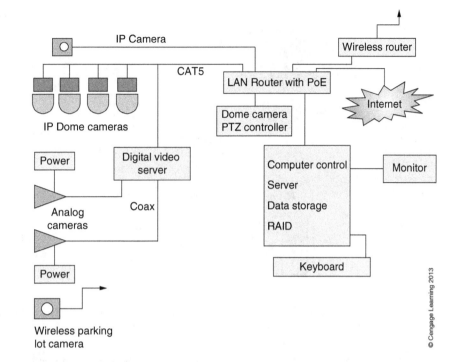

Figure 14–12
Networked camera system block diagram.

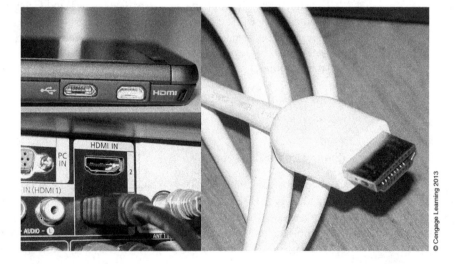

Figure 14–13
HDMI connector and cable.

Type B is a 29-pin connector, approximately 21 × 4.5 mm, and is currently not used on any consumer products. It can carry double the video bandwidth of Type A and was developed for very high-resolution displays, such as 3840 × 2400.

Type C is a miniconnector intended for portable devices, approximately 10.42 × 2.42 mm, with the same 19-pin configuration of Type A, but with slightly different pin assignments. Where connecting between Type C-Mini and Type A, an adapter or Type A-to-Type C cable is required.

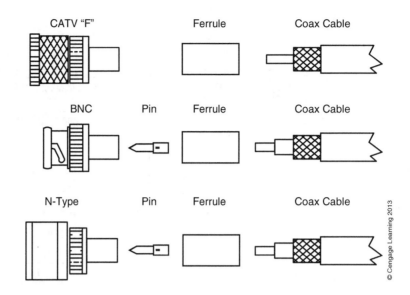

Figure 14–14
Coaxial cable connectors. BNC, British Naval Connector; CATV, community antenna television.

Type D is a microconnector having the same 19-pin configuration as Type A and C, but smaller in size, similar to that of a micro-USB connector. Type D measures approximately 2.8 × 6.4 mm.

Type E is used on automotive connections and equipment.

HDMI cables fall into two categories: Category 1 and Category 2. Category 1 cables are considered Standard HDMI cables, certified and tested to 74.5 MHz. Category 2 are considered high-speed cables, certified and tested to 10.2 Gbps or 340 MHz, for high-resolution, 3D, and Deep Color systems.

HDMI cable lengths are not subject to the same issues as FireWire or USB. Category 1 cables can be found to lengths of 5 m (16 ft), and depending on the quality of construction and wire gauge, HDMI cables can reach up to 15 m (49 ft) in length.

Coaxial Cable Connectors

Common coaxial cable connectors include CATV "F," BNC (British naval connector), and N-Type (Figure 14–14). Where making connections, the ferrule in the illustration is crimped around the cable as a means of securing the connector. Special crimp tools are required for each style of connector. Some connectors do exist, however, that use a special twist-on ferrule, which does not require the use of a crimp tool.

SEC 14.8 *NATIONAL ELECTRICAL CODE ARTICLE 820* REQUIREMENTS

The *Code* requirements for CCTV systems are not very different from those for communications systems. When comparing *Article 800* to *Article 820* of the *NEC*, there is very little difference, and they are virtually identical except for the following items:

For coaxial cables entering buildings from underground, the cables must be permanently separated by a fixed barrier from those of Class 1 conductors. In addition, underground burial of coaxial cables requires a minimum 12-in. (304.8 mm) separation from buried power lines or Class 1 cables that are not already installed in conduit (*NEC 820.47(A)* and *(B)*). The 12-in. (304.8 mm) separation is not required if the coaxial cables are installed inside of a raceway or have a metallic armor. For direct burial, coaxial cables also must be listed and labeled for such purposes. A specific burial depth is not required.

Where grounding coaxial cables associated with CCTV systems, as long as the outer conductive shield on the cable is grounded, no other ground is required. When installing a separate ground rod for coaxial cable systems, the ground rod needs to be 8 ft in length, as opposed to the minimum 5 ft length allowed in *Article 800*. For additional information, please refer to Chapter 3 in this book for the details of grounding and bonding coaxial cables.

Power Limitations

Coaxial cables are permitted to deliver power along the central conductor associated with distributed RF systems, provided the voltage does not exceed 60 V and that the current is supplied through a power-limited transformer or device. The power shall also be blocked from devices on the network that are not intended to be powered by the coaxial cable.

Table 820.179 of the *NEC* shows the required cable markings for CCTV system cable, and *Table 820.154(b)* details a list of cable substitutions.

The 2011 edition of the *NEC* has also added an expanded section, *820.113 Installation of Coaxial Cables*, which details the types of cables that can be used in fabricated ducts, plenums, risers, risers in metal raceways, risers in fireproof shafts, risers in one- and two-family dwellings, cable trays, distributed frames and cross-connects, multifamily dwellings, and other building locations. *Table 820.154(a)* was also added, which gives a detailed listing of Applications for Listed Coaxial Cables in Buildings. When deciding on which cable to use for a given location, refer to these tables and sections of the *NEC*.

Support of Conductors

Coaxial cables are not allowed to be strapped, taped, or attached to the exterior of any conduit or raceway as a means of support (*Article 820.133(B)*). One exception does exist, however: the exterior mast or raceway used to support overhead (aerial) cables that are outside and entering buildings. Once the cable has entered the building, attachment to any conduit or raceway is prohibited.

CHAPTER 14 FINAL QUESTIONS

1. What are the five main components of a CCTV system?
2. The video tube of an analog tube camera is a type of _____, such as used in older style television sets.
3. What are the two types of CCD technology?
4. Which is more sensitive to light, CMOS or CCD?
5. A light meter can be calibrated to read in _____ or _____.
6. What is the formula for calculating the amount of light at the camera?
7. Express 1526 ft-candles in lux.
8. What is the aspect ratio?
9. For facial recognition, it is recommended that HD cameras have _____ pixels per foot.
10. Explain what will happen if a large-format lens were used on a small-format camera?
11. Explain the difference between a fixed-focus and a varifocal lens.
12. What is the average focal length of the human eye?
13. How does an *f*-stop of 4 compare with an *f*-stop of 11? What is the difference in light level between the two settings?
14. How does an increase of shutter speed affect exposure and how is it related to *f*-stop?
15. What is depth of field?
16. When the *f*-stop is increased, what happens to the depth of field?
17. You are using a 0.25-in. camera format, with a FL of 35 mm. You want to view a conference table that has a horizontal span of 5 ft. How far must the camera be located from the table for you to be able to see a viewable picture on the monitor?
18. Explain the purpose of a multiplexer, and give an example of where one might be used.
19. What is the disadvantage of using a time-lapse video recorder?
20. What are the advantages to using a digital video recorder?
21. A PoE switch cannot output more than _____ V and _____ W.
22. Which mode of PoE delivers power to the spare pairs on a CAT5 cable?
23. What standard defines PoE?
24. The original standard for broadcast analog video in the United States was _____.
25. What standard replaced the analog television standard in 1995?
26. What are the three most common vertical resolutions for HD television that use an aspect ratio of 16:9?

27. When designing a network-based CCTV system, what is the most important detail?
28. In a modern-day CCTV system, the central computer controls every aspect of the system through _____ software.
29. What is required to be able to connect an analog video camera to a digital network?
30. How many categories of HDMI cable exist, and which can be used for high-speed, 10.2 Gbps transmission?
31. Is HDMI compatible with DVI?
32. What is the minimum distance requirement for a direct buried coaxial cable and a direct buried power line?
33. The central conductor of a coaxial cable can provide device power for distributed RF equipment up to a maximum of _____ V.
34. Where grounding coaxial cable, a minimum _____ ft rod is required.

Glossary

active filter   An amplified filter network that contains some form of internal electronic feedback to help stabilize the signal. In some cases, the parameters of active filters are user defined and adjustable, making them ideal for devices such as crossovers and equalizers.

active infrared sensor   An active IR sensor emits a continuous infrared light beam from an IR-LED. A detected break in the IR beam alerts the control panel of a security breach and triggers an alarm.

acoustic glassbreak sensor   A sensor that detects the unique frequency response of glass when it breaks. Regardless of the type of glass—plate, tempered, or safety—the sensor is usually able to decipher the difference and trigger the alarm. Installation and placement of the acoustic glassbreak sensor is typically on a wall within proximity of the window. Once installed, the sensor is tested for performance with a handheld, electronic glassbreak simulator. The simulator mimics the sound of many varieties of breaking glass to be able to activate the sensor and trigger an alarm.

advanced mobile phone service (AMPS)   An analog cellular phone service developed in the early 1990s. Within a single cell, AMPS can authorize up to 395 duplex analog voice channels operating in a frequency band between 824 and 894 MHz.

american wire gauge (AWG)   The standardized system in the United States for specifying the size of wire. An increase of three gauge numbers doubles the area and weight of a conductor and simultaneously cuts the dc resistance in half.

amplitude modulation   Varying the amplitude of a carrier frequency (vertical deflection of the sine wave) at a rate equal to the frequency of the superimposed information signal.

anaerobic   A non-heat curing method of connecting optical fiber connectors that uses a Loctite® 648 adhesive and a primer for curing.

analog-to-digital converter (A/D)   An electronic circuit that converts an analog signal into that of a binary digital signal. Such a conversion is necessary when interfacing an analog sensor or input device to a computer or microprocessor.

annexes   Located in the back of the *National Electrical Code (NEC)*, between *Chapter 9* and the *Index*. The annexes are a good reference; however, they are not an official part of the *NEC*. The annexes are included for informational purposes only; they are not valid *Code* references.

annunciator   See *annunciator panel*.

annunciator panel   Visually and audibly announces the location of a fire alarm. The panel is often segregated into building, floor, or zone. The zones correspond to the various sections of the building or floors of a protected structure. The specified zones visually and audibly indicate their alarm status on the front panel through an alphanumeric display, printed output, colored LEDs, back-lit indicator labels, or graphics.

antenna   A metallic conductor used to transmit or receive radio signals. An antenna can be a simple loop of wire, a multi-element device, or a parabolic dish, depending on the type of transmission, wavelength of the carrier signal, and bandwidth of the tunable receiver.

area code   The first three numbers of a telephone number, which identify the geographical region of a telephone subscriber.

articles   Individual subjects covered within the *NEC*. There are approximately 140 articles contained within the *NEC*. Article numbers directly relate to the chapter in which they are found. For example, *Chapter 1* contains *Articles 100* through *199, Chapter 2* contains *Articles 200* through *299*, and so on.

attenuation   The amount of voltage lost to a transmitting signal as it propagates through a cable or an electronic circuit. Attenuation measurements are calculated and listed in decibels.

attenuation crosstalk ratio (ACR)   A cable specification measurement similar to signal-to-noise ratio. High levels are therefore preferred. The perfect cable would have low attenuation and low crosstalk, thus allowing for maximum signal transmission down the line.

audio mixer   Used to combine signals from multiple sources, either to a single output channel or to multiple output channels, as selected by the user. Mixers also provide separation between individual signals so that they do not interfere with or load each other when combining.

authority having jurisdiction (AHJ)   The electrical inspector. The job of the AHJ is to ensure that the installation is *Code* compliant.

autotransformer   A single winding transformer, where the secondary and primary share the same coil. Because of this, an autotransformer is unable to provide electrical isolation between the input and output.

balanced   See *balanced signal or balanced line.*

balanced signal or balanced line   Ungrounded and measures an equal level of impedance from either side of the electrical connection to ground. A balanced connection is desirable because it helps reduce the amount of noise and interference from stray EMF or electrostatic interference.

band-pass   Only passes frequencies to the output within a specified range between the lower and upper cutoff frequencies. Everything between the lower and upper cutoff frequencies is considered the pass-band. The cutoff frequencies measure 3 dB below the peak voltage of the center frequency.

bandwidth   The separation between the highest and lowest producible frequency within an electronic system or cable. For example, if a piece of audio equipment has a frequency response between 100 Hz and 1000 Hz, then the bandwidth equals 900 Hz.

base unit   The hard-wired connection point of a wireless system, whether it is a cordless telephone, wireless microphone, or wireless network. The base unit is typically powered by ac line voltage and provides the electronic circuitry necessary to modulate and demodulate signals traveling over the wireless connection.

battery   A dc power source that converts chemical energy into electrical energy. A simple battery cell consists of two dissimilar metals separated by an electrolyte. The combination of dissimilar metals determines the amount of cell voltage.

bit rates   Specifies the maximum transmission speeds for data, or the number of bits per second, over specific lengths of cable or through electronic devices. As an example, 10 Mbps indicates data traveling at 10 mega-bits-per-second (Mega = 10^6 or million).

bluetooth   A wireless standard that provides a short-range communication link between peripheral devices and a computer, typically 10 cm to 10 m apart. Bluetooth was originally designed as a communication link between small, inexpensive devices such as notebook computers, PDAs, cellular phones, peripheral devices, modems, and loudspeakers. Bluetooth is similar to the IEEE 802.11 standard, but it operates in the 2.4 GHz band.

bonded/bonding   Connected to establish electrical continuity and conductivity.

bonding conductor   A reliable conductor to ensure the required electrical conductivity among metal parts required to be electrically connected.

breakout cable   Optical cables that combine multiple tightpack bundles under a single jacket. For this reason, they are stronger, larger, and more expensive. Breakout cables typically are used for long conduit runs and riser or plenum applications.

breakout kit   Prior to installing connectors on loose-tube cables, a breakout kit is required to protect the individual strands of fiber once the outer jacketing has been removed. Breakout kits are loose tube coverings that are placed over the bare fibers of loose-tube cables to separate and protect the individual strands while terminating connectors.

bridge   A network device that is used to connect and disconnect computer clusters operating in two separate domains. The two domains are normally not networked and are only temporarily connected through the bridge when communication is required.

buck-boost transformer   A type of ac line regulator that uses a special type of transformer that allows inductive feedback to either increase (boost) or decrease (buck) the level of voltage on the output. A semiconductor control circuit continuously monitors the output for voltage variations. The control circuit then adjusts the level of feedback accordingly to help maintain a constant output voltage.

bus topology   A method of computer networking that connects each node of the network to a common line (bus). The bus represents a broadcast environment where all devices on the bus hear all transmissions simultaneously.

busy signal   Detected when a telephone line is in use. The busy signal operates at approximately 480 Hz. It is a choppy, noncontinuous signal that alternates between on for 0.5 second and off for 0.5 second.

camera   Collects reflected light images from objects in the environment. The job of the camera is to capture the visual image of an event. Video cameras convert the light images into electronic signals that typically measure a maximum of 1 Vp-p (volt, peak-to-peak) for transmission through cables and to be displayed on monitors.

camera resolution   The quality of an image or the ability to see fine details in an image. Camera resolution is the measure of a camera's ability to distinguish between objects in close proximity to each other.

camera sensitivity   The minimum amount of light that the camera requires to produce a full video image.

cardioid A microphone sensitivity pattern that is unidirectional in nature; it can better detect sounds in the forward direction but has a greatly reduced sensitivity toward the rear.

carrier frequency Used in wireless communications to carry an information signal across a desired channel. The carrier is a high-frequency electromagnetic sine wave that radiates out over the channel. The channel is the air space between the transmitter and receiver antennas. To receive a communication, the receiver must be tuned to the carrier frequency of the transmitter.

cathode ray tube (CRT) Used in older style video cameras and television sets to record and display an image electronically through a finely focused electron beam. To record an image, the beam would scan a light-sensitive target inside the camera. The presence of light through the camera lens caused the light sensitive target to become more conductive, increasing the level of current flow through the tube. To view the image on a monitor, the process was reversed as electrons were aimed at a phosphorescent screen, causing it to glow when struck by electrons, thereby recreating the original presence of light. To create the illusion of motion and animation, a continuous stream of images is quickly scanned, synchronized, and rescanned, line by line, frame by frame, at a rate of 30 frames per second (25 frames per second in Europe). Because 30 frames per second is faster than human vision can perceive, the images appear as a steady stream of fluid motion, rather than a series of electronic flash cards.

central office (CO) The local telephone company that provides service to the customer.

central station An organization owned and operated by a private company that is responsible for monitoring, managing, and reporting the occurrence and status of the security or fire alarm systems of the various clients and facilities within a municipal area. Examples of central station service organizations include ADT and SIMPLEX.

channel In a hard-wired communications link, such as in a landline telephone system, the channel is the connecting wire between the sender and the receiver. In a wireless system, the channel is the air space between the transmitter and receiver antennas. See *carrier frequency*.

chapters The nine major categories within the *NEC* are known as chapters. Each chapter contains chapter-specific articles.

characteristic impedance The level of charge resistance a cable exhibits when transmitting high frequency signals between two points of a communication link. Also referred to as nominal impedance, the characteristic impedance must match the source and the load impedance or standing waves result. Characteristic impedance is a constant value; the length of the cable does not determine it. Instead, its value is determined by the size of the conductors, spacing, and type of insulation.

charge-coupled device (CCD) cameras Cameras that use a semiconductor target instead of a vacuum tube to record an image. CCD cameras are generally designed to be small in size and can suitably record images in environments where light levels are low and variable.

chassis ground Indicated on a schematic when the electronic signal ground or circuit common of the equipment and the earth ground of the electrical system are attached to the outer metal enclosure of the equipment. The reason for a chassis ground is two-fold. First, the chassis ground provides safety from ground-faults, and possible short circuits. And second, the metallic chassis provides a large conductive surface area for the electronic signal ground or common to help minimize ground resistance and reduce noise on the internal circuit.

circuit common A common ground point in an electrical circuit from which voltage measurements are taken.

circuit integrity (CI) A special type of listed and labeled fire alarm cable, consisting of solid, copper conductors and a lightweight fire resistant insulation that exhibits a 2-hour rating when burned. The insulation of CI cable remains flexible and easy to install at room temperature, but when exposed to high temperatures, it changes state, becoming a 2-hour fire-resistant protective barrier to the central conductors.

circular mil (c-mil) A term used to define the cross-sectional area of a conductor. One circular mil is equal in area to a 1/1000-in. diameter circle. As an example, a 0.25-in. diameter conductor equals 25 mils. To find the value of circular mil, the mil diameter must then be squared: $25^2 = 625$ c-mil. As the example shows, a conductor having a diameter of 0.25 in. equals 625 c-mil.

cladding A material layer placed between the core and the outer coating of an optical cable. Mechanically, the layer of cladding placed around the core acts as a reflecting mirror. Because the cladding is made of material with a different density than the core, light waves refract and reflect as they transmit though the optical cable. Without the layer of cladding, the light waves do not transmit through the cable.

Class 1 circuit Defined by *Article 725* of the *NEC*. There are two types of Class 1 circuits: Class 1 remote control and signaling and Class 1 power-limited.

Class 1 remote control and signaling has no power limit and can operate up to 600 V. Class 1 power-limited is limited to 30 V and 1000 VA.

Class 2 circuit   Defined by *Article 725* of the *NEC*. In general, Class 2 circuits are power-limited up to 100 VA. Class 2 circuit conductors must be rated up to 150 V. Current limitations do change based on voltage levels (see *Chapter 9*, *Table 11(A)* and *(B)* of the *NEC*, which defines the specific voltage ranges and current limitations of inherently limited and non-inherently limited, ac and dc, Class 2 circuits and power supplies).

Class 3 circuit   Defined by *Article 725* of the *NEC*. In general, Class 3 circuits are power-limited up to 100 VA. Class 3 circuit conductors must also be rated up to 300 V and must not be smaller than 18 AWG. Current limitations do change based on voltage levels (see *Chapter 9*, *Table 11(A) and (B)* of the *NEC*, which defines the specific voltage ranges and current limitations of inherently limited and non-inherently limited, ac and dc, Class 3 circuits and power supplies).

client/server   A client/server network designates one node of the system as the main file server and system resource for all other connecting nodes of the network.

closed circuit television (CCTV)   Isolated television systems that are primarily used in surveillance and security systems. A common CCTV system consists of a camera, a transmission link (wiring), a monitor, and optional peripherals such as video or time-lapse recorders.

closed loop   Indicates that circuit contacts are closed, providing continuity, and a low resistance path for the source.

coating   Also known as plating, refers to an optional layer of tin, silver, or nickel that covers the exterior of most copper conductors. The plating helps eliminate problems when soldering, allows for more simplified stripping and removal of insulation, and helps prevent the oxidation and deterioration of the central conductor at high temperatures.

coaxial   Implies that the conductors share the same central axis point in the cable. Coaxial cable consists of a solid central conductor (stranded central conductors are also available) surrounded by a polyethylene insulator with a concentric outer conductor, or shield, consisting of multiple braided strands. Coaxial cables are traditionally used on unbalanced circuits to reduce the amount of noise and interference on the conductors.

coaxial cable   See *coaxial*.

coaxial speakers   Sepakers with two or three diaphragms mounted inside a single speaker housing, all driven by one voice coil. The different sized diameter cones are mounted on top of each other, terminating back to one central point at the voice coil. The three separate diaphragms then move simultaneously with the voice coil, thereby generating a full range of sound frequencies within a single space, all radiating out from a central point.

coherence   An in-phase relation between two signals. When two signals are in-phase and mixed, they positively combine to increase the total amplitude at the output, resulting in an increased signal level.

coherent bundle   Optical fiber cables can be manufactured in randomized or coherent bundles. Coherent bundling means that the individual glass fibers of the cable are carefully lined up end-to-end in such a way that an image can be transmitted through the cable and displayed in a viewable form at the opposite end.

common ground   Also referred to as circuit common, it represents a common point in an electrical circuit from which voltage measurements are taken.

common-mode signal   A signal present on both sides of a power line, measuring equal in level to ground.

communication circuit   Communication circuits and systems are covered by *Articles 800, 810, 820,* and *830* of the *NEC*. They include telecommunication circuits, security and alarm circuits, radio and TV antenna systems, CCTV and radio distribution systems, and network-powered broadband.

compensator   A portion of the telephone that compresses the dc or the loudness of the voice signals passing over the local loop. The compensator helps to regulate and maintain voice signals at an average level regardless of how loud or soft they may be.

compliance   A speaker specification that is measured in Newtons per meter (N/m), indicating the ease with which a speaker's diaphragm moves.

composite optical fiber cable   Composite cables, also known as hybrids, come in one of two varieties: those that contain both optical fibers and copper conductors under a single jacket and those that contain multimode and single-mode optical fiber under a single jacket. Composite cables are commonly used for special applications where the need exists to transmit between systems over multiple types of cable.

condenser microphones   A microphone made from a capacitor. In early times, capacitors were referred to as condensers, hence the name condenser microphone. The diaphragm of a condenser microphone attaches to one of the plates of the capacitor. As sound pressure from the atmosphere pushes on the diaphragm, the spacing between the metal plates changes. The overall capacitance increases as the plates move closer together and decreases as the plates move farther apart. As capacitance changes, so too does the quantity of electrostatic charge stored inside the device, developing a direct analog to the changing sound pressure on the diaphragm. Condenser microphones require a power source to operate.

conductive optical fiber cable   Often includes a variety of metallic strength members, metallic vapor barriers, and metallic armor or sheathing. Although these metallic components are not meant to carry current, they are still conductive in nature and must be grounded when installed.

conduit fill   Where installing electrical wiring within a raceway, the number of conductors to be run should be limited to a certain percentage of fill. Limiting the number of conductors within a raceway allows for the easy addition and removal of future wires to avoid damaging those already existing. In addition, not stuffing the raceway to the maximum fill allows room for the dissipation of heat. To calculate conduit fill, refer to *Chapter 9, Tables 1* through *5* of the *NEC*.

constant-voltage transformer regulator   Uses a saturating transformer and the coupling of a resonant tank circuit to achieve regulation. The resonant tank circuit is made from the internal inductance and capacitance of the transformer. The main concept is that the resonant circuit provides the additional output when the input decreases during voltage lulls. Also, during voltage spikes or unexpected supply increases, the saturating core has the counteracting effect of reducing output voltages back to normal levels.

control unit   The brain of the fire alarm system. It monitors all input sensors and initiating devices, such as smoke or heat detectors or radiant energy devices, and also controls alarm and signaling devices, such as horns, strobes, ventilation dampers, elevator recall, and automatic doors. The control unit also performs supervisory services by continually monitoring the functionality of the system, connecting wires, power, and active devices. During an alarm, the control unit will also seize control of the telephone line and dial out to the central station command center, alerting authorities to potential threat of fire at the specified location.

copper-clad   Copper-clad wire is available in a variety of types. The most common examples include copper clad aluminum and copper-clad steel. Copper-clad aluminum fuses an outer layer of copper to a central core of aluminum. Copper-clad steel fuses an outer layer of copper to a central core of steel.

copper-weld   Fuses a thin coating of hard-drawn, bare copper to a steel core. Although copper-weld is only 40% conductive at low frequencies, it has a significantly higher level of tensile strength, making it nearly impossible to stretch. Copper-weld is ideal for high-frequency applications, especially shortwave transmitting aerials and the construction of antenna systems.

core   The central part of an optical fiber cable, typically made from glass strands or plastic.

critical angle   The numeric aperture of an optical fiber cable. Any light waves attempting to enter the cable outside of the critical angle are rejected. Essentially, the critical angle is the angle of acceptance for all light passing through the optical cable.

crossovers   Filter networks that are used to separate and direct a complex mixture of audio frequencies to their required sized loads. For example, a low-pass filter is used to pass low-frequency bass signals to a woofer; a band-pass filter is used to pass midrange frequencies to a midrange diaphragm; and a high-pass filter is used to pass high frequencies to a tweeter. These filters make up the crossover.

crosstalk   The unwanted transfer of signals from one conductor to another through the process of electromagnetic induction or capacitive coupling.

damping   To restrain or reduce in amplitude. When designing speakers, damping refers to mechanical restraint that helps to prevent unwanted bounce on the speaker diaphragm during the absence of signal. The level of damping affects the transient response of the diaphragm and the resonant frequency of the speaker.

decibel (db)   The common unit of measurement for sound, power attenuation or gain, and voltage attenuation or gain. dB stands for 1/10 of a Bel on a logarithmic scale. The power dB formula is

$$dB = 10 \log(P/P_R)$$

where P represents the measured power and P_R represents the reference power.
The voltage dB formula is

$$dB = 20 \log(V_2/V_1)$$

where V_1 represents an initial voltage measurement across a specific resistance (before a changing condition), or it can represent a constant voltage reference. V_2 represents a voltage measurement from the same point in the circuit after a changing condition.

delay skew   Timing differences among the four pairs of a communication or network cable resulting from the varying twist rates of pairs and varying propagation delays. The timing of delay skew is usually measured in nanoseconds (a billionth of a second, or 10^{-9} second).

definitions   The *NEC* contains a glossary of terms referred to as definitions found in *Article 100*. All definitions found in *Article 100* apply to multiple

NEC articles. Definitions supplied in *Article 100* help to clarify the initial intent of the *Code*. *Article 100* does not contain all of the definitions found within the *NEC*. Some definitions only pertain to specific articles. In such cases, those definitions are article specific, meaning that they apply only to that article. Definitions may also be found inside of sections; again, those definitions only apply to the specific section for which they are found.

demarcation point The junction of the main service wiring and the premise wiring. The main service wiring is controlled and maintained by the service provider, whereas the customer is responsible for and maintains the premise wiring.

depth of field Relates to the measured distance of focus in front of and behind the main subject of an image. As an example, when the camera is focused on a spot in the viewing frame, there is a measured distance in front of and behind the spot which remains in focus.

dial tone The telephone company provides a dial tone on the line to indicate that the phone is ready for use. In the United States, the tone is made by modulating a 350 Hz tone with 440 Hz. The tone is also continuous at a level of −12 dBm.

diffraction The bending of a waveform around an object.

digital subscriber line (DSL) An alternative to a standard dial-up modem Internet connection because it allows high-speed computer data to share the telephone line with the analog voice communication. The service provider of a DSL system installs a special DSL modem that filters and splits the analog voice signal from the high-speed digital data. The speed of DSL can operate at nearly 1.5 Mbps, up to 50 times faster than dial-up modems. DSL comes in six varieties, all varying in speed, upstream and downstream symmetry, and maximum distance. They include asymmetrical digital subscriber line (ADSL), symmetric digital subscriber line (SDSL), rate-adaptive digital subscriber line (RADSL), high-bit-rate digital subscriber line (HDSL), and very-high-bit-rate digital subscriber line (VHDSL).

digital-to-analog (D/A) converter An electronic circuit that converts a binary digital signal into an analog signal. Such a conversion is necessary when interfacing the output of a computer or microprocessor to the input of an analog device. An example of an analog device is an audio loudspeaker.

digital video recorder (DVR) Uses a hard drive-type storage device for recording images. The DVR looks similar to a typical VCR except that there is no tape. DVR is capable of high-resolution picture playback and also gives the user the ability to almost instantly access a particular event without having to scan through the entire recording in a linear manner, as with videocassette tapes. It is also possible to view recorded images remotely through a secure connection using a DVR.

dissimilar metals The joining of unlike metals such as copper and aluminum.

domain name server (DNS) Converts the numbered Internet address code of a URL to a series of recognizable words. As an example, the URL address of an Internet pet shop may be 145.22.78.82. The DNS then converts the coded URL into a recognizable format such as http://www.petshop.com/newpets/blacklabs.html.

downlink The process of receiving a satellite transmission from outer space on the ground.

dual-tone multifrequency (DTMF) A two-tone signal developed by a telephone company to facilitate touch-tone dialing. Each number of a touch-tone keypad generates a unique dual-tone frequency combination. Equipment at the TELCO then decodes the frequency combinations and connects the desired call.

duplex coil Provides an auditory feedback from the mouthpiece of a telephone to the earpiece. It is important to control the amount of voice signal feeding back to the earpiece of the handset so speakers can hear a portion of their sound as they speak.

dynamic microphones Operate under the principle of induction, as sound pressure variations cause a coil of wire to move within a magnetic field. The moving coil then produces electrical impulses at the input of the mixer and amplifier. Dynamic microphones tend to be quite durable, and unlike condenser microphones, they do not require a power source to operate.

dynamic range The maximum span, or range, between the loudest and softest level of sound an object is capable of producing. An object can be a person, a musical instrument, or any mechanical device.

earpiece The earpiece of a telephone handset is made up of an inexpensive, small, 8 Ω speaker. The speaker vibrates in the presence of an ac voice signal produced by the mouthpiece. The vibration recreates the voice signals sent by the person on the other side of the line over the local loop.

earth ground A circuit ground point that is connected to the earth

echo See *reverberation*.

effective ground-fault current path An intentionally constructed, low-impedance electrically conductive path designed and intended to carry current under ground-fault conditions from the point of a ground-fault on a wiring system to the electrical supply source. The purpose is to facilitate the operation of an overcurrent protective device or ground-fault detectors on high-impedance ground systems.

electric strike Used in security systems as a door latching mechanism. They typically operate from 12 or 24 V, but higher voltage strikes are available. The strike may run on ac, dc, or both, depending on the make and model. An electric strike uses a pivoting ramp plate to allow or deny access through the door. The device is mounted in the door frame where the door latch normally resides. When activated in fail-safe mode, an electronic solenoid within the strike latches the pivot plate, making it impossible to swing freely or move and thereby securing access through the door. Once deactivated, the internal solenoid releases the pivot plate, allowing it to swing clear as the door is pushed open. Electric strikes can be configured as fail-safe or fail-secure.

electromagnetic wave Light and radio frequencies travel in the form of electromagnetic waves; that is, the wave emits both electric and magnetic fields as it moves through space. The fields vibrate at right angles to the direction of movement and at right angles to each other. The amount of energy in a light or radio wave is proportionally related to its frequency; high-frequency waves have high energy and low-frequency waves have low energy. When traveling through a vacuum, electromagnetic waves can move at speeds of more than 186,000 miles/sec, or 300,000,000 m/sec.

electronic key system (EKS) Telephone system for small businesses that allows them to share multiple lines from the CO without having to go through a PBX. The system is essentially a series of electronically controlled switches and relays that multiplex lines to specific station sets as needed. Customers can not only make calls between station sets locally, but also share incoming lines from the CO.

emergency situation Defined as any event that is a possible threat to life safety. The possibility for loss of human life is always a primary concern in any emergency situation, with property loss and damage ranking as a secondary concern.

emergency telephone Isolated, separate system that provides two-way communications or party-line connections to multiple handsets throughout a building. They are not connected to or part of the main public telephone system. Emergency telephones are used by the fire department to communicate between fire command centers and building stairwells. Emergency telephones also do not need to be dialed because connection to the communication loop is made by simply lifting the handset. All parties can communicate simultaneously without the risk of external interference or line loss.

end-of-line device Fire alarm and security systems often use end-of-line devices on their loop circuits to help supervise the presence of system activity. End-of-line devices typically consist of resistors, relays, or diodes, depending on the design of the system.

end-of-line relay Used on fire alarm or security system loop circuits to activate or deactivate power to a secondary system when an alarm has been activated. An example would be to turn off the blower fans of an HVAC system during a fire alarm to prevent toxic smoke from blowing through the entire building.

equalizers Provide line-level audio processing control of specific frequencies in an audio system. An equalizer typically includes a series of active low-pass, high-pass, band-pass, and notch filters that are used to control individual frequencies of an audio signal. The two most common varieties are graphic equalizers and parametric equalizers. Graphic equalizers are easy to use and display the graphical curve of the adjusted frequency spectrum on the front panel display of the device. Parametric equalizers are more complex and require a high-level knowledge of frequencies and filtering to be able to use them.

equipment grounding conductor (EGC) The connective path(s) installed to connect normally non-current-carrying metal parts of equipment together and to the system earth ground.

Ethernet The first computer networks, known as Ethernets, ran at maximum speeds of 2.94 Mbps on coaxial cable and were developed in the 1970s by Xerox. Xerox soon joined with Intel and Digital Equipment Corporation to develop a standard for a 10 Mbps Ethernet, known as DIX (Digital, Intel, Xerox). In 1987, the development of the 10Base-T Ethernet communicated at speeds of 10 Mbps on a twisted pair line; such networks could reliably transmit over a 100 m run.

exceptions Alternatives to specific sections of *NEC Code*. There are many exceptions listed throughout the *NEC*. Often, the *NEC* states a rule to follow, then clarifies that rule by saying "except." In such cases, it is important to read all of the exceptions to be sure of the alternative options.

exchange The middle set of three numbers in a telephone number represent the exchange (612-555-1234). The exchange is a switch that decodes incoming tones or pulses to another inside TELCO. The last four numbers represent the exact phone line.

fail-safe A device that deactivates when power is removed.

fail-secure Implies that a device remains activated when power has been removed. An example may be door locks to a secured area that continue to maintain their latching mode during a power failure, thus maintaining a level of security to the area at all times.

far-end crosstalk (FEXT) Signal distortion caused by the retransmission of a received signal at the far end of the line. The destination of a transmission not only contains the intended signal but also the retransmitted receive signal from the other end. The mixing of transmitted and retransmitted data undoubtedly distorts the intended transmission, as well as generates unwanted errors.

federal communications commission (FCC) The federal agency that regulates the standards, licensing, and use of radio frequencies in the United States.

fiber distributed data interface (FDDI) Uses optical fiber to transport high-speed data over a computer network. Originally developed in 1987 for token ring networks, FDDI was able to transmit data at speeds of up to 100 Mbps. Today, FDDI still exists as the main backbone for many installations requiring long-distance runs including universities and large corporate complexes. FDDI is also commonly used to interconnect two or more LANs because optical fiber inherently transmits a higher quality signal over a greater distance than copper.

fiber-optic cable Transmits a light signal through glass or plastic conductors, as opposed to an electrical signal through copper wire.

field coil The outer coil of a speaker or generator that generates a stationary magnetic field from the presence of an applied dc current flow. The stationary magnetic field is then used to counteract the changing magnetic field of the speaker voice coil or motor armature. Motion results from the attraction of opposing magnetic fields and the repulsion of like fields.

field of view (FOV) An external measurement of image coverage, relating to the maximum range of horizontal or vertical visibility as seen by the camera.

file transfer protocol (FTP) Used to upload or download files to or from network servers. FTP is built into most Unix- and Windows-based operating systems and Web browsers. When files need to be transferred between servers or network nodes, FTP is engaged to perform the desired task and is often transparent to the user's viewpoint. FTP is how files move around a network or the Internet.

Fine Print Notes (FPNs) Can only be found in editions of the *NEC* prior to 2011. They are included within the *NEC* for clarification purposes. FPNs are not enforceable and are not a requirement of the *Code*. In 2011, the *NEC* relabeled the FPN as an Informational Note. Words such as *may* or *might* are used in FPNs, but words such as *must* or *should* are never used.

fire alarm circuit Authorized *Code* for fire alarms and fire alarm circuits are published by the National Fire Protection Agency. NFPA 72 is the *National Fire Alarm Code Handbook,* and NFPA 70, or the *NEC,* outlines the wiring of non–power-limited and power-limited fire alarm circuits in *Article 760.*

fire command center An approved, designated area, typically located at the entrance to a building, that the fire department uses as a central communication command post. In the event of an actual emergency, the fire command center is used by the fire department to direct and manage all activities related to building evacuation, automated building controls, emergency communications, and fire fighting. Large structures may have multiple command centers operating simultaneously, but only one can be designated as the central command center.

firewall Prevents unauthorized individuals from gaining access to a local intranet or to a private network of computers. Access to such computers is only granted to specified users with correct authentication codes such as a valid username and password.

FireWire Also known as i.Link, FireWire is a high-bandwidth, serial communication standard developed by IEEE (IEEE 1394). The connection allows a maximum of 63 devices with a data transfer rate of 400 Mbps. Although FireWire is ideal for the transfer of video and audio information, it is also commonly used to connect hard drives and computer storage devices because of its high bandwidth and speed. Maximum cable lengths for FireWire are limited to 14 ft.

focal length (FL) A measurement internal to the camera. It defines the distance from the focal plane of the camera (θ) to the lens when the lens is focused on infinity.

foot-candle A standard unit of measurement for most light meters. A foot-candle is the amount of illumination produced by a standard candle at a distance of 1 foot.

frequency Measured in Hertz (Hz), frequency represents the vibrational rate of an alternating waveform in cycles per second. 1 Hz represents 1 cycle per second.

frequency division multiple access (FDMA) A modulation technique that controls the allocation of channels to cell phone subscribers. It allows customers to be reassigned frequencies as needed due to problems of interference and availability at cell towers.

frequency modulation (FM) Involves varying the carrier frequency of a radio signal at a rate equal to the superimposed information signal. When a 92.5

MHz carrier is FM modulated by a 1 kHz audio tone, the carrier frequency shifts 1 kHz above and 1 kHz below 92.5 MHz (92.499–92.501 MHz). The absence of a modulating signal causes the carrier to remain at a constant 92.5 MHz.

frequency response   The range of producible frequencies for an object, subject, instrument, or piece of audio amplification equipment. The frequency response of most audio amplification equipment is 20–20,000 Hz, which is the range of human hearing.

frequency shift keying (FSK)   A type of frequency modulation that causes the carrier frequency to vary with respect to two separate tones to represent the highs and lows of a binary number. For example, the two tones may be 1kHz and 10 kHz. The 1 kHz tone can represent a binary 0, whereas the 10 kHz can represent a binary 1. By alternating the sequence of the modulating tones, the carrier can then be made to transmit a direct digital code.

full-duplex   A method of communication where both parties of a communication link can simultaneously transmit and receive because they are operating on separate channels. Such a link allows an individual to hear a response while in the process of responding. A standard telephone connection is an example of a communication link that operates in full-duplex mode.

gateway   Used to interconnect computer systems, e-mail systems, or any network devices that would otherwise be unable to communicate directly with each other. Such examples would include Macintosh to IBM PC or even different varieties of mainframe computer.

generator   Provides electrical power to a circuit or load, whether it is ac or dc. A generator comes in two varieties, those driven by a motor and those driven by an engine. Engine-driven ac generators are often used as a source of backup power in cases where the main power line has failed. An ac motor-driven generator provides a source of isolation from the main utility power line. The main utility power line often serves as the source of power for the motor generator. Motor-driven generators have the added advantage of being able to provide ac-to-ac conversion or dc-to-ac conversion, depending on the type of motor at the front end of the system. Ac-to-ac conversion also can include a frequency conversion such as 60 to 400 Hz.

geosynchronous satellite (GEO)   A multichannel, multifrequency transponder that rotates 35,880 km (22,300 miles) above the Earth in a geosynchronous orbit. The distance from the Earth to the geosynchronous orbit is the exact location where the centrifugal forces pushing the satellite out of orbit and the gravitational forces pulling the satellite down toward Earth are equal and balanced. The achievement of a geosynchronous orbit usually guarantees that the satellite remains in a fixed elevation above Earth while simultaneously appearing to be stationary from a ground perspective because of the speed of orbital rotation.

glassbreak sensor   Comes in two varieties, acoustic glassbreak sensors and shock-type glassbreak sensors. The acoustic glassbreak sensors detect the sound vibrations of breaking glass. Shock-type glassbreak sensors are mounted directly on the glass to detect the physical shock vibration of the breaking glass.

global system for mobile (GSM)   A version of PCS used in Europe, most of Asia, and parts of North America. GSM is a type of TDMA, but it uses a Gaussian minimum-shift keying method of modulation. Such a method reduces the overall bandwidth between the two shift-key modulation frequencies, making the system far more economical than those of standard FSK methods.

goosenecks   A flexible metal conduit that is available in various lengths for mounting and connecting a microphone to a podium or countertop. The flexibility of the device allows for multiple positioning angles by the user to achieve optimum microphone performance. A secondary benefit of the flexible metal conduit is that it provides an added level of shielding to the internal wires, as well as a physical barrier to protect the device from possible damage.

graded index   A type of multimode optical fiber (MMF). Graded index MMF comprises many different layers of glass, all chosen with a different index of refraction to produce an overall index profile approximating that of a parabola. Because light travels faster through a lower index of refraction, the light travels faster along the outer edges of the cable and slower closest to the core. Light traveling at higher modes through graded index fiber has an overall travel time similar to those of the straight-line axial rays along the central core, thus allowing for minimal higher mode signal dispersion and higher bandwidths. Graded index fiber can provide nearly 100 times the bandwidth of step index fiber, and it is traditionally used in LED-driven data communications systems.

graphic equalizer   Equalizers which use a combination of active filters separated by equal bandwidth, which is sometimes as narrow as 1/6 of an octave along the entire audio spectrum, to help adjust and change the frequency response of a line-input audio signal. The graphic equalizer gets its name from the front panel display that graphically illustrates the adjusted frequency response of the output signal. The position of the front panel faders, when adjusted, graphically displays the curve of the adjusted frequency response of the signal.

ground   Connected to the earth. An earth ground in an electrical circuit is used to direct lightning into the earth to protect wiring and cables from high-voltage surges. Since the Earth's atmosphere regularly develops static electrical charges, which occasionally discharge in the form of lightning bolts, a protective, conductive pathway must be provided to help draw the energy away from buildings to safely dissipate it into the earth.

ground fault   An unintentional, electrically conducting connection between an ungrounded conductor of an electrical circuit and the normally non-current-carrying conductors, metallic enclosures, metallic raceways, metallic equipment, or earth.

ground-fault circuit interrupter (GFCI)   A device intended for the protection of personnel that functions to de-energize a circuit, or portion thereof within an established period of time when current to ground exceeds the values established for a Class A device. The established over-current value that triggers a ground-fault circuit interrupter is typically 5 mA.

grounded conductor   A system or circuit conductor that is intentionally grounded. The grounded conductor provides current to the load.

grounded/grounding   Connected to ground or to a conductive body that extends the ground connection.

grounding electrode   A conducting object through which a direct connection to the earth is made.

grounding electrode conductor (GEC)   A conductor used to connect the system grounded conductor or the equipment to a grounding electrode or to a point on the grounding electrode system.

grounding electrode system   All the grounding electrodes that are present at a building or structure that are bonded together to form the grounding electrode system. The grounding electrode system can include inter-bonded combinations of the following: metal underground water pipe, metal frame of the building or structure, concrete-encased electrode, ground ring, rod, pipe or plate, or other listed grounding electrodes permitted by the authority having jurisdiction.

ground loop   Results from unwanted ground currents flowing between two low-impedance ground points, which typically results from 60 Hz power being referenced to the earth at multiple locations. The presence of unwanted currents along the ground plane can develop stray ground voltage and unwanted noise along the connecting ground resistance. Once present, the ground loop can also act as an inductive pickup for any other stray magnetic or electromagnetic fields in the environment. When present, a ground loop generates noise spikes, voltage fluctuations, and a 60 Hz hum.

ground start   A method of establishing a call connection when dialing through a PBX. To seize the line, the PBX momentarily grounds the "ring" conductor of the telephone circuit. At the CO end of the circuit, the ring is connected to -48 V dc. As soon as the PBX attaches the "ring" to ground, the CO senses the flow of current and responds by grounding the "tip" conductor. The PBX then senses the flow of current along the tip, and responds by removing the "ring" ground. The loop circuit is now connected and dial tone is activated. To disconnect the circuit, the CO removes ground from the tip conductor, indicating to the PBX that the call has ended.

half-duplex   An alternative to full-duplex, in half-duplex communication the transmitter and receiver share the same frequency carrier or channel. In such cases, the receiver is disabled when the transmitter is operating and vice versa. Communication protocols must also be established when using a half-duplex system to help minimize confusion or interference between listeners on both ends of the link.

handset   The mobile piece of the cordless phone. It does not have anything physically tying it down to a specific location, unlike the base unit. The handset also contains circuitry to transmit and receive radio signals. The handset contains a speaker, microphone, keypad, and rechargeable batteries.

harmonic filters   Used to filter out unwanted harmonics from electronic circuits, power supplies, and power lines. The presence of unwanted harmonics can cause signal distortion, power supply noise, the overheating of cables and conductors, and the possible threat of fire.

harmonics   Direct multiples of a base frequency. If the base frequency is equal to 1000 Hz, then the second harmonic is 2000 Hz, the third harmonic is 3000 Hz, the fourth harmonic is 4000 Hz, and so on. The first harmonic is always equal to the base frequency.

heat detector   Designed to detect abnormally high temperatures within a given space or area. Heat flow or heat transfer can occur in two ways, as either convective or radiant heat. Because of the differing methods of heat transfer, a heat detector may operate under any of the following three principals: fixed temperature detector, rate compensation, or rate-of-rise detector. The detector can be designed in any of the three manners, or they can be all inclusive, depending on the manufacturer.

high-definition multimedia interface (HDMI)   An audio/video interface used for transmitting digital data between compatible devices, computers, and monitors. HDMI supports 8 channels of audio, compressed or uncompressed, and any uncompressed format of TV or PC video. HDMI is also electrically compatible with Digital Visual Interface (DVI) when a DVI to HDMI adaptor is used. DVI was an earlier style connector for digital video, which is still found on some equipment.

high-pass   The high-pass filter passes a band of frequencies on the upper end of the spectrum. Compared with the low-pass filter, the high-pass filter has the -3 dB cutoff point on the low side of the pass-band. Any frequencies below the cutoff are significantly attenuated.

hook switch   Indicates whether the phone is "on hook" or "off hook." The switch can be easily identified as the plastic button depressed by the earpiece of a standard phone when it is on the hook. The switch is open when the handset of the telephone is on the hook, indicating an open circuit drawing no current. When the handset is off the hook, the switch closes and the line draws approximately 10–90 mA of dc current, depending on the resistance of the local loop. Current draw on the loop notifies TELCO to provide a dial tone and make the line ready for use. The typical off-hook resistance of a telephone is about 400 Ω.

horizontal wiring   The wiring of a specific floor represents horizontal wiring, from the intermediate distribution frame (IDF) to the specific user. Horizontal wiring does not run from floor to floor, but is instead contained to a specific horizontal plane.

horns   Horns are audible alarm indicator devices for fire alarm systems. Horns provide a high-power, high-decibel, audio output, necessary to cover large spaces or to help cut through noisy environments during a fire alarm.

Hotmelt®   Hotmelt® connectors are a 3M product for optical fiber cables. The connector is prefilled with an epoxy which must be heated in a Hotmelt® oven for about 2 minutes. Once the epoxy is softened by the heat, the unjacketed, unbuffered fiber is inserted into the connector. The connector is then air cooled prior to polishing.

hub   Connects multiple user stations of a computer network through multiple ports. The hub does not make any decision or determination on the data, but instead acts as a multiport repeater. This device fits into layer 1 of the OSI model.

index   The index is located in the back of the *NEC*, organized in alphabetical order from keywords found within the electrical *Code*. The index is the best reference for finding multiple occurrences of specific words and phases within various articles and sections of *Code*.

index of refraction   A ratio comparing the speed of light in a vacuum with that of any other material or medium. In a vacuum, light travels at its maximum speed of 186 miles/sec. Through any other material, the speed of light slows down. In a vacuum, the index

of refraction is equal to 1. As a result, all other materials or mediums exhibit a higher index of refraction due to the slower speed of travel. As light travels between materials having different index of refraction levels, the light wave bends. The material makeup of an optical fiber cable uses differing index of refraction levels for the core and the cladding to get the light wave to bend as it travels through the cable.

Informational Note   Prior to the 2011 edition of the *NEC*, the Informational Note was a Fine Print Note (FPN). Informational Notes are used within the *Code* book to provide additional information and reference sources for a specific section of code.

inherently limited   An inherently limited power supply is clamped internally and unable to deliver more than a specific amount of energy to a load. Any attempt to push an inherently limited source past its maximum limit causes it to shut down or self-destruct in a safe manner.

initiating device circuit (IDC)   Initiating devices are input sensors on fire alarm circuits that are used to trigger alarms. Examples of initiating devices include smoke detectors and heat detectors. Initiating device circuits connect to initiating devices and are divided into the following classifications: Class A, style D or E and Class B, style A, B, and C. Class A circuits use a 4-wire connection and Class B, a 2-wire connection.

instrumentation tray cable (ITC)   Covered by *Article 727* of the *NEC*. ITC specifications are limited to 150 V and 5 A for sizes 22 to 12 AWG wire. Size 22 AWG is limited to 150 V and 3 A. ITC can only be used and installed in industrial establishments that are maintained and supervised by qualified personnel. It is not to be used for general-purpose installations or as a substitute for Class 2 wiring. (Class 2 wiring is covered by *Article 725* of the *NEC*.)

integrated amplifier   A self-contained audio-processing device that provides multichannel input selection, tone control, power amplification, and connection to speaker outputs. An integrated amplifier is essentially a combination preamplifier/power amplifier.

integrated services digital network (ISDN)   A dial-up service to the Internet. It does not offer a dedicated or always-on connection to a network like DSL. In some respects, ISDN is similar to a standard dial-up modem connection except that the ISDN has the ability to share the line simultaneously with the telephone, fax, and computer.

intermediate distribution frame (IDF)   A free-standing or wall-mounted rack for managing and interconnecting the telecommunications cable between telsets, fax machines, modems, or answering machines. The IDF is used in tall buildings where placing the main distribution frame (MDF) in the basement or first floor would cause cable runs in excess of thousands of feet. IDFs are typically located on every floor or every other floor, as deemed necessary by the designer.

internet protocol (IP)   A routing protocol that contains the addresses of the sender and the intended network destination. IP addresses can be permanently or temporarily assigned, depending on how the computer is connected to the network.

intersystem bonding termination   A device that provides a means of connecting bonding conductors from communications systems to the grounding electrode system. In most cases the device is either a multi-connection terminal bar, or copper plate.

inverse square law   States that the measured level of signal intensity decreases by the square of the distance, or 6 dB for every doubling of distance from a signal source. Likewise, at one-half the distance from the source, the measured level of signal intensity increases by 6 dB. The inverse square law works for all types of waveforms radiating out into a three-dimensional space, including electromagnetic waves (radiofrequency), sound waves, and photons of light. The following formula is used to calculate the expected loss or gain in signal level from a measured distance of d to the source:

$$dB \text{ loss or gain at distance } d = 20 \log (\text{Reference Distance}/d)$$

isolated ground   Used on technical circuits as a means of isolating and separating sensitive circuits from electrical noise commonly found on the chassis or equipment ground of an existing ground plane. Isolated grounds use a single-point grounding scheme that connects back to a main ground point within a building.

isolation transformer   Used to isolate a piece of electrical or electronic equipment from the main power line. A 1:1 relation exists between the primary and the secondary coils of the isolation transformer. Although the output has the same voltage and current flow as the input, it is electrically isolated, having only a magnetic coupling between the two sides of the transformer.

labeled   Equipment that has been proven safe and acceptable for use by a certifying laboratory, such as Underwriters Laboratories (UL), and is physically labeled to indicate its certification.

LAN extender   Remote-access, multilayer switches that are used to filter out and forward network traffic coming from a host router. Although they can be used to help filter out unwanted traffic by address, they are unable to segment or create security firewalls.

lapel   See *lavaliere.*

lavaliere   A lavaliere is often referred to as a lapel microphone. It is like a condenser in nature and, as referenced by the alternate name, is usually clipped to the lapel of a shirt or worn in close proximity to an individual's chest to achieve optimum signal response.

lens   The lens of a camera is a transparent optical device used to converge or diverge transmitted light to form images. Images pass through the lens to the internal focal plane or film plane of the camera to be recorded.

light-emitting diode (LED)   An optical signal source. LEDs are low-current, low-voltage, semiconductor devices, typically operating at 2 V and 20 mA. On an atomic level, the light is produced as electrons traveling between the negative and positive side of the diode fall into the vacant holes within the crystal structure of the covalent bond. The LED is similar to a standard diode in that it only conducts current in one direction (from negative to positive). The difference, however, is the colored plastic lens built around the exterior of the LED to help concentrate the light as it radiates out from the junction of the device. LEDs are available in an assortment of colors.

lightning arrestor   Primary protectors that are used to limit lightning and switching transients to safe levels. Lightning arrestors appear as insulators or open circuits to all voltages below a specific trigger threshold. Once the threshold has been reached, the internal resistance of the device decreases significantly, providing surge currents an alternate path to ground, safely bypassing the connecting loads. As line voltages return to their normal levels, dropping back below the trigger threshold, the device turns off, preventing the flow of current to ground. Such devices can be used repeatedly without risk for damage or need of replacement.

line pad   Used to help reduce signal levels, if necessary, to avoid overloading the channel inputs of mixers or amplifiers. Line pads can be made switchable, often in the form of −20 and −40 dB pad switches found on most microphone mixer inputs, or they can be in-line, meaning that they connect directly to microphone or line-level cables as needed to achieve a desired signal attenuation.

listed   Equipment materials, or services have been tested as safe by an organization acceptable to the AHJ. An example of a listing laboratory would include UL. Such an agency evaluates and tests a piece of equipment, guaranteeing that it meets required safety standards and that it is appropriate for public use. All equipment, materials, or services must be tested to ensure that they meet appropriate safety standards.

They must also be found to be suitable for a specific purpose or use. After having passed a series of inspections and tests, the equipment, materials, or services are then placed on a list that is published by the listing agency.

listing and labeling laboratory   Performs the evaluation and testing of equipment, materials, and services to ensure that they are suitable and safe for public use. Examples of listing and labeling laboratories include UL and The Canadian Standards Association (CSA).

local area network (LAN)   The linking of computers and users in a small geographical area such as an office complex, school, or college campus.

local loop   The connection from your house to the telephone company. Originally, the local loop was designed for analog voice transmission over twisted pair cable. Today, because of the use of computer modems, the local loop uses both analog and digital transmission technology and is now sometimes referred to as the "subscriber loop."

loop start   A loop start circuit is what residential telephone systems use to establish a call. The trunk is idle until the station set goes off hook. An off-hook configuration connects the loop circuit, and current flows from ring to tip.

loose-tube cable   A type of optical fiber cable designed for outside applications; that is, the cable is typically exterior to the structure and is subject to environmental elements not normally seen on the interior of a building. In this design, color-coded plastic tubes house and protect the optical fibers. Because moisture is a concern in outdoor installations, this cable is designed with a gel filling compound to impede water penetration. To protect the fibers from damage during installation or environmental loading, a steel central member also serves as an antibuckling element.

loudspeaker   Converts an electrical current back into a mechanical vibration to reproduce sound. The term *direct radiator* is also used to describe a loudspeaker because it can directly affect the movement of air molecules, thereby generating sound waves and moving wave fronts within the atmosphere. A basic speaker consists of a voice coil, a magnetic field, and a mechanical cone or diaphragm.

low-earth-orbit (LEO) satellite   LEO satellites are not stationary or geosynchronous. Instead, they are moving continuously around the north and south geographical poles of the Earth in a low orbit, often at an altitude of 300 to 1500 km (186–932 miles). The benefits of using LEO include lower power and shorter signal delays, making them ideal for voice communications and cellular phone usage.

low-noise block converter (LNB)   The receiving element for a satellite dish antenna. The purpose of the LNB is to provide an impedance match between the dish and the 75 Ω coaxial cable, frequency conversion to a lower value (necessary for connection to the input to the satellite receiver), and signal preamplification.

low-pass   A low-pass filter passes a band of frequencies on the low end of a frequency spectrum until the -3 dB cutoff is reached. Any frequency past the -3 dB cutoff point on the upper end of the spectrum is effectively filtered out from the signal.

lux   A unit of illumination for light calibration equal to 1 lumen/m^2. Manufacturers often specify light calibration in lux or foot-candles. To convert between the two standards of measurement, 1 lux is equal to 0.0929 foot-candles.

maglock   A magnetic locking mechanism for a door. A maglock typically operates from a 12 to 24 V dc power supply. All magnetic locks are fail-safe. This means that they need a constant source of current to remain locked. If power is removed, the lock opens.

magstripe card   A type of access-control card that has a black magnetic stripe along the back side which is read by sliding the card through a magstripe card reader.

main distribution frame (MDF)   The place where all individual wire pairs from the PBX of a commercially installed telephone system terminate. The MDF is typically located in the basement or on the first floor of a building. It is used to cross-connect any outside line with any inside line of the telephone system.

marginal notations   The *NEC* is revised every three years. When this is finished, all changes are noted in the margins by the presence of a vertical line. These changes are referred to as marginal notations. The vertical line indicates that the section has been revised since the last edition.

master box   The communication link or junction point between the fire alarm control panel and central station or fire station.

matching transformer   Provides an impedance match between differing electronic circuits. For example, when a high-impedance source needs to connect to a low-impedance load, it must do so through a matching transformer to achieve maximum power transfer. To make the connection without the use of a matching transformer causes loss of signal and power. The turn ratio of the transformer directly determines the impedance match between the primary input and the secondary output of the device.

medium-earth-orbit (MEO) satellite   MEO satellites operate somewhere between LEO and GEO satellites, typically at altitudes of 8000 to 20,000 km (4970–12,427 miles). As would be expected, because of the higher orbit and larger orbital circumference, MEO satellites require more devices in orbit than those of LEO to accomplish a smooth communication link between parties. Power requirements for MEO satellite transmissions are also higher because of the higher altitude, forcing hardware to be larger and less portable.

microphone paging   Results from connecting a microphone directly to an audio system equipment rack for distribution through the network of building speakers. A paging microphone differs from a standard microphone in that it requires the activation of a muting switch prior to paging. The muting switch provides two functions: It turns on the microphone and sends a muting control signal out to the audio mixer to turn off any music programming that may be playing prior to a page.

microwave motion sensor   Motion sensors that operate in a similar manner to ultrasonic but at a higher frequency. Microwave sensors are highly sensitive and can pick up movement through glass. Therefore, the placement of such sensors must be carefully thought through to prevent false alarms caused by movement on the other side of a window.

midrange   See *midrange speakers.*

midrange speakers   Midrange speakers typically range in size between 2 and 8 in. and are used to recreate a midrange of frequencies between 500 and 8000 Hz.

mineral-insulated cable (MI)   A type of fire-rated cable that is capable of resisting the spread of flame and fire for up to 2 hours (see *NEC Article 332*). MI cable is made of solid copper conductors, insulated with magnesium oxide, covered over by a rigid copper sheath. The cable can be either single conductor or multiconductor, depending on the application. Previously, MI cables were widely used on emergency generators, fire pumps, low-voltage fire protection circuits, and emergency voice systems. Such systems require the use of a fire-rated cable to help minimize damage to circuits and guarantee the functionality of an emergency system, especially during the initial hours of an evacuation.

modem   A device that allows a computer to communicate over an analog telephone line. The word *modem* is a conjunction of *mod*ulation and *dem*odulation. Modulation places an intelligent signal on to a communication carrier for transmission. An intelligent signal may be voice, data, or video. The carrier then carries the signal through a designated channel to the recipient of the communication. Demodulation is the opposite of modulation; it removes the intelligent signal from the carrier so that the recipient

can interpret the message. A dial-up modem also provides a digital-to-analog conversion so that the analog phone line can use and pass the signal.

modulation   Places an intelligent signal on to a communication carrier for transmission. Examples include amplitude modulation (AM), frequency modulation (FM), phase modulation (PM), pulse modulation, and frequency shift keying modulation (FSK).

monitor   Used to display the video image of a CCTV signal. A video monitor only accepts a composite or component video signal. Because a monitor does not contain an adjustable receiver, it is unable to tune in a standard radio broadcast frequency or television station.

mouthpiece   The part of the telephone that communicates the audible voice of an individual to the other side of the line. Contained within the mouthpiece is a small carbon microphone. When an individual speaks into the microphone, the sound pressure created by the person's voice vibrates carbon granules between two, thin-metal plates. The applied sound pressure changes the internal resistance of the mouthpiece, thereby modulating the dc current passing through the telephone handset.

multiconductor   The number of insulated conductors in a multiconductor cable can vary from 3 to 50, depending on the type of cable and application. Multiconductor cables are available in a variety of sizes (solid or stranded), voltage and temperature ratings, and insulation types; they also can be shielded or unshielded. Typical applications for multiconductor cables include communication, instrumentation, control, audio, and data transmission.

multimode   See *multimode fiber.*

multimode fiber   A type of optical fiber cable. Multimode gets its name from the various light waves that are dispersed into numerous paths, or modes, as they travel through the cable's core, which is typically 850 or 1300 nm in diameter. The signal can therefore enter and exit the fiber at different angles. The main disadvantage of multimode is that over long cable runs (>3000 ft [914.4 m]), the multiple paths of light can cause signal distortion, resulting in unclear or incomplete data transmission at the receiving end of the line. Also, signals traveling at higher modes can become attenuated as they travel down the fiber because of absorption of light into the cladding and also through scattering, as the light it becomes bent at angles outside the critical angle. There are two basic varieties of MMF: step index and graded index.

multiplexer   An electronic circuit that combines multiple data sources into a single transmission line.

multiplexing   The mixing of several communication channels from multiple sources into a single, large-capacity transmission channel.

muting control Allows for selected inputs of amplifiers or mixing consoles to be shut down or muted during a microphone or telephone page. The intent is to interrupt or reduce the intensity of a music-programming source during the period of an announcement or page.

National Electrical Code (*NEC*) The *NEC* is sponsored by the National Fire Protection Agency (NFPA); it is also known as NFPA Document 70. The purpose of the *NEC* is to ensure the promotion of electrical standards and the safety of individuals. The *NEC* is not intended as an installation manual or how-to guide. The NFPA has published the *NEC* since 1911; revised editions are published every 3 years., with 2011 being the latest edition of the electrical *Code*.

National Fire Protection Agency (NFPA) Oversees the national fire code for the United States and, in addition, publishes the standards for the *NEC*. The NFPA is also the source of many other documents and publications dedicated to the safety of individuals.

near-end crosstalk (NEXT) Results from high levels of transmission signal at each end of a communication link interfering with the reception of an intended signal. The interference from near-end transmissions only serves to degrade the quality of reception. High NEXT levels (in decibels) are desirable.

network Refers to the interconnection of computers for the purpose of intercommunication and the sharing of information.

Newtons A metric unit of force. (International System of Units).

nonconductive optical fiber cable Contains no conductive elements. During the manufacturing process, the cable is constructed from glass or plastic and a variety of insulators. Nonconductive optical fiber cable is incapable of any type of electrical conduction accidentally short-circuiting.

non–power-limited circuit A circuit with no current limitations that is permitted to draw whatever power is necessary from the supply. Although such circuits are not limited from a design point of view and may be made to draw any required level of current flow, they still require overcurrent protection as a means of protecting the connecting cables and supply from a possible short circuit. Power-limited technicians are restricted from installing non–power-limited circuits because such circuits fall under the authority of a journeyman or master electrician.

non–power-limited fire alarm circuit *Article 760* of the *NEC* details the installation requirement of non–power-limited fire alarm circuits. Although the design parameters of non–power-limited fire alarm circuits are not restricted to specific current levels, they still require overcurrent protection as a means of protecting the connecting cables and supply from a possible short circuit. Power-limited technicians are restricted from installing non–power-limited fire alarm circuits because such circuits fall under the authority of a journeyman or master electrician.

notch filter Used to remove a specified frequency or band of frequency from a desired signal. It is the opposite of a band-pass filter. Therefore, the only frequencies to make it through a notch filter are those outside of the filtered bandwidth that are above and below the cutoff frequencies.

notification appliance circuit (NAC) A circuit that connects an alarm indicator device that provides an audible or visual indicator output for a fire alarm circuit. A horn/strobe is an example of a NAC.

octave The doubling of a base frequency or half the value of a base frequency. Therefore, if 1000 Hz represents a reference base frequency, then 2000 Hz is the first upper octave, and 500 Hz is the first lower octave. Multiple octaves away from 1000 Hz can be calculated by multiplying or dividing the base frequency by 2^n, where n represents the level of octave to be calculated.

omnidirectional An almost perfect 360° transmission or reception pattern around a device. As an example, an omnidirectional microphone is sensitive to sounds from the front, sides, and rear of a diaphragm, in a spherical, three-dimensional pattern.

open loop An open-loop circuit or device indicates that circuit contacts are open, and measure infinite resistance from the source. Current does not flow in an open loop configuration.

open systems interconnection (OSI) The primary architectural model for intercomputer communications. The OSI reference model divides the movement of communications between computers into seven tasks or layers: application, presentation, session, transport, network, data link, and physical. The OSI is a conceptual model. Conceptually dividing the model into seven self-contained tasks helps to reduce the complexity of the network. Smaller tasks are ultimately easier to understand and manage.

operating system (OS) The focus of a computing experience. It is the first software you see when you boot up a computer, the software that guides you through your applications, and the software that safelys shuts down the computer when you are ready to end a session. Modern OSs generally use a graphical user interface (GUI), meaning they use pictures and icons to interact with the user. The OS is the software that allows programs to be launched and used and that organizes and controls the hardware.

optical fiber cable   Communicates using light energy, not electrons. Compared to standard copper conduction, optical fiber cables provide wider bandwidths and faster communication speeds over longer distances.

overcurrent   See *overcurrent protection.*

overcurrent protection   Protects cables and circuits from excessive levels of current flow that could result in a potential fire hazard. Examples of overcurrent protection include fuses, circuit breakers, or active electronic feedback circuits.

panic button   A hidden alarm button within a security system that, when pressed, silently notifies security personnel of an apparent emergency, without simultaneously tipping off the thieves.

parallel communication   Sends multiple binary bits, all at one time, between two points in a communication system. In most cases, entire binary words can be sent simultaneously, helping to reduce the total transfer time and making parallel communications much faster than serial (in actuality, eight times faster).

parametric equalizer   Most commonly used by audio studio engineers. They offer continuous variability and adjustment to frequencies and bandwidths, thus allowing the user to fine-tune the signal precisely to any desired shape or curve. The parametric equalizer is not as user-friendly as the graphic equalizer, and a more detailed knowledge of signals and filtering is needed to operate the device.

parts   Many articles within the *NEC* are broken into parts. Parts are used to subdivide articles into simpler topics of focus. Parts are displayed as roman numerals. Most articles contain three to four different parts; however, *Article 230* is a good example of an article subdivided many times. *Article 230* contains eight parts.

passive infrared (PIR) sensor   An indoor/outdoor device that is capable of detecting motion through variations in temperature. Security systems often use PIR sensors to detect unexpected movement within a given space, often from the radiating body temperature of an intruder. In addition to security alarms, the sensor can also be wired to trigger the operation of cameras, recording devices, and lighting.

personal communication system (PCS)   Digital PCS represents the second generation of cellular telephone operating within six frequency bands between 1850 and 1980 MHz. The development of portable, low-power, digital handheld phones was the initial intent of the PCS system, replacing the previous analog version of cellular, AMPS.

phantom power   An external power source for condenser microphones, typically supplied by the mixing console through the connecting wires of the microphone. The phantom power acts as a replacement for the internal batteries of the microphone. Most manufacturers use a level of 48 V dc. A button at each individual channel of the mix console is available for turning the phantom power on or off, as desired.

phase   The synchronization of multiple ac waveforms traveling along the same channel, cable, or electronic circuit. Waveforms that are in-phase and moving in the same direction, positive or negative, are additive; out-of-phase waveforms, moving in opposite directions, subtract. In-phase and out-of-phase also relate to coherence and incoherence.

phase modulation (PM)   A type of modulation similar to FM except that the phase of the carrier signal is made to shift at a rate equal to the superimposed information signal. PM modulation is used in many types of digital communication circuits.

phon   Used to measure the perception of loudness for the human ear. The phon is equal to the decibel level of a 1000 Hz reference tone within a bandwidth of frequencies. The perception levels of frequency bands, ranging from 20 to 20,000 Hz, tend to be quite uneven at low phon levels but tend to flatten at higher phon levels.

photoelectric sensor   Detects the presence of light. The output switches states between open or closed, based on a predetermined internal calibration and the level of available light.

photoelectric smoke dectector   See *smoke detector.*

physical addressing   Different from network addressing, physical addressing defines how devices are addressed at the data link layer; it is similar to a zip code as opposed to an actual house address.

picocells   Small-range cellular repeater systems intended for indoor use that are installed to help bridge the gap between an outdoor radio signal and a shielded building. Elements of a picocell include an outdoor antenna, multiple indoor antennas, and a central repeater through which all signals pass. Picocell systems are often quite expensive to purchase and install, depending on the number of channels needed and the desired signal throughput.

ping   A standard troubleshooting utility available on most network OSs. Besides determining whether a remote computer is "alive," ping also indicates something about the speed and reliability of a network connection. Ping is like the game of Ping-Pong (table tennis): You hit the ball to the other side and it should be hit back. When you use ping, you are actually sending a request directly to another computer to respond.

pink noise   Pink noise generators are perceived as equal in energy per octave band. Pink noise is white noise that has been filtered -3 dB per octave. Each octave within the 20 to 20,000 Hz bandwidth is 3 dB lower. Therefore, 40 Hz is 3 dB less than 20 Hz, and 80 Hz is 16 dB less than 20 Hz. Pink noise generators are used to balance the frequency response of electronic equipment and the acoustics response of a room.

plating A thin layer of tin, silver, or nickel that is applied to the exterior of conductors during the manufacturing process. There are three essential reasons why plating of conductors is necessary. First, plating helps to eliminate problems when soldering; second, it allows for a more simplified stripping and removal of insulation; and third, it prevents deterioration of the central conductor at high temperatures.

plenum The space above a suspended ceiling where ductwork is run. The plenum space is used by the HVAC system to circulate environmental air throughout the building.

polar patterns Used to graphically illustrate the sensitivity of microphones. A polar response is measured by rotating the microphone through a 360° arc while keeping the sound source constant, both in level and direction. The most common varieties of polar patterns are omnidirectional, bidirectional, unidirectional or cardioid, super-cardioid, hyper-cardioid, and shotgun.

polarization A technique known as signal polarization is used to help minimize interference between adjacent transponders. Polarization can take the form of either a linear vertical or a linear horizontal orientation. The rotational angle of the pickup element inside of the antenna feed horn determines the polarity of the uplink or downlink. In most cases, antennas have dual elements installed, one being horizontal and the other vertical, making polarity selection a simple matter of switching between the two, as needed. As long as the polarity of the antenna element matches the polarity of the radiated microwave signal, communication between devices is possible. Mismatched polarities, however, cause confusion and loss of signal.

post office protocol 3 (POP3) Defines the method and protocol for retrieving e-mail from an e-mail server.

power amplifier The last stage of amplification between the signal input and the output speakers. Power amplifiers can be multichannel or monaural, and they must be rated to deliver all the necessary power required by the load. A power amplifier provides wide-band amplification across the entire audio spectrum from 20 to 20,000 Hz. They do not offer any tone control or signal-processing functionality other than adjustable volume control for the main signal outputs. Power amplifiers provide connection to 4, 8, or 16 Ω loads or to 25, 70, or 100 V matching transformers.

power-limited circuit Power-limited circuits are limited in output and capacity. Power-limited circuits fall into three varieties: Class 1, Class 2, and Class 3. Class 1 power-limited is limited to a maximum output power of 1000 VA and 30 volts. The limitations of Class 2 and Class 3 power-limited circuits are defined by *Chapter 9, Table 11(A)* and (*B*) of the *NEC*. In general, Class 2 and Class 3 power-limited circuits must not exceed 100 VA. The allowable voltages for Class 2 and Class 3 circuits may go as high as 150 volts, with maximum current levels based on the level of supply voltage.

power-limited fire alarm circuit (PLFA) *Article 760* of the *NEC* covers the installation and *Code* requirements for power-limited fire alarm circuits. Power-limited fire alarm circuits shall be rated as Class 3, having a maximum power rating of 100 VA and a maximum cable rating of 300 V.

power limited tray cable (PLTC) A special type of listed and labeled, nonmetallic-sheathed cabling that is intended for use in cable trays of factories or industrial establishments. *Table 725.61* of the *NEC* also states that PLTC may be used as an approved substitute for Class 2 or 3 wiring in general-purpose or dwelling locations. PLTC is rated for 300 V, 100 VA and can be purchased in sizes 22 AWG through 12 AWG.

power line conditioner Incorporates combinations of regulation, filtering, and surge protection within a single unit. Such units, however, are not able to provide protection against power line abnormalities such as variations in frequency or power blackouts. Most conditioners also provide a form of current limiting to limit the maximum load current deviations to a range of 150% to 200% at full voltage.

power over Ethernet (POE) A system technology used to pass electrical power safely, along with data, on Ethernet cabling. PoE is an IEEE standard, currently 802.3at-2009. PoE is used, for instance, to power CCTV cameras by connecting the camera through a CAT5 or CAT6 cable to a PoE Ethernet switch.

power sum (PSUM) A test procedure where the same signal is applied simultaneously to three pairs of a cable, with the resultant measurement taken from the fourth. All pair combinations are tested and averaged for a final power sum value, which is measured in decibels.

preamplifier Line-level signal amplifiers that provide tone control and level adjustment for input signals before final-stage amplification. Output connections can be either 150 or 600 Ω balanced or 10,000 to 35,000 Ω unbalanced. Preamplifiers are not designed to drive 8 Ω loudspeakers.

precedence A signal that has priority over all other signals. An example of a signal with precedence would be a fire alarm signal or emergency paging signal; in an emergency, a fire alarm signal or emergency paging signal would have precedence over all other input signals to a mixer or amplifier. All signals not having precedence, such as music programming, are muted or shut down during an emergency broadcast.

primary cell   Batteries that are not rechargeable. Once the internal energy of the cell has been depleted, it must be thrown away. Primary cells include carbon-zinc, alkaline, button, or lithium-type cells.

private branch exchange (PBX)   A telephone system within a business that switches calls between users on local lines while allowing all users to share a certain number of external phone lines. The main purpose of a PBX is cost saving. The business is charged a flat fee for the service as opposed to charging per line, per use. The PBX is operated and owned by the business, not the telephone company. The PBX is typically located in the basement or on the first floor of the establishment.

proprietary station   Security or fire alarm reporting stations that are privately owned and operated by the property owners of the protected premises. The structure can vary from a single-building complex, such as a shopping mall or high-rise, to several buildings within a corporate or collegiate complex. Properties can be contiguous or noncontiguous, provided they are all communicating to a supervised central station that is properly staffed by qualified and trained personnel.

protected premises   The location that is being protected by a supervisory fire alarm system or security alarm.

protocol   Communication between computers is made possible through a set of protocols. Protocols are a formal set of rules governing the transfer and exchange of data through the network medium and individual layers of the OSI. A wide variety of protocols exist, such as LAN or WAN protocols, network protocols, or routing protocols.

proximity   See *proximity effect.*

proximity effect   Many cardioid microphones exhibit a phenomenon known as the proximity effect. The proximity effect produces an increased bass response as the microphone moves closer to the sound source.

proximity cards   Proximity readers and cards use contactless technology to communicate; physical contact does not have to occur when using proximity technology. Proximity access cards contain a tightly wound wire embedded inside the card. This wire acts as an antenna and only needs to come within the proximity of the reader to be read. Proximity cards can be broken down into three general

categories: immediate proximity cards that require a 1 mm distance, close proximity cards that require 10 cm of distance from the reader in a specific orientation, and vicinity cards that can function at a range 30 to 70 cm from the reader in any orientation.

public fire alarm reporting system   The method by which a fire alarm is reported to authorities. Examples include pull boxes, line seizure from the fire alarm control panel, and the master box.

pull station   Public fire alarm reporting systems that are manually activated by an individual during a fire emergency. The pull station is essentially nothing more than a shorting switch contact. When the contact closes, the supervised alarm notification circuit is shorted, causing the fire alarm control panel to activate all alarm notification devices such as horns and strobes.

pulse modulation   Involves varying the amplitude, duration, or time interval between pulses to represent a formatted coded signal. The pulses are then superimposed on the carrier frequency for transmission.

qualified person   According to the *NEC Article 100, Definitions,* a qualified person must be familiar with the aspects of electricity but, most importantly, must have had some form of electrical safety training.

radiation patterns   Relate to how a specific radiated waveform propagates or travels through a known medium. Examples of radiated waveforms include light, sound, and radio frequency signals. The shape or footprint of a radiated waveform is typically plotted around a 360° axis, and graphed out in polar form.

radiators   A device that radiates a type of waveform; examples may include a transmitting antenna and a loudspeaker.

radio frequency identification (RFID)   A communication technology that communicates over radio waves between an electronic tag and a reader. The tag, also known as a label, is composed of an embedded, digital radio circuit and an antenna. The reader, referred to as an interrogator, provides a radio frequency carrier signal, necessary to activate the tag. Once activated, the digital information embedded within the tag superimposes itself on the RF carrier signal and reflects back over the antenna to the reader as a modulated response.

radiofrequency interference (RFI)   A type of noise signal that is often inductively coupled into electrical conductors and circuits from the unwanted presence of electromagnetic waves. Common sources of electromagnetic waves are high-frequency transmitters and antenna systems, fluorescent lights, motors, and generators.

randomized bundle   The central core of an optical fiber cable can be organized in coherent or randomized bundles. Randomized bundles means that the glass strands running through the center of the cable core are not organized in any special order; for this reason, the cables are easier to manufacture and are less expensive. Randomized cables are used in most cases where the application requires only the transmission of on/off light pulses. In such cases, the organization of individual strands within the bundle is not as critical as it is for transmitting actual images.

RCA connectors   Commonly used to connect the inputs and outputs of unbalanced audio circuits and composite video.

reflection   Occurs when a waveform strikes an object and bounces off, sending it back from where it came or off in a different direction of travel. Reflections can cause echoes and reverberation.

regulator   Electronic circuits use regulators to help maintain outputs at a constant level. Voltage regulators are most commonly used on ac and dc power supplies. The voltage regulator uses a form of negative feedback to automatically adjust for and counteract sudden or variable changes to the input. Current regulators are similar in operation to voltage regulators, but they are more commonly used on circuits that require a constant flow of current. A battery charger is an example of a circuit that uses a current regulator to trickle charge a cell over a period of many hours.

remote-control circuit   Circuits that control other circuits. Control voltages often can vary from as high as 600 V to as low as 5 V, depending on the type of system. As an example, motor control circuits typically use 120 V on starter coils, but many other types of control circuits operate from 24 V or less. Coiled relays, transistors, and silicon-controlled rectifiers (SCRs) often are used to accomplish the task. Common uses for remote-control circuits are motor controls, elevators, conveyer systems, automated processes, and garage door openers.

remote station   Remote station fire alarm systems are used when a protected premises does not have access to a public fire alarm system or does not use a central station service or proprietary supervising station system. Remote stations are not certified systems, and companies installing them do not need to be UL certified. Often, landlords with building structures in isolated locations use remote systems as a way to link multiple structures to a single, centralized, manned control center. Remote stations often transmit fire alarm signals by means of a private radio, dedicated phone circuit, one-way phone, or any approved method deemed acceptable by the AHJs. Monitoring of supervisory or trouble alarm signals from a remote protected premise can

be achieved through a central control unit, fire command center, or supervising station, provided they are continually staffed and attended by qualified response personnel.

repeater   Provides the function of passing data or radio transmissions from one system or antenna cell to another. The primary function of a repeater station is to receive signals and then pass them on, unaltered, to the next destination, acting much like an electronic data relay.

resolution   See *camera resolution*.

reverberation   Echo and reverberation are a result of reflection. A desired sound wave may take a direct path to the ear of the listener, but it may also bounce around the room, off the walls, floor, ceiling, and furnishings, before eventually arriving at the final destination. An echo is an example of an early reflection arriving just after the direct sound wave. Reverberation is considered a later reflection because it arrives some time after the echo, due to the more indirect pathway it takes from the source to the ear of the listener.

ribbon fiber-optic cable   Constructed by bonding up to 12 coated fibers into a flat, horizontal geometry. Multiple ribbons can be stacked on top of each other within a single cable, which more effectively uses the size of a given space. A ribbon-style cable can pack more than 100 fibers into a 0.5-in. square, which is a more efficient use of the geometry of a small space and allows for higher density installations and lower costs from using less cable.

ring   The individual wires in a twisted pair telephone line are called the tip (green) and the ring (red). These terms date back to when telephone operators would "patch" calls manually at a switchboard with a tip-ring-sleeve patch cord. The tip connects to ground and the ring to the battery at TELCO.

ring-back signal   A telephone company uses the ring-back signal when a customer places a call to a line that is not currently being used. It represents the sound of a bell ringing at 440 Hz. In the United States, the format for a typical ring back signal consists of 6 seconds on, 2 seconds off.

ringer equivalence number (REN)   The telephone company only guarantees to ring up to three ringers on a single phone line. More ringers would require additional current, which may or not be available. The REN is used to calculate how many phones can be connected to a single phone line. Every phone should have a Federal Communications Commission registration label that contains the REN. The format of the REN is standard across all phones. It contains NUMBER, then LETTER. The NUMBER is a decimal number representing how much power the ringer requires as compared with a standard phone. A standard phone is defined as a standard gong ringer as supplied in a standard telephone company desk telephone. The number 1.0 represents that the phone uses 100% of the power of a standard phone. A number of 3.2, for example, indicates that the telephone uses the power of 3.2 standard phones. The LETTER represents the frequency the telephone requires. The letter *A* signifies that a 20 Hz ring signal is required. The letter *B* signifies that the telephone rings at any frequency.

riser   An architectural term referring to the vertical column or shaft between floors in a building that is typically used for plumbing, air ducts, and electrical wiring.

roaming   The act of traveling and communicating between adjacent cells within a cellular telephone system.

router   A network bridge that is used to link entirely separate networks. A bridge connects clusters of computers within the same network, whereas a router is used to connect isolated networks. As an example, it would be used to link separate entities such as the University of California to the University of Minnesota. The router still operates from a list of programmed addresses; however, the address is targeting a location inside the network layer of the OSI by using IP addressing instead of the data link layer, where CSMA/DC manages network traffic.

satellite   Used as a means of transmitting communication signals through wireless technology over microwave. Three basic types of satellites exist, depending on their function and use: geosynchronous satellites (GEOs), low-earth-orbit satellites (LEOs), and medium-earth-orbit satellites (MEOs).

secondary cell   A rechargeable type battery. A NiCd battery is an example of a secondary cell.

sections   Articles and parts within the *NEC* are broken down into sections. A section is assigned a corresponding article number and "dot" notation, followed by the section number (e.g., *800.1*). This example indicates that we are referencing *Article 800, Section 1*. Sections increase chronologically, but like articles, some section numbers are skipped to allow for future expansion.

sensitivity   See *camera sensitivity.*

sensor   Supplies an analog or digital input to an electronic circuit.

serial communication   Sends data bits one bit at a time between two points in a computer network or system. Serial communication is the slowest form of communication between computers.

server   The center of a multi-station computer network. The server is the main network resource where commonly used application programs reside, data is stored, and network activity is monitored.

shielded twisted pair (STP)   Character impedance of STP is typically 150 Ω. Although STP can provide more protection against noise and EMF, it is not widely used anymore because of its incompatibility with most of the new modular connectors and terminators currently used in the industry.

shielding   Used on electrical cables and electronic equipment to isolate them from sources of electromagnetic interference and to prevent the emission of unwanted electromagnetic radiation into the surrounding atmosphere. Examples of shielding include aluminum foil, braids, and steel enclosures.

shock-type glassbreak sensor   The sensor uses a piezo electric transducer which is tuned to the shock frequency of breaking glass. The shock-type glassbreak sensor is mounted directly on the glass to detect the mechanical shock of the breaking glass.

shotgun   A type of microphone that has a highly selective and directional polar pattern, allowing for the detection of signals toward the front axis of the diaphragm, and greatly attenuating signals that enter from the sides or rear of the device.

shutter speed   The speed of a camera shutter. Shutter speed is adjusted to prevent image blurring and to adjust for differing levels of light and exposure. A doubling of shutter speed cuts an exposure by half; reducing the shutter speed by half doubles the exposure.

signal level   Another term for the voltage of an input signal.

signal source   Provides the input signal to an electronic circuit or system. The signal source is not the power supply. The power supply is referred to as the power source.

signaling circuit   Defined in *Article 725* of the *NEC*, signaling circuits are circuits that activate notification devices. Examples include lights, doorbells, buzzers, sirens, enunciators, and alarm devices.

signaling line circuit (SLC)   Fire alarm circuits and security systems use SLCs. The SLC is an addressable system used for connecting and communicating directly between circuit devices and the control panel.

simple mail transport protocol (SMTP)   A network protocol that allows the sending of e-mail and messages from a user station to an e-mail server. The process is one-directional. As a result, SMTP cannot request mail, only send it.

simple network transfer protocol (SNTP)   Exists as a means of monitoring computers, routers, switches, and network usage.

simplex cable   A type of fiber-optic cable. It is a jacketed, single-fiber, tight-buffered cable that uses Kevlar for strength and reinforcement. Simplex cables are typically used as patch cords and in backplane applications. The backplane refers to the socketed interconnection of multiple cables and circuit cards for purposes of routing and expansion.

single-mode   See *single-mode fiber.*

single-mode fiber (SMF)   Provides only one mode of transmission through a single strand of glass fiber, typically having a core diameter of 8.3 to 10 μm.

skin effect   For high-frequency signals, the skin effect causes current flow to travel along the surface of a conductor, resulting in no center conduction. Skin effect is responsible for signal attenuation at high frequencies and for the increased resistance of the conductor.

smoke detector   A device that detects visible or invisible particles of combustion. Types of smoke detectors include ionization, photoelectric, air sampling-type, projected beam-type, and spot-type.

solidly grounded   Connecting to ground without inserting any resistor or impedance device; directly connected to ground.

speaker efficiency   Measured as the ratio of output power to input power. Input power is electrical, whereas output power is radiating, acoustical sound pressure. A speaker's efficiency is usually measured (in decibels) at a distance of 1 m, and it is derived from a level of electrical input power of 1 W.

spectrum analyzer   Graphically illustrates the full range of signal level along a frequency spectrum. For an audio system, an audio spectrum analyzer measures and displays the entire audio spectrum from 20 to 20,000 Hz.

spider mount   The voice-coil of a speaker mounts to the spider mount. The spider mount acts as a shock absorber to help dampen the coil movement during the absence of current flow.

standards agency   Promotes, approves, and monitors industry practice and standards.

standing waves   The reflected waves of energy along a transmission line due to a mismatched source, line, or load impedance.

star topology   Connects individual workstations or computers in a network through a central device that acts as a hub to the surrounding satellites. Each of the network connections is considered a node to the central hub.

step index   A type of multimode optical fiber that has a core composed of one type of glass. The light can travel through the fiber in two ways: in a straight line along the central axis or through a more indirect path, in modal form, reflecting back and forth along the core and cladding. Signals traveling along the more direct, straight-line axis transmit faster and achieve higher bandwidths; those traveling at higher modes have a longer distance and travel time, resulting in lower transmission bandwidths. Modal travel may also be subject to greater levels of signal dispersion and attenuation. For this reason, step index fiber typically is used over short, low-speed data links in situations where higher bandwidths are not critical.

strobe   Visual alarm indicator devices. The central control unit of a fire alarm or security system outputs a control signal to activate strobes during an alarm status. Strobes visually flash synchronized indicator lights throughout the protected area, similar to the way a police light operates, indicating the presence of an alarm.

structural return loss   The digital equivalent of a standing wave on the transmission line.

supervisability   Refers to an electronic circuit that has the ability to self-monitor its functions and reliability through a control loop. Fire alarms and security systems are examples of supervised circuits.

supervisory station   A command center, occupied by trained personnel 24 hours a day, that monitors the status of a fire alarm or security systems. All reports of actual alarms and trouble alerts are sent automatically to the supervisory station by the control panel of the fire alarm or security system.

supplemental electrode   The use of an additional grounding electrode to lower ground resistance when a single electrode does not suffice. According to the *NEC*, the resistance to earth for a grounding electrode shall be 25 Ω or less.

surge protector   Also called lightning arresters, transient suppressors, surge arresters, or transient voltage suppressors. The purpose and use of such devices is twofold: to limit peak voltage surges and to divert power surges and transients to ground.

survivability   Refers to the ability of a circuit to operate during an actual fire. Notification appliances, such as horns and strobes, must be able to maintain their functionality during the process of an evacuation, allowing occupants the needed time to withdraw safely from the building or structure. To ensure the survivability of the fire alarm circuit, cables and assemblies typically have a survivability rating of 2 hours.

tables   Tables within the *NEC* are used to display *Code* rules in a logical and easy-to-read order. Each table found within an article is article specific. When referencing a table within an article, the reference must include the word *Table* and the section number (e.g., *Table 110.31*).

TELCO   The telephone company.

telephone paging   An audio signal sent from the main phone system of a building structure to an audio system equipment rack for distribution through the network of building speakers.

tight-buffered cable   A type of optical cable more commonly used inside of building structures. It comes in two varieties: single-fiber and multifiber tight-buffered designs. In the tight-buffered design, the buffering material is in direct contact with the fiber, making it less heat and moisture resistant than the loose-tube cable. The tight-buffered cable lacks a central steel member. It is most often used for intrabuilding, risers, general building, and plenum applications. More specifically, single-fiber tight-buffered cables are used as pigtails, patch cords, and jumpers, whereas multifiber tight-buffered cables are used primarily for alternative routing and handling flexibility for installations in tight areas.

tightpack cable   Optical fiber cables that are used as distribution cables and contain several tight-buffered fibers bundled under a single jacket. Tightpack cables also use Kevlar for strength and reinforcement of the bundle. The individual fibers, however, are not reinforced separately and therefore must be terminated inside of patch panels or junction boxes. A typical use for tightpack cables is for a short, indoor, plenum, or riser installation.

time division multiple access (TDMA)   A method of cellular communication that involves dividing channels into time slots rather than assigning a subscriber an individual frequency, as is commonly done with FDMA. Analog signals are first converted into digital and then assigned time slots among multiple subscribers. The main difference is that subscribers in the TDMA system share pieces of a communication channel with other subscribers, but they do not have access to the entire channel as is allowed in FDMA. Digital audio signals are organized into data packets and then assigned to an allocated time slot within the complete data stream. Control data packets are also included in the mix.

time-lapse video recorder   Similar to a typical videocassette recorder (VCR) that uses a tape to record events. However, a time-lapse recorder is unique because it can take hundreds of hours of viewing and put it on a 3-hour tape. It does this by skipping frames to save space. Viewing a recorded image from a time-lapse recorder is different from viewing a real-time image, because frames have been left out randomly. For this reason, the picture tends to look jerky. In some cases, particular event details may even be missing. Time-lapse recorders are often connected to a motion sensor. In such cases, the recordings only take place when a movement has been detected in the viewing frame.

tip   The individual wires in a twisted pair telephone line are called the tip (green) and the ring (red). These terms date back to when telephone operators would "patch" calls manually at a switchboard with a tip-ring-sleeve patch cord. The tip connects to the ground and the ring to the battery at TELCO.

tip-ring-sleeve   A type of patch cord connector that is typically used on balanced audio circuits. Tip-ring-sleeve connectors can also be used for the connection of unbalanced stereo connections; an example would include the connection of headphones to the output of an audio device. Tip-ring-sleeve connectors come in 1/4-inch, 1/8-inch, metric, and micro sizes.

token ring network (TRN)   A type of network connection where each node of the network is connected in series with the next through a connecting cable. Communication is achieved by passing a short message in the form of binary data packets, called a token. The token passes continually around the ring, from node to node, in the form of a message relay.

topology   Defines how computer networks are connected. Examples include token ring, bus, star, and tree.

transceiver   A combination transmitter and receiver.

transducers   A device that converts one form of energy into a secondary form. An electrical transducer converts mechanical force into an electrical impulse; an example would include a pressure sensor or a microphone.

transmission   Involves the sending of a communication signal. Therefore, a transmitter transmits a communication signal to a receiver.

transmission link   Defines the type of connection between two points in a communication link. Examples include wireless radio frequency, coaxial cable, optical fiber, or a network category 5 (CAT5).

transmitter   The job of a transmitter is to send a signal source over a communication channel to the receiver. In telecommunications, the communication channel is the connecting wire or fiber-optic cable between the caller and the recipient. In radio communications, the channel is the actual airspace between the transmitting antenna and the receiving antenna.

transponder   In satellite transmission, a transponder is a repeater that responds to incoming transmissions of narrow-beam, radiated microwave. The transponder not only transmits and receives signals but also frequency shifts signals to achieve full-duplex communication between an uplink and downlink.

transport control protocol (TCP)   Works with Internet protocol (IP) as a set of protocols to manage the transfer of data across a network. TCP operates from the transport layer at both ends of the communication link. Its capabilities include data stream transfer management, flow control, full-duplex operation, and communication multiplexing. TCP is mainly responsible for the sorting and reassembling of data packets; error correction; and the detecting, requesting, and resending of lost packets. TCP does not concern itself with the destination routes of the various packets; instead, it simply operates from the end points of the communication link, spending the majority of its time sorting and reorganizing packets as they arrive.

tree topology   A type of bus topology that connects nodal points of a computer network to multiple branches.

triaxial   A type of coaxial cable that has an added second layer of braided shielding. The second shielding layer is separated electrically from the first by an additional insulation barrier. Functionally, having two layers of shielding allows for the broadcast of multiple signals over the same cable.

trunk   Similar to, but different from, a telephone line. A telephone line connects between an individual phone and the CO switching equipment. A trunk connects switching circuits together, such as between the CO and a PBX, or from one CO switch to another CO switch.

tweeters   High-frequency speaker elements. Tweeters are small in diameter, typically less than 2 in., and more often resemble a flat diaphragm rather than a cone.

twinaxial   Differs from coaxial cable in that it uses two central conductors, allowing for a low-noise, crosstalk-free, balanced transmission signal. Twinaxial is more expensive than coaxial cable and is most often used by the broadcast industry.

twin-lead   Cable originally designed for older style, balanced, television and radio antenna systems. It is comprised of two conductors running parallel through an insulator, separated by a fixed distance. In most cases, the characteristic impedance of twin-lead cables is 300 Ω, as determined by the conductor spacing and the gauge of the wire, 20 to 26 AWG being the most common. Notably, manufacturers of new style televisions, VCRs, and satellite receivers have stopped including 300 Ω inputs on many of their products, replacing it with the now standard 75 Ω coaxial "F" connector. When it becomes necessary to connect a 300 Ω, twin-lead cable to a 75 Ω coaxial input, a matching transformer or balun is needed. Not using a matching device introduces an impedance mismatch into the circuit, causing signal loss and a degradation of quality.

twisted pair   Twisted pair cables offer improved noise immunity over multiconductor cables. The twist in the cable helps to attenuate high-frequency noise transmission caused by the increased inductance, and because the conductors are grouped into pairs providing equal impedance to ground, they are more often used by the telecommunications industry where balanced transmission lines are needed, helping to provide common mode rejection and reducing noise and crosstalk between neighboring conductors. As a result, a twisted paired cable usually offers greater rates of data transmission or bit rates than a multiconductor cable and has become the standard for communication cables in the industry. Twisted pair cables are available in a variety of sizes (solid or stranded), voltage and temperature ratings, and insulation types; and they can be of the shielded or unshielded variety. Examples include unshielded twisted pair (UTP), screened twisted pair (ScTP), and shielded twisted pair (STP).

ultrasonic motion sensor   Used to detect motion. An ultrasonic sensor emits a high-frequency sound wave into a given field. When a person or object enters the field, the sound waves bounce off the object and back to the sensor's receiver at a lower frequency. Only an object with movement transmits a lower frequency signal. With objects in a field that stay stationary (table, file cabinet, desk, etc.) the sound waves bounce off the object but return to the sensor at the same frequency that they were transmitted.

unbalanced   An unbalanced line uses a single-ended input from a grounded source to a grounded load. Because one side of the circuit is grounded and unable to balance off the main conductor, a magnetic field radiates out from the main line, thus acting as a potential source of interference to other conductors. A shielded or coaxial cable should be used in such instances to help reduce the effects of unwanted radiation and to prevent the pickup of stray electromagnetic field and electrostatic interference.

ungrounded   Not connected to ground or to a conductive body that extends the ground connection.

unidirectional microphone   Unidirectional is a type of microphone polar pattern, often called the cardioid. The unidirectional polar pattern can better detect sounds in the forward direction but has a reduced sensitivity toward the rear of the diaphragm.

uniform resource locator (URL)   All available resources or files on the Internet must have a unique address, known as a URL, or uniform resource locator. The URL provides the necessary protocols for finding and accessing Web sites or information files residing on the Internet. URLs are registered through network providers to ensure that they are unique and available for use, because two individual Web sites are not allowed to have the same URL.

uninterrupted power supply (UPS)   An electronically controlled power system that provides ac regulation and filtering to loads requiring consistent, uninterrupted, steady-state power during a partial or even total failure of the main supply input. A variety of UPS systems are available that offer a wide range of configuration options, often based on the design and size of the system. Additional options often include power line conditioning, lightning protection, EMI protection, and redundancy. The size and capacity of a UPS can range from less than 100 W to as high as several megawatts. Single- or three-phase outputs also are available, with typical frequencies of 50, 60, or 400 Hz. The output waveform also can vary from that of a near-perfect sine wave, or square wave, depending on the design of the system.

units of measurement   When using the *NEC*, it is essential to understand how the *Code* uses units of measurement. References to measurements are always listed as International System (SI) units first, and the inch-pound conversion is listed in parentheses; for example, 2.5 m (8.2 ft).

universal serial bus (USB)   A high-bandwidth data communication standard that can support multiple devices within an entire system. The maximum number of devices is limited to 127, with a maximum data rate of 12 Mbps. A revision to the USB standard, known as USB 2.0, can now support data transfer speeds of 480 Mbps.

unshielded twisted pair (UTP)   Solid conductor UTP has become the standard for most data and telecommunication installations because of its relative low cost and ease of installation. The characteristic impedance of UTP is typically in the range of 100 Ω. UTP offers improved noise immunity over that of multiconductor cables. The twist in the cable helps to attenuate high-frequency noise transmission caused by the increased inductance, and because the conductors are grouped into pairs providing equal impedance to ground, they are more often used on balanced transmission lines, helping to provide common mode rejection and reduce noise and crosstalk between neighboring conductors.

uplink   Refers to the transmission of a satellite communication signal from a ground-based transmitter to an orbiting satellite transponder.

User Datagram Protocol (UDP)   A connectionless network protocol that does not send back any type of acknowledgment to the originating transmitter regarding the status of data packets. A UDP transmission therefore must rely on upper-layer protocols for reliability, because data checks are not built into the communication.

varifocal   A type of camera lens that is often referred to as an auto-iris lens. It is used in situations where variable focal lengths (wide angle to telephoto) with the ability to adjust focus is a requirement. It is also a good choice of lens when lighting conditions can vary from full light to low light, from minute to minute, or over the course of a day. The varifocal lens automatically adjusts the diameter of the iris by using a small forward and reversible motor.

varistor   A solid-state, voltage-dependent, variable resistor that is used to protect circuits and line voltages from power surges. The most common type of varistor used in surge protection is the metal oxide varistor; in addition, other types of varistors can provide high current protection in the range of 70,000 A for power applications. Varistors can be made to provide surge protection at all levels.

velocity of propagation   A cable specification that refers to the charge delay of the transmitted signal as it moves down the cable to the load. The rate of transmission is compared with the speed of light, 300 million meters per second, or 186,000 miles per second. Cable specifications usually specify the velocity of propagation as a percentage called the velocity factor. As an example, an 86% velocity of propagation means that the signal is traveling at 86% the speed of light or $300 \times 10^6 \times 0.86 = 258$ million meters per second. The rate of propagation is almost entirely dependent on the dielectric rating of the insulation material used in the construction of the cable.

video motion detection   A software driven method of detecting movement within a video image. The software is sensitive to changing video patterns. If the video pattern remains static or unchanging, then there is no perceived motion. Any nonrepetitive movement within a field of view is detected as an intruder. CCTV systems can use video motion detection to help alert security to an apparent threat.

wavelength   The wavelength (λ) of a waveform can be defined as the velocity, v (feet per seconds), divided by the frequency in Hertz, f (cycles per second).

$$v \div f = \lambda$$

white noise   White noise generators generate equal intensities of all frequencies in the range of 20 to 20,000 Hz. The effect is similar to the sound of a vacuum cleaner. White noise generators are used to flatten the response of a room; they also are used as noise-masking devices to help break the dead silence of a quiet room.

wide area network (WAN)   Interconnects multiple LANs over a wide geographical area, such as a city or state. A WAN achieves a communication link over long distances through a network line brought in by a service provider.

Wiegand technology Security systems use Wiegand technology in certain types of access cards and access card readers. Wiegand technology uses short lengths of small-diameter, specially treated wire with distinctive magnetic properties. The wire is embedded into a code strip within an access card. This wire, when presented to a changing magnetic field provided by a Wiegand reader, represents either a 1 or 0 of a binary code. The access-control system grants the user access if the binary code on the card matches up with the pre-programmed user code.

wireless communication Involves the sending of signals by air instead of through a hardwired connection or cable. Examples include radio, satellite, microwave, infrared, and ultrasonic.

woofers Large-diameter speaker cones, typically larger than 8 in. They are used to recreate low-frequency bass signals.

XLR Commector commonly used to connect balanced, low-impedance circuits, including microphones and line-level audio signals, to processing equipment, and amplifiers.

zipcord A two-fiber version of simplex, optical fiber cable.

Index

10Base-T Ethernet, 167–168
25 twisted-pair color codes, 367–368

A

Abandoned cables, 27
ac. *See* Alternating current
Access-control systems, security alarms and, 377–402
Acoustic glassbreak sensor, 380–381
ACR. *See* Attenuation/crosstalk ratio communication cable
Active crossovers, 131
Active filters, 152
Active infrared sensor, 379
A/D. *See* Analog-to-digital converters
Addressable systems, 321–322
Addressing, computer and internet, 206–207
Address Resolution Protocol (ARP), 206
ADSL. *See* Asymmetrical Digital Subscriber Line
Advanced Mobile Phone Service (AMPS), 426–428
Advanced Research Projects Agency Network (ARPANET), 201
Advanced Television System Committee (ATSC), 475
AHJ. *See* authority having jurisdiction
Air, wavelength and sound traveling in, 104–105
Air sampling-type smoke detectors, 299–300
Alarm, 110
　false, 324
　indicator devices, 303–304
　relay, 399
Alarm communications. *See* Voice/Alarm communications
Alarming, purpose of, 378
Alarm-initiating devices, 291–302
Alternating current (ac)
　generator, 231–233
　systems less than 50 V, 80–81

Alternating current (ac) voltage regulators, 224–226
Aluminum, 19, 75
AM. *See* amplitude modulation
Amateur transmitting and receiving stations, 456
American National Standards Institute (ANSI), 10–11
American Standard Code for Information Interchange (ASCII), 181
American Wire Gauge (AWG), 20–21, 145, 261
Ampacity, 37
　ratings, conductor, 261
Amplification, 107–160
Amplifier
　bandwidth, 98
　electrical properties, 140–149
　headphone, 141
　integrated, 149
　power, 134, 142–144, 149
Amplitude modulation (AM), 410, 411, 416
AMPS. *See* Advanced Mobile Phone Service
Anaerobic, 342–343
Analog
　addressable systems, 321–322
　cameras, 461
　digital *vs.* 412–413
　video formats, 474–476
Analog-to-digital converters (A/D), 413–414
Annunciator display panels, remote fire alarm, 289–290
ANSI. *See* American National Standards Institute
Antenna, 408, 415–424
Application and process layers, 203–205
Area code, telephone dialing and, 354
ARP. *See* Address Resolution Protocol
ARPANET. *See* Advanced Research Projects Agency Network
ASCII. *See* American Standard Code for Information Interchange

Associated systems within enclosures, 269–270
Asymmetrical Digital Subscriber Line (ADSL), 170
Asynchronous transfer mode (ATM), 199–201
ATM. *See* Asynchronous transfer mode
ATSC. *See* Advanced Television System
    Committee
Attenuation
    cutoff frequency and signal, 42–43
    dB signal, 91
    skin effect and signal, 38–39
Attenuation/crosstalk ratio (ACR) communication
    cable, 53
Audible alarm indicator devices, 303–304
Audio
    cable, 27
    circuits, Class 2 or 3 conductors and, 273
    mixers, 150–151
    physics, 89–105
    signal-processing, 107–160
    system circuits
        and PLFA circuits, 313
    transformer outputs, 145
Authority having jurisdiction (AHJ), 4, 281
Automatic extinguishing systems, 301
Autotransformers, 223–224
Auxiliary fire alarm system, 284
AWG. *See* American Wire Gauge

B

Balanced circuits, 119
Balanced *vs.* unbalanced transmission line,
    120–121
Band-pass filters, 129–130, 134
Bandwidths, 42, 46, 97, 98, 199
BAP. *See* Battery-assisted passive tags
Base unit, cordless phone, 355
Batteries. *See* Storage batteries
Battery-assisted passive (BAP) tags, 393
Bell, 110, 154, 303
    wiring, 388
Bidirectional polar pattern, 114–115
Biometric readers, 393
Bit rates, 41
Block diagram
    completed system, 154–156
    fire alarm circuit, 286
Bluetooth, 449–451
Bonding
    conductors, 74, 78–80
    effective ground-fault current path, 78

electrical system purpose, 77–78
    grounding and, 63–85
    ground rods, 78
    Intersystem termination, 83
    jumpers, 72–74
    NEC and, 81–85, 157
Braid shields, 33–34
Breakout cables, 336
Breakout kits, 342–343
Bridge, 192–194
Bridge rectifiers, 238–240
Broadband antenna, 420–421
Broadcast, 189
Buck-boost transformer regulators, 225
Bus network, 185
Busy signal, telephone, 353
Bytes, 175

C

Cables, 308. *See also* Fiber cables;
    Fiber-optic cable
    category 3, 4, 5, 5e, 6, and 7, 49–51
    classified types, 29–32
    classified *vs.* listed, 28–32
    coaxial, 18, 22, 48, 54–56, 191, 472–473, 480
    communication, 27, 52–53, 211–212
    composite, 272
    composite optical fiber, 346
    conductive and nonconductive optical
        fiber, 272, 345, 346
    construction and insulation, 17–24
    designs, 333–334
    electrical conductors and, 15–59
    electrical properties, 35–37, 39, 41–47, 54,
        189–192
    fire alarm types, 311–312
    instrumentation tray, 266
    modems, 171–172
    multiconductor, 17, 47, 48, 310–311
    network category 5, 473
    optical fiber, 27, 174, 272, 328–344
    ratings, 374
    shielded *vs.* unshielded, 18
    signal conductor, 47
    single conductor, 17, 47
    solid *vs.* stranded, 17, 22
    special multipurpose hybrid, 57
    support of conductors and, 273–274
    trays, 27, 28, 265–266
    triaxial and twinaxial, 48, 56

twin-lead, 18, 56–57
twisted pair, 48–49, 166–167, 367–368
types of, 47–57
Cable connectors
  allowable numbers of power supply, 311
  coaxial cable, 480
  fiber-optic types of, 340–344
  line input, 141–142
  stranded *vs.* solid, 17, 22
  terminal, 51–52
  termination methods, 343
  troubleshooting, 211–213
Cable television systems, 452–454
Cabling
  premise wiring, 356
  residential connections, 355–358
Cameras
  analog, 461
  CCD, 462–464
  CCTV, 273, 381, 459–481
  Newvicon, 462
  sensitivity, 464
  tube, 461–462
  ultricon, 462
  vidicon, 461–462
Canadian Standards Association (CSA),
    12, 16, 28
Capacitance, cable, 41
Capacitor filter, 237–238
Carbon monoxide (CO) detectors, 301
Cardioid microphones
  polar patterns, 114–117
  proximity effect, 119
Carrier frequency, 409
Carrier sense multiple access with collision
    detection (CSMA/CD), 185–188
Cathode ray tube (CRT), 461
CATV. *See* Community antenna and television
    distribution
CB. *See* Citizen band radio
CCD. *See* Charge-coupled device
CCTV. *See* Closed circuit television camera systems
CDDI. *See* Copper distributed data interface
CDMA. *See* Code division multiple access
Cellular phone communication, 424–432
Cellular systems, PCSs, 425, 428–429
Centimeter-gram-second (CGS) system, 90
Central fire alarm control units, 284–285, 288,
    290–291, 301–304
Central office (CO), 350

Central station service, 287–288
CGS. *See* centimeter-gram-second
Channels, 406
  of satellite communications, 437, 439
  yagi antenna, 419
Characteristic impedance, 43–44, 54
Charge-coupled device (CCD) camera, 462–464
Charging cautions and concerns, storage
    batteries, 246–247
Chassis ground, 66
CI. *See* Circuit integrity cable
Circuits. *See also* Communication circuit
  audio, Class 2 or 3 conductors, 273
  audio systems, 313
  balanced, 119
  Class 1, 257, 259–261
  Class 2, 27, 257, 261–262
  Class 2 and 3, 264–273
  Class 3, 27, 257, 262
  common ground, 66
  door status, 398–399
  electrical types, 256–257
  initiating device classification, 315–318
  multiple Class 2 and 3 installation requirements,
      272–273
  NAC, 318–319
  NEC classification of, 30, 255–276
  NPLFA, 308–310
  PLFA, 272, 312–314
  power-limited Class 2 and 3 separation
      requirement, 261–271
  remote-control, 257–259
  separation of unrelated, 259–260
  SLC, 258, 322
  switching, 196–197
Circuit identification
  Class 1, 2, and 3 markings and, 267
  fire alarm, 307
Circuit integrity (CI) cable, 308
Circular mil (CM), 21–22
Citizen band (CB) radio, 122
CL2. *See* Class 2 cable
CL3. *See* Class 3 cable
Cladding, coating, and core, 330–331
Class 1, 2, and 3 circuit identification and
    markings, 267
Class 1 circuits, 257, 259–261
Class 2 and 3 circuits, 264–273
Class 2 and 3 power sources, 263–264
Class 2 (CL2) cable, 28

Class 2 circuits, 27, 257, 261–262
Class 3 (CL3) cable, 28
Class 3 circuits, 27, 257, 262
Classification of circuits, *NEC Article 725*, 30, 255–276
Classified cable, types of, 29–32
Client/server networks, 195–196
Closed circuit television (CCTV) camera systems, 273, 381, 459–481
   components, 461, 470–472
   NEC Table 820 requirements, 480–481
   purpose, 460
   recording devices, 476–480
   system specifications, 464–470
   transmission links, 472–474
   viewing monitors, 461, 474–476
Closed loop, 270–271, 383, 386–387
CM. *See* Circular mil; Communications cable
CMOS. *See* Complementary metal-oxide semiconductor
CMP. *See* Communications plenum
CO. *See* Carbon monoxide; Central office
Coating, core, and cladding, 330–331
Coaxial cable, 22, 48, 54–56, 191, 472–473
   connectors, 480
   twin-lead cable *vs.* 18
Coaxial speakers, 132–133
Code division multiple access (CDMA), 429–432
Coherence of signal, 100–101
Coherent bundle, 330
Collision detection, 185–192
Combinational topology, 185
Combination shields, 34–35
Combination varistor/spark gap suppressor, 228
Commercial systems, 358–360
   fire alarm, 284–288
Common-mode signal, 221–222
Communication
   cables, 27, 52–53, 211–212
   emergency voice/alarm, 304
   methods, 176–178
   OSI and, 181–184
   parallel, 178
   satellite, 432–444
   serial, 176–177
   wireless, 405–457
Communication circuit, 257, 258, 273, 370–375
   cable ratings, 374
   installation methods within buildings, 373–374

   protection devices, 372–373
   separation of, 374–375
   wires and cables outside and entering buildings, 370–372
Communications (CM) cable, 28
Communications plenum (CMP), 28
Communication system grounding means, 84
Community antenna and television distribution (CATV), 28, 171, 258
Compensator, telephone, 351–352
Complementary metal-oxide semiconductor (CMOS), 463
Compliance, speaker, 131–132
Composite cables, 272
Composite optical fiber cables, 346
Computer network
   addressing, 206–207
   wireless, 444–452
Concrete-encased electrode, permitted for grounding, 70–71
Condenser microphones, 112–113
Conductance, cable, 41–42
Conductive optical fiber cables, 272, 345, 346
Conductor. *See also* Electrical conductors
   ampacity ratings, 261
   AWG, 261
   bonding, 74, 78–80
   coatings, 20
   number of, 157, 275–276
   in raceway, number and size of, 275–276
   shielding, 32–35
   size for PLFA, 314
   support of, 273–274
   types, 18–20, 75
Connectors. *See also* Cable connectors; Optical fiber connectors
   Hotmelt, 343
   line input, 141–142
   microphone, 124–125, 141–143
   power supply, 311
   pressure, 73
   RJ, 51, 209–210
Constant-voltage transformer (CVT) regulators, 226
Control panels
   fire alarm types, 154, 283
   keypads, 383, 389
Control systems. *See* Access-control systems
Control units
   alarm-initiating devices and, 291–302
   central, 284–285, 288, 290–291, 301–304

Control valve supervisory signal-initiating device, 302
Copper distributed data interface (CDDI), 167
Cordless telephones, 354–355
Core, cladding, and coating, 330–331
Critical angle, 333
Crossovers
active and passive, 131
electrical properties, 129–130
filters, 134
Crosstalk shield, 35
CRT. *See* Cathode ray tube
CSA. *See* Canadian Standards Association
CSMA/CD. *See* Carrier sense multiple access with collision detection
Cutoff frequency and signal attenuation, 42–43
CVT. *See* Constant-voltage transformer regulators

D

D/A. *See* Digital-to-analog converters
DACT. *See* Digital alarm communicating transmitter
Damping level, 128
DARPA. *See* Defense Advanced Research Project Agency
Data circuit-terminating equipment (DCE), 200
Data errors, 198–199
Datagrams, 198
Data link layer of OSI method, 182
Data terminal equipment (DTE), 200
dB. *See* Decibels
DBS. *See* Direct broadcast satellite
dc. *See* Direct current power supply
DCE. *See* Data circuit-terminating equipment
Decibels (dB), 90–95
Defense Advanced Research Project Agency (DARPA), 197
Delay skew
communication cable, 53
terminal connectors, 52
DEOL. *See* Double end-of-line
Depth of field, 470–472
Detection
collision, 185–192
range of heat, 293–294
Detectors
carbon monoxide, 301
fire, 292–293
heat, 292–296
radiant energy, 300
smoke, 297–300

Devices
alarm indicator, 303–304
configuration, 386–387
control units and alarm-initiating, 291–302
drivers, 213
end-of-line, 320
infrared, 451–452
network, 192–195
overcurrent protection, 257, 260–261
recording, 476–480
reset *vs.* nonrestorable initiating, 294
sprinkler water flow alarm-initiating, 301
supervisory signal-initiating, 301–302
Dialing, 353–354
area code, 354
telephone, 352
Dial tone, telephone, 351
Dial-up modem, 168, 169
DID. *See* Direct inward dialing
Diffraction, 104–105
Digital, analog *vs.* 412–413
Digital, Intel, Xerox (DIX), 164
Digital alarm communicating transmitter (DACT), 323
Digital Equipment Corporation, 164
Digital-processing equipment, 153
Digital Subscriber Line (DSL), 165, 168–170
Digital television standards, ATSC, 475
Digital-to-analog (D/A) converters, 413–414
Digital video recorders (DVRs), 477–478
Dipole antennas, 419
Direct broadcast satellite (DBS), 172–173
Direct current (dc) power supply, 235–243
Direct inward dialing (DID), 363
Direct radiator, 127
Direct-sequence spread spectrum (DSSS), 430
Dish size, satellite, 436–437
Display and reporting systems, 383–385
Dissimilar metals, 57–59
Distribution trunk and feeders, 361–364
DIX. *See* Digital, Intel, Xerox
DNS. *See* Domain Name Server
Domain Name Server (DNS), 207
Door
hardware, 394–396
releases, 320
status circuit, 398–399
Double-conversion UPS, 234–235
Double end-of-line (DEOL), 387
Downlink antenna, 433

Dry-cell electrolytes, 245
DSL. *See* Digital subscriber line
DSSS. *See* Direct-sequence spread spectrum
DTE. *See* Data terminal equipment
DTMF. *See* Dual-tone multifrequency
Dual-technology motion sensor, 379–380
Dual-tone multifrequency (DTMF), 352
Duplex coil, telephone, 351
DVRs. *See* Digital video recorders
Dynamic microphones, 111–112
Dynamic range, 96–97

E

Earpiece, telephone, 351
Earth ground, 66
EBCDIC. *See* Extended Binary Coded Decimal
    Information Code
Echo, 135
Effective ground-fault current path, 78
EGC. *See* Equipment grounding conductor
EIA. *See* Electronic Industries Alliance
EIA/TIA. *See* Electronic Industries Alliance/
    Telecommunications Industry Association
EKS. *See* Electronic key systems
Electrical circuits, types of, 256–257
Electrical conductivity heat detectors, 296
Electrical conductors
    cables and, 15–59
    cable construction and insulation, 17–24
    cable selection, 16–17
    cable types, 47–57
    classified *vs.* listed cables, 28–32
    conductor shielding, 32–35
    dissimilar metals, 57–59
    electrical properties of cables, 35–47
    plenum *vs.* riser, 24–27
Electrical equipment, access to behind panels, 307
Electrical properties
    of amplifiers, 140–149
    of cables, 35–37, 39, 41–47, 54, 189–192
    of sound systems, 111–127
    of speakers, 127–139
Electrical service entrance, 64–66
Electric strike, 395–396, 398
Electrodes permitted for grounding, 70–72
Electrolytes, wet-or-dry cell, 245
Electrolytic corrosion, 58
Electromagnetic interference (EMI), 174
Electromagnetic wave, 328
Electronic Industries Alliance (EIA), 11

Electronic Industries Alliance/Telecommunications
    Industry Association (EIA/TIA), 210, 290
Electronic key systems (EKS), 364–365
e-mail, 165
Emergency systems, 248–250
Emergency telephones, 288, 305–306
Emergency voice/alarm communications, 304
EMI. *See* Electromagnetic interference
E+M tie line, 363–364
Encapsulation layer of OSI method, 183–184
Enclosures
    associated systems within, 269–270
    loudspeakers, 154
    raceways within, 269
    with single openings, 270
    speaker, 137–139
End-of-line (EOL)
    configurations, 387–388
    devices, 320
    relays, 320
    resistor, 386–387
Engine-driven ac generator, 233
EOL. *See* Rnd-of-line
EPDM. *See* Ethylene-propylene-diene elastomer
Equalizers, 152–153
Equipment
    digital-processing, 153
    networking and information technology,
        163–214
    reproduction, 91, 107–160
    signal-processing, 91, 108, 153
Equipment grounding conductor (EGC), 68
    bonding conductors, 74
    clamps and hardware - listing and labeling, 74
    equipment connection method, 72–73
    grounded conductor *vs.* 69
    grounding electrode conductors, 74
ETA/TIA. *See* Electronic Industries Alliance/
    Telecommunications Industry Association
Ethernet, 164
    10Base-T, 167–168
    evolution, 208–209
    PoE, 473–474
    thick *vs.* thin, 190
Ethylene-propylene-diene elastomer (EPDM)
    insulation, 23
Eutectic salts and solders, 294–295
Exchange, telephone, 354
Exchange messaging (e-mail), 165
Exothermic weld, 74

Expanding air volume, 295
Expanding bimetallic components, 294
Expanding liquid volume, 295
Extended Binary Coded Decimal Information Code
     (EBCDIC), 181

F

Fail-safe magnetic locks, 394–395
Fail-secure magnetic lock, 395–396
False alarms, 324
Far-end crosstalk (FEXT) communication cable,
     52–53, 211–212
FCC. *See* Federal Communications Commission
FCS. *See* Frame check sequence
FDDI. *See* Fiber distributed data interface
FDMA. *See* Frequency division multiple access
Federal Communications Commission (FCC),
     210–211, 407–408
FEXT. *See* Far-end crosstalk communication cable
Fiber
     classification of, 345–346
     varieties of, 340–344
Fiber cables
     composite optical, 346
     conductive and nonconductive optical,
          272, 345, 346
     optical, 27, 174, 272, 328–344
     styles of optical, 334–337
Fiber distributed data interface (FDDI), 166, 167
Fiber-optic
     construction, glass, 330
     fiber color codes, 344
     splices, 343–344
Fiber-optic cable and *NEC Article 770*, 85,
     328–344. *See also* Optical fiber cable
     applications, 338
     concepts of light, 328–329
     connector types, 340–344
     fiber classifications, 345–346
     fiber varieties, 339–340
     optical fiber cable, 329–338
     ribbon, 333–334
Field coil, 127
Field of view (FOV), 467, 469
File Transfer Protocol (FTP), 203–204
Filters
     active, 152
     band-pass, 129–130, 134
     capacitor, 237–238
     crossover, 134

harmonic and noise suppression, 229–230
     high-pass, 129, 134
     low-pass, 129, 134
     notch filter, 130, 134
     power amplifier, 134, 142–144, 149
Fine Print Notes (FPNs), 8
Fire
     low-voltage wiring and, 27
     plenum space and, 25
Fire alarm
     cables, 311–312
     Class 1 conductors, non-power-limited
          and, 311
     control panels, 154, 283
Fire alarm circuit
     block diagram, 286
     identification, 307
Fire alarm systems
     commercial, 284–288
     electrical requirements of, 307–314
     household, 284
     NFPA definition of, 282
     OSHA requirements for, 306
     public, 288
     purpose of, 280–282
     remote station, 288
     single-stage, 284
     supervisability of, 282
     troubleshooting, 324–325
     types of, 284–288
     wiring of, 315–321
Fire alarm systems defined
     fire alarm control panels, types of, 154, 283
     fire alarm systems, types of, 284–288
     fire command center, 288–289
     remote fire alarm annunciator display
          panels, 289–290
Fire command center, 288–289
Fire detectors (heat and smoke), 292–293
Firewalls, 194, 207
FireWire, 176, 178
Fixed-focus lens, 468–469
Fixed-temperature heat detector, 292
FL. *See* Focal length
Flexible coaxial cable, 54–56
Flux, 59
FM. *See* Frequency modulation
Focal length (FL), 469–470
Foil braid shield, 34
Foil shields, 33

Formats, viewing video, 474–476
Fourth generation of wireless, 432
FOV. *See* Field of view
FPL. *See* Power-limited fire protective
FPNs. *See* Fine Print Notes
Frame check sequence (FCS), 194
Frame relay, 200
Frame transfer CCD camera, 463–464
French braid shields, 34
Frequency, 95–96
  carrier, 409
  cutoff, 42–43
  high, 43
  low, 42, 46
  radiation patterns of different, 105
  radio, 392–393, 408
  response, 97, 123
  satellite, 142
Frequency division multiple access (FDMA), 428
Frequency modulation (FM), 410, 411
Frequency shift modulation (FSK), 410, 411
FSK. *See* Frequency shift modulation
FT4/IEEE 1202 vertical flame test, 28
FTP. *See* File Transfer Protocol
Full-duplex communication, 409
Full-wave rectifiers, 238–239
Full-wave *vs.* bridge rectifiers, 239–240
Fusible links, 294–295

G
Gateways, 195
GEC. *See* Grounding electrode conductor
Generators, 230–233
  ac motor, 231–233
  engine-driven AC, 233
  pink noise, 103–104
  white noise, 103
GEOs. *See* Geosynchronous satellites
Geosynchronous satellites (GEOs), 433, 436
Glassbreak sensor, 380–381
Glass fiber-optic construction, 229–330
Global positioning satellite (GPS), 407
Global system for mobile (GSM), 429, 430
Gooseneck, 126
GPS. *See* Global positioning satellite
Graded index MMF, 340
Graphic equalizers, 152
Ground
  chassis, 66
  earth, 66

EGC, 68
faults, 75–78
grounding electrode conductor, 68–69
loops, 121–122
NEC definition, 66, 68–69
neutral or circuit common, 66
resistance supplemental electrodes, 71–72
ring, 71
rod, 64
solidly grounded, 69
start, 361
symbols, 67
ungrounded conductor, 69
Grounded communication installation, 82
Grounded conductor, 68–69
Grounding
  electrical systems purpose, 64
  electrode, 68
  electrode system, 70–71
Grounding and bonding, 63–85
  alternating current systems less than 50 V, 80–81
  aluminum and aluminum grounding conductors, 75
  bonding, 77–80
  bonding jumpers, 72–74
  connecting EGCs, 72–74
  electrical service entrance, 64–66
  ground definition, 66–67
  ground faults, 75–77
  grounding electrical systems purpose, 64
  grounding electrode system, 70–71
  ground resistance and supplemental electrodes, 71–72
  lightning protection systems, 75
  NEC ground definitions, 66, 68–69
  NEC grounding articles, 81–85
Grounding electrode conductor (GEC), 69, 371
GSM. *See* Global system for mobile

H
Half-duplex communication, 410
Handset, cordless phone, 355
Hardware
  door, 394–396
  EGC, 74
  media, 208–213
  microphone, 124–125
Harmonic and noise suppression filters, 229–230
Harmonics *vs.* Octaves, 99

HDMI. *See* High-definition multimedia interface
HDSL. *See* High-Bit-Rate Digital Subscriber Line
HDTV. *See* High-definition television standard
Headphone amplifiers, 141
Hearing
    graph, 101
    sound and, 101–105
    threshold of, 90
Heat detection, range of, 293–294
Heat detector
    fixed-temperature, 292
    location, 296
    rate compensation detector, 292
    rate-of-rise, 293
    technologies, 294–296
Heating, ventilation, and air conditioning (HVAC)
        system, 58
Helical antennas, 423
High-Bit-Rate Digital Subscriber Line (HDSL), 170
High-definition multimedia interface (HDMI),
        478–480
High-definition television standard (HDTV), 475
High-frequency transmission, 43
High-pass filter, 129, 134
History
    mobile radiotelephone, 425–429
    of networks, 164–168
    telephone, 350
    wireless communication, 406–408
Hoistways, 271
Hook switch, telephone, 351
Horizontal wiring, 360
Horns, 133, 303
Hotmelt connectors, 343
Household fire alarm systems, 284
HTML. *See* Hypertext markup language
HTTP. *See* Hypertext Transport Protocol
Hub, 192
Human voice, 99
HVAC. *See* Heating, ventilation, and air
        conditioning
Hybrid cables, special multipurpose, 57
Hypercardioid polar patterns, 116–118
Hypertext markup language (HTML), 203
Hypertext Transport Protocol (HTTP), 203–204

I

IBC. *See* International Building Code
ICEA T-29 520 vertical flame test, 29
ICMP. *See* Internet Control Message Protocol

IDC. *See* Initiating device circuits
Identification
    Class 1, 2, and 3 circuit markings and, 267
    fire alarm circuit, 307
IDF. *See* Intermediate distribution frame
IEC 323-3 vertical flame test, 29
IEEE. *See* Institute of Electrical and Electronics
        Engineers
IMAP. *See* Internet Mail Access Protocol
Impedance, 39
    characteristic, 43–44, 54
    microphone, 119
    nominal, 43, 45
    speaker, 128–129
Improved Mobil Telephone Service (IMTS), 425
IMTS. *See* Improved Mobil Telephone Service
Index of refraction, 332–333
Information technology equipment, networking
        and, 163–214
Infrared
    devices, 451–452
    sensor, 379
Inherently limited Class 2 power source, 263
Initiating device circuits (IDC), 315–318
Installation requirements
    for communication circuits, 373–374
    for multiple Class 2 and 3 circuits, 272–273
Institute of Electrical and Electronics Engineers
        (IEEE), 11, 164
    383 vertical flame test, 29
Instrumentation tray cable (ITC), 266
Insulated cables. *See* Mineral-insulated cables
Insulation
    cable construction and, 17–24
    Class 1 conductor requirements, 261
    Class 2 and 3 circuits requirements, 265
    types, 23–24
Insurance company, 281
Integrated amplifiers, 149
Integrated services digital network (ISDN), 165,
        171, 200
Integrated services digital network digital subscriber
        line (ISDL), 171–174
Intercoms. *See* Public address systems/intercom
Interconnection of power sources, 264
Interior transmitting stations, 457
Interline transfer CCD camera, 463
Intermediate distribution frame (IDF), 359–360
International Building Code (IBC)/Municipal and
        Local Building Codes, 281

International Organization for Standardization
(ISO), 11, 179
International System (SI) of Units, 4
International Telecommunications Union (ITU), 432
Internet access
dial-up modem, 168, 169
DSL, 165, 168–170
ISDN, 165, 171, 200
optical fiber, 27, 174, 272, 328–344
satellites, 432–444
wireless networks, 405–407
Internet addressing, 207
Internet Control Message Protocol (ICMP), 206
Internet Explorer, 203
Internet Mail Access Protocol (IMAP), 206
Internet Protocol (IP), 175, 202
Internet Service Provider (ISP), 173, 182
Intersystem bonding termination, 83
Inverse square law, logarithms and, 94–95
Ionization smoke detectors, 298
IP. *See* Internet Protocol
ISDL. *See* Integrated services digital network digital
subscriber line
ISDN. *See* Integrated services digital network
ISO. *See* International Organization for
Standardization
Isolated ground, 157–159
Isolation transformers, 220–223
common-mode signal, 221–222
traverse-mode noise, 223
ISP. *See* Internet Service Provider
ITC. *See* Instrumentation tray cable (ITC)
ITU. *See* International Telecommunications Union

J

Jitter data error, 198

K

Keypads/control panels, 383, 389

L

Labeling laboratories, listing and, 12
Laboratories, listing and labeling, 12
LANs. *See* Local area networks
Latency data error, 198
Lavaliere or lapel microphones, 123
Layers
application and process, 203–205
data link, 182
network, 182

physical, 183
presentation, 181
session, 181
transport, 181–182
Lead-acid batteries, 246
Least significant digit (LSD), 175
LED. *See* Light-emitting diode
Lens, 467
LEOs. *See* Low-earth-orbit satellites
Levels
damping, 128
line, 141
microphone, 140
power amplifier, 142–144
signal, 131, 140–141
sound, 153–154
Light, concepts of, 328–329
Light-emitting diode (LED), 331
Lightning arrestors, 228
Lightning protection system, 75
Line
input connectors, 141–142
levels, 141
to load, matching, 43, 45
pads, 141
transmission, 39–41, 55, 120–121
Listed cables, classified *vs.* 28–32
Listed clamp, 74
Listing and labeling laboratories, 12
LNB. *See* Low-noise block converter
Load, voltage loss at, 36–37
Local area networks (LANs), 164–165,
167, 184, 194
Local loops, 350
Locations
heat detector, 296
smoke detector, 300
Logarithms, 90–95
Loop
closed, 270–271, 383, 386–387
ground, 121–122
local, 350
open, 383, 386–387
start, 362–363
Loose-tube cable design, 333
Lost packets data errors, 199
Loudspeakers, 111
damping level, 128
direct radiator, 127
enclosures, 154

field coil, 127
radiation, 138
spider mount, 127–128
Low bandwidth data error, 199
Low-earth-orbit (LEOs) satellites, 433, 436
Low-frequency transmission
and bandwidth, 42, 46
Low-noise block (LNB) converter, 422
Low-pass filter, 129, 134
Low-power network-powered broadband
communication, 273
Low-voltage
residential network applications, 214
wiring, 27
LSD. *See* Least significant digit

M

MAC address, *See* Media access control address
Maglocks. *See* Magnetic locks
Magnetic door/window contacts, 382–386
Magnetic locks (Maglocks), 394–398
Magnetic stripe, 392
Main distribution frame (MDF), 358–360
Manholes, 270
MANs. *See* Metropolitan area networks
Master box, 285
Matching transformers, 144
Maximum power transfer, 144–145
MDF. *See* Main distribution frame
Media
connections, hardware, and installation
techniques, 208–213
connector troubleshooting, 211–213
modular wire-mapping options, 209–211
types and ethernet evolution, 208–209
Media access control (MAC) address, 175
Medium-Earth-orbit (MEOs) satellites, 433, 436
Megger ground resistance tester, 71–72
Melting insulators, 295
MEOs. *See* Medium-Earth-orbit satellites
Metal frame of building, electrode permitted
for grounding, 70
Metal underground water pipe
electrode permitted for grounding, 70
supplemental electrodes and, 72
Metropolitan area networks (MANs), 164
Microphone, 111–121
Microphone connectors
miniconnectors, 124–125, 142–143
monoplug, 125

Tip-ring-sleeve, 124, 141–142
XLR, 124, 141
Microwave/radar sensor, 379
MIDI. *See* Musical instrument
digital interface
Midrange cones, 128
Mineral-insulated (MI) cables, 308
Miniconnector, microphone, 124–125, 142–143
MMF. *See* Multimode fiber
MMTA. *See* Multimedia Telecommunications
Association
Mobile radiotelephone history
AMPS, 426–428
PCSs cellular systems, 425, 428–429
roaming, 426
Modem, 165, 168, 169, 171–172
Modular patch panels, 366–367
Modular wire-mapping options, 209–211
Modulation, 409, 410–413, 416
Monitors, viewing, 461, 474–476
Monoplug microphone connector, 125
Monopole, quarter-wave, 419
Most significant digit (MSD), 175
Mouthpiece, telephone, 351
Mozilla FireFox, 203
MP. *See* Multipurpose cable
MPP. *See* Multipurpose plenum
MSAU. *See* Multistation access unit
MSD. *See* Most significant digit
Multicast, 189
Multiconductor cables, 47, 48
NPLFA, 310–311
twisted pairs cables *vs.* 17
Multimedia Telecommunications Association
(MMTA), 11
Multimode fiber (MMF), 336–337, 340
Multi-National Harmonized Communication Cable
Standards, 28
Multiple Class 2 and 3 circuits, installation
requirements for, 272–273
Multiple octaves, 98
Multiplexers, 475–476
Multiplexing, 414
Multipurpose (MP) cable, 28
Multipurpose plenum (MPP), 28
Multistation access unit (MSAU), 185
Musical instrument digital
interface (MIDI), 153
Muting, 110
switch, microphone, 126

N

NAC. *See* Notification appliance circuits
National Electrical Code (NEC), 2–12
  Article 90, 3
  Article 90.3, 6
  Article 90.9, 4
  Article 100, definition, 3–4, 8
  Article 110.5-100.11, 16
  Article 110.6, conductor sizes, 20–21
  Article 110.7, wiring integrity, 22–23
  Article 110.11, deteriorating agents, 59
  Article 110.14, electrical connections, 57–58
  Article 640, audio signal processing, amplification
    and reproduction equipment, 91, 108
  Article 640.4, protection of electrical
    equipment, 108
  Article 640.5, access to electrical equipment
    behind panels, 109
  Article 640.6, mechanical execution of
    work, 109
  Article 700, emergency systems, 248–250
  Article 725, circuit classification, 30, 255–276
  Article 760, 30–31
  Article 770, fiber-optic cables and raceways, 85,
    328–346
  Article 770.6, optical fiber cables, 338
  Article 800, communication circuits, 31, 81–84,
    349–375, 402
  Article 800.44, telecommunications, 7
  Article 810, Radio and television requirements,
    454–457
  Article 820, CATV, 32, 85, 480–481
  Article 830, 188–189
  Article 830.90, related to protection, 189
  Article 830.179, related to Broadband Equipment
    and Cables, 189
  Definition 640.2, loudspeaker, 127–128
  grounding and bonding articles, 81–85
  on ground definitions, 66, 68–69
  requirements, 156–160, 275–276
  Table 310-104 (A) conductor applications
    and insulations, 23
National Fire Protection Agency (NFPA), 2, 23,
  281–283
National Testing Association (NTA), 12
NC. *See* Normally closed
Near-end crosstalk (NEXT) communication cable,
  52, 211–212
NEC. See National Electrical Code
Neoprene insulation, 23

Network
  addressing, computer, 206–207
  application, low-voltage residential, 214
  bus, 185
  category 5 cable, 473
  client/server, 195–196
  computer, 206–207, 444–452
  devices, 192–195
  history, 164–168
  ISDL, 171–174
  ISDN, 165, 171, 200
  LANs, 164–165, 167, 184, 194
  layer of OSI method, 182
  peer-to-peer, 195
  protocols and functions, 199–206
  ring, 184–185, 187
  software management, 195–196
  star, 184–185
  topology, 164, 184–185
  wireless, 405–407
  wireless computer, 444–452
Networking
  challenges, 175
  information technology equipment and, 163–214
Networking, architecture and topology
  byte, 175
  circuit switching, 196–197
  collision detection, 185–188
  communication and OSI model, 181–183
  communication methods, 176–178
  data errors types, 198–199
  datagrams, 198
  internet access, 168–174
  network devices, 192–195
  network history, 164–168
  networking challenges, 175
  network software management, 195–196
  network topology, 184–185
  NIC, 174–175
  OSI seven layers, 179–180
  packet switching, 197–198
  point-to-point links, 196
  transmission methods, 188–189
  unicast, broadcast, and multicast, 189
  velocity of propagation and, 45–47, 189–192
Networking and information technology equipment
  computer network addressing, 206–207
  low-voltage residential network applications, 214
  media: connections, hardware and installation
    techniques, 208–213

networking, architecture and topology, 164–199
network protocols and functions, 199–206
other concerns, 213–214
Network interface card (NIC), 174–175
Network operating system (NOS), 184
Neutral ground, 66
Newtons per meter, 131
Newvicon camera, 462
NEXT. *See* Near-end crosstalk communication cable
NFPA. *See* National Fire Protection Agency
NIC. *See* Network interface card
Nickel-cadmium batteries, 247–248
NO. *See* Normally open
Nominal impedance, 43, 45
Nonconductive optical fiber cables, 272, 345, 346
Nonpower-limited fire alarm (NPLFA) circuit,
    308–310
Nonpower-limited fire protective (NPLF) cable, 28
Normally closed (NC) or open-loop switch, 383
Normally open (NO) or closed-loop
    switch, 383, 386
NOS. *See* Network operating system
Notch filter, 130, 134
Notification appliance circuits (NAC), 318–319
Notification appliances
    audible alarm indicator device, 303–304
    emergency telephones, 288, 305–306
    emergency voice/alarm communications, 304
    strobes, 304–305
NPLF. *See* Nonpower-limited fire protective cable
NPLFA. *See* Nonpower-limited fire alarm circuit
NTA. *See* National Testing Association

O

Occupational Health and Safety Administration
    (OSHA), 306
Octave, 98
    harmonics *vs.* 99
Ohm's law, 36
Omnidirectional polar pattern, 113–114
Open loop, 383, 386–387
Open source systems, 400–401
Open Systems Interconnection (OSI), 199
    communication and, 181–184
    protocols, 180
    router protocols, 180
    seven layers, 179–180
Operating systems (OSs), 195–196, 213–214
Optical fiber cable, 174, 328–344. *See also*
    Fiber-optic cable

abandoned cable, 27
breakout cables, 336
cable designs, 333–334
composite, 346
conductive and nonconductive, 272, 345, 346
core, cladding, and coating, 330–331
glass fiber-optic construction, 229–330
index of refraction, 332–333
MMF, 336–337, 340
NEC 770, 329–339
NEC 770.6, 338
signal loss, 337–338
simplex cables, 335
SMF, 336–337, 339–340
styles of, 334–337
tightpack cables, 336
transmission, 331–332
Optical fiber connectors, 341–344
Orbital slots, 435–436
OSHA. *See* Occupational Health and Safety
    Administration
OSI. *See* Open Systems Interconnection
OSs. *See* Operating systems
Overcurrent protection devices, 257, 260–261

P

Packet switching, 197–198
Packet-switching exchange (PSE), 200
Paging, 110, 149
PAL. *See* Phase Alternating Lines
Panels
    control, 154, 283, 383, 389
    remote fire alarm annunciator display, 289–290
Panic button, 385
Parabolic dish antennas, 422
Parallel communication, 178
Parametric equalizers, 152
Passive crossovers, 131
Passive infrared (PIR) sensor, 379
PBX. *See* Private Branch Exchange
PC. *See* Personal computer
PCSs. *See* Personal communication systems
Peer-to-peer networks, 195
Personal communication systems (PCSs),
    425, 428–429
Personal computer (PC), 165
Phantom power, 113
Phase Alternating Lines (PAL), 475
Phase-array antennas, 424
Phase modulation (PM), 410, 412

Phase of signal, 100
Phon, 101–103
Photoelectric
    beam, 381
    smoke detectors, 298–299
Physical addressing, 182
Physical layer of OSI method, 183
Picocells, 432
Ping, 213
Pink noise generators, 103–104
PIR. *See* Passive infrared
Plain old telephone system (POTS) splitter, 169
Plate electrodes, permitted for grounding, 71
Plating, 20
Platinum-resistive thermometer (PRT), 296
Plenum, 24–25
Plenum *vs.* riser, 24–28
PLFA. *See* Power-limited fire alarm
PLTC. *See* Power-limited tray cable
PM. *See* Phase modulation
PoE. *See* Power over Ethernet
Point-to-point links, 196, 435
Polarity, 437, 439
Polarization, 439
Polar patterns
    bidirectional, 114–115
    omnidirectional, 113–114
    shotgun, 116
    supercardioid and hypercardioid, 116–118
    unidirectional or cardioid, 114–117
Polyethylene insulation, 24
Polypropylene insulation, 24
Polyvinyl chloride (PVC) insulation, 24
POP3. *See* Post Office Protocol 3
Post Office Protocol 3 (POP3), 203, 205
POTS. *See* Plain old telephone system splitter
Power
    access-control systems wiring, 396–397
    amplifier, 134, 142–144, 149
    dB and, 91
    phantom, 113
    transfer, maximum, 144–145
Power-limited circuit, 257, 258
    specifications, Class 1, 260
Power-limited Class 2 and 3 circuits, separation
        requirements for, 261–271
Power-limited fire alarm (PLFA)
    audio system circuits and, 313
    circuits, 272, 312–314
    conductor size for, 314

Power-limited fire protective (PLF) cable, 27
Power-limited tray cable (PLTC),
        27, 28, 265–266
Power line conditioners, 228–229
Power over Ethernet (PoE), 473–474
Power sources, for PLFA circuits, 313–314
Power sum (PSUM) communication cable, 53
Power supply
    connectors, allowable numbers of, 311
    dc types, 235–243
    regulated *vs.* unregulated, 242–243
    selection, 241–242
    UPS, 233–235
Preamplifier, 126, 151
Precedence, 110
Premise wiring, 356
Presentation layer of OSI model, 181
Pressure connector, 73
Pressure supervisory signal-initiating device, 302
Primary cells, 245–246
Private Branch Exchange (PBX),
        358–360, 361
    trunk services, 361–364
Projected beam-type smoke detector, 299
Propagation, velocity of, 45–47, 189–192
Propagation times, 442–443
Proprietary station systems, 287
Protected premises, 283
Protection devices, 372–373
    overcurrent, 257, 260–261
Protocols
    IP, 175, 202
    network, 199–206
    OSI, 180
    TCPs, 182
Proximity cards, 392
Proximity effect, 119
PRT. *See* Platinum-resistive thermometer
PSE. *See* Packet-switching exchange
PSUM. *See* Power sum communication cable
Public address systems/intercom, 110
Public fire alarm system, 288, 291–292
Pull stations, 291–292
Pulse modulation, 410, 412
PVC. *See* Polyvinyl chloride insulation

Q

Quad shield, 34–35
Quarter-wave monopole antenna, 419

R

Raceways
within enclosures, 269
number and size of conductors and, 275–276
Radiant energy detectors, 300
Radiation patterns of different frequencies, 105
speaker, 138–139
Radiators, 127
Radio and television, *NEC Article 810* requirements
for, 454–456
Radio frequency identification (RFID), 392–393
Radio frequency interference (RFI), 408
Radiotelephone history, mobile
AMPS, 426–428
PCSs cellular systems, 425, 428–429
roaming, 426
RADSL. *See* Rate-Adaptive Digital Subscriber Line
Randomized bundle, 330
Rate-Adaptive Digital Subscriber Line (RADSL), 170
Rate compensation heat detector, 292
Rate-of-rise heat detector, 293
RCA connector, 141–142
Reader connections, 397
Receivers, satellite, 442
Reclassification guidelines of Class 2 and
3 circuits, 266–273
Recorders, time-lapse, 476
Recording devices, 476–480
Rectifiers, 236–237
bridge, 192–194
full-wave, 238–239
full-wave *vs.* bridge, 239–240
Reduce nuisance alarms. *See* False alarms
Reflection, 104–105
Refraction, index of, 332–333
Registered jack (RJ) terminal connector, 51
Regulated *vs.* unregulated power supplies, 242–243
Relative voltage measurement, 92
Relays
alarm, 399
end-of-line, 320
frame, 200
Remote-control circuits, 257–258
Remote fire alarm annunciator display panels,
289–290
Remote station fire alarm system, 288
REN. *See* Ringer Equivalence Number
Repeater, 192
Reporting systems, display and, 383–385
Reproduction equipment, 91, 107–160

Request to exit (RTE), 396, 399
Reset *vs.* nonrestorable initiating devices, 294
Residential
connections, 356–358
network applications, low-voltage, 214
Residential cabling
premise wiring, 356
residential connections, 356–358
tip and ring, 355, 368–369
Residential systems, 355–369
Resolution, 466–467
Resonance, speaker, 128–129
Response, frequency, 97, 123
Reverberation, 135
RFI. *See* Radio frequency interference
RFID. *See* Radio frequency identification
Ribbon fiber-optic cable design, 333–334
Rigid coaxial cable, 54
Ring-back signal, telephone, 353
Ringer Equivalence Number (REN), 353
Ringers, telephone, 352–353
Ring network, 184–185, 187
Riser, 25–26
plenum *vs.*, 24–28
RJ. *See* Registered jack
RJ-11 connector, 209
RJ-14 connector, 209–210
RJ-21X, 365–366
RJ-45 connector, 210
Roaming, 426
Rod and pipe electrodes, permitted for
grounding, 71
Room temperature supervisory signal-initiating
device, 302
Router, 194
RTE. *See* Request to exit
Rubber or synthetic-based rubber insulation, 23

S

Satellite, 172–174
dish and alignment, 440–442
frequency, 442
geosynchronous, 433, 434, 436
low-Earth-orbit, 444
medium-Earth-orbit, 444
receivers, 442
Satellite communications, 432–444
channels of, 437, 439
dish size, 436–437
orbital slots, 435–436

Satellite communications (*Continued*)
   polarity types, 437, 439
   polarization, 439
   propagation times, 442–443
   satellite dish and alignment, 440–442
   satellite frequencies, 437, 439
   satellite receivers, 442
Satellite communications
   service areas, 439–440
   signal strength, 436–437
   transponders, 433, 435
   uplink/downlink configurations types, 433, 434–435
Screened twisted pair (ScTP), 48–49
ScTP. *See* Screened twisted pair
SDSL. *See* Symmetric Digital Subscriber Line
Secondary cells, primary *vs.* 245–246
Security alarms
   access-control systems and, 377–402
   alarming purpose, 378
   display and reporting systems, 383–385
   keypads/control panels, 383, 389
   magnetic door/window contacts, 382–386
   sensors, 378–382
Security alarm system wiring, 385–390
   electrical code, 402
   power, 385–386
   required wire, 386–389
Seebeck Effect, 58
Semiconductor intensified tube (SIT), 462
Sensitivity
   camera, 464
   microphone, 113
Sensors, 378–380
SEOL. *See* Single end-of-line
Separation of communication circuits, 374–375
Serial communications, 176–177
Server networks. *See* Client/server networks
Servers, 165
Service areas of satellites, 439–440
Session layer of OSI model, 181
SFF. *See* Small form factor
Shielded twisted pair (STP), 48–49, 167
Shielded *vs.* unshielded cables, 18
Shielding, conductor, 32–35
Shields, 34–35
Shock hazard, 221
Shock mount, 125
Shock-type glassbreak sensor, 380–381
Shotgun polar pattern, 116
Shutter speed, 470

SI. *See* International System of Units
Signal attenuation
   cut off frequency and, 42–43
   dB and, 91
   skin effect and, 38–39
Signal coherence, 100–101
Signal-conductor cable, 47
Signaling line circuits (SLC), 258, 322
Signaling tones, 353
Signal-initiating devices, supervisory, 301–302
Signal level, 131, 140–141
Signal loss, 337–338
Signal-processing
   audio, 107–160
   equipment, 91, 108, 153
Signal sources, 154
   of audio mixer, 150–151
Signal strength, 436–437
Signal transmission, 330
Silicone insulation, 23
Simple Mail Transport Protocol (SMTP), 203, 205
Simple Network Transfer Protocol (SNTP), 205
Simplex cables, zip cord, 335
Single-channel yagi antennas, 419
Single conductor cables, 17, 47
Single-conversion UPS, 235
Single end-of-line (SEOL), 387
Single-mode fiber (SMF), 336–337, 339–340
Single-stage fire systems, 284
SIT. *See* Semiconductor intensified tube
Skin effect, 20, 38–39
SLC. *See* Signaling line circuits
Small form factor (SFF) connectors, 342
SMF. *See* Single-mode fiber
Smoke detectors, 297–300
SMTP. *See* Simple Mail Transport Protocol
SNTP. *See* Simple Network Transfer Protocol
Software management, network, 195–196
Solder, 59
Solidly grounded, 69
Solid *vs.* stranded cables, 17, 22
Sound
   creation, 96
   hearing and, 101–105
   traveling in air, 104–105
Sound level analyzers/spectrum analyzers, 153–154
Sound systems, components and electrical
     properties, 111–127
   ground loops, 121–122
   lavaliere or lapel microphones, 123

microphone frequency response, 123
microphone hardware, 124–125
microphones, 111–121
windscreens, 125–126
wireless microphones, 122
Sound wave physics, 95–101
    amplifier bandwidth, 98
    bandwidth, 97
    dynamic range, 96–97
    examples, 90–91
    frequency, 95–96
    frequency response, 97
    harmonics, 99
    human voice, 99
    multiple octaves, 98
    octave, 98
    phase and coherence of signal, 100–101
    sound waves, 95
Speaker
    coaxial, 132–133
    compliances, 131–132
    efficiency, 132
    electrical properties and, 127–139
    enclosures, 137–139
    placement, 134–136
    resonances and impedance, 128–129
    transformers, 146–148
Special multipurpose hybrid cable, 57
Spectrum analyzers, 153–154
Spider mount, 127–128
Spiral/serve shields, 34
Spot-type smoke detector, 300
Sprinkler water flow alarm-initiating devices, 301
Standards and standards agencies, 10–12
Standing waves, 43
Star network, 184–185
Star-wired bus, 185
Steiner Tunnel Test, 28
Step index MMF, 340
Storage batteries, 243–248
STP. *See* Shielded twisted pair
Stranded *vs.* solid conductors, 17, 22
Strobes, 304–305
Structural return loss, 45
    communication cable, 53
Subpart B, separated by barriers, 269
Subpart C, raceways within enclosures, 269
Subpart D, associated systems within enclosures, 269–270
Subpart E, enclosures with single openings, 270

Subpart F, manholes, 270
Subpart G, closed-loop and programmed power distribution, 270–271
Subpart H, hoistways, 271
Subpart I, all other applications, 271
Supercardioid polar patterns, 116–118
Supervisability of fire alarm system, 282
Supervisory signal-initiating devices, 301–302
Supervisory station, 285
Supplemental electrodes, 71–72
Surge protectors, 226–227
Survivability, 307–308
Switches, 126, 194
Symmetric Digital Subscriber Line (SDSL), 170
System
    access-control, 377–402
    addressable, 321–322
    associated, 269–270
    automatic extinguishing, 301
    cable television, 452–454
    CCTV, 273, 381, 459–481
    commercial, 284–288, 358–360
    display and reporting, 383–385
    emergency, 248–250
    fire alarm, 279–325
    near water, 156
    PCSs cellular, 425, 428–429
    programming, 389, 398–400
    public address, 110
    residential, 355–369
    sound, 111–127
    wiring of fire alarm, 315–324
Système Électronique pour Couleur Avec Mémoire (SECAM), 475

T

T1 line, 364
Tamper, 387
Tap-changing regulators, 224–225
TBB. *See* Telecommunications bonding backbone
TCP. *See* Transfer Control Protocol
TCP/IP. *See* Transfer Control Protocol/Internet Protocol
TCPs. *See* Transport control protocols
TDD. *See* Time division duplex
TDMA. *See* Time division multiple access
Technical ground, 159
Technology equipment, networking and information, 163–214
Teflon insulation, 24

TELCO, 350
Telecommunications, *NEC Article 800* and, 349–375
Telecommunications bonding backbone (TBB), 80
Telecommunications Industry Association (TIA), 11
Telephone, 110, 149, 350–355, 388–389
Telnet, 203–205
Temperature-sensitive
  resistors, 296
  semiconductors, 295
Terminal bars, 73
Terminal connectors, 51–52
Terminations and color codes, 365–370
Thermistors, 295
Thermopiles, 296
Third generation of wireless, 430
Threshold of hearing, 90
TIA. *See* Telecommunications Industry Association
Tight-buffered cable design, 333
Tightpack cables, 336
Time division duplex (TDD), 431
Time division multiple access (TDMA), 429–430
Time-lapse recorders, 476
Tip and ring residential cabling, 355, 368–369
Tip-ring-sleeve microphone connector, 124, 141–142
Token ring network (TRN), 165–167
Tones, 110, 154
Transceiver, 409
Transducers, 111, 413
Transfer Control Protocol (TCP), 202–203
Transfer Control Protocol/Internet Protocol (TCP/IP), 201–206
Transformer. *See also* Autotransformers
  isolation, 220–223
  outputs, audio, 145
  volume control, 148
Transient protection, 398
Transmission
  high-frequency, 43
  LED, 331
  links, 472–474
  methods, 188
  optical fiber, 331–332
  photoelectric sensor, 331
  signal, 330
Transmission line
  balanced *vs.* unbalanced, 120–121
  components, 39–41, 55
Transmitter, 408
Transmitting stations, 456, 457

Transponders, 433, 435
Transport control protocols (TCPs), 182
Transport layer of OSI model, 181–182
Traverse-mode noise, 223
Tree topology, 185
Triaxial cable, 48, 56
Tri-shield, 34
TRN. *See* Token ring network
Trunk services, 361–364
Tube cameras, 461–462
Tweeters, 128
Twinaxial cable, 48, 56
Twin-lead cable, 18, 56–57
Twisted pair cable, 48–49, 166–167, 367–368
Two-stage fire systems, 287

U

UDP. *See* User datagram protocol
UL. *See* Underwriters Laboratories
UL-1581 vertical flame tests, 28
Ultrasonic sensor, 379
Ultricon camera, 462
Unbalanced transmission line, 120–121
Underwriters Laboratories (UL), 4, 12, 16, 26, 28
Ungrounded conductor, 69
Unicast, 189
Unidirectional microphone, 114–116
Unidirectional wireless system, 408
Uniform resource locator (URL), 206–207
Uninterruptible power supply (UPS), 233–235
Unit equipment, 249
Units of measurement, 4
Universal serial bus (USB), 171, 177–178
Universal Service Order Code (USOC), 210–211
Unregulated power supplies, regulated *vs.* 242–243
Unrelated circuits, separation of, 259–260
Unshielded twisted pair (UTP), 48–49, 166–167
Uplink antenna, 433
Uplink/downlink configurations, 435
UPS. *See* Uninterruptible power supply
URL. *See* Uniform resource locator
USB. *See* Universal serial bus
User datagram protocol (UDP), 203
USOC. *See* Universal Service Order Code
UTP. *See* Unshielded twisted pair

V

Varifocal lens, 469
Varistor-type lightning arrestors, 228
VCR. *See* Videocassette recorder

Velocity of propagation, 45–47
    related to collision and CSMA/CD, 189–192
    wavelength, 45
Vertical flame tests, 28–29
Very-High-Bit-Rate Digital Subscriber Line
        (VHDSL), 170
Very small aperture terminal (VSAT), 435
VHDSL. *See* Very-High-Bit-Rate Digital
        Subscriber Line
Videocassette recorder (VCR), 476
Video formats, 474–476
Video motion detection, 381
Vidicon camera, 461–462
Virtual local area network (VLAN), 194
VLAN. *See* Virtual local area network
Voice, human, 99
Voice/alarm communications, 304
Voltage. *See also* Low-voltage
    Class 2 and 3 circuits requirements, 265
    loss at load, 36–37
    regulators, 224–226, 237
VSAT. *See* Very small aperture terminal

W

WANs. *See* Wide area networks
Water, systems near, 156
Water flow alarm-initiating devices, sprinkler, 301
Water-level supervisory signal-initiating
        device, 302
Water temperature supervisory signal-initiating
        device, 302
Wavelength, 45, 104–105
Wet- or dry-cell electrolytes, 245
White noise generators, 103
Wide area networks (WANs), 165, 166, 199, 200
Wiegand technology, 392
Window contacts. *See* Magnetic door/window
        contacts
Windscreens, 125–126
Wire
    gauge, 20–21
    resistance, 35–37
Wireless communications, 405–457
    antennas, 415–424, 433

cable television systems, 452–454
cellular phone communication, 424–432
fundamentals of, 408–414
history of, 406–408
NEC Article 810 requirements for, 454–457
satellite communications, 432–444
wireless computer networks, 444–452
Wireless computer networks, 444–452
Wireless LANs, 444–445
Wireless microphones, 122
Wireless networking, components of, 445–452
    402.11n, 449
    802.11 (Wi-Fi), 447
    802.11a, 449
    802.11b, 449
    802.11g, 449
    Bluetooth, 449–451
    DSSS, 448–449
    FHSS, 447–448
    infrared devices, 451–452
    miscellaneous devices, 451
Wiring
    access-control system, 396–400
    bell, 388
    fire alarm systems, 315–324
    horizontal, 360
    integrity, 22–23
    low-voltage, 27
    methods for Class 2 and 3 circuits, 264–265
    premise, 356
    security alarm system, 385–390
    zone, 386
Woofers, 128

X

X.25 Packet-Network Protocol, 199–200
Xerox, 164
XLR microphone connector, 124, 141

Z

Zener diode, 237
Zipcord simplex cable, 335
Zone wiring, 386

CPSIA information can be obtained
at www.ICGtesting.com
Printed in the USA
BVHW010035211020
591458BV00005B/31